HANDBOOK ON COMPLEXITY AND PUBLIC POLICY

HANDBOOKS OF RESEARCH ON PUBLIC POLICY

Series Editor: Frank Fischer, *Rutgers University, New Jersey, USA*

The objective of this series is to publish *Handbooks* that offer comprehensive overviews of the very latest research within the key areas in the field of public policy. Under the guidance of the Series Editor, Frank Fischer, the aim is to produce prestigious high quality works of lasting significance. Each *Handbook* will consist of original, peer-reviewed contributions by leading authorities, selected by an editor who is a recognized leader in the field. The emphasis is on the most important concepts and research as well as expanding debate and indicating the likely research agenda for the future. The *Handbooks* will aim to give a comprehensive overview of the debates and research positions in each key area of focus.

Titles in the series include:

International Handbook on Ageing and Public Policy
Edited by Sarah Harper and Kate Hamblin

Handbook on Complexity and Public Policy
Edited by Robert Geyer and Paul Cairney

Handbook on Complexity and Public Policy

Edited by

Robert Geyer

Professor of Politics, Complexity and Policy, Department of Politics, Philosophy and Religion, Lancaster University, UK

Paul Cairney

Professor of Politics and Public Policy, Division of History and Politics, University of Stirling, UK

HANDBOOKS OF RESEARCH ON PUBLIC POLICY

Cheltenham, UK • Northampton, MA, USA

© Robert Geyer and Paul Cairney 2015
Chapter 8 © Eve Mitleton-Kelly 2015

All rights reserved. No part of this publication may be reproduced, stored in a retrieval system or transmitted in any form or by any means, electronic, mechanical or photocopying, recording, or otherwise without the prior permission of the publisher.

Published by
Edward Elgar Publishing Limited
The Lypiatts
15 Lansdown Road
Cheltenham
Glos GL50 2JA
UK

Edward Elgar Publishing, Inc.
William Pratt House
9 Dewey Court
Northampton
Massachusetts 01060
USA

A catalogue record for this book
is available from the British Library

Library of Congress Control Number: 2014957096

This book is available electronically in the Elgaronline
Social and Political Science subject collection
DOI 10.4337/9781782549529

ISBN 978 1 78254 951 2 (cased)
ISBN 978 1 78254 952 9 (eBook)

Typeset by Servis Filmsetting Ltd, Stockport, Cheshire
Printed and bound in Great Britain by T.J. International Ltd, Padstow

Contents

List of contributors		viii
List of abbreviations		xiii

1	Introduction	1
	Paul Cairney and Robert Geyer	

PART I THEORY AND TOOLS

2	Complexity, power and policy	19
	Graham Room	
3	Complexity and real politics	32
	Adrian Little	
4	Critical Legal Studies and a complexity approach: some initial observations for law and policy	48
	Thomas E. Webb	
5	'What's the big deal?': complexity versus traditional US policy approaches	65
	Michael Givel	
6	Can we discover the Higgs boson of public policy or public administration theory? A complexity theory answer	78
	Göktuğ Morçöl	
7	The policymaker's complexity toolkit	92
	Jim Price and Philip Haynes with Mary Darking, Julia Stroud, Chris Warren-Adamson and Carla Ricaurte	
8	Effective policy making: addressing apparently intractable problems	111
	Eve Mitleton-Kelly	

PART II METHODS AND MODELLING FOR POLICY RESEARCH AND ACTION

9	Complexity theory and political science: do new theories require new methods?	131
	Stuart Astill and Paul Cairney	
10	Complexity modelling and application to policy research	150
	Liz Johnson	

v

vi *Handbook on complexity and public policy*

11 Policymaking as complex cartography? Mapping and achieving probable
    futures using complex concepts and tools   171
    Kasey Treadwell Shine

12 The role of models in bridging expert and lay knowledge in policy-making
    activities   190
    Sylvie Occelli and Ferdinando Semboloni

13 Modelling complexity for policy: opportunities and challenges   205
    Bruce Edmonds and Carlos Gershenson

14 Using agent-based modelling to inform policy for complex domains   221
    Mirsad Hadzikadic, Joseph Whitmeyer and Ted Carmichael

PART III APPLYING COMPLEXITY TO LOCAL, NATIONAL AND
          INTERNATIONAL POLICY

15 Local government service design skills through the appreciation of
    complexity   245
    Catherine Hobbs

16 Managing complex adaptive systems to improve public outcomes in
    Birmingham, UK   261
    Tony Bovaird and Richard Kenny

17 Brazil and violent crime: complexity as a way of approaching 'intractable'
    problems   284
    Kai Enno Lehmann

18 Educating for equality: the complex policy of domestic migrants' children in
    China   299
    Qian Liu

19 The emergence of intermediary organizations: a network-based approach to
    the design of innovation policies   314
    Annalisa Caloffi, Federica Rossi and Margherita Russo

20 Complexity theory and collaborative crisis governance in Sweden   332
    Daniel Nohrstedt

21 Going for Plan B – conditioning adaptive planning: about urban planning
    and institutional design in a non-linear, complex world   349
    Gert de Roo

22 Complexity and health policy   369
    Tim Tenbensel

23 A case study of complexity and health policy: planning for a pandemic   384
    Ben Gray

Contents vii

24 How useful is complexity theory to policy studies? Lessons from the climate
change adaptation literature 399
Adam Wellstead, Michael Howlett and Jeremy Rayner

25 Agent Based Modelling and the global trade network 414
Ugur Bilge

26 The international financial crisis: the failure of a complex system 432
Philip Haynes

CONCLUSION

27 Conclusion: where does complexity and policy go from here? 457
Paul Cairney and Robert Geyer

Index 466

Contributors

Stuart Astill is principal consultant at IOD PARC and a Research Associate of the Public Policy Group at LSE. He has spent over twenty years at the heart of analytical, strategic, policy and operational thinking in government, consultancy and academia, creating innovative analysis and insight with practical applications.

Ugur Bilge is a modelling expert who designs and develops Data Mining and Agent Based Simulation tools for applying complexity thinking to real world problems. He set up SimWorld Ltd in 1998 and provides innovative solutions for a range of projects in finance, retail, logistics, economics, governance and healthcare.

Tony Bovaird is Professor of Public Management and Policy at INLOGOV and Third Sector Research Centre, University of Birmingham and Director of Governance International. Tony Bovaird is a specialist in strategic management, evaluation, performance management and user and community co-production of public outcomes. He is co-author of *Public Management and Governance* (3rd edn, 2014).

Paul Cairney is Professor of Politics and Public Policy in the Division of History and Politics, University of Stirling. His research interests are in comparative public policy, including: comparisons of policy theories (*Understanding Public Policy*, 2012); policy outcomes in different countries (*Global Tobacco Control*, 2012), Scottish politics and policy (*The Scottish Political System Since Devolution*, 2011), and UK and devolved government policymaking.

Annalisa Caloffi is Assistant Professor, Department of Economics and Business, University of Padua (Italy). Annalisa Caloffi's main research interests include innovation networks, R&D consortia and industrial and innovation policies. Her works have been published in peer-reviewed journals, books and other international outlets.

Ted Carmichael is an Affiliated Assistant Research Professor at UNC Charlotte, and is the Senior Research Scientist at TutorGen, Inc. His research has focused on novel computer simulations of complex systems, and he has created agent-based models for many different domains, including: biology, ecology, economics, philosophy, sociology and political science.

Mary Darking is Senior Lecturer in Social Policy and Informatics, School of Applied Social Science, University of Brighton. Dr Darking is currently researching digital health services policy and practice. She has recently published a co-authored protocol for 'practice-centred' health IT evaluation in BMC Health Services Research.

Gert de Roo is Professor in Spatial Planning, University of Groningen, the Netherlands. Gert de Roo is President of the Association of European Schools of Planning (AESOP, 2011–15). He has been coordinator of AESOP's thematic group on complexity and planning and is editor of Ashgate's series 'New Directions in Planning Theory'. His books on complexity and planning are titled *A Planner's Encounter with Complexity* (2010) and *Planning & Complexity: Systems, Assemblages and Simulations* (2012).

Bruce Edmonds is Professor at the Centre for Policy Modelling, Manchester Metropolitan University. He is Director of the Centre for Policy Modelling. His interests cross the boundaries of complexity science, computer science, sociology, philosophy and policy.

Carlos Gershenson is Head of the Computer Science Department, Instituto de Investigaciones en Matemáticas Aplicadas y en Sistemas, Universidad Nacional Autónoma de México. He is interested in self-organizing systems, complexity, urbanism, artificial life, evolution, cognition, artificial societies and philosophy.

Robert Geyer is Professor of Politics, Complexity and Policy in the Department of Politics, Philosophy and Religion, Lancaster University. His research interests include all aspects of policy and complexity and his major publications include: *Complexity and Public Policy* (2010) (co-authored with Samir Rihani) and *Complexity Science and Society* (2007) (co-edited with Jan Bogg).

Michael Givel works at the Department of Political Science, University of Oklahoma. In 2009, Michael Givel was the first US Fulbright in Bhutan.

Ben Gray, MBChB MBHL FRNZCGP, is Senior Lecturer, Department of Primary Health Care and General Practice, University of Otago Wellington School of Medicine and Health Sciences. Ben Gray works part time as a GP. His academic interests include cross-cultural care, cultural competence and the use of interpreters. He has recently completed a Masters' in Bioethics and Health Law with a dissertation on how the concept of cultural competence affects the practice of bioethics and health law.

Mirsad Hadzikadic is Professor at the Department of Software and Information Systems, and Director of the Complex Systems Institute, College of Computing and Informatics at UNC Charlotte. Mirsad Hadzikadic's research interests include data mining, health informatics, complexity theory, brain informatics, and a systems view of policies in financial services, economics, defence, healthcare and political science.

Philip Haynes is Professor at the School of Applied Social Science, University of Brighton. Philip Haynes researches and teaches the application of complexity theory in the applied social sciences. His recent publications include: *Managing Complexity in the Public Services* (2014, 2nd edn) and *Public Policy Beyond the Financial Crisis: An International Comparative Study* (2012).

Catherine Hobbs is a PhD researcher at the University of Hull. In pursuit of a compelling interest in local government reform, Catherine Hobbs completed an MSc in Local Governance at Birmingham University (2008). Her research interests include a fascination with systemic thinking to address the challenge of fundamental reform through leadership and collaboration; her PhD in Systems Science is exploring this.

Michael Howlett is Burnaby Mountain Chair in the Department of Political Science at Simon Fraser University, Vancouver, Canada, and Yong Pung How Chair Professor in the Lee Kuan Yew School of Public Policy at the National University of Singapore. He specializes in public policy analysis, political economy and resource and environmental policy.

x   *Handbook on complexity and public policy*

Liz Johnson is a researcher at the Complex Systems Institute, University of North Carolina, Charlotte. She co-founded and serves as the Managing Editor of the *Journal on Policy and Complex Systems*. Her recent publication is 'Applying complexity to qualitative policy research: An exploratory case study', *Journal of Social Science for Policy Development*.

Richard Kenny, BA (Hons), MSc, MBA (Warwick), Warwick Institute for Science of Cities, University of Warwick and Centre for Geographical Economic Research, University of Cambridge. Richard Kenny is Head of Strategic Development at Birmingham City Council, supporting the strategic development of the council and its city. He is a thought leader and strategist for cities and experienced director of programmes. He has held senior positions at national, regional and local levels for major UK cities and local government.

Kai Enno Lehmann is Professor Doutor, Institute for International Relations, University of São Paulo, Brazil. Kai Lehmann works on the application of complexity to issues of public policy and regionalism, with specific focus on the European Union. He is currently also working on a collaborative research project led by Professor Thomas Diez on the EU's actions in regional conflicts outside its neighbourhood.

Adrian Little is Professor of Political Theory and Head of the School of Social and Political Sciences, University of Melbourne. He has published several books and many journal articles on contemporary political and social theory with many works informed by complexity theory. His most recent book is *Enduring Conflict: Challenging the Signature of Peace and Democracy* (2014).

Qian Liu is Associate Professor, Institute of Anthropology, Renmin University of China. Qian Liu's research field is Anthropology of Education and Anthropology of Medicine. She is Vice General-Secretary of the China Union of Anthropological and Ethnological Sciences.

Eve Mitleton-Kelly is Founder and Director of the Complexity Group at London School of Economics. She is a member of the World Economic Forum, Global Agenda Council on Complex Systems. She is editor and co-author of five books and has led or participated in 30 research projects. She has developed a theory of complex systems and a methodology to address apparently intractable problems.

Göktuğ Morçöl is Professor of Public Policy and Administration, The Pennsylvania State University. Göktuğ Morçöl is Editor-in-Chief of the journal *Complexity, Governance & Networks*, author of *A Complexity Theory for Public Policy* (2012), and co-editor of *Challenges to Democratic Governance in Developing Countries* (2014).

Daniel Nohrstedt is Associate Professor of Political Science at Uppsala University, Sweden. Daniel Nohrstedt studies the policy process in the context of crisis management focusing on interorganizational collaboration, learning and performance. His recent articles appear in *Public Administration*, *Policy Studies Journal*, *Administration & Society* and *Public Management Review*.

Sylvie Occelli is Senior Research Fellow at the Istituto di Ricerche Economico Sociali del Piemonte (IRES), Turin, Italy. She has worked in various fields of regional science, ranging from housing, transportation, mobility, innovation, urban modelling and spatial analysis. She is currently in charge of a research unit aimed at developing model-based socio-technical systems in policy practices.

Jim Price, MA FRCP FRCGP FAcadMEd, is a GP and Principal Lecturer in Medical Education. He is the Medical Education Research Lead and Course Leader for the MSc and Postgraduate Certificate in Medical Education at Brighton and Sussex Medical School. His academic interests are in complexity science as it relates to medical and health education, public policy and health systems, leadership and change management.

Jeremy Rayner is Professor, Johnson-Shoyama Graduate School of Public Policy, University of Saskatchewan. Jeremy Rayner is currently working on sustainable energy transitions and problems of reflexive long-term policy design.

Carla Ricaurte is Associate Professor in Tourism at the Escuela Superior Politécnica del Litoral, Ecuador. Carla was a consultant involved in public tourism planning in her country before completing a PhD in the Centre for Tourism Policy Studies in the University of Brighton. Her research interests are tourism public policy, planning and governance, including decentralization, local government tourism management and participatory tourism planning.

Graham Room is Professor of European Social Policy, University of Bath. He is author, co-author or editor of twelve books, the most recent being *Complexity, Institutions and Public Policy: Agile Decision-Making in a Turbulent World* (Edward Elgar Publishing, 2011). He was Founding Editor of the *Journal of European Social Policy* and is a member of the Academy of Social Sciences.

Federica Rossi is a lecturer at the Department of Management, Birkbeck, University of London. Federica Rossi has worked as researcher and consultant for the OECD, the EC/Eurostat and regional and local development agencies. She has written on the economics and management of intellectual property, the economics of higher education, innovation networks, and science and technology policy.

Margherita Russo is Full Professor of Economic Policy, Department of Economics, University of Modena and Reggio Emilia, Italy. Analysis and policies of local development and innovation are Margherita Russo's main research interests. Since 2000, she has been the head of the university research-action, Officina Emilia, whose main goal is to foster in young people an active knowledge of the local context in the global scenario through hands-on and multimedia laboratories.

Ferdinando Semboloni is Assistant Professor in Town Planning and Design, and member of the Center for the Study of Complex Dynamics, University of Florence. His main expertise relies on the study of the urban and regional system, and on the methods for the steering of its evolution.

Kasey Treadwell Shine currently works in the Activation and Employment Policy, Department of Social Protection, Dublin. Her responsibilities include research, policy analysis and monitoring progress on Irish and EU poverty targets. She is the national expert on the EU Indicators Sub-Group of the SPC. She previously had responsibility for child poverty policy and for the development of the Irish child poverty target. She xii *Handbook on complexity and public policy*

has worked on a number of European projects and authored several articles exploring policymaking and management using complex concepts and tools.

Julia Stroud is a social work academic at the School of Applied Social Science, University of Brighton and her main research area is child homicide. She has employed complexity theory in understanding issues in child protection. Her recent publication is 'Multi-agency child protection: can risk assessment frameworks be helpful?', *Social Work and Social Sciences Review* (2013).

Tim Tenbensel works at the Health Systems Section, School of Population Health, University of Auckland. He researches and teaches in health policy. He has published widely on the interplay of hierarchies, markets, networks and communities in health policy, and is currently involved in studies of the implementation of health targets in New Zealand.

Chris Warren-Adamson is a qualified social worker and practice teacher in child and family social work, and has taught and researched at Sussex, Southampton and Brighton Universities since 1992, with a special interest in community and child and family centres, increasingly drawing on complexity theory as a theoretical framework for practice and research. Much of his work on centres in child and family practice, in collaboration with Anita Lightburn, has been under the auspices of the International Association for the Evaluation of Outcomes in Child and Family Services, of which he is a founder member.

Thomas E. Webb is a Faculty Academic Fellow in Law at Lancaster University. His publications include 'Exploring system boundaries' (2013, *Law and Critique*), and 'Tracing an outline of legal complexity' (2014, *Ratio Juris*). His other work concerns constitutional theory and administrative decision-making and review processes.

Adam Wellstead is Assistant Professor, Department of Social Sciences, Michigan Technological University, USA. Climate change adaption policy is one of Adam Wellstead's current research areas. The former government of Canada social scientist also undertakes empirical research on policy capacity issues.

Joseph Whitmeyer is Professor of Sociology, Department of Sociology, UNC Charlotte. Joseph Whitmeyer studies mathematical models of social processes. His recent publications are in *Markov Processes and Related Fields*, *Stochastics and Dynamics*, and *Managing Complexity* (ed. Hadzikadic et al., 2013).

Abbreviations

ABM	Agent Based Model/Agent Based Modeling
ABS	Agent Based Simulations
ACSES	Actionable Capability for Social and Economic Systems
BMI	Body Mass Indicator
CA	Cellular Automata
CAS	Complex Adaptive System
CERN	European Organization for Nuclear Research
CIAM	*Congrès Internationaux d'Architecture Moderne*
CLS	Critical Legal Studies
COR	Club of Rome
CPI	Consumer Price Inflation
CPT	Complex Practice Toolkit
CSTK	Brighton Complex Systems Toolkit
DARPA	Defense Advanced Research Projects Agency
DCYA	Department of Children and Youth Affairs
DEFRA	Department for the Environment, Food and Rural Affairs
ECB	European Central Bank
EE	Enabling Environment
EIPM	Evidence-informed Policymaking
ERDF	European Regional Development Fund
ESRC	Economic and Social Research Council
EU	European Union
FCIC	Financial Crisis Inquiry Commission
FDA	United States Federal Drug Administration
FDI	Foreign Direct Investments
GA	Government Agency
GC project	Game Changers project
GDP	Gross Domestic Product
GTNS	Global Trade Network Simulator
HMN literature	Hierarchies, Markets and Networks/heterarchy
IBPS	Brazilian Institute of Social Research
IFs	International Futures
IIASA	International Institute for Applied Systems Analysis
IMF	International Monetary Fund
IPCC	Intergovernmental Panel on Climate Change
IT	Information Technology
KIBS	Knowledge-Intensive Business Service providers
MSB	Swedish Civil Contingencies Agency
NAPinclusion	National Action Plan for Social Inclusion
NEF	UK New Economics Foundation

NGO	Non-Governmental Organization
NHS	UK National Health Service
NPM	New Public Management
OECD	Organisation for Economic Co-operation and Development
PCC	*Primeiro Comando da Capital*
PIs	Performance Indicators
PM	Military Police, Brazil
QE	Quantitative Easing
RANs	Random Agent Networks
RPIA-ITT-2002	The Regional Programmes of Innovative Actions ('*Innovazione Tecnologica in Toscana*' 2001–2004)
RPIA-VINCI-2006	Regional Programmes of Innovative Actions and 'Virtual Enterprises' 2006
SNA	Social Network Analysis
SOPs	Standard Operating Procedures
SM	Specificity of Manipulation
SMEs	Small and Medium-sized Enterprises
SPD	Single Programming Document
SR	Specificity of Reference
TQM	Total Quality Management
UPPs	*Unidades Policiais Pacificadoras*, Pacifying Police Units
WB	World Bank
WEO	IMF World Economic Outlook

1. Introduction
Paul Cairney and Robert Geyer

A NEW DIRECTION IN POLICYMAKING THEORY AND PRACTICE?

The aim of this Handbook is to improve the theory and practice of policymaking by drawing on the theory, concepts, tools and metaphors of complexity. In both theory and practice, the key aim is to advance 'complexity thinking', which describes a way to understand and explain the policymaking world, act accordingly, and invite others to do the same. To do this, the Handbook brings together a wide range of specialists to address these issues from different angles: disciplinary specialists examining how complexity thinking influences the study of topics such as the law, philosophy and politics; interdisciplinary teams examining how best to model or describe complex systems; case study specialists explaining the outcomes of real world events; and scholars and practitioners examining how to 'translate' complexity theory into 'simple' policymaking advice.

This is an ambitious project which applies a new theoretical approach to the philosophy, methodology and real world case studies and practice of politics and policymaking. Its ambition is in keeping with the approach of many well-established complexity theorists. It is common in the complexity theory literature to make bold claims about its novelty, reach and explanatory power: to say that it is radically new; a scientific revolution that will change the way we think about, and study, the natural and social world.

Yet it is not necessary, or sensible, to reject the past in such a wholesale way. Rather, we feel that complexity builds on earlier traditions and bridges many others. In theoretical and empirical studies, the identification of complex systems (and associated behaviour and outcomes) may further aid our understanding of policymaking and help select the most useful methods of study. In normative studies, it may add weight to a well-established focus on topics such as the unintended/ adverse consequences of 'top-down' policymaking.

To clarify these points we will briefly discuss some of complexity theory's main features (particularly relevant to policy) and explore how much is really new. Following this, we outline the basic structure of the book and provide brief chapter reviews. The book is divided into three main parts:

- Theory and tools;
- Methods and modelling for policy research and action;
- Applying complexity to local, national and international policy.

As with any edited book, not all of the chapters fall into exact categories. Most of the chapters are a mix of theory and application. Moreover, we tried to make the Handbook as wide-ranging and accessible as possible. Hence, there is no single unifying policy area or theme running through the chapters. We are well aware that editing a large volume

2 *Handbook on complexity and public policy*

is like trying to manage a loosely structured complex system. We had a clear idea of the direction in which we wanted to go and the type of book we wanted to see, but the exact nature of each of the chapters involved a range of unpredictable developments and interactions. Nevertheless, we are very pleased with the result and hope that you agree.

COMPLEXITY THEORY'S MAIN FEATURES

Advocates of complexity theory describe it as a new scientific paradigm providing new ways to understand, and study, the natural and social worlds (Mitchell, 2009: x; Mitleton-Kelly, 2003: 26; Sanderson, 2006: 117). Its point of departure is 'reductionism', or the attempt to break down an object of study into its component parts. Complexity theory suggests that reductionism is fatally flawed because complex systems are greater than the sum of their parts. Elements interact with each other to produce outcomes that cannot simply be attributed to individual parts of a system. Consequently, it suggests that we shift our analysis from individual parts of a system to the system as a whole, as a network of elements that interact and combine to produce systemic behaviour. The metaphor of a microscope or telescope, in which we zoom in to analyse individual components or zoom out to see the system as a whole, sums up this potential to shift our focus and approach.

The potential reach of complexity theory is remarkable. It is described by its advocates as a way to bring together the study of the natural and social world – by identifying the same broad features in each object of study and, as far as possible, using the same understanding and methods to study markedly different phenomena. Complexity theory has been applied to a wide range of activity, from the swarming behaviour of bees, ant colonies, the weather and the function of the brain, to social and political systems. The argument is that all such systems have common properties, including:

1. A complex system is greater than the sum of its parts; those parts are interdependent – elements interact with each other, share information and combine to produce systemic behaviour.
2. Some actions (or inputs of energy) in complex systems are dampened (negative feedback) while others are amplified (positive feedback). Small actions can have large effects and large actions can have small effects.
3. Complex systems are particularly sensitive to initial conditions that produce a long-term momentum or 'path dependence'.
4. They exhibit 'emergence', or behaviour that results from the interaction between elements at a local level rather than central direction.
5. They may contain 'strange attractors' or demonstrate extended regularities of behaviour which may be interrupted by short bursts of change (punctuated equilibrium).

Since complexity is a theory of many things, the language is inevitably broad and often vague. It is not always clear why we would want to compare potentially similar processes in the natural and social world, or why scholars would collaborate to do so. There are some visible tensions in the field regarding what complexity theory is, what its terms mean, how it can be used to explain outcomes and which theorists are most worthy of study (Cairney, 2012: 354; Paley, 2010).

Introduction 3

At the very least, complexity theory needs some interpretation in each discipline or field. That interpretation process may be usefully done by interdisciplinary teams, each conscious of their own terms and biases. Or, single disciplinary studies may produce distinctive contributions to complexity thinking. For example, consider the idea of 'emergence' in the absence of central direction. This may be a straightforward idea if we are studying single cells interacting to produce outcomes in the absence of a central brain. The interaction may follow simple rules, but the rules are determined locally and without reference to a wider 'order' determined by a central, rule-making body. This is not a straightforward idea in politics and policymaking, where the 'centre' (usually central government) exists and sets many rules for local actors to follow. In this context, 'emergence' may refer to behaviour which results from local interaction *despite* central government policies or rules, not in their absence. It is more difficult to explain outcomes with reference to local actors who interact with each other (and their environments) with no reference to, or awareness of, attempts by the centre to produce a different kind of outcome or order.

In broader terms, we should be careful about comparing the natural, physical and social worlds – at least unless we are content with: (a) the use of complex systems simply as metaphors with no direct source of explanation (as is common with studies using terms associated with evolution); or (b) the same use of mathematical terms and formulae, to maintain the same level of abstraction when producing explanations. This is difficult to do in studies of the social world, where there is greater potential to 'psychologize' processes such as attraction, order and self-organization to produce explanations with no equivalent in the physical world (Paley, 2010; 2011).

Even in the absence of this tendency, we need to consider how our objects of study differ and how we should respond. Human behaviour, or 'the capacity to reflect and to make deliberative choices and decisions among alternative paths of action', makes the social world an unusual object of study (Mitleton-Kelly, 2003: 25–6; see also Padgett and McLean, 2006: 1464; Steinmo, 2010: 13). In particular, it makes it difficult to follow studies of the physical systems which produce 'deterministic' arguments: 'if the complex *system* is predominantly the causal factor then we lose sight of the role that policy makers play; there may be a tendency to treat the system as a rule-bound structure that leaves minimal room for the role of agency' (Cairney, 2012: 353, emphasis in original).

Disciplinary adaptation may also involve drawing on 'old' or established concepts in each field – providing some way to compare new insights with the knowledge, gathered in the past, that we may still find useful (Cairney, 2012; 2013a).

NEW THEORY OR OLD WINE IN NEW BOTTLES?

Although complexity theory is often described as radical and new, many of its insights can already be found in policy studies. For example, focusing on theoretical and empirical studies of complexity, Klijn (2008: 302) describes complex systems as a series of policy 'subsystems', all of which have their own rules of behaviour and external forces to deal with (see also Bovaird, 2008: 321). This has been the general approach to policy studies for decades, at least since Heclo (1978: 94–7) highlighted a shift from the simple 'clubby days' of early post-war politics to 'complex relationships' at multiple levels of

4 *Handbook on complexity and public policy*

government and among a huge, politically active population (compare with Sabatier, 2007: 3–4; Jordan, 1981: 98; Richardson, 2000: 1008; McCool, 1998: 554–5).

Consequently, our task may be less about declaring theoretical or empirical novelty, and more about detailed comparisons between old and new concepts and forms of explanation. Relevant texts in the empirically-orientated policymaking literature include the following.

Path dependence and historical institutionalism. Path dependence suggests that when a commitment to a policy has been established and resources devoted to it, over time it produces 'increasing returns' (when people adapt to, and build on, the initial decision) and it effectively becomes increasingly costly to choose a different path (Pierson, 2000; compare with Room, 2011, 7–18). Pierson (2000: 253) and Room (2011: 16) adopt the same language (the 'Polya urn') and examples (such as the QWERTY keyboard) to describe the unpredictability of events and initial choices followed by subsequent inflexibility when the rules governing systemic behaviour become established and difficult (but not impossible) to change.

Punctuated equilibrium theory. Jones and Baumgartner (2005: 7; see also Baumgartner and Jones, 1993[2009]) argue that policymakers are surrounded by an infinite number of 'signals'; they must simplify their decision-making environment by ignoring most (negative feedback) and promoting few to the top of their agenda (positive feedback). Negative feedback may produce long periods of equilibrium since existing policy relationships, rules and responsibilities are more likely to remain stable, and policy is less likely to change, when the issue receives minimal attention from policymakers. Positive feedback may produce policy 'punctuations' because, when policymakers pay a disproportionate amount of attention to an issue, it is more likely that policy will change dramatically. The 'selective attention' of decision makers or institutions explains why issues can be relatively high on certain agendas, but not acted upon; why these powerful signals are often ignored and policies remain stable for long periods. Change often requires a critical mass of attention to overcome the conservatism of decision makers and shift their attention from competing problems (Jones and Baumgartner, 2005: 19–20; 48–51). Information processing is characterized by 'stasis interrupted by bursts of innovation' and policy responses are unpredictable and episodic rather than continuous (Jones and Baumgartner, 2005: 20).

Lipsky's idea of 'street-level bureaucracy'. Lipsky (1980) suggests that there are so many targets, rules and laws that no public agency or official can be reasonably expected to fulfil them all. In fact, many may be too vague or even contradictory, requiring 'street-level bureaucrats' to choose some over others. The potential irony is that the cumulative pressure from more central government rules and targets effectively provides implementers with a greater degree of freedom to manage their budgets and day-to-day activities. Or, central governments must effectively reduce their expectations by introducing performance measures which relate to a small part of government business. Consequently, we can explain much 'street-level' behaviour with reference to how local actors interact with each other and their environment.

Hjern and Porte's (1981) focus on intra-departmental conflict, when central government departments pursue programmes with competing aims, and *interdependence,* when policies are implemented by multiple organizations. Programmes are implemented through 'implementation structures' where 'parts of many public and private organizations cooperate in the implementation of a programme'. Although national governments create the overall framework of regulations and resources, and there are 'administrative imperatives' behind the legislation authorizing a programme, the main shaping of policy takes place at local levels.

Governance. A lack of central control has prompted governments in the past to embrace New Public Management (NPM) and seek to impose order through hierarchy and targeting. However, local implementation networks (with members from the public, third and private sectors) have often proved not be amenable to such direct control (see for example, Rhodes, 1997; Bevir and Rhodes, 2003; Kooiman, 1993).

Many of these texts also have an implicit or explicit argument about the limits to central control and the potential advantage to more local or less radical forms of policy-making. This compares to older work, with a relatively explicit message for policymakers, which seems to anticipate complexity themes:

> Making policy is at best a very rough process. Neither social scientists, nor politicians, nor public administrators yet know enough about the social world to avoid repeated error in predicting the consequences of policy moves. A wise policy-maker consequently expects that his [sic] policies will achieve only part of what he hopes and at the same time will produce unanticipated consequences he would have preferred to avoid. If he proceeds through a succession of incremental changes, he avoids serious lasting mistakes (Lindblom, 1959: 86).

Consequently, we should reject the idea of novelty for novelty's sake. The value of complexity theory is not that it is new – it is that it allows us to use our knowledge of the natural and social world to understand and influence real world problems in a particular way. This may involve building on, rather than rejecting, the existing literature – or at least demonstrating (rather than merely asserting) why it should replace previous insights.

The former requires careful thought, to deal with concepts that may mean different things in studies of the natural and social world, and in studies of complexity which may or may not describe the same processes as established concepts (Cairney, 2013b). For example, part of the explanation for terminological similarities may be the tendency for intuitive terms to cross disciplines. It is possible that some terms – such as path dependence and sensitivity to initial conditions – may be used in areas such as computing science or systems biology to explore unpredictability, but in political science to explain long-term inertia and stability. Positive and negative feedback may be used differently in computing and political science, particularly when policymaking studies discuss the intentional, psychological process behind paying no/ high attention to issues. Some terms, such as 'first order', refer to major change in the physical world but minor change in policymaking (Hall, 1993). 'Chaos' may relate to a deterministic process in physics but be used, in common parlance and social science, to describe unpredictability. 'Local' outcomes may emerge in the absence of a centre in the physical world but despite the actions of the centre in the policymaking world.

Ultimately, complexity theory will only be valuable if we can produce some results. To a large extent, this involves the meaningful application of 'complexity thinking' to the study of real world problems by using case studies or other approaches. By 'meaningful', we mean that those studies do more than add a complexity gloss to the analysis. They use complexity theory to produce a broad understanding of the world, to guide the way in

6   *Handbook on complexity and public policy*

which we seek to generate knowledge of it, and to guide their choice and use of methods. If the complex system underpins explanation, we would expect studies to identify the complex system, how it operates (which might include a discussion of the rules used by agents within it) and what outcomes emerge from the interaction between agents, the rules they follow, and their environment. Or, studies might identify particular aspects of a complex system and examine, for example, how agents adapt to their environments or how we can identify and explain the generation of rules within systems. In other words, they do more than just state that the world is complicated.

The case has to be made for the value of complexity theory as a way to organize the study of policy and policymaking. Ideally, we should be able to adopt the same language in a wide range of different case studies, to allow us to compare the results and consider the extent to which the same concepts explain behaviour in many cases. This is a high bar that is not always met, even in well-established policy-relevant theories such as punctuated equilibrium, the advocacy coalition framework and the institutional analysis and development framework (Cairney, 2013a; Cairney and Heikkila, 2014). However, in each of those cases there has been a marked attempt to set a common theoretical and methodological agenda, to allow the comparison of results and the accumulation of knowledge. This Handbook is an attempt to add to this process and eventually meet that 'high bar'.

PART I: THEORY AND TOOLS

These themes, and many others, are taken up in a variety of ways by the authors in Part I, 'Theory and tools'. For example, Graham Room argues that a major weakness in complexity is its limited discussion of power. This is partly because complexity theory and approaches developed in the natural sciences and other fields where this would not be a key issue. However, for the social, political and policy fields, it is essential. To respond to this, Room explores classic discussions of power in the social sciences, the insights that we can generate when we consider power and complexity, and the extent to which we can hold actors such as (but not exclusively) policymakers to account for policy outcomes when they depend on the interactions among many people in a complex system.

Adrian Little draws on complexity themes to engage with a fundamental question to philosophy and social science: how can we understand and conceptualize the real world? He challenges the way in which reality is conceptualized in international relations scholarship, and encourages political theorists to engage in a more meaningful way with politics by removing artificial barriers between their analysis of the real world and what political reality should be. Complexity theory suggests that what we call 'real' is a brief snapshot of a world that is always in flux. Consequently, to advance our study of policymaking in a complex world, we need to understand the problematic ways in which policymakers see and respond to it. This is an actor-centred account which suggests that what is 'real' to people is in fact what they need to see as reality in order to allow them to operate effectively within it.

Shifting into the field of law, Thomas Webb highlights the potential for complexity theory to prompt new thinking in legal studies. He draws on complexity terms, such as emergence and contingency, to argue that the legal process cannot be boiled down to a set of simple laws and rules to be implemented by government bodies such as the courts. Rather, people interpret rules and interact with each other to produce outcomes that are difficult to predict with reference to the statute book. Webb identifies this broad argument in historical debates initiated by the Critical Legal Studies Movement in the US, but finds that they often became stalled when debates focused on the ideological biases of each position. He explores the ability of complexity thinking to take us beyond the politics of particular countries and produce a set of ideas to guide legal thinking.

Michael Givel explores the ability of complexity theory to advance, or provide an alternative to, US-dominated theories of policy process development. He identifies three main methodological positions relevant to political science – positivism, post-positivism and critical realism – to argue that complexity theory represents a major problem for positivist studies driven by a belief in objectivity and linear cause and effect. His critique is that, while many policy theories identify several aspects of complex policymaking systems, they still rely largely on positivist methods. His proposed solution is that policy scholars not only recognize, fully, the nature of complex systems, but also that they change their research assumptions and methods to allow them to understand such systems (compare with the chapter by Wellstead et al., which is more critical of complexity research and more hopeful about the contribution of established policy theory).

Following on this policy-oriented theme Göktuğ Morçöl examines the extent to which we can produce a general, unifying theory of public administration. He uses an analogy with the physical sciences – CERN and the Higgs boson particle – to make three key points. First, even in the physical sciences, with general laws regarding the make-up of the physical world and billions of dollars of research, scientists are still uncertain about their results. Second, although the natural and social sciences share a common focus of study, their *object* of study is different. Third, the social sciences involve a much higher degree of uncertainty because each individual is a complex system, not a constant. Consequently, it is less appropriate to think in terms of general theory and traditional methods. In that context, Morçöl examines the value of methods such as agent-based modelling and case study research.

Jim Price and Philip Haynes, with Mary Darking, Julia Stroud, Chris Warren-Adamson and Carla Ricaurte, shift the discussion from theory to tools by providing a 'toolkit' for policymakers and practitioners when dealing with complex policymaking and social systems. The seven key tools in their 'Brighton Complex Systems Toolkit' are: identify the properties and members of the system; think of leadership as the actions of many people, not just a CEO; encourage a sense of 'self-organization' in systems rather than seeking top-down control; accept that people must use short cuts to gather information and make decisions; develop appropriate ways to scan for information; experiment with policy interventions rather than seeing policy as key events; and evaluate policies regularly to pursue the pragmatic mantra, 'do more of what works and less of what doesn't'. Overall, these tools add up to a common theme in complexity theory: that we should identify a coherent alternative to the ideal-type of hierarchical corporate management headed by a single executive.

In the concluding chapter in this part, Eve Mitleton-Kelly presents the 'EMK methodology', a toolkit for policymakers and practitioners to use when trying to solve problems within complex systems. Building on 20 years of experience and 30 different projects, she has developed ten complexity principles, including the identification of interconnectivity

8 *Handbook on complexity and public policy*

and interdependence between actors, and the properties of systems (feedback, emergence, self-organization, far from equilibrium) that can be applied to various policy situations. In her chapter she applies these to the case study of an Indonesian government agency, showing that they can be used to facilitate extensive conversations within an organization to identify problems in performance. For example, the process unearthed important concerns about staff attitudes to strong leadership, nepotism and individual behaviour.

PART II: METHODS AND MODELLING FOR POLICY RESEARCH AND ACTION

One consequence of the broad nature of complexity theory, the tension regarding its meaning, and the potential for many approaches in many disciplines, is that there is no inevitability about which method or modelling approach to use. As Stuart Astill and Paul Cairney discuss, we may decide that complexity requires new methods derived from other disciplines, or simply incorporate an understanding of complexity in well-established social science methods (perhaps including a mixed-methods approach). The same issue can be seen in relation to the level of simplicity we seek. For example, any method is a 'model' to a greater or lesser degree. To 'model' is to turn a system, whose complexity we cannot fully understand in its entirety, into a smaller number of key elements which interact to produce a complex system's outcomes. As they demonstrate, the key is to examine the assumptions we make when we select and use any method – a task that may be particularly useful in fields which focus more on the sophistication of modelling methods at the expense of a 'first principles' discussion of why they use those methods in the first place.

Given Astill and Cairney's argument that the methodological field for complexity is wide open, the following chapter by Liz Johnson takes this forward and provides an excellent introduction and overview of some of the core research and methodological tools of complexity and a robust justification for their importance to policymaking and policy actors. Stressing a goal of trying to 'keep complexity simple', she begins with a thorough review of some of the core vocabulary of complexity before walking the reader through introductions to network and agent-based modelling. Throughout, she argues that general policy theory does not adequately take into account how policymaking interrelates and intersects with complexity in the real world. Complexity theory and methods provide the means to move from research that describes *what is* to further inquiry into *what could be*. For Johnson, complexity methodologies can be tools for improving perspective and the practice of policymaking and policy research.

Kasey Treadwell Shine's chapter tries to create a complex cartography of the policymaking process. The chapter does this through identifying key 'first principles' (such as emergent behaviour and complicity) and exploring the implications of social complexity and its ethical consequences, where policymakers act as 'compass bearers', correcting and re-orienting emergent ordering and pathways. It then goes on to explore these principles and concepts through two case studies, child poverty and social inclusion/exclusion policy, arguing that complex policymaking helps policymakers and others to decide when to 'let go' and let many hands make light work; when to be vigilant (for example at transition points); and when to 'intentionally disrupt' atypical pathways, to reconnect to pathways of educational advantage. Complex policymaking demands a new direction away from evidence-based policymaking and towards evidence-informed policymaking (EIPM). EIPM asks, not 'what works', but 'what are the best possible solutions for all': recognizing and embracing change, over time and space, as the potential to achieve desired future(s).

Mirroring the earlier discussion in Astill and Cairney, Sylvie Occelli and Ferdinando Semboloni argue that in increasingly ICT-mediated human organizations, models can be effective vehicles for promoting dialogue between scientific experts and stakeholders and facilitating participative policymaking. Building on research carried out in Italy over the last decade, they examine how the functional and cognitive mediator roles of models in policymaking are addressed. They suggest that both design of models and the implications for the community using them should be given foremost attention and argue that by providing an epistemology for understanding social processes, models can help leverage the organizational capability of communities. To take advantage of models, however, co-evolutionary changes should take place: (a) in the encoding–decoding processes underlying model activities; and (b) in the ways in which policymaking activities are implemented in real governmental systems.

Bruce Edmonds and Carlos Gershenson review the purpose and use of models from the field of complex systems and, in particular, the implications of trying to use models to understand or make decisions within complex situations, such as those that policymakers usually face. This includes a discussion of the different dimensions of formalized situations, the different purposes for models and the different kinds of relationships they can have with the policymaking process. This is followed by an examination of the compromises forced by the complexity of the target issues. Several modelling approaches from complexity science are described (system dynamics, network theory, information theory, cellular automata, and agent-based modelling) with brief discussions of their abilities and limitations. Following some examples of policy case studies, they conclude by outlining some of the major pitfalls facing those wishing to use such models for policy evaluation.

To bring this part to a close, Mirsad Hadzikadic, Joseph Whitmeyer and Ted Carmichael discuss the strengths and weaknesses of applying agent-based modelling (ABM) to the domain of emergent social behaviour. They do this through a review of complex adaptive systems and ABM, looking in particular at some of the main aspects of ABM (agents, attributes, rules, adaption, fitness function and so on). In their case study of citizen allegiance during the Taliban insurgency in Afghanistan, funded by the Defence Advanced Research Project Agency, one of the most interesting aspects they found was of the geographic 'clustering tendency' of agents (both Taliban and pro-government forces) when the model was initiated. Despite this finding, and others, they stress that in many ways, designing agent-based models is 'more art than science' and at best they are aids to decision-making and not answers in themselves.

PART III: APPLYING COMPLEXITY TO LOCAL, NATIONAL AND INTERNATIONAL POLICY

In the policymaking field, the identification of a complex system is used to make suggestions which often represent a mix of descriptive and prescriptive elements, telling us how the world works and how we should respond. For example, a common argument is that law-like behaviour is difficult to identify because elements of the system interact

10   *Handbook on complexity and public policy*

in different ways. Sometimes they reinforce each other, but sometimes they cancel each other out. The methodological conclusion is that we need to find ways to identify and measure this behaviour, which may defy traditional quantitative methods (or at least those which seek to produce 'linear' results). The broader philosophical conclusion is that any method produces evidence of snapshots in time – prompting us to consider how much we can generalize and how many claims we can reasonably make from limited data.

The practical conclusion is that a policy that was successful in one context (one place, one time) may not have the same effect in another. In some cases, we may be able to identify the elements which interacted to contribute most to policy success, but not that those elements will interact in the same way in a different time or space. Since policymaking systems change so quickly, and are so difficult to predict and control, policymakers should not be surprised when their policy interventions do not have the desired effect. Instead, they should adapt quickly and not rely on a single policy strategy.

This major limit to our knowledge of, and ability to control, policymaking systems often produces a 'bottom up' or local approach to practitioner advice, with arguments including:

1. Rely less on central government driven targets, in favour of giving local organizations more freedom to learn from their experience and adapt to their rapidly-changing environment.
2. To deal with uncertainty and change, encourage trial-and-error projects, or pilots, that can provide lessons, or be adopted or rejected, relatively quickly.
3. Encourage better ways to deal with alleged failure by treating 'errors' as sources of learning (rather than a means to punish organizations) or setting more realistic parameters for success/ failure.
4. Encourage a greater understanding, within the public sector, of the implications of complex systems and terms such as 'emergence' or 'feedback loops'.

In other words, this literature tends to encourage a movement away from rigid governmental hierarchies, top-down policymaking, centrally driven targets and performance indicators – all based on the idea that a policymaking system can be controlled, that policymakers can impose order (Geyer and Rihani, 2010). A policy approach based on this belief in order, determined by the centre, may simply result in policy failure and demoralized policymakers. In that context, complexity theory often represents a counter-narrative set up to challenge the mechanical, 'state in control', approach in which governments go through a periodic process of failure, 'learning lessons', reform and failure because they support a governing system doomed to fail and produce blame (Geyer, 2012). This is not a straightforward task, particularly in countries with a 'Westminster model' understanding of accountability and responsibility, based on the centralization of power in government to ensure accountability to the public through Parliament. In that context, a counter-narrative based on our analysis of the real world may have to be accompanied by a narrative of democracy and accountability.

We have set out this advice in relatively straightforward terms, but this is not always a feature of complexity theory-derived practitioner advice. There is great potential for the advice to be too jargon-filled and unattractive to busy practitioners with their own language (unless practitioners are trained in relevant policy skills – see Hallsworth and Rutter, 2011: 30). Or, the translation may produce rather banal advice. The onus is on complexity theorists to provide understandable, meaningful and valuable advice that practitioners cannot get elsewhere. For example, it has proved possible to make these arguments against centralization without reference to complex systems, and some advice (such as 'use trial-and-error as a strategy') could be described as little more than common sense (Geyer and Rihani, 2010: 186). There is an onus on complexity theorists to demonstrate that their advice is more than a fashionable way to repackage old advice.

These themes are taken up in Part III, 'Applying complexity to local, national and international policy' where we have asked a mixture of practitioners and academics to explore the implications of complexity for distinct policy areas. Catherine Hobbs embodies this combination of 'pracademic', with over 20 years in UK local government and seven years of research activity. In her chapter, she argues for a radically 'complex' approach to local government in the UK. At present, pressures for budget cuts, increasing public demands and 'getting more for less' put intolerable strains on local government where retrenchment seems to be the only answer. She contends, following an excellent review of the complexity and local policy literature, that the current juncture can also be an opportunity to apply the broad and growing body of literature on complexity and policy to open up new policy debates, structures and actions. This should help local policy actors to accept the reality of complexity and learn how to make it work for, rather than against, them. She concludes that only if researchers and policy actors continue to 'polish the gem of complexity' will they be able to open up a new, positive and more flexible approach to local government.

Also in the UK, Tony Bovaird and Richard Kenny examine how a recent project exploring the cause-and-effect chains connecting government agency initiatives in Birmingham with their public outcomes has revealed a number of service areas and interventions which were not susceptible to conventional cause-and-effect analysis, despite cost-effectiveness claims by the stakeholders involved. Bovaird and Kenny then go on to explore these areas, first by seeking to classify them within the CYNEFIN framework (Snowden and Boone, 2007) to establish the knowledge domain in which the various stakeholders believe they are working and then by identifying and studying some of these service areas and interventions as potential complex adaptive systems. They demonstrate that there are areas of public sector intervention where the complexity lens is appropriate, that some of the current claims to cost-effectiveness of these interventions cannot be justified and that these interventions would be more appropriately managed if a complexity perspective were applied.

Exploring local policy changes in a very different part of the world, Kai Lehmann brings together an international and local dimension to his chapter by examining the implications for complexity thinking on the 'intractable' problems of violent crime in São Paulo and Rio de Janeiro. Both cities had extremely high homicide rates in the 1990s – responding with traditional strategies of 'containment' and 'confrontation'. However, in the 2000s both cities began to look at the problem of violent crime in new, more flexible and adaptive ways. Without being aware of the work on complexity, they began to adapt strategies that mirrored much of what complexity thinkers would suggest: recognize that the violence is part of an evolving complex process and that the police and local governments are 'interacting' with that process and not something external to it. As he

12 *Handbook on complexity and public policy*

points out, in many ways changing the perception of the police (from 'killing bad guys' to 'calming the violence') had a fundamental impact on the actions of the communities and the perpetrators. For Lehmann, recognizing the underlying complexity of the problem is the key to maintaining the successful strategies of these two cities.

Continuing the international-local theme, Qian Liu explores the complex interaction of migration and education policy in China by examining the policy changes and pressures on local authorities and policy actors in Beijing to respond to the changing educational needs of internal migrant children. With the massive growth in urbanization in China during the past 20+ years, large numbers of migrants have moved into the cities. Previously, the children of these migrants had very limited rights to education in these urban areas, leading to the rise of various types of self-organized schools and educational institutions in response to local education needs. Recent central government policy changes have begun to improve the education rights of these children, but this has put huge pressures on local schools and existing semi-legal migrant schools. From a study of one particular area in Beijing, Qian Liu argues that implementing these changes will continue to be a complex adaptive process where the various stakeholders will play a key role.

Using a more quantitative approach, Annalisa Caloffi, Federica Rossi and Margherita Russo explore an area with strong linkages to complexity: social network analysis (SNA). Using SNA, they argue that our understanding of what network configurations contribute to innovation and how they do so is still very limited. In order to stretch these limits they focus on the role of 'intermediaries' in innovation networks and argue that their role is much more important than just being an organizational 'matchmaker'. Using a quantitative study of innovation networks created by EU regional funding in the Tuscany region in Italy in the 2000s, they analysed the networking roles of 1366 different organizations. Their general conclusions were that intermediaries were primarily local governments, though intermediary roles were played by a wide variety of institutions and varied from project to project. Moreover, while 'brokers' (linking unconnected agents) tended to form in turbulent environments, 'intercohesive agents' (bridging cohesive communities of network agents) primarily operated in more stable environments.

Continuing the theme of the importance of local networks, Daniel Nohrstedt conducts an empirical study of adaptive capacity in local-level emergency preparedness collaborations in Sweden. Given the uncertainty of risks and threats, local managers in emergency preparedness generally engage in networking to mobilize knowledge and other resources from multiple stakeholders. But even if collaborative management is elevated as a precursor to effective crisis management, the effects of networking in this area are still poorly understood. The chapter illustrates how complexity theory and concepts can be applied to structure empirical analysis of collaborative performance in crisis governance. Using survey data on emergency preparedness in Swedish municipalities (n = 290) over a period of four years (2009–12), the chapter investigates the relationship between adaptive capacity and outcomes (capacity to respond to crisis) in Sweden and draws some implications for complexity theory more generally.

Shifting to a broad-ranging discussion of planning, Gert de Roo begins his chapter with a detailed history of the main approaches to urban and spatial planning since WWII. He charts the rise and fall of centralized planning based on the ideas of certainty, linear science and 'planning for control'. This was followed by the rise of the 'communicative turn' in planning that recognized the uncertainty and contingent nature of the planning process and strived to create an 'agreed reality' between planning actors and society. However, for de Roo the communicative turn is not enough to embrace the increasingly non-linear complex reality. For a non-linear world, planning must embrace 'adaptive planning' where the planner must act as a manager of change trying to enhance the positive effects of change for communities and societies as a whole and minimize its possible negative effects. This 'post-normal' scientific perspective radically challenges traditional interpretations of and policy strategies for all forms of planning.

Focusing on health policy in general, Tim Tenbensel examines the use of complexity theory in the literature on healthcare policy. This has two main elements: the idea of 'upstream' health policy, which focuses on the complex causes of health outcomes; and 'downstream', which examines the complex interplay among actors in health services. Focusing primarily on the latter, he identifies a general use of complexity as a metaphor for healthcare systems. Most intriguingly, Tenbensel makes a distinctive argument about the normative implications of complexity theory for healthcare management. A general argument in this literature is that we should encourage the kinds of 'bottom up' inter-action and outcomes that emerge in the absence of central coordination. However, he challenges this account, arguing that it relied on a caricature of 'top down' and hierarchi-cal, command and control policymaking that no longer exists in modern governments. Instead, he stresses the importance of new forms of organization which combine govern-ment, market and network solutions.

Ben Gray carries on the focus on healthcare by providing an in-depth perspective of a busy practitioner trying to make sense of his work within a complex healthcare system. He argues that doctors are commonly confronted by organizations (such as pharmaceutical companies and some governments) promoting simple interventions and ethics scholars advocating simple principles for all doctors to follow. However, neither is adequate in real world situations. Building on the difference between simple, compli-cated and complex problems in healthcare, Gray explores how to seek pragmatic solu-tions when faced with limited information, complexity, and the need to negotiate medical decisions with a wide range of colleagues and patients.

Moving from health to climate policy, Adam Wellstead, Michael Howlett and Jeremy Rayner identify an important irony in the complexity literature: complexity thinking is used increasingly to identify problems with climate change, and our need to adapt to climactic complex systems, but it is not used to understand the complex policymaking system itself. Instead, there is too much reliance on a 'black box' or func-tional understanding of policymaking in which policy simply reflects what is required of it. In this context, the ability of climate scientists to influence the policy process is limited, because they present vague or unrealistic recommendations to policymakers and do not know enough about the policy process to engage with its key actors. Wellstead et al. argue that, for complexity theory to become a useful bridge between natural and political science, its advocates need a greater understanding of 'meso level' political processes such as governing institutions and the networks between government and non-government actors.

Keeping the focus at the global level, Ugur Bilge provides an in-depth case study to demonstrate the value of agent-based modelling to examine scenarios in complex systems. Drawing on work commissioned by bodies such as the Finland government,

14 *Handbook on complexity and public policy*

he produces a model of global trade to simulate a range of scenarios over the next three decades. The main focus of this ABM simulation is on the role of the US and China, simulating what would happen if their relationships changed markedly in favour of one country or the other. The added-value to this discussion is the ability of an ABM to examine the role of major shocks to systems and even to explore potential 'butterfly effects' (how a small country could cause a major global economic crisis) on these shocks and responses.

Concluding the global policy theme, Philip Haynes draws on complexity themes and concepts to explore the international financial crisis in the middle to late 2000s. He argues that the global financial system has the attributes of a complex social system. Consequently, it cannot be controlled easily using traditional economic levers by individual countries, particularly since major companies are mobile and able to move money across the world. Haynes identifies sources of attractors and instability in this system and explores the ability of governments to influence its outcomes, comparing the limited influence of domestic measures with the potential for more effective measures based on international collaboration.

Finally, to end the Handbook, Cairney and Geyer will assess what this volume says about the current state of complexity, where complexity may be going from here and if a complexity informed policy process is really something new and possible.

ACKNOWLEDGEMENT

The volume editors give special thanks to Nicola Mathie for all her help with and input into this book.

REFERENCES

Baumgartner, F. and B. Jones (1993[2009]), *Agendas and Instability in American Politics*, Chicago, IL: Chicago University Press.
Bevir, M. and R.A.W. Rhodes (2003), *Interpreting British Governance*, London: Routledge.
Bovaird, T. (2008), 'Emergent Strategic Management and Planning Mechanisms in Complex Adaptive Systems', *Public Management Review*, 10(3), 319–40.
Cairney, P. (2012), 'Complexity Theory in Political Science and Public Policy', *Political Studies Review*, 10(3), 346–58.
Cairney, P. (2013a), 'Standing on the Shoulders of Giants: How Do We Combine the Insights of Multiple Theories in Public Policy Studies?', *Policy Studies Journal*, 4(1), 1–21.
Cairney, P. (2013b), 'What is Evolutionary Theory and How Does it Inform Policy Studies?', *Policy and Politics*, 41(2), 279–98.
Cairney, P. and T. Heikkila (2014), 'A Comparison of Theories of the Policy Process', in P. Sabatier and C. Weible (eds), *Theories of the Policy Process*, 3rd edn, Chicago, IL: Westview.
Geyer, R. (2012), 'Can Complexity Move UK Policy beyond "Evidence-Based Policy Making" and the "Audit Culture"? Applying a "Complexity Cascade" to Education and Health Policy', *Political Studies*, 60(1), 20–43.
Geyer, R. and S. Rihani (2010), *Complexity and Public Policy*, London: Routledge.
Hall, P. (1993), 'Policy Paradigms, Social Learning, and the State: The Case of Economic Policymaking in Britain', *Comparative Politics*, 25(3), 275–96.
Hallsworth, M. and J. Rutter (2011), *Making Policy Better*, London: Institute for Government.
Heclo, H. (1978), 'Issue Networks and the Executive Establishment', in A. King (ed.), *The New American Political System*, Washington, DC: American Enterprise Institute.

Hjern, B. and D. Porter (1981), 'Implementation Structures: A New Unit of Administrative Analysis', *Organizational Studies*, 2, 211–27.

Jones, B. and F. Baumgartner (2005), *The Politics of Attention*, Chicago, IL: University of Chicago Press.

Jordan, G. (1981), 'Iron Triangles, Woolly Corporatism and Elastic Nets: Images of the Policy Process', *Journal of Public Policy*, 1(1), 95–123.

Klijn, E. (2008), 'Complexity Theory and Public Administration: What's New?' *Public Management Review*, 10(3), 299–317.

Kooiman, J. (1993), 'Socio-political Governance: Introduction', in J. Kooiman (ed.), *Modern Governance*, London: Sage.

Lindblom, C. (1959), 'The Science of Muddling Through', *Public Administration Review*, 19, 79–88.

Lipsky, M. (1980), *Street-Level Bureaucracy*, New York: Russell Sage Foundation.

McCool, D. (1998), 'The Subsystem Family of Concepts: A Critique and a Proposal', *Political Research Quarterly*, 51(2), 551–70.

Mitchell, M. (2009), *Complexity*, Oxford: Oxford University Press.

Mitleton-Kelly, E. (2003), 'Ten Principles of Complexity and Enabling Infrastructures', in E. Mitleton-Kelly (ed.), *Complex Systems and Evolutionary Perspectives of Organisations*, Amsterdam: Elsevier.

Padgett, J. and P. McLean (2006), 'Organizational Invention and Elite Transformation: The Birth of Partnership Systems in Renaissance Florence', *American Journal of Sociology*, 111(5), 1463–568.

Paley, J. (2010), 'The Appropriation of Complexity Theory in Health Care', *Journal of Health Services Research & Policy*, 15(1), 59–61.

Paley, J. (2011), 'Complexity in Health Care: A Rejoinder', *Journal of Health Services Research & Policy*, 16(1), 44–5.

Pierson, P. (2000), 'Increasing Returns, Path Dependence, and the Study of Politics', *The American Political Science Review*, 94(2), 251–67.

Rhodes, R.A.W. (1997), *Understanding Governance*, Buckingham: Open University Press.

Richardson, J.J. (2000), 'Government, Interest Groups and Policy Change', *Political Studies*, 48, 1006–25.

Room, G. (2011), *Complexity, Institutions and Public Policy*, Cheltenham, UK and Northampton, MA, USA: Edward Elgar Publishing.

Sabatier, P. (2007), 'The Need for Better Theories', in P. Sabatier (ed.), *Theories of the Policy Process*, Cambridge, MA: Westview.

Sanderson, I. (2006), 'Complexity, "Practical Rationality" and Evidence-based Policy Making', *Policy & Politics*, 34(1), 115–32.

Snowden, D. and M. Boone (2007), 'A Leader's Framework for Decision Making', *Harvard Business Review*, November, 69–76.

Steinmo, S. (2010), *The Evolution of Modern States*, Cambridge: Cambridge University Press.

PART I

THEORY AND TOOLS

2. Complexity, power and policy
Graham Room

INTRODUCTION

This chapter will pose three questions: how should complexity thinking handle 'power'; how can this enrich larger debates in political and social science about power; and how, finally, should this affect our approach to policy analysis and policy making?

The complexity literature generally lacks much discussion of 'power', notwithstanding its centrality to sociological theory. This lacuna is evident for example within agent-based modelling (ABM), one of the most popular methodological derivatives of complexity analysis. It is also evident in attempts to apply evolutionary approaches to social and economic systems – evolutionary models being a prime example of complex systems (Room, 2012).

This gap attests to the origins of complexity thinking in the natural and informational sciences and it vitiates our efforts to analyse and illuminate policy making. One consequence is that in applying complexity perspectives to societal change, there is in general little reference to the distribution of power in the society concerned, and to the institutional mechanisms through which power is exercised.

This chapter takes stock of the debates on power within the sociological and political science literature and it considers how complexity perspectives might bring additional analytical insights. Such perspectives will, it argues, need to be combined with the insights of political economy, if they are to provide real value-added, in terms of both research and policy guidance.

COMPLEXITY AND SOCIAL SYSTEMS

The last quarter century has seen a noteworthy coming together around a new paradigm of 'complexity'. One facilitating factor was the establishment of institutes specifically committed to the multi-disciplinary treatment of complex systems, notably that at Santa Fé (http://www.santafe.edu/). Another was the development and ready availability of computer power, with which complex processes could be simulated and explored in ways hitherto impossible.[1]

The new complexity paradigm grows out of earlier currents of thought (Room, 2011: Ch. 2). Among the intellectual currents that it has woven into a new synthesis are thermodynamics and 'far from equilibrium' systems; neo-Darwinian evolutionary biology (bringing together Darwin's interest in variation and selection with more recent molecular and genetic biology); information science and systems of distributed intelligence; heterodox economics and its treatment of increasing returns and the growth of knowledge.

Complexity science takes us far from the classical focus on equilibrium and stability. It presents us instead with a world where open systems re-shape their environment and

20 *Handbook on complexity and public policy*

its dynamic development. Here are path-dependent trajectories, with positive feedback, lock-in and ratchet effects. Here also, from the local interactions of micro-entities, emerge macro-level patterns, forms of 'self-organization' that are non-linear and counter-intuitive.

Applying complexity thinking to human societies, institutions matter because they articulate and enforce the rules of interaction among social agents. Such rules are socially and politically constructed. Nevertheless, much of the complexity literature that deals with human societies gives only limited attention to such institutions and rules. Thus, for example agent-based modelling can incorporate forms of social action that are oriented to shared norms and understandings (see for example Cliffe et al., 2007). Even here, however, the focus is on how norms may develop endogenously, as the emergent product of micro-interaction. There is little acknowledgement of the pre-existing normative and institutional structures that confront and dominate our lives, lending or denying authority to the choices we make.

Institutions constrain and channel agent interactions; however, they can also be subverted from below or reformed from above, as actors lift their gaze, reflect upon the overall socio-economic system in which they live and reinforce or reshape the rules and architectures of those systems. This is the stuff of politics and political choice. It suggests an agenda for complexity research in social science which builds upon, but is distinct from, that in the natural sciences. It accepts the potential value of modelling social dynamics as a self-organizing system, analogous to those in the natural sciences. On the other hand, it insists that social science must also be centrally interested in the socio-political processes by which these dynamics are re-shaped. It is to such an institutionally grounded complexity perspective that this chapter contributes (for an elaborated treatment see Room, 2011).

TWO CONCEPTS OF POWER

The literature on complex systems offers powerful insights into the dynamics of physical and biological phenomena and those of human societies. Nevertheless, in presenting human societies as *systems*, it also enters long-running disputes over the role of human agency, reflection and purposefulness. These disputes also bring competing understandings of what we mean by power.

One of the classic treatments was Talcott Parsons' sociological essay 'On the concept of political power' (1963). He posed two questions in particular. First, is power zero-sum (as for example argued by Weber and Wright Mills) or non-zero-sum (as Parsons himself tended to argue)?[2] Second, should we conceive of power as the generalized capacity of a political system to 'get things done' (Parsons); or as the capacity of one individual or group to secure their purposes, even against the resistance of others (Weber and Wright Mills)? Nevertheless, as Parsons concludes, on neither of these questions may it be a simple matter of one or the other.

Notice that Parsons' preferred position (non-zero-sum, generalized capacity) is that of someone conceptualizing society as a system, as against those taking human agency as the starting point. The same conceptual tension provided the focus for a subsequent essay by Dawe, 'The two sociologies' (Dawe, 1970). One sociological tradition – he instances Durkheim, Spencer and Parsons – is centrally concerned with 'the problem of order': how social systems avoid descending into chaos. Here individuals tend to be seen as adjusting to the social structures in which they find themselves enmeshed.

The other tradition – Dawe instances Weber – is concerned in contrast with 'the problem of control': the efforts of individual actors to give coherent meaning to their biographies and to fulfil the projects which will deliver this. This involves an action frame of reference which brings individual definitions of the situation centre-stage; as also therefore the contestation of such definitions and the social interactions – cooperative, competitive or conflictual – which ensue. Social systems here appear as the outcome – to some extent unanticipated – of social interactions among a myriad of individuals, amidst the institutional legacy of previous struggles.

From this second standpoint, the first tradition is guilty of 'reification' – treating society not as a human creation but as a 'thing' to which humans must adapt. It therefore implies that the task of the sociologist is to analyse the social structures and constraints in which humans find themselves, so that they may better understand and adjust to such necessities. Against this, the second tradition aims to 'demystify' such reification and to reveal the human and historical origins of social and political arrangements. By illuminating the social world, moreover, it aims to enable social and political actors to take responsibility and change it. And, not least, it aims to hold to account those who exercise power, seen as zero-sum capacity to overcome the resistance of others.

There have been a variety of efforts to transcend this gulf. These include, for example, the 'social constructionism' of Berger and Luckmann (1967) and the 'structuration' of Giddens (1984). Many of those writing on complex systems would see their work as a further contribution – think for example of agent-based modelling and the 'emergent' properties of social systems that this can reveal. So would those writing in the 'realist' tradition, many of whom take inspiration from complexity thinking (Byrne and Ragin, 2009).[3] Nevertheless, if we re-visit the concept of power, and how this is treated in the complex systems literature, we may be disappointed. It is to this that the present chapter aims to contribute.

POWER, COMPLEXITY AND EMERGENCE

Across wide swathes of the literature on complex systems, as applied to human societies, the concept of power – in the sense used by sociologists and political scientists – is largely absent. We look in vain among some of the classics of the literature, including social scientists such as Schelling and Arthur.

The same goes for writers who apply evolutionary models – an important example of complex systems – to social change (Room, 2011: Ch. 3).[4] Thus, for example, Potts and Dopfer, who have made important contributions to evolutionary economics, acknowledge the importance of politics and power differentials, and they recognize that evolutionary dynamics may produce 'exclusion, extinction, domination or enslavement' (Dopfer and Potts, 2008: 75, fn. 43). Nowhere, however, do these come to the centre of their attention. They defend their approach in terms of an intellectual division of labour: evolutionary economics focuses upon choice, whereas an evolutionary approach to sociology or politics would emphasize power and lack of choice (Dopfer and

22 *Handbook on complexity and public policy*

Potts, 2008: 40). Nevertheless – and notwithstanding their protestations to the contrary (Dopfer and Potts, 2008: 96, para. 1.3) – they have not entirely escaped from the intellectual attractors within which their own journey began: Hayekian laissez-faire economics and post-Darwinian evolutionary theory. This leaves them with a view of history – and of public policy – that underplays fundamental political choices.

Evolution by natural selection is a blind process, effected through differential population dynamics. This is a fruitful perspective to apply to the waves of innovation that characterize market economies, as Dopfer and Potts demonstrate. This is, however, only part of the picture. In human societies, people also in some degree make their own history. They probe and they experiment, not randomly but by systematic testing and learning (Bronowski, 1981: Chs 2–4). They seek strategically to re-shape the technologies and institutions of their world. They develop thereby their understanding and their capacities; their control over their lives; their positional advantage and leverage. This brings interests and power and politics centre-stage (Room, 2012).[5]

Among writers on agent-based modelling, Squazzoni is one of those most sensitive to these sociological debates (Squazzoni, 2012). He argues the need 'to embed ABMs within the entire set of empirical methods for social science', including qualitative interviews and questionnaires and large-scale statistical surveys (Boero and Squazzoni, 2005: 5.3). Even so, 'power' is not given any significant attention in his work or in the great mass of writings on ABM. This is perhaps no accident. ABM is concerned with patterns that emerge bottom-up from the interactions among agents. However, patterns are also imposed top-down by more powerful actors, as they mobilize the institutional legacy of the past. This is missed.

The principal – if limited – reference to 'power' in the literature on complex systems is in the context of 'emergent' phenomena. It is in this vein that Elder-Vass (2010) entitles his 2010 book *The Causal Power of Social Structures: Emergence, Structure and Agency*. However, this might seem to be a notion of 'power' quite different from the sociological concepts with which this chapter is principally concerned.

Elder-Vass is an exponent of 'realism', as developed by philosophers of science such as Harré (1972: Ch. 4). Realism insists that if we seek to develop explanations of phenomena, it is not enough to establish correlations of independent and dependent variables by appropriate statistical techniques. Explanation – in both the natural and the social sciences – must also include an account of the processes, the 'generative mechanisms', which produce the patterns we observe. Ontology matters.

For Harré, these 'generative mechanisms' involve potentialities or powers that are unlocked or closed down by different contextual conditions. Explosives such as gunpowder and dynamite provide an example he frequently cites. The chemical composition of the explosive provides the capacity or power to explode: but whether or not it does so depends on such factors as the absence of damp, the presence of oxygen, the ambient temperature and so on. Causal analysis of generative mechanisms aims at revealing such contingencies.

It is in this spirit that Elder-Vass investigates the causal powers of entities in the physical, biological and social worlds: the generative mechanisms which produce the patterns we observe, which may be unanticipated, counter-intuitive and non-linear. His interest is in those generative mechanisms which operate only when the different elements of a system come together – and which are therefore to be seen as powers of that whole. Thus does Elder-Vass connect causal power with the notion of 'emergence' in the literature on complex systems.

At first glance, we have here a distraction from our concern with the sociological debates on power with which we started – a false trail prompted by the exposition of 'causal powers' offered by Harré and his successors. Elder-Vass in his closing pages (Elder-Vass, 2010: 204–5) recognizes that there is still a need to theorize power – as distinct from causal powers – and its relationship to structure and agency, and that he has at most provided an ontological foundation for this. Nevertheless, building on this recognition, we can return to some of our earlier questions about power in our sociological sense and show how a complexity perspective can enhance and illuminate those debates.

This realist concern with 'causal powers', far from being a distraction, can directly inform our sociological quest. As we have seen, Harré would have us think of causal processes as releasing or blocking potentialities (Harré, 1972: 121–2). In human affairs such blockages – and their release – are constructed institutionally and historically. The same indeed goes for the potentialities themselves: here the parallel with explosives such as gunpowder and dynamite breaks down, for in their case the potentialities are fixed by their chemistry.[6] Applied to human societies, any realist account of potentialities created, unlocked or closed down is an account of this contingent historical and political struggle.[7]

THE DYNAMICS OF SEGREGATION AND POLARIZATION

We now take the conceptual distinctions made by Parsons as the standpoint from which to re-examine the contribution of the complexity literature to the sociological study of power. We recall first the questions around power seen as a zero-sum concept, and involving the capacity of one individual or group to secure their purposes, even against the resistance of others.

In recent decades, Lukes (1974; 2005) has provided the most prominent point of reference in sociological and political science debates around such a zero-sum concept of power. Much of this has been concerned with the three 'dimensions' of power that Lukes distinguishes and, in particular, the extent to which his third dimension can be operationalized empirically.[8] Lukes has nuanced his original argument in a number of ways, in response to the critics, but has retained the core of his original argument (see for example the 2006 *Political Studies Review* symposium and the response that Lukes (2006) makes there to his critics).

Lukes is, however, concerned not only to analyse the exercise of power by reference to these three dimensions, but also to hold to account those who exercise power. In the afore-mentioned symposium, Hayward (2006) addresses this challenge, of fixing responsibility and holding powerful actors to account. She insists, however, that this should apply not just to individual power holders, but also to interconnected arrays or communities of actors whose actions impinge destructively on the interests of others. After all, no individual power holder is entirely autonomous: a wider array of actors is invariably complicit and they too should be held responsible.

The specific example that Hayward and Lukes debate is that of inner city flight: when middle class households flee to the suburbs, leaving a shrinking tax base and crumbling

24 *Handbook on complexity and public policy*

public services for the disadvantaged households who remain, instead of staying and using their political muscle to resist and reverse that downward spiral. No one household can be said to be individually responsible for pushing the inner city beyond some 'tipping point'; nevertheless, Hayward wants them to be held collectively responsible, held to account for their collective exercise of power, their destruction of the hopes and living conditions of their erstwhile poorer neighbours. After all, she argues, when a larger array or community of actors behaves in ways that have destructive consequences, it is rarely the case that none of them could have acted otherwise and that all of them were wholly and necessarily ignorant of those consequences. Lukes seems undecided as to whether this is appropriate.

In an early piece on 'The concept of power', Dahl (1957: 80) describes his 'intuitive idea of power' as 'something like this: A has power over B to the extent that he can get B to do something that B would not otherwise do'. This definition assumes that we can readily identify A and B and the effect of A's actions specifically on B. Hayward asks that even without this specific attribution, this chain of effects, we should identify such white flight with an exercise of power. The argument has obvious parallels with attempts in jurisprudence to hold collectivities – corporations, nations – responsible for their actions, not just as organized hierarchies under the command of particular leaders, but as dispersed communities, all of whose members played some part in producing the negative outcome in question.

Now consider Schelling (1978: Ch. 4), one of the classics of complexity writing, who uses cellular automata to illuminate the dynamics of racial zoning in cities. His cellular automata live at different addresses on a grid or lattice. Schelling then posits a 'tolerance schedule', a preference not to have within one's immediate neighbourhood more than a certain threshold proportion of another race; the general racial composition of the larger city is, however, of no concern. Starting with an initial random distribution of households across the city, Schelling conducts repeated simulations, as households walk step-wise to an adjacent address, whenever their immediate neighbours exceed this racial threshold. He shows that even a rather mild level of racial antipathy – and concerned only with immediate neighbours – will quickly generate zones of racial segregation across the city, and to an extent that is much greater than simple extrapolation from the thresholds of antipathy at the micro-level might lead us to expect.

'Macro-behaviour' – in the form of a racially segregated city – is here the emergent result of 'micro-motives'. How far should we regard the processes of racial segregation embodied in Schelling's model as involving the exercise of power? The limitation is that in Schelling's model, members of the two racial groups are assumed to be equal bearers of the mild intolerance that sets the process in motion, from which high levels of segregation eventually emerge. His model omits an additional key element of Hayward's concerns: the way that racial and other groups become segregated as one group seeks to occupy and control particular scarce resources and opportunities. Thus for example Schelling's model assumes city residents can move to an adjacent property without cost or restriction: real world housing markets operate very differently. What is more, as particular social or racial groups coalesce in particular neighbourhoods, this in itself tends to affect the desirability and pricing of that housing, in ways that can reinforce the zoning. To this extent Schelling's model ignores the dynamics of cumulative disadvantage that develop in the inner city as a result of 'white flight' to the suburbs.

Nevertheless, we can still pose the question: when more affluent and advantaged residents use their resources to move out of mixed neighbourhoods and into the suburbs, can the 'emergent' results of their myriad actions be regarded as an exercise of power, allowing them to secure their purposes, even against the resistance of others? And for this should they be deemed collectively responsible and held to account? To answer in the affirmative – as Hayward would wish – suggests that complexity analysis can play a central and essential role in the analysis of power dynamics: in revealing the emergent processes by which micro-behaviours unwittingly produce such destructive and cumulative effects.

It is in this sense that Elder-Vass's concern with the 'causal power of social structures' is relevant to our own interest in power. Both Elder-Vass and Schelling are interested in the generative processes that can develop in complex systems, producing the macro-patterns we observe. To repeat, however, Schelling ignores the efforts of different groups to occupy and control particular resources and opportunities, and the dynamics of cumulative disadvantage that then develop. As well as patterns that emerge bottom-up from the interactions among agents, we have here patterns imposed by groups of powerful actors acting top-down to shape and reinforce their positional advantage, by excluding others.

In terms of the policy implications that one might draw from Schelling, one seems clear: to rely on encouraging general multiracial tolerance is unlikely to secure racial mixing. What Schelling omits is the institutional and policy context in which racial segregation develops: the housing market but also the labour market and city policies on zoning and development. To these questions we will return.

THE DYNAMICS OF ECONOMIC DEVELOPMENT

We turn now to power as the non-zero-sum generalized capacity of a political system to 'get things done'. We introduced this originally by reference to Parsons. He treats power conceived in this way as a property or resource of a social system: indeed, he presents it as analogous to the role of money in the economic system, a medium of exchange and a standard of value, an indispensable enabler of economic life.[9] This is, however, a rather static notion of a social system and its capacity to 'get things done'.

Among the complexity writers of recent times, Arthur et al. (1997) has provided one of the best-known accounts of the economy as an evolving complex system. It echoes Marshall, Young, Myrdal and Kaldor, scholars of an earlier generation who highlighted the dynamic relationship between innovation and the expansion of markets; the path dependency of economic development; and the processes of cumulative change that this involves (Toner, 1999). It requires that we recognize economic activity as fatefully re-shaping its environment and producing self-reinforcing feedback effects. This makes for increasing returns of a sort fundamentally at odds with academic economic orthodoxy, in the Anglo-Saxon world at least, over the last century. Nevertheless, Arthur offers little in the way of policies for purposive economic development: for this we must look elsewhere.

One such writer is Hirschman (1958), concerned with the generalized capacity of a political system to 'get things done', in particular, to 'do' economic development.

26 *Handbook on complexity and public policy*

Hirschman does not believe in 'balanced' economic development across the various sectors that will make for a developed economy. Instead he advocates development policies that will generate and maintain 'tensions, disproportions and disequilibria' (Hirschman, 1958: 66). Investment on one front, if carefully chosen, will provide leverage for accelerated development on other fronts. At each stage, the State must combine and mobilize resources so as generate the increasing returns and positive feedback effects to which Arthur points. This will build the generalized capacity of the economic and political system to get things done: or, in the language of Elder-Vass, the generative mechanisms from which economic prosperity can emerge, by pulling together the elements of that system, so as to realize its causal powers.

Hirschman takes for granted that the State is the prime mover in development. Nevertheless, investment by the State will open up market opportunities and incentives for further investments by private entrepreneurs, and those investments will then induce further rounds of investment by the State.[10] He sees the task of development policy as steering the most effective co-evolution of the two.

Hirschman tells us (1958: 34) that in advanced capitalist economies there is a great supply of entrepreneurs 'who are especially trained in the art of perceiving and ferreting out economic opportunity'. In under-developed countries their numbers may be lower, although once the development process starts, they fairly quickly emerge. But what does this art of the entrepreneur typically involve, 'ferreting out' such opportunities? Here Hirschman is somewhat vague and we need to fill out the argument.

The birds follow the gardener, as s/he breaks up the soil and disrupts the cosy shelter afforded to beetles and worms (Room, forthcoming). The small fry nuzzle up to the big players, because these shakers and movers disrupt their larger environment, as they pursue their own projects. Nuzzling provides safe ground and reduces uncertainty, but it also allows selective probing for new possibilities, amidst the interstices of the big actors.[11] It is in this that the 'art of perceiving and ferreting' surely consists (see for example Gans and Stern, 2003). Even so, most entrepreneurs whose quest bears fruit are gobbled up by the big movers; most fail even before they are worth gobbling and only rarely do they gobble the gobblers.

Hirschman – who in many ways anticipated later writers on complex systems – thus has a much more dynamic approach to the development of generalized capacity than Parsons offered. As that capacity develops, through the 'tensions, disproportions and disequilibria' that Hirschman describes, there is an endless struggle, among the ferreting entrepreneurs, to capitalize on such tensions and probe for new opportunities. It is here that zero-sum power struggles are fought – but on an evolving, not a static landscape.

Hirschman does not, however, explicitly address power as a zero-sum struggle. He seems not to recognize that as the State builds generalized capacity, this will affect different groups of the population in different ways. Development is hedged with uncertainty, but this is not uncertainty for everyone equally. It is driven disproportionately by the powerful: their projects generate disruption and the burden of uncertainty is displaced disproportionately onto those who are weaker (Marris, 1996: Ch. 7). Indeed, Hirschman (1958: 59) explains the success of capitalist societies, in terms of this offloading of the costs of change, from the capitalist onto other social groups. Nobody holds them to account or reins them in.

Hirschman, as we have seen, assumes that the State is the prime mover in development, and that its investments will in turn open up a range of new market opportunities for private entrepreneurs. However, those entrepreneurs include not only the small fry but the large multinational corporations. With his focus on the development dynamics within individual countries, Hirschman can perhaps be forgiven for reducing his focus to the State and the indigenous entrepreneurs he hoped would emerge (Adelman, 2013). Nevertheless, it is multinational corporations who today in considerable measure drive development in both the advanced and developing economies. These – and not just the State – are the big actors amidst whose interstices the small fry must carve their niches.

POWER AND SELF-ORGANIZATION

Writers on complexity examine the patterns that 'emerge' bottom-up from micro-interactions among the elements of a system, producing different forms of 'self-organization'. This is true of Schelling and Squazzoni and other exponents of agent-based modelling; Elder-Vass, concerned with the causal powers that systems develop, as their various elements combine; and evolutionary economists such as Dopfer and Potts, analysing the economy as a complex system of co-evolving elements.

The preceding discussion builds on this literature. In reference to Lukes, we have argued that an analysis of emergence and self-organization can illuminate the exercise of power within dispersed communities. It can help in tracing how such communities of individuals, responding to the social and market signals they individually confront, together set in motion a larger process that despoils particular communities of their fellows. In orthodox economic analysis, individuals who respond to market signals and incentives are not changing or reinforcing the system, nor are they exercising power. In a complex system in contrast, individuals respond to system signals; they do so individually, but the result of these myriad responses can be to shift the system rapidly to a quite different state.

This is 'emergence'. It develops 'bottom-up'. It displays clear parallels with processes of self-organization within physical and biological systems. This is, however, only half of the picture. In human societies, actors at all levels seek to extend their capacities and their positional advantage within the complex world they inhabit. They not only observe local signals; in varying degrees they monitor the larger and non-linear transformations that are under way, and these they seek to re-organize and steer.

This may involve particular social or racial groups coalescing in particular neighbourhoods and then using their collective resources to exclude other groups, something that is absent from Schelling's model. It may also involve powerful interest groups intervening to re-organize the institutional and policy context in which racial segregation develops – the housing market but also the labour market and city policies on zoning and development. Here the social researcher has the task of penetrating the veil of secrecy that surrounds the actions of powerful interest groups, including for example industrial lobbies. This is essential to any understanding of zero-sum power and the assignment of responsibility for its consequences.

If Lukes and Hayward want to assign responsibility for the consequences of the exercise of power, Hirschman assigns credit. He celebrates the way that the State, the prime mover, combines and mobilizes resources, thereby opening up opportunities and

28 *Handbook on complexity and public policy*

incentives for further investments by private entrepreneurs. Between them, they can make use of 'tensions, disproportions and disequilibria' to leverage accelerated development. This is a process of co-evolution, albeit one which must be steered and cultivated by the State.

This means that self-organization is also central to Hirschman (albeit not a term he himself uses). Development strategy mobilizes such processes of self-organization, so as to develop the generalized capacity of the society in question and the non-zero-sum power that this entails. Nevertheless, as we have suggested, this leaves Hirschman somewhat neglecting power as a zero-sum struggle. In addition, he looks to the emergence of indigenous entrepreneurs, but neglects the part played by large multinational corporations in shaping the direction of development, in both the advanced and developing economies.

Here are strategic agents with the power to shape and dominate a zero-sum struggle, and to write the rules of the game so as to cream off the lion's share of any enhancement of a society's generalized capacity (Lieberson, 1987). The claim that the market is a self-organizing system then risks becoming an ideological and self-serving mask. Again therefore, this underlines the significance not only of bottom-up self-organization of any complex system, but also the efforts by strategic agents to re-organize and re-direct that system.

In short: the order and regularities of social life may attest in part to its self-organization as a complex system, but they attest no less to the exercise of power, and the success of some social actors in negotiating or imposing that order on others. This brings interests and power and politics centre-stage: complexity analysis must be combined with political economy (Room, 2011).

CONCLUSION: COMPLEXITY, POWER AND POLICY

Lukes and Hayward distinguish two ways in which we can hold power-holders to account. There is on the one hand the 'backward-looking' attribution of blame or credit for the present situation. It is with this that we have so far been primarily concerned. There is also, however, the 'forward-looking' attribution of responsibility for addressing that situation henceforth, on the part of 'politically responsible agents, in strategic positions, who are able to make a difference' (Lukes, 2006: 172). This is the realm of public policy.

To identify such agents and to suggest how they might 'make a difference' is no easy task. As we have seen, complexity science alerts us to a world of path-dependent trajectories, with positive feedback, lock-in and ratchet effects, and with forms of 'self-organization' that are non-linear. Schelling's citizens may express only mild levels of preference to live alongside neighbours of their own ethnicity, but from such micro-motives they find that they have collectively produced a city of sharp racial divisions. This 'macro behaviour' is as counter-intuitive and unintended as the descent into war in 1914 seems to have been to many of the governments involved (Macmillan, 2013). This is, more generally, a feature of what have come to be described as 'wicked' policy problems, involving complex trade-offs which allow no easy resolution and may involve path dependencies and lock-ins which are difficult to escape.

Nevertheless, Hirschman reminds us that the State need not be a mere bystander. It can, in some degree, expánd the generalized capacity of the society and shift the ground on which zero-sum struggles are fought out. Not even the most wicked of policy problems is wholly inescapable. However, the task of leveraging 'tensions, disproportions and disequilibria', so as to have public and private stakeholders working in tandem, is perhaps more of a challenge than Hirschman suggests.

The policy maker must be alert to how, in Elder-Vass's words, to combine different elements of a social system so as to enhance its emergent powers. In a complex world, the conventional orthodoxy of 'evidence-based policy making' and 'what works', with a preference for randomized controlled trials, is unlikely to be of much help. Policy interventions are not isolated: they are launched into a crowded policy ecosystem. The impact of any intervention is generally contingent on the various synergies that it develops – or fails to develop – with other interventions and on the ways that those affected lever it to their own particular ends. What matters are the transformative synergies that develop *among* these interventions and their respective stakeholders. In all this, interests and power and politics are centre-stage (Room, 2013).

For Lukes, this forward-looking responsibility for public policy is meant to address the wanton damage that the exercise of power can produce. In a complex social system it is never wholly possible to allocate clear responsibility for such damage: it is necessary for the State to take ultimate responsibility for solidarity and compensation across the society as a whole, if those damages are not to lie where they fall. This was a central tenet of Titmuss, perhaps the most authoritative of the post-war generation of scholars who mapped the contours of social policy (Titmuss, 1968: Ch. 11).

Nevertheless, this forward-looking responsibility is not just about repairing damage: it also, as Hirschman demonstrates, involves expanding capacities and horizons. Public policy – and the implicit social contract or settlement that it embodies between government and governed – has to combine social investment and social compensation. For this, it is insufficient to look to the economy's self-organizing propensities. Civility does not 'self-organize'; it must be politically constructed, and we cannot escape the social and political choices of our time.

ACKNOWLEDGEMENTS

I am grateful to Jean Boulton, Graham Brown and Steven Lukes for comments on an earlier draft of this chapter. It draws upon a programme of work supported by the Economic and Social Research Council (Award RES-063-27-0130) and my book *Complexity, Institutions and Public Policy: Agile Decision-Making in a Turbulent World* (2011).

NOTES

1. There is a range of good texts for the general reader – and for the social scientist approaching these matters for the first time – which can be consulted for elaboration of the exposition offered here (Ball, 2004; Buchanan, 2000; Johnson, 2001; Waldrop, 1992). The paradigm is admittedly somewhat inchoate,

30 *Handbook on complexity and public policy*

even among the Santa Fé community (Cowan et al., 1994), as perhaps befits any lively and novel enterprise; nevertheless the elements highlighted here would probably command general assent. See also the opening chapter of this volume.

2. Power is said to be zero-sum if the more power that I have in my relationships, the less by definition do the others involved in those relationships. The same goes for any cake that is being divided up: the more one person has, the less there is for the others. Even if a bigger cake is brought, the portions that are made available must still add up to one cake.

3. Compare this chapter's discussion of the realist tradition in the philosophy of science with the realism discussed in international relations and political theory in Adrian Little's chapter.

4. It is important to distinguish these from those writers who purport to explain social phenomena by reference to human propensities – including the pursuit and exercise of power – which may have roots in evolutionary biology. What such attempts at explanation cannot do is explain the great diversity of social phenomena produced by biologically identical human beings: to this extent, their efforts are largely a waste of time.

5. The same dynamic interrelationship of blind evolution and strategic choice is evident in Darwin's own thinking (Room, 2012). His account of natural selection began with his observation of husbandry and 'artificial selection', as practised by the pigeon breeders and horticulturalists of his day (Darwin, 1859: Ch. 1). They looked out for novel characteristics in the offspring of each new generation that would better meet their requirements. These superior varieties were selected for breeding, so as to combine and progressively accentuate these advantages. From here Darwin made the mental leap to posit 'natural selection', with the harsh struggle for scarce sustenance culling the less fit as rigorously – albeit over a much longer time period – as the breeder or horticulturalist.In adapting his model of evolution and natural selection to human societies, we move back into the practices of active husbandry from which Darwin began. These are the arts of civilization. Instead of blind adaptation of a population to different environments, they involve reflection, learning, experimentation, collaboration and the growth of knowledge. This is true of the husbandry of pigeons and livestock and plants: it is also true of the 'cultivation' by entrepreneurs of new technologies and institutions. Nevertheless, processes of evolution continue, with an endless dance between natural and artificial selection, as humans cultivate but also respond to the wild.

6. Indeed, even in the case of explosives, practical actors are likely to be interested not in gunpowder per se, but in the weapons technologies of which it is no more than a component, and whose potentialities, far from fixed, are the stuff of desperate arms races.

7. Marris (1996: Ch. 7) offers a similar account of 'power as the mastery of contingencies rather than the accumulation of assets'.

8. Lukes starts from the assumption, made by American political scientists such as Dahl, that power within democratic political institutions is wielded by voting strength within deliberative assemblies. This is what Lukes describes as the first dimension of power. Against this, he sets the power to shape and limit what issues get onto the agenda of those assemblies: this is the second dimension. But he then appeals to such writers as Gramsci, to argue that the cultural power of particular social groups may be such as to hide from the disadvantaged the way that their interests and needs are systematically disregarded. This is what Lukes describes as the third dimension of power. It is around the conceptual definition of this, and its empirical investigation, that much of the subsequent debate has centred. To repeat, Lukes took as his starting point some central tenets of American liberal democratic theory, subjecting them to fundamental criticism. This was appropriate to its times: the late 1960s and early 1970s saw vigorous re-examination of that tradition, partly in response to the upheavals of the civil rights movement and the Vietnam War. But this does mean that care must be taken that those origins do not limit the discussion of power today. It was with his third dimension that Lukes rested his case: he had done enough to undermine his chosen target. Subsequent writers such as Foucault have developed a broader account of power in capitalist societies, one which embraces Lukes' three dimensions but also the positive notion of power as generalized capacity to get things done (Gaventa, 2003). Foucault's ambition therefore has parallels with Berger and Luckmann and with Giddens, mentioned earlier in this chapter, and with earlier writers such as Marcuse (1964). See Lukes (2005: 89–107, 139–43) for his own assessment of both Foucault and Bourdieu, in relation to these conceptual and methodological debates.

9. Just as we earlier noted the context in which Lukes wrote his classic text on power, so here we need to recall the context in which Parsons wrote. The Western economies had failed to 'get things done' during the 1920s and 1930s: the economic and political failure of the Depression was still alive in public memory.

10. These linkages and inducements hold echoes of Keynes' investment accelerator; they form a key part of Hirschman's argument, but the argument remains rather general.

11. Burt (2003) uses network analysis to depict the role of brokers who span gaps across unconnected regions of the network, merchants, idea-brokers and so on who see an opportunity for profit or arbitrage, at least while they monopolize the connection in question. It is often by spanning such connections between big actors that innovators develop their own niche.

REFERENCES

Adelman, J. (2013), *World Philosopher: The Odyssey of Albert O. Hirschman*, Princeton, NJ: Princeton University Press.

Arthur, W.B., S.N. Durlauf and D.A. Lane (eds) (1997), *The Economy as an Evolving Complex System II*, Boulder, CO: Westview.

Ball, P. (2004), *Critical Mass: How One Thing Leads to Another*, London: Heinemann.

Berger, P.L. and T. Luckmann (1967), *The Social Construction of Reality*, Harmondsworth: Penguin.

Boero, R. and F. Squazzoni (2005), 'Does Empirical Embeddedness Matter? Methodological Issues on Agent-Based Models for Analytical Social Science', *Journal of Artificial Societies and Social Simulation*, 8(4).

Bronowski, J. (1981), *The Ascent of Man*, London: Futura.

Buchanan, M. (2000), *Ubiquity*, London: Weidenfeld and Nicholson.

Burt, R.S. (2003), 'Structural Holes and Good Ideas', *American Journal of Sociology*, 110(2), 349–99.

Byrne, D. and C. Ragin (eds) (2009), *The Sage Handbook of Case-Based Methods*, London: Sage.

Cliffe, O., M. De Vos and J.A. Padget (2007), 'Specifying and Reasoning about Multiple Institutions', in J. Vazquez-Salceda and P. Noriega (eds), *Lecture Notes in Computer Science*, Berlin: Springer, 4386: 63–81.

Cowan, G.A., D. Pines and D. Meltzer (eds) (1994), *Complexity: Metaphors, Models and Reality*, Reading, MA: Addison-Wesley.

Dahl, R.A. (1957), 'The Concept of Power', *Behavioural Science*, 2(3), 201–5.

Darwin, C. (1859), *The Origin of Species*, (reprinted 1998), London: Wordsworth.

Dawe, A. (1970), 'The Two Sociologies', *British Journal of Sociology*, 21(2), 207–18.

Dopfer, K. and J. Potts (2008), *The General Theory of Economic Evolution*, London: Routledge.

Elder-Vass, D. (2010), *The Causal Power of Social Structures: Emergence, Structure and Agency*, Cambridge: Cambridge University Press.

Gans, J.S. and S. Stern (2003), 'The Product Market and the Market for "Ideas": Commercialization Strategies for Technology Entrepreneurs', *Research Policy*, 32(2), 333–50.

Gaventa, J. (2003), *Power after Lukes: A Review of the Literature*, Brighton: Institute of Development Studies.

Giddens, A. (1984), *The Constitution of Society*, Cambridge: Policy Press.

Harré, R. (1972), *The Philosophies of Science*, Oxford: Oxford University Press.

Hayward, C.R. (2006), 'On Power and Responsibility', *Political Studies Review*, 4(2), 156–63.

Hirschman, A.O. (1958), *The Strategy of Economic Development*, New Haven, CT: Yale University Press.

Johnson, S. (2001), *Emergence: The Connected Lives of Ants, Brains, Cities and Software*, Harmondsworth: Penguin.

Lieberson, S. (1987), *Making it Count: The Improvement of Social Research and Theory*, Berkeley, CA: University of California Press.

Lukes, S.M. (1974), *Power: A Radical View*, London: Macmillan.

Lukes, S.M. (2005), *Power: A Radical View*, 2nd edn, Basingstoke: Palgrave Macmillan.

Lukes, S.M. (2006), 'Reply to Comments', *Political Studies Review*, 4(2), 164–73.

Macmillan, M. (2013), *The War that Ended Peace: How Europe Abandoned Peace for the First World War*, London: Profile Books.

Marcuse, H. (1964), *One-Dimensional Man*, London: Sphere Books.

Marris, P. (1996), *The Politics of Uncertainty*, London: Routledge.

Parsons, T. (1963), 'On the Concept of Political Power', *Proceedings of the American Philosophical Society*, 107(3), 232–62.

Room, G. (2011), *Complexity, Institutions and Public Policy: Agile Decision-Making in a Turbulent World*, Cheltenham, UK and Northampton, MA, USA: Edward Elgar Publishing.

Room, G. (2012), 'Evolution and the Arts of Civilisation', *Policy and Politics*, 40(4), 453–71.

Room, G. (2013), 'Evidence for Agile Policy Makers: The Contribution of Transformative Realism', *Evidence and Policy*, 9(2), 225–44.

Room, G. (forthcoming), 'Nudge, Budge or Nuzzle?'

Schelling, T.C. (1978), *Micromotives and Macrobehaviour*, London: W.W. Norton.

Squazzoni, F. (2012), *Agent-Based Computational Sociology*, Chichester: Wiley.

Titmuss, R.M. (1968), *Commitment to Welfare*, London: Unwin.

Toner, P. (1999), *Main Currents in Cumulative Causation: The Dynamics of Growth and Development*, London: St Martin's Press.

Waldrop, M.M. (1992), *Complexity: The Emerging Science at the Edge of Order and Chaos*, London: Viking.

3. Complexity and real politics
Adrian Little

INTRODUCTION

The invocation to 'get real' is now commonplace in the social sciences. Across the disciplines that are part of the broad social scientific family, we see appeals to relinquish abstract philosophy that is not applied to particular real world problems or to focus our attention on the practical policy implications of theoretical arguments. More prosaically, academic researchers are encouraged to demonstrate the 'impact' of their work in practical terms. In terms of making some of the less engaged, academic work relevant to contemporary debates, these calls are not without merit especially if they encourage us to base research agendas around particular problems rather than articulating answers (often reflective of particular ideologies) before we have even identified what the problems are. However, the notion of 'getting real' is also problematic in itself in so far as it embodies the assumption that there is a 'real world' that is agreed, incontrovertible and easily delineated by 'facts' that we just read off from analysis of the prevailing conditions. An alternative view, and one that will be pursued in this chapter, is that the constitution of the real world is much more complicated than the 'fact-derived' account outlined above indicates (Little, 2015b). Indeed, the major contention is that by considering the temporal and epistemological dynamics of complexity theory, we need to think in a more critical fashion about how we construct the real world that more abstract social scientific theorizing is supposed to relate to in concrete policy terms.

In the last decade there has been a shift in political theory in particular to rethink the nature and meaning of realism. One reason for this is a critical reaction to the use of the term 'realism' in the international relations (IR) literature (Der Derian, 1995; Walker, 1993; Guzzini, 2004). Although IR realism, often associated with Kenneth Waltz and his ilk, is highly diverse, it does tend towards a combination of hard-headed practical wisdom, advocacy of scientific positivism, calculations of rational self-interest and a focus on the operation of power. But, as Guzzini (2004: 538–48) argues, practical wisdom and scientific positivism are often in tension or contradiction with one another and advocacy of the former often negates the need for the latter. Moreover, Guzzini also points to the rather one-dimensional understanding of power in IR realism which fails to comprehend the diffuse and relational ways in which power affects political actors.

The second main reason for the emergence of debates on realism focuses on political theory and its role in relation the practical world of politics. In particular, it is a debate that has focused on the distinction between theory and practice and the extent to which it is useful to conduct theoretical research in abstract terms which do not relate to the real world. Where commentators such as Swift and White (2008) maintain a distinction between the short-term, strategic considerations of political actors and the more abstract and rarefied activities of political philosophers, others such as Philp (2010) and Williams (2005) are less willing to distinguish between the two in such clear-cut ways. This has

Complexity and real politics 33

spawned a substantial debate in political theory, with Baderin (2014) identifying two main camps: 'detachment realists' who focus on the distance between abstract political philosophy and the practical realm of political decision-making, and 'displacement critics' who suggest that political theory threatens real politics.

While discussing briefly both the debates in IR and political theory on realism, this chapter focuses most attention on two further approaches to these problems. The first is the emergence of complexity theory in politics and the ways in which this unsettles the established narratives around what constitutes realism. The second is phronetic social science, which invites political analysts to engage in highly contextualized case study work if we want to have anything meaningful to say about the real world. This chapter contrasts the picture of the real that has emerged from complexity theory and phronetic social science, and suggests that phronesis has at least two major shortcomings in conceptualizing real politics that are highlighted by complexity.

Although realism can take on many different forms in politics, it is understood here as usually predicated on two dimensions: first, a belief in the existence of self-evident or incontrovertible 'facts' that are 'read off' to provide the background conditions for political action; second, realism then tends to make assumptions about the motivations of political actors based on their pursuit of particular (and often self-interested) rationalities.

In the last decade, the limitations of this conception of realism have been highlighted in various ways by political theorists including Bernard Williams (2005), Raymond Geuss (2008), Mark Philp (2010) and William Galston (2010). However, none of these commentators have interrogated realism through the lens of complexity theory. Their focus has been on the flawed interpretation of realism as a form of political strategy and the methodological limitations that emanate from 'the seductive power of normative ethical thinking' (Freeden, 2014: 1). Thus, there has been less attention paid to the ways in which a focus on complexity theory and attendant concepts such as path dependence, emergent properties, and adaptive systems might make us reconsider what is actually considered to be 'real'. That is, the established debates have failed to grapple with the *constitution of the real* that realism purports to address (Little, 2015b). This chapter examines some of the key dimensions of complexity theory, the ways in which they contribute to a reconsideration of the constitution of the real and the implications of this rethinking for how social scientists should conceptualize realism.

In the first section of the chapter, I outline the new debate on realism, the ways in which it challenges more traditional conceptions of realism and the continued blindspots in both of these ways of theorizing the implications of realism for politics. In the second section, the chapter goes on to examine the specific challenges for realism of complexity theory especially in relation to realism's methodological assumptions. The third section focuses on the epistemological shortcomings of realism that complexity theory highlights as well as the impact of complexity on the constitution of what we take to be the real. The fourth section of the chapter shifts the focus to the emergence of phronetic social science as an alternative account of how to understand real politics through using the Aristotelian concept of practical wisdom. I then compare and assess these reinterpretations of real politics and the foundational building blocks they provide for political realism, before concluding the chapter with a brief comment on their implications for political action.

34 *Handbook on complexity and public policy*

THE NEW DEBATE ON REALISM

The discourse of realism has become a subject of renewed interest in contemporary political theory. While it can be used descriptively to identify those approaches to politics which are grounded in the immediate practicalities of political decision-making (Swift and White, 2008), it is frequently employed to distinguish practical politics from a number of other ways in which political arguments can be constructed and justified, including abstract moral philosophy, normative ethical claims, constructivist models of desirable alternatives, ideology and advocacy. To be a realist, by this understanding, is to be able to analyse politics and decision-making on the basis of the real conditions which prevail and which inform the strategic actions of decision-makers. The justification for this approach is – at least partially, if not mostly – driven by a perception of the unrealistic assumptions of the alternative approaches listed above. That is, more abstract and less applied political theories routinely underestimate the structural constraints that inhibit the choices of political actors, or the kinds of strategic calculations that inform their mindset which do not concentrate necessarily on what is right or good.

Complexity theory in the social sciences emphasizes the simultaneously orderly and disorderly dimensions of the social conditions that affect perceptions of the boundaries of possible political action (Urry, 2003). Significantly, theorists such as William Connolly have pointed to the ways in which these understandings of the realm of possible action are subject to temporal change, which means that perceived opportunities for political action also expand and contract over the course of time (Connolly, 2010). This conception of complexity makes 'fact-derived' accounts of realism problematic precisely because the 'facts' upon which these perspectives are based are understood to be transient. At the very basic level, 'fact-derived' accounts of the real assume that what is real is self-evident and can merely be read off from events by suitably informed political analysts. The idea of realism suggests that the analysis of political problems and options available to decision-makers should focus on what is actually feasible rather than being sidetracked by the pursuit of particular ethical outcomes or ideological objectives. Realists then can concentrate on what is genuinely possible within prevailing constraints rather than being drawn into a potentially unrealistic politics of what might be desirable. So realism is supposed to be focused on the facts of a particular political matter rather than less practical concerns about how things might be differently organized.

Complexity theory challenges these assumptions in a number of ways but, as we will see, most significantly on epistemological and motivational grounds. But there is a key point at stake before assessing these objections: the debates about realism focus on what it means to conduct 'realistic politics' without attending to a much more fundamental methodological problem of how political theorists define what the real actually is. In fact, this chapter suggests that what actually constitutes the real is transient and largely undefined due to a number of factors associated with complexity analyses. The challenge for political theorists in trying to demonstrate the applicability of their work to real politics is to think more clearly about what constitutes the 'real world' of politics. Instead, too often, we see theorists – typified by Swift and White (2008) – unnecessarily falling into the trap of distancing theory and real politics. This is an exceptionally narrow view of real politics that contrasts with the 'political turn' in political theory in recent years (Freeden, 2013). Swift and White construct an artificial divide that assumes that 'real

Complexity and real politics 35

politics' is largely self-constituting without establishing the epistemological foundations of this assumption. This then perpetuates the division between theory and real politics which, as Raymond Geuss makes clear, lets realists off the hook of having to actually explain why their interpretation of the real is preferable to alternative accounts (Geuss, 2008: 59–60). Thus, realism actually becomes a licence to avoid difficult questions around both the foundations of knowledge and the ontologies of political actors.

The emergence of new debates on realism, in which Geuss has been a prominent figure, challenges the sustainability of uncritical approaches to the constitution of real politics. Others, such as Mark Philp, have attempted to establish a stronger connection between political theory and the practice of politics by focusing on the behaviour of political actors and suggesting that political theorists should take more considered, realistic judgements on political action (Philp, 2010). William Galston, on the other hand, outlines several different variants of political theory and uses that to formulate an argument that some theoretical approaches are much more realist than others (especially those which he challenges for adopting utopianism) (Galston, 2010). And, Bernard Williams contributed to this debate by focusing attention on basic legitimacy as one of the cornerstones of realist political theory which he differentiated from forms of liberalism that employed political moralism as their foundation (Williams, 2005). In all of these accounts, however, the focus is on how political theorists should address the reality of political practice, whereas the argument here is focused on how they interpret what actually constitutes the space of real politics.

This chapter suggests that paying greater attention to the challenges raised by complexity theory could place political theorists in a stronger position to identify some key foundational questions in real politics. Rather than relying on a one-dimensional notion of the immediately identifiable 'facts' as the real, it could allow us to outline the epistemological and ontological shortcomings of these narrow understandings of real politics: a kind of 'mono-realism' in which only readily observable and computable 'facts' from pre-established horizons of interpretation can be construed as real. From a complexity theory perspective, this unnecessarily limits the field of possibilities. Ultimately, the existence of gaps in the epistemological and ontological foundations of realism – that is that we can never actually *know* what constitutes the real or how human actors will interpret phenomena – opens up space for a reconsideration of realist politics (Little, 2015b). The epistemological gaps in any construction of the real open up opportunities for alternative accounts to mount hegemonic projects for the reconsideration of realism. If the real is fluid and dynamic (in often unforeseen ways) and acted upon by multiple actors in a complex fashion, then there is always a battle to be forged on what constitutes reality. In the next section, then, I use some of the features of complexity theory to examine the epistemological limitations of realism and some of its flawed ontological assumptions about political actors.

CHALLENGING REALISM: COMPLEXITY, EPISTEMOLOGY, ONTOLOGY

The starting point in understanding the ways in which complexity challenges realist modes of political theory is to make clear that *political* theory needs to provide more than abstract philosophical argument about how the world should be. *Political* theory

36   Handbook on complexity and public policy

involves analysis of structures of power and the institutional and contextual factors which may act as impediments to the operationalization of particular normative ideals. Political theory needs to shed some light on the possibilities for political action, even if it can sometimes only identify and examine the obstacles to realizing particular normative goals (Philp, 2010: 478). Realism, based on the existence of incontrovertible facts that constrain the field of action, operates to marginalize views which challenge established models and the boundaries they create for what is deemed to be politically feasible. This prevents political theorists from contributing to a debate on what might be *actually* possible in terms of real political action.

Complexity theory on the other hand suggests at least two fronts on which this process of reassessing realism can be opened up:

- first, the limitations of existing systems and institutions in dealing with the political uncertainty engendered by complexity;
- second, the weakness of traditional constructions of realism in understanding the emotive and psychological dimensions that may affect decisions about political action.

First, then, the growing body of literature on complexity theory has fundamental implications for the epistemological foundation on which we comprehend the real in social and political terms (Cilliers, 1998; Geyer and Rihani, 2010; Harrison, 2006; Urry, 2003). Complexity theory focuses on the chaotic blend of orderly and disorderly phenomena in social and political life and this suggests that the actual state of reality, even if it is in some way discernible, will always be in flux. Complex environments are dynamic, with a plurality of actors engaging on issues in a multiplicity of different ways and generating effects that are unknown to each other. Moreover, the effects of what we do are not linear, given that multiple other parties are acting on the same issue simultaneously. So although we might act politically with specific outcomes in mind, we can never guarantee the intended result. The actions and behaviours of other interested parties may interrupt the linearity between action and outcome. This ensures that the actual nature of things is dynamic and always in a stage of development – at best, depictions of reality are momentary snapshots.

The complicated interaction of a plurality of actors on a social phenomenon will always generate emergent dimensions of the real that we may not fully comprehend. Part of the reason for this epistemological gap is that we tend to be beholden to established modes of understanding in making sense of emerging entities. This path dependence constrains our horizons and can bind us to particular methodologies for understanding real politics. In addressing social issues, political actors are always playing catch up as they are partly bound by established techniques of political management while trying to deal with emerging problems. On this account, merely depicting the obvious as the real is deeply problematic because it ties us even further into existent, path-dependent modes of understanding. Thus, the epistemological blind spots and sheer unknowability of all relevant information in complex environments should require us to take more risks in the way we organize institutional politics rather than merely being beholden to our historical inheritance (Little, 2012).

All of this makes it rather difficult to discern what is definitively 'real' at any given point of time. Indeed, even if we could attain a definitive understanding of dynamic entities in the now, it would soon be superseded by new emergent properties. In a complex society, then, the dynamic nature of the environment and the emerging and unforeseen problems it generates needs to be understood alongside the more static nature of political institutions and systems. The dynamic interplay between a plurality of actors creates a flexible, changing and sometimes disorderly context to which institutions and systems need to adapt constantly. From this perspective, systems are always struggling reactively to contain issues that are constantly changing and evolving in relation to a multiplicity of actors engaging with them. In short, this complexity engenders an epistemological gap that challenges the operation of political institutions and undermines the realist claim to a patent on the comprehension of what is *actually* real.

The second major challenge to the construction of real politics does not emanate directly from complexity theory but does resonate with some of its insights. This ontological challenge, primarily articulated by Raymond Geuss, focuses on the limits of our comprehension of the human motivations that contribute to possible forms of political action. Geuss (2008) highlights the challenge for human beings in trying to act in a systematic fashion when faced with a complex array of competing options, their membership of multiple collectivities, and the shifting nature of the issues they have to address. Here Geuss introduces an emotive dimension to our understanding of action that he juxtaposes with the construction of politics as mere applied ethics (see also Philp, 2010; Williams, 2005). He argues that human beings have to make decisions in possession of incomplete knowledge and with very little comprehension of the outcomes of their actions in the longer term. Furthermore they may not actually know what they want or which course of action is going to further their interests (Geuss, 2008: 2–3). This approach resonates with some of the more traditional debates in public policy around sub-optimality, 'satisficing' and 'bounded rationality' (Little, 2012; Simon, 1982; Browne and Wildavsky, 1984; Lindblom, 1959).

From this foundation, Geuss rejects what he sees as 'ethics-first idealism' in political philosophy and makes the case that, if we take the ontological challenge seriously, then realism needs to focus just as much on the obstacles to the realization of normative ethical arguments as the establishment of those ethical positions (Geuss, 2008: 9). Realism also implies a theory of political action – political theory needs to give an action-based account of problems (Philp, 2010) and a historically located account of the structural context in which we put theoretical propositions into practice (Williams, 2005:13). Importantly, then, realist politics is not just about purely strategic action from individual or collective utility maximizers, Geuss sees it as a skilled art or craft where political actors take account of their contextual and structural constraints and work out their favoured options on that basis. Reproducing theories in practice is never straightforward. So, the distinction between political theory and political practice is dangerously reinforced by theoretical approaches which can only be understood as analytically coherent if they are abstracted from conditions that may pertain in actual political practice. In this vein Geuss argues that it is problematic to approach politics through focusing 'abstractly [on] the good, the right, the true, or the rational in complete abstraction from the way in which these items figure in the more motivationally active parts of the human psyche, and particularly in abstraction from the way in which they impinge, even if indirectly, on human action' (Geuss, 2008: 28).

A further question that Geuss asks us to bear in mind when considering political reality

38   *Handbook on complexity and public policy*

is the relationship between the ontology of political actors and the actions that they take. The reason for this is that he conceives politics as a relational matter. Politics is never a straightforward matter of simply identifying and enacting the correct course of action. For Geuss, the decision to act in one way is also a decision to not act in another way. Thus, the choices we have to make are not between isolated options but options that are related to one another. Humans always act in relational terms and, rather than following a definitive path between right and wrong or good and bad, we act in preferential ways depending on our particular calculation of what is at stake in considering different options: 'politics is not about doing what is good or rational or beneficial *simpliciter* . . . but about the pursuit of what is good in a particular concrete case by agents with limited powers and resources, where choice of one thing to pursue means failure to choose and pursue another' (Geuss, 2008: 30–31).

Relational choices about political action are significant in complex systems because the decision to do one thing over another may foreclose other options or potentially create path-dependent conditions in which alternative courses of action may become much more perilous. So, to decide on a particular course of action at a given time is not without long-term effects – this makes the preferential process of one action rather than another an incredibly loaded decision in real politics. This necessitates an approach to the real which goes beyond a short-term focus on the here and now and narrow interpretations of the strategic considerations of political actors. While short-termist, calculating responses from political actors to specific problems in practical situations may be understandable, they do not constitute a persuasive interpretation of the real in and of itself and they certainly do not address the kinds of epistemological and ontological questions that theories of complexity generate.

COMPLEXITY THEORY'S EPISTEMOLOGICAL AND CONSTITUTIVE CHALLENGES TO REALISM

Given the epistemological constraints articulated by complexity theory highlighted in the previous section, the chapter now turns to the question of what implications this epistemological uncertainty has for the constitution of 'real politics'. Part of the reason for revisiting this constitutive element of the real is that so much of what is usually thought to constitute real politics is filtered through the mechanisms and institutions of existing political systems. Each of these systems has limits that make its operation comprehensible amid the disorder and unpredictability of conditions of social complexity. These limits of political systems attempt to facilitate political action and a degree of certainty in environments of openness and transience. Thus, as Human and Cilliers make clear, limits make an economy or a system possible but the effect is that complex systems which are often 'conceptualized as being open to their environments' are simultaneously 'operationally closed' (Human and Cilliers, 2013: 28).

This generates considerable epistemological challenges and has an influential bearing on what we are able to make sense of within the limits that enable us to comprehend phenomena as a part of real politics. Thus, the limits that systems establish help us to act on things by foreclosing their inevitable openness. At the same time, however, they inhibit us from fully grasping the permeable and transient nature of the limits we establish.

Complexity and real politics 39

These epistemological shortcomings of political systems have potentially dramatic consequences in the understanding of real politics because of the longevity and reproductive capacity of systems. Once a system has been established according to a specific logic, it becomes considerably more difficult to view it through a different lens or in an alternative way. And the most adaptive systems are highly adept at incorporating alternatives at the margins of intelligibility within their systemic rationalities. This scenario means that epistemological limitations are not just impediments with regard to what we know; they also have fundamental ramifications for how we conduct politics, make sense of political issues and build political institutions. In this sense, epistemological imperfection cascades through political systems. At one and the same time, epistemological imperfection leaves open the possibility of political change given the flawed logic underpinning political systems, while closing off opportunities to act upon these possibilities in a politically coherent way:

> that which is excessive, that which is excluded, stands outside the particular logic or reason of that epistemology. In order to ensure the rational representation of what is investigated, that which is excluded is often depicted as being 'noise' or 'inconsequential' . . . It is by exclusion, the setting of limits, that the economies become useful to us. These limits are productive not only because they are constitutive, but also because they allow the very antagonisms within the models to function. That which is excluded makes possible the debates or differences found inside the system (Human and Cilliers, 2013: 29).

Building on the earlier work of Paul Cilliers on the relationship between postmodernism and complexity theory (Cilliers, 1998), Human and Cilliers have recently advanced this particular confluence of two distinctive modes of political theory. Augmenting the argument above, that exclusion and the drawing of firm limits enables a particular economy of thought to make sense (and therefore facilitates the operation of political systems), Human and Cilliers make a crucial intervention in distinguishing between heterogeneity and difference. They label as *heterogeneous* 'that which is perceived as noise from the perspective of the system or the model itself' but, crucially, they argue that heterogeneity is 'not noise or a mystical force but simply that which does not make sense from the limited perspective of *this* model' (Human and Cilliers, 2013: 30). *Difference*, on the other hand, 'refers to the discrimination which can be made from the perspective of the model under consideration. Differences can be recognized only in terms of a common frame' (Human and Cilliers, 2013: 30). In other words, then, contemporary political systems can accommodate difference as long as it is formulated within the acceptable boundaries within which the particular economy of a system continues to make sense. The heterogeneous is that which sits beyond the limits of a particular economy and challenges the boundaries of political systems; it is the constitutive outside that enables the system to make sense at all.

The implications of this distinction between heterogeneity and difference for the constitution of real politics are quite profound. It suggests that what is considered 'real' is that which is credible according to the prevalent economy of a system. To be clear, the systems that are best able to reproduce in a complex environment are those which are highly adaptive. In practice, what this means is that those systems which are capable of incorporating extraneous and unforeseen elements within the boundaries of their economy, are those which are most adaptive. And, in a complex terrain, adaptivity is

40　*Handbook on complexity and public policy*

of course a key attribute of political systems. Robust political systems are highly reliant on the capacity to incorporate a variety of different perspectives within their logic or economy. At the same time, however, they must also be prepared to identify that which is beyond the pale precisely because heterogeneity defines the boundaries of the economy.

From this foundation, however, it is important to introduce a temporal dimension because where the lines are drawn between heterogeneity and difference will shift as time passes. The economy that underpins any particular system will evolve over the course of time in reaction to the changing circumstances in which the system operates. As a result the boundary between what is acceptable difference and what is heterogeneous noise will also move. Importantly, then, at any given point in time there will be limits to an economy so we can distinguish between difference and heterogeneity but, as economies develop, the limits move, providing the potential for what was once noise to become part of the accepted discursive structure of a particular economy. As a result, complex systems are characterized by the evolution of the phenomena that constitute what is real – the real is not settled but evolves as the structuring limits of political discourse transform over the course of time. What this means, of course, is that 'realism' has to shift as well. A realist politics needs to take account of the fact that any version of 'how things really are' is only that which is discernible within the prevailing structures of political discourse in the now and that they too are in a state of flux. At all times then, when we discuss what is real in politics, we must do so from a foundation of epistemological uncertainty in which we are open to the temporal dynamics that ensure that any depiction of the real is partial and ephemeral.

An extension to this argument deployed by Human and Cilliers involves Edgar Morin's distinction between 'restricted complexity' and 'general complexity'. For Morin, 'restricted complexity' refers to the classical scientific practice of understanding complexity in a circumscribed way which still enables the development and application of scientific principles. Indeed, it is only because complexity is restricted through a principle of 'simplification' leading to techniques of 'epistemological determinism, reduction, and disjunction' (Human and Cilliers, 2013: 31) that scientific practice is enabled. In Morin's work this is contrasted with a very different understanding of 'general complexity' based on:

> an epistemology of complex systems which examines the *relationships* between the parts as well as the parts themselves . . . the focus of analysis shifts away from the parts to a consideration of the contingent sets of relationships between the parts. It is this contingency which denies simple and universal models. The best we have are models which are partial and provisional (Human and Cilliers, 2013: 32).

This is a relational model of complexity that implies that there is not a simple distinction to be made between the real and the unreal or imagined. It highlights the ways in which difference can be incorporated within existing complex systems (understood restrictively) but that a more general account of complexity needs to recognize the ways in which heterogeneous elements intervene across the borders of general complexity. That is, the entities which comprise spheres of difference and heterogeneity are not distinct from one another but exist instead on a relational plane. Where the limits are placed between the acceptable and the heterogeneous moves around on these relational planes. Moreover, because these relations are not strictly linear but are subject to cross-cutting interventions of other relational interactions between actors and concepts, the general condition of complexity is much more uncertain, disorderly terrain and is substantially different from the much more static and linear imagination of 'restricted complexity'. Restricted models may be necessary to enable practical political action but they are thereby highly susceptible to unforeseen elements and thus have a greater propensity for failure and error (Little, 2012).

THE EMERGENCE OF PHRONESIS AS A NEW FORM OF REAL POLITICAL ANALYSIS

In connecting this analysis of general complexity with an account of real politics, then, it is perhaps useful to turn to the re-emergence of the Aristotelian concept of 'phronesis' in contemporary social science debates (Flyvbjerg, 2001; Flyvbjerg et al., 2012a). Phronesis refers to attempts to enhance 'a socially relevant form of knowledge' which is interpreted as 'practical wisdom on how to address and act on social problems in a particular context' (Flyvbjerg et al., 2012b: 1). Unlike the decontextualized practice of the natural sciences (akin to the discussion of restricted complexity above), phronesis develops a broader Aristotelian perspective which incorporates local context, human interaction and values into the analysis of social phenomena. As such, as a method focused on the significance of practical wisdom in the face of contingency and uncertainty, advocates of phronesis suggest that it is superior to the restricted analyses of the natural sciences in addressing social obstacles and comprehending localized problems. Unlike the realist turn in contemporary political philosophy, then, phronesis tries to develop a much stronger sense of some of the motivational complexities which affect human action in its advocacy of a 'praxis-oriented epistemology, theory of science and methodology which makes it particularly effective in dealing with issues of power' (Flyvbjerg et al., 2012b: 6).

The complexity of the human subject and the shifting object of study over time in the social sciences are features that characterize both complexity and phronetic theories. However, where the former seeks to borrow from the emergence of complexity theory in the natural sciences, the latter seeks to distinguish the quality of social scientific knowledge from that of the natural sciences. So part of the distinction between the two approaches is that complexity focuses on the capacity of the natural sciences to inform social scientific research at least in terms of methodology, whereas phronetic social science distances itself from what it sees as the flawed approach of the natural sciences. For example, phronetic social scientists such as Schram deride the translation of the methodology of the natural sciences into the social sciences as 'decontextualised universal rationality stated in abstract terms of false precision' (Schram, 2012: 17). This depiction of the natural sciences is contested, however, with some complexity theorists in the social sciences drawing explicitly on Foucault's critique of the epistemological foundations of this history of science to demonstrate the highly contested nature of knowledge in the natural sciences (Little, 2012). The debate around contested epistemology highlights how it permeates both the natural and social sciences in ways that call into question the distinction that phronetic social scientists like Schram employ to emphasize the distinctiveness of their approach.

42   *Handbook on complexity and public policy*

So the social scientific use of complexity theory challenges some of the clear distinctions and binaries that appear in the phronesis literature. At the same time, however, it does so from a basis of alignment with the general thrust of that literature and there is much shared ground. For example, Schram (2012: 18–19) highlights four key aspects of phronetic social science that accord with many of the assumptions of complexity theory in the social sciences. These are:

- the critical assessment of 'values, norms and structures of power' above general laws that are used to predict action;
- a subjective form of practical reason which highlights dialogue and judgement over disinterested research;
- relinquishing objective views of absolute truth to emphasize 'a contextual notion of truth that is pluralistic and culture-bound, further necessitating involvement with those we study';
- a methodological outlook which recognizes that the position of the interpreting researcher is power-laden.

This outline implies that we need to develop practice-based and practice-led approaches to the problems and issues that we address as social scientists, and indeed Schram cites Ian Shapiro in making the case for problem-driven research that is driven by political contingency rather than ethics-first idealism (see Shapiro and Bedi, 2007). Phronetic social science makes much of the methodological issues in social scientific research although it is fair to say that it is not beholden to one particular approach. This is an important point because, rather than just assuming the real is comprised of what is immediately obvious to us, phronesis encourages the use of deep, contextual knowledge to get closer to understanding practical problems. While there is an underpinning commitment to mixed methods research in phronesis rather than solely positivist or interpretive perspectives in isolation, case study research is particularly prominent in the phronesis literature because it allows for greater depth of understanding and is framed in the context of highly specific and localized epistemologies. Phronetic social science also relates strongly to ethnographic techniques as a potential means of researchers acquiring deep, contextual knowledge about grass roots, local issues. It is clear then that social scientific approaches to phronesis have devoted considerable thought and attention to methodological issues in a way that might be commonly implied by elements of complexity theory but that is less frequently worked through systematically.

A further area of methodological–theoretical consideration is the emergence of narrative approaches to particular social and political problems (Little, 2014; Bevir, 2006; Moon, 2008; Patterson and Monroe, 1998) because narratives are central components of phronetic research. For example, Todd Landman argues for the usefulness of narrative analysis (focusing, in particular, on truth commissions and transitional justice initiatives) as a methodological approach that can attend to both macro- and micro-level events:

> such events carry with them other analytical features, such as a start date and time (also known as an *onset*) and an end date and time, where the difference between the two is understood as the event *duration*. They also have various dimensions of *magnitude* and *size*, including the number of actors involved . . . the types of things that actors do and the types of things that happen to them (Landman, 2012: 29).

Complexity and real politics 43

The value of focusing on narratives is derived at least in part from the way in which these articulations of events and problems can help to generate theories that are attuned to practical issues at a local level. Rather than developing fully formed theoretical models and then imposing them on specific problems in their local conditions, phronesis suggests that we need to try to build theory from understanding the constraints of these practical conditions at the outset. Thus, Flyvbjerg et al. contend that, in order to act appropriately on social and political problems, we require:

> a knowledge of context that is simply not accessible through theory alone. In phronetic social science, 'applied' means thinking about practice and action with a point of departure not in top-down, decontextualized theory and rules, but in 'bottom-up' contextual and action-oriented knowledge, teased out from the context and actions under study by asking and answering the value-rational questions that stand at the core of phronetic social science' (Flyvbjerg et al., 2012c: 284–5).

It should be fairly clear that complexity theory and phronesis share scepticism about the way in which the 'reality' of politics is constituted and used as a technique which justifies some forms of political action and not others. However, while complexity theory points to epistemological uncertainties and conflicting ontologies as a general condition of the field of human action, phronesis implies that different methods of accruing knowledge and closer attention to the narrative claims of those in local contexts (often those with the least power) can lead us towards a better comprehension of social and political problems and opportunities to act upon them. In this sense, while complexity implies that the real is a shifting, transient construction that is always illusive to some extent, phronesis suggests that we can get closer to what is actually real by developing better methodological approaches focused on acquiring deeper knowledge at the local and deeply contextual level. In the final section, I address these claims and assess their implications for the reconstitution of notions of real politics.

TOWARDS A COMPLEX ECONOMY OF THE REAL

Flyvbjerg et al. conclude their book with the statement that 'social science is at a critical juncture. The momentum lies with alternatives to the dominant scientistic social science that unreflectively seeks to apply the models of the natural sciences to the social world' (Flyvbjerg et al., 2012c: 296). This cuts to the first of the two major divisions between phronetic social science and complexity theory. Where phronesis rejects the 'scientistic' attempt to employ a natural sciences mentality in approaching the social sciences, complexity theory points to multiple methodologies within the natural sciences in ways that give distinctive understanding to the meaning of science. So where the view from phronesis owes at least something to C.P. Snow's argument about 'two cultures' (Snow, 2013) and seeks to make capital out of this division, complexity theory highlights the emergence of complexity within the natural sciences and tries to employ that critique of overly or exclusively positivist and empirical approaches to the social sciences. Therefore, while phronesis wants to accentuate the differences between the social and natural sciences, complexity theory recognizes the differences within the latter category and seeks to

44 *Handbook on complexity and public policy*

employ the insights of complexity in the natural sciences to the ways in which knowledge is constructed in the social sciences.

So, while much of the natural science literature may adopt a more narrowly positivistic framework, complexity theory does not, and challenges positivism's methodological assumptions. Moreover, even if natural scientific methodology does purport to be about the establishment of verifiable 'truths', it would be odd, given the history of science, not to recognize that those 'truths' can only be based on what we happen to know at a given point of time (Foucault, 1998; Little, 2012). Over the course of time scientific knowledge changes and develops especially as sciences with different economies engage with one another. Either theoretically or practically the natural sciences are not quite as monolithic as Flyvbjerg et al. suggest, even if they are quite rightly sceptical of laboratory-based methodologies (which are, of course, problematic from a complexity perspective too – see Latour and Woolgar, 1986). Ultimately, following Foucault, we need to understand the contested nature of scientific epistemes and the ways in which our knowledge about social phenomena is similarly polysemic.

The second key point of difference between phronesis and complexity is the way in which complexity theory in the social sciences uses the idea of an 'economy' and how that informs its understanding of phronetic knowledge. To reiterate, 'the idea of economy allows us to begin to say something about the nature of the relationships "inside" a complex system whilst acknowledging the openness of a system's boundaries' (Human and Cilliers, 2013: 39). While defining the boundaries of an economy is always elusive for complexity theorists, it highlights the existence at any point in time of a boundary between the acceptably different and the heterogeneous outside. For Human and Cilliers, this has several implications but, most significantly for the argument here, it implies that there is always a wealth of possibilities in acting on a particular political issue but that many of them sit beyond the constructed limits of intelligibility of our political systems:

> the excess of the general economy of complexity implies a normative dimension to our engagement with the world . . . The normative dimension rests in the fact that we have to choose what we include in the economy, but there is no objective means of doing so. We cannot escape this normative dimension when dealing with complexity as we are forced to make exclusions in order to maintain the coherence of our economies. Our actions are meaningful precisely because they close down other possibilities (Human and Cilliers, 2013: 40).

This helps to make clear the second dimension of difference between complexity and phronesis. Where complexity sees the creation of restricted economies in all modes of social life, phronesis accords a degree of privilege to practical wisdom based on deep engagement with particular cases and contexts (and preferably mixed methods techniques) to get closer to the *actual reality*. While intuitively this phronetic approach seems attractive because it is focused on deep, contextual knowledge derived from the bottom up rather than imposed or received ideas, from a complexity perspective, it is another form of economy. While perhaps immune to some of the more abstract ethics-first approaches to political questions, phronesis still requires boundaries around the intelligible to formulate coherent forms of knowledge. Whether these boundaries are established by lines between local and more global forms of knowledge, or between different methodological techniques for generating knowledge, phronesis still needs the

Complexity and real politics 45

boundaries of its economy to make sense of itself. From a complexity perspective, this will of course involve difference within the boundaries but, given those boundaries are shifting and permeable within a complex environment, we also need to pay attention to the heterogeneous arguments which sit outside the frames of intelligibility which give phronetic knowledge its meaning.

So, in terms of understanding the relationship between complexity and real politics, phronetic social science helps us to renegotiate the terms of the real incorporating a much more contextual, bottom-up approach. However, it does not provide all of the answers to the two challenges laid out in the first part of the chapter. If our understanding of the constitution of the real in complex societies is complicated by epistemological uncertainty coupled with temporal change, and further uncertainty about the human motivations underpinning political action, then phronetic knowledge is undoubtedly significant but is subject to the same constraints as purported realism. As alluded to above, Human and Cilliers discuss a complex order in terms of an economy in which the relational dimension is clear. This suggests that we focus on 'the *play* between parts of a system' as well as the 'limits or constraints determined by the relationships between the components in the system' (Human and Cilliers, 2013: 27). This has implications for how we conceive real politics (conceived in phronetic terms or otherwise) because it suggests that the reality is not settled or given. Instead, the real emerges from the interplay of a range of actors, social forces, and material structures in many different ways. The real then will always be interpreted differently especially as the adjudication of the reality takes place at different temporal stages when the interaction of multiple social forces may be quite dramatically different.

Therefore, in complex environments, because the boundaries of the real may be quite transient, the capacity of political systems to adapt to these changes is paramount. This requires that the logic that underpins a particular system must be capable of subsuming some of the more marginal elements of thought or action within its limits or economy. In other words, the limits of intelligibility within social and political paradigms are not fixed but can shift according to relational interactions between a wide range of variables. In this scenario what makes a system resilient is the capacity to subsume some of the ideas and practices that are closer to the margins of the limits into the paradigmatic logic. If the constitution of the real is dependent on these kinds of processes, then our understanding of what real politics involves is reliant on the adaptive capacity of our institutions. However, whenever there is a rupture in this systemic capacity to subsume and make sense of sufficient difference, paradigm shifts can take place that involve a redrawing of the limits of an economy so that different relations and interactions can be established. This in turn gives new meaning to what constitutes the real.

CONCLUSION

This chapter has examined recent debates on realism in the social sciences and found them wanting in terms of addressing a preliminary question: *what is the constitution of the real?* While the emergence of realism as a topic of concern in political theory is to be welcomed, the contributors fall into some of the traps evident in the international relations literature in assuming that the real from which realism derives its rationale is

46 *Handbook on complexity and public policy*

self-evident. The argument here challenges this assumption by teasing out some of the foundational arguments from complexity theory about epistemology and temporality to suggest that the real is in fact a highly contingent and transient phenomenon. It contrasts the complexity approach with the recent emergence of phronesis as an alternative way of addressing key social scientific questions in the real world. But, whereas phronesis advocates practical wisdom derived from getting closer to the contextual substance of the real, complexity implies that such techniques merely create a new economy of the real (albeit one that is much more applied than abstract interpretations of the real world).

All of this suggests that an understanding of real politics that is attuned to the shifting dynamics of social complexity needs to be capable of providing a suitably flexible account of the relationship between reality and temporality. This is a challenge for political analysis because so much of it is conducted in time-bound terms and with a need for immediacy. This pressing nature of political decision-making gives excessive weight to restricted accounts of what is real at a particular point in time and thus politics tends to be conducted from within a very time-bound mindset – we have to act now, if we do x then y will ensue, if we don't do x then z will ensue, and so on. While this immediacy is perhaps inevitable, it draws attention from the much more processive and incremental nature of political development (Little, 2015a). And it binds us to 'in the now' considerations of what is real that are not sustainable if we take the epistemological and ontological critiques from complexity theory seriously. Therefore, it is important not to reproduce the artificial schisms between practice and theory, and both complexity theorists and phronetic social scientists alike have helped to make this clear. However, as outlined above, only complexity theory provides us with the tools to understand the contingency and transience of the constitution of the real and its implications for political action.

REFERENCES

Baderin, A. (2014), 'Two Forms of Realism in Political Theory', *European Journal of Political Theory*, 13(2), 132–53.

Bevir, M. (2006), 'Political Studies as Narrative and Science, 1880–2000', *Political Studies*, 54(3), 583–606.

Browne, A. and A. Wildavsky (1984), 'What Should Evaluation Mean to Implementation?', in J. Pressman and A. Wildavsky (eds), *Implementation* (3rd edn), Berkeley, CA: University of California Press.

Cilliers, P. (1998), *Complexity and Postmodernism*, London: Routledge.

Connolly, W. (2010), *A World of Becoming*, Durham, NC: Duke University Press.

Der Derian, J. (ed.) (1995), *International Theory: Critical Investigations*, New York: New York University Press.

Flyvbjerg, B. (2001), *Making Social Science Matter*, Cambridge: Cambridge University Press.

Flyvbjerg, B., T. Landman and S. Schram (eds) (2012a), *Real Social Science: Applied Phronesis*, Cambridge: Cambridge University Press.

Flyvbjerg, B., T. Landman and S. Schram (2012b), 'Introduction: New Directions in Social Science', in B. Flyvbjerg, T. Landman and S. Schram (eds), *Real Social Science: Applied Phronesis*, Cambridge: Cambridge University Press.

Flyvbjerg, B., T. Landman and S. Schram (eds) (2012c), 'Important Next Steps in Phronetic Social Science', in B. Flyvbjerg, T. Landman and S. Schram (eds), *Real Social Science: Applied Phronesis*, Cambridge: Cambridge University Press.

Foucault, M. (1998), 'Life: Experience and Science', in M. Foucault and J.D. Faubion (eds), *Aesthetics, Method and Epistemology*, New York: New Press.

Freeden, M. (2013), *The Political Theory of Political Thinking*, Oxford: Oxford University Press.

Freeden, M. (2014), 'Editorial: The "Political Turn" in Political Theory', *Journal of Political Ideologies*, 19(1), 1–14.

Galston, W. (2010), 'Realism in Political Theory', *European Journal of Political Theory*, 9(4), 385–411.

Geuss, R. (2008), *Philosophy and Real Politics*, Princeton, NJ: Princeton University Press.

Geyer, R. and S. Rihani (2010), *Complexity and Public Policy*, London: Routledge.

Guzzini, S. (2004), 'The Enduring Dilemmas of Realism in International Relations', *European Journal of International Relations*, 10(4), 533–68.

Harrison, N. (2006), *Complexity in World Politics*, Albany, NY: SUNY Press.

Human, O. and P. Cilliers (2013), 'Towards an Economy of Complexity: Derrida, Morin and Bataille', *Theory, Culture & Society*, 30(5), 24–44.

Landman, T. (2012), 'Phronesis and Narrative Analysis', in B. Flyvbjerg, T. Landman and S. Schram (eds), *Real Social Science: Applied Phronesis*, Cambridge: Cambridge University Press.

Latour, B. and S. Woolgar (1986), *Laboratory Life: The Construction of Scientific Facts* (2nd edn), Princeton, NJ: Princeton University Press.

Lindblom, C. (1959), 'The Science of "Muddling Through"', *Public Administrative Review*, 19(2), 79–88.

Little, A. (2012), 'Political Action, Error and Failure', *Political Studies*, 60(1), 3–17.

Little, A. (2014), *Enduring Conflict: Challenging the Signature of Peace and Democracy*, New York: Bloomsbury.

Little, A. (2015a), 'Performing the Demos: Towards a Processive Theory of Global Democracy', *Critical Review of International Social and Political Philosophy*, forthcoming.

Little, A. (2015b), 'Reconstituting Realism: Feasibility, Utopia and Epistemological Imperfection', *Contemporary Political Theory*, forthcoming.

Moon, C. (2008), *Narrating Political Reconciliation: South Africa's Truth and Reconciliation Commission*, Lanham, MD: Lexington Books.

Patterson, M. and K.R. Monroe (1998), 'Narrative in Political Science', *Annual Review of Political Science*, 1, 315–51.

Philp, M. (2010), 'What is to be Done? Political Theory and Political Realism', *European Journal of Political Theory*, 9(4), 466–84.

Schram, S. (2012), 'Phronetic Social Science: An Idea Whose Time Has Come', in B. Flyvbjerg, T. Landman and S. Schram (eds), *Real Social Science: Applied Phronesis*, Cambridge: Cambridge University Press.

Shapiro, I. and S. Bedi (eds) (2007), *Political Contingency*, New York: New York University Press.

Simon, H. (1982), *Models of Bounded Rationality*, Cambridge, MA: MIT Press.

Snow, C.P. (2013), 'The Two Cultures', accessed 24 October 2013 at http://www.newstatesman.com/cultural-capital/2013/01/c-p-snow-two-cultures.

Swift, A. and S. White (2008), 'Political Theory, Social Science and Real Politics', in D. Leopold and M. Stears (eds), *Political Theory: Methods and Approaches*, Oxford: Oxford University Press.

Urry, J. (2003), *Global Complexity*, Cambridge: Polity Press.

Walker, R.G.B. (1993), *Inside/Outside: International Relations as Political Theory*, Cambridge: Cambridge University Press.

Williams, B. (2005), *In the Beginning was the Deed*, Princeton, NJ: Princeton University Press.

4. Critical Legal Studies and a complexity approach: some initial observations for law and policy
Thomas E. Webb

INTRODUCTION

Complexity theory is not the first movement to attempt to teach those with an interest in law and legal processes to think differently (see further Fischl, 1987: 510–13). The purpose of this chapter is first to introduce complexity theory to the understanding developed by the Critical Legal Studies movement (CLS), and its proponents 'the Crits', and position complexity theory as a logical development from the CLS. The second aim is to provide some initial indication of how this interfacing can enhance understandings of law, policy, and their interaction. There are several tensions to navigate in such an exploration, particularly, for example, where a reader might be unfamiliar with one or other of the theoretical perspectives. As such, I deal only with the possibility of communication between CLS and complexity theory, and from this make some general observations concerning the implications for law and policy, and leave other matters open to future exploration. It is proposed to confine discussion in this regard to the consideration of connected concepts found in CLS and the complexity approach, indeterminacy and destabilization, contingency and emergence, and ongoing critique and self-reflexivity.

It should also be borne in mind that the construction and interpretation of law and policy in the modern administrative state is a polycentric process involving many actors operating in a wide range of contexts. Consequently, while in places I might imply a bright-line distinction between law and policy, this is not the case in practice. It is true to say that law grants the power to implement policy, and that law may subsequently be altered in accordance with the experience of implementation. However, it is also the case that not all policy derives its power from law, not all policy thought to be lawful is actually so, that substantial changes in policy may often require no amendment to enabling legislation, and that the nature of policy and its instruments is varied taking, for example, the form of guidelines, strict rules and open-textured frameworks. Thus, while law and policy are implicated in the same reflexive processes, this is a complex, rich and varied relationship. It is the assumptions, analyses and interpretation of these relationships which CLS and complexity theory speak to.

As I discuss in the following section, a critical account of complexity theory (see further Cilliers, 1998) has the capacity to expose the limitations of models constructed to understand how different aspects of society approach law and policy (see Geyer and Rihani, 2010: 102–103). It is argued that the approach offers lawyers and others a new way of thinking, in terms of both theory and practice, about how legal measures interact with, and are constructed by, the areas they target (see Cilliers, 2001: 141). At the same time, this account can be enhanced by a consideration of the CLS critique of law, politics and society made during the 1970s and 1980s in America. I propose to discuss

Critical Legal Studies and a complexity approach 49

the understanding of the critical complexity approach employed here and its relatively limited appearance in law, and then to give an overview of the origins of the CLS in Legal Realism, and the shape of CLS thinking. From there, it will be possible to contrast the concepts of the complexity and CLS approaches.

LAW AND A CRITICAL COMPLEXITY APPROACH

Complexity theory can be used to examine the consequences of interactions between the multiplicity of unique individuals in society. Where society is viewed as a complex system, individual participants (people, or collections of people; organizations, states, businesses, NGOs) each construct a unique understanding of reality to cope with novelty (Waldrop, 1994: 177; Holland, 1995: 31–4). These models, like their creators, are interfaced with the environment (Webb, 2013), and are situated in a particular spatial (Webb, 2005: 235; Cilliers, 1998: 92) and temporal context (Cilliers, 1998: 4). This generates a view of the world which is contingent upon that context (Richardson et al., 2001:12; Webb, 2005: 235, 241). Any account will be coloured by the approach which the individual chooses to interpret the bombardment of information and events in society, in concert with how the system, society or some aspect of it, is presented to the individual (Cilliers, 1998; 2001: 141; Webb, 2005: 237 and n. 43, 237). Furthermore, a definitive and objective account of society and its constituent elements is not possible because of the sheer quantity of available data (Rescher, 1998: *xiv–xv*; and see also Cilliers, 2001: 137), and the inability of the individual to view information from a position other than their own (Richardson, 2004: 77). These models and their creators cannot be isolated from society, they must interact and argue, compete, and adapt in response to change elsewhere. At the same time, the individual's adaptation demands a reaction from the rest of society. The collective interaction of this vast range of behaviours produces emergent consequences, outcomes arising out of the quality of the interactions rather than the characteristics of the individual parts, which must be interpreted and incorporated into the models.

This account should not be treated as being representative of ontological reality. While complexity theory is a powerful analytical tool capable of exposing problematic assumptions and unexplored aspects of interpersonal and inter-institutional relationships, it is only one lens for viewing the world. It is self-limiting through its assertion that context is crucial, and that competing perspectives are legitimate and may offer useful alternative understandings of the same subject matter (Mauthe and Webb, 2013a: 257–9). Moreover, much of the complexity literature originates beyond law and related considerations, and so the limits of translation should be borne in mind. Complexity is, above all, modest about its claims.

In addition to this reading of complexity theory there are several understandings of the approach in legal scholarship (see further Webb, 2014). The most developed account is that offered by Ruhl (for example, 1996a; 1996b; 1997; 2008; 2014; Ruhl and Ruhl, 1997; Ruhl and Salzman, 2002; 2003), which focuses on the utility of complexity to the regulation of environmental law in the United States and wider administrative aims. Other views have taken research in the direction of chaos and complexity via Deleuze and Guattari (Murray, 2006; 2008; 2010), complexity theory as an offshoot of the dominant

50 *Handbook on complexity and public policy*

autopoietic systems tradition in law (Webb, 2005), constitutional theory and adjudication (Vermeule, 2012), property and resistance (Finchett-Maddock, 2012), and comparisons between complexity theory and other systems approaches (Mauthe and Webb 2013a; Webb, 2013; 2014). The complexity approach in law outlined below is developed from Cilliers' critical complexity approach (1998), but it can be supplemented by these other perspectives. By way of introduction to the concepts discussed in this chapter, it is proposed to outline in greater detail two elements of the critical complexity approach: emergence and contingency.

EMERGENCE

At its core, emergence is the idea that the whole is greater than the sum of its parts. More specifically, it explains how it is the contextually situated interactions between the parts of the system which generate its character, not the individual nature of the parts themselves (Cilliers, 1998: 1–2; Heylighen et al., 2007: 118–19; Mitleton-Kelly, 2003: 40; Murray, 2008: 231, 239; Richardson, 2004: 77; Vermeule, 2012: 50–51; Waldrop, 1994: 63–6; Webb, 2005: 231). From the interactions of system parts emerge behaviours and models used by the participant to cope with novelty in the world. Emergence in this sense implies a form of self-organization within the system. This process involves the appearance of order from interactions (Heylighen et al., 2007: 126; Waldrop, 1994: 124) without direct, external or centralized influence (Cilliers, 1998: 12; Waldrop, 1994: 145), which instead relies on many local decisions (Urry, 2003: 13; see also Goldstein, 1999; 2000; Holland, 1995). At the same time, there are limits to the interactive possibilities open to system components (Cilliers, 2001: 141; Goldstein, 2000: 10; see also Webb, 2013), many of which are, ironically, an emergent product of social interaction, for example, technological or value-derived restrictions (Waldrop, 1994: 119–20; Webb, 2005: 239). Although awareness of these interactive limits may allow humans to influence boundaries (Goldstein, 2000: 19; Ruhl, 1996b: 867–8), it is not the case that humanity can dictate the structure of society. Individual elements of an organization may have ordering preferences, but non-linear causality, the non-proportionate relationship between cause and effect, coupled with the contingency of assessments, means that the individual's actions may have a large or negligible effect (Cilliers, 1998: 4; Goldstein, 1999: 60). While the extent of effect is largely beyond direct individual control, powerful individuals may have a greater degree of influence (Cilliers, 1998: 4).

The converse of emergence is reductionism. The critical complexity approach can be employed to confront reductionist assumptions (Webb, 2013: 141–5); for example, a belief that more specific rules lead to better control (Ruhl, 1996b: 860), and that the character and capabilities of rule-making and rule-administering bodies can be reduced to the sum of their parts (Vermeule, 2012: 50–51). Ruhl has discussed the problems of reductionism in relation to environmental law in America, particularly with regard to its efforts to obtain certainty by the proliferation of granular rule-structures (Ruhl, 1996b: 853), and to avoid risk through detailed regulation (2014: 602–603). His work focuses on three related themes concerning the administrative state: understanding rule-making processes and goals; explaining regulatory failures; and patterns of legal development and stagnation. Central to understanding the difficulties of rule-making, and the beliefs

Critical Legal Studies and a complexity approach 51

about what rules can achieve, is the pervasive influence of reductionism on both the academic and professional communities of law (Ruhl, 1996b: 859–60). In their efforts to control rule-outcomes Ruhl argues that lawyers indulge in 'excessive doses of hyper-detailed regulation' and have developed an 'ingrained reliance on top-heavy administrative structures' to manage them (1996b: 860). Similar trends can be seen in the United Kingdom in the proliferation of rules, and efforts to centralize control of some elements of the state, for example persistent attempts to reorganize the National Health Service (Geyer and Rihani, 2010: 98–103).

Ruhl argues that, in the American context, whereas the basic organizing principles and values of the common law, the system of law developed over time through case law, anticipate unpredictability but allow a degree of flexibility to cope with new circumstances (see Webb, 2005: 235), the burgeoning volume of 'regulatory micromanagement' has left the system unable to deal with novelty and shock (Ruhl, 1996b: 860, and see 2014). Although the purpose of generating more and more specific rules may be intended to serve an overriding principle of law, legal certainty, the specificity of the rules may actually limit the ability of administrators, tribunals and courts to fit them into unusual cases. Where the rules run out, a decision-maker might appropriately consider the general aims of the rules, or apply relevant legal principles to reach a defensible judgment (notwithstanding my comments below, see discussion of Dworkin and Hart on pp. 53–4). However, in the face of highly specified rules, it is not clear that such approaches are legitimate, as the rule-makers may not have intended a situation to be covered by otherwise very detailed provisions. It should also be remembered that the vast majority of decisions never see a court or tribunal involved, and administrators will be forced to act within the gaps between the rules, whether this is legitimate or not. Thus, specificity runs paradoxically counter to the principle of legal certainty, and raises questions over the legitimacy of decision-making processes in this context.

Reductionism is also pervasive and can be seen in the compositional and divisional fallacies used to conceptualize institutions which administer and enforce law (see Vermeule, 2012). A compositional fallacy is the assumption that the characteristics of the whole institution or group are derived from the nature of the parts in isolation. Conversely, a divisional fallacy is the assumption that if a whole institution or group possesses a certain set of characteristics, then those traits must originate in its component parts (2012: 9, 15–16). Democracy is, like other values, an emergent phenomenon (Vermeule, 2012: 50–51, 169–70; Webb, 2005: 239); a system is democratic because of how its parts interact and behave, not because each part is democratic. Consider, for example, the European Union (EU) which is said to be undemocratic because its various elements are undemocratic (Walker, 2006: 76; 2010: 229). Viewed through the lens of emergence, it would be incorrect to assume that because parts of the EU are deemed to be undemocratic, the overall arrangements must themselves be undemocratic. This is not to say that the EU is or is not democratic, but that to assume this on the basis of the character of the individual parts, rather than in the manner in which they interact and are appraised by participants, may lead to flawed conclusions. In this context it is important to examine actual practice and the motivating beliefs and actions of the individuals comprising such organizations.

This aspect of reductionist thinking has implications for how lawyers explain and attempt to resolve issues of (non-/partial-) compliance with necessarily contingent rule

52   *Handbook on complexity and public policy*

and value structures. If this is considered in terms of, for example, removing one rule from a given system, the consequences are unlikely to be limited to the area of competence it dealt with. An individual rule is not hermetically sealed from the rest of the system; the rule and the character of the system are a consequence of their interaction. As such, a reductionist understanding is likely to lead to poor decision-making and reform efforts, because these activities will be premised on the idea that the whole is merely the sum of its parts.

CONTINGENCY

The question of how best to approach law- and policy-making, even if emergent consequences are considered, remains contingent. Contingency acknowledges the impossibility of objective, final positions, because everything remains open to contestation (Cilliers and Preiser, 2010: 290). Explanations of regulatory failure, or poorly deployed policy, which claim to have objectively derived the root cause of a difficulty must be viewed sceptically. Contingency requires self-criticality and the consideration of the range of alternative outcomes possible within the wider nexus of assumptions underpinning the conclusions drawn. Similarly, it implies that the wider the range of positions taken into account, the more likely that an effective position can be adopted. In this context, although disagreement will be ongoing, the complexity approach indicates that disagreement itself, discourse (Cilliers, 1995: 129; 1998: 120, 123; Preiser and Cilliers, 2010: 270), may be the key to effective outcomes. Conversely, constraining decision-making under strict sets of rules premised on a limited understanding of the system will damage the resilience of the system against novelty by starving it of the energy which competing views represent.

If legal rules and the system in general are only comprehended in terms of a limited reductionist perspective which suppresses the notions of emergence and contingency, then this increases the likelihood of a number of negative consequences, including unexpected interaction and conflicts between rules (Ruhl and Salzman, 2003: 807), the increase in volume of rules seeking finer levels of control and the corresponding difficulty of tracing compliance and conflict (ibid.: 810), challenges for rule-enforcement over conflicting pressures or standards (Ruhl, 1996b: 908–10), and regulatory failure. Ruhl observes that the natural inclination of lawyers and administrators in the United States to correct law with more law and regulatory guidance on the basis of 'causal intuition' alone conceives of the problem in atomistic terms; if Law A is wrong, Law B can be used to correct it (Ruhl, 1996b: 919, see also 2014; and Vermeule, 2012: 64). However, if the understanding of the current law and its replacements are based on a reductionist understanding of the law-and-society system, they will tend to lock-in the negative 'social defects . . . spawned in the previous generation of laws' (Ruhl and Ruhl, 1997: 413). If law- and policy-makers present a problem as simple or clearly defined, the risk is that the cause of the problem is viewed as being that we have failed to understand at a sufficiently detailed level what are *believed* to be the precise, reductively derived, predictable mechanisms in operation. On this approach there is no recognition that the method employed may be intrinsically flawed, and that the subject matter is quite *un*predictable (see Richardson, 2005: 617). The natural outcome of such a view is to attempt to combat

Critical Legal Studies and a complexity approach   53

the problem with solutions produced on the same defective premises. This position is not entirely new: similar difficulties have been experienced for much of the last century, with the challenge to Legal Formalism by the American Legal Realists, and the subsequent move by the CLS to overcome the flaws of the liberal legal response that followed.

FROM AMERICAN LEGAL REALISM TO THE CLS

In order to understand the conceptual similarities between the complexity approach and CLS, some examination of the historical background is important. CLS was active in the United States during the 1970s and 1980s (on the introduction of CLS to England see Goodrich, 1992) and sought to undermine the prevailing liberal legal tradition. The movement developed and expanded the earlier critique of Formalism by the American Legal Realists undertaken in the 1920s and 1930s (see further Tushnet, 1986; Unger, 1983; White, 1986; see also Mauthe and Webb, 2013b), although other movements such as the less significant Law, Science and Policy school and the much larger Law and Society Association bridged the gap between Realist and CLS discourse (White, 1986: 827, 830–36). The Formalists argued that, by confining the discretion of a decision-maker to a set of rationally determined rules, it was possible to discern the objectively correct outcome to a given case, interpretation of the law, or policy question (Tushnet, 1984: 239; Hasnas, 1995: 86–7; Fischl, 1987: 510–13; Mauthe and Webb, 2013b: 30). The Legal Realists, interested in the processes by which decisions were reached, undermined the Formalists' claim by demonstrating that the outcome determined by any process was inevitably bound up with the character of the decision-maker (Tushnet, 1984: 239). This was not to say that the Realists did not value empirical data; they wanted scientific evidence to support judicial decision-making, but their understanding of legal reality was to be derived from *social*, rather than natural, scientific methods (Fischl, 1987: 521). Whereas the Formalists argued that the key to legal decision-making was the use of the deductive and reductionist scientific methodology to produce objectively certain outcomes, the Realists showed that because many laws within a given legal system were often in conflict, and possessed only indeterminate scope and meaning, it remained open to the decision-maker to decide how they would interpret the particular combination of legal rules and facts; this prevented the objective prediction of decisions (Hasnas, 1995: 86–7, 88–9; Fischl, 1987: 513; Llewellyn, 1930: 464; 1931: 1241–2).

The essence of the Realists' account was that the meaning of law, particularly its application in concrete situations, was indeterminate, and that the decision-maker's own perspective would heavily inform the conclusions drawn. The Realists' critique challenged a central belief of contemporary lawyers: that law had a discernible natural-scientific meaning which could be used to make objectively 'right' decisions (Fischl, 1987: 515). Given the importance of the belief in the objectivity of, and the possibility of reaching single answers to legal questions, it is unsurprising that, rather than accept the conclusions of the Realists, legal theorists sought to reclaim their lost certainty, and 'recover from the shock of realism' (Hunt, 1986: 5; see also White, 1986: 831). The authors of seminal jurisprudential texts examined the nature of indeterminacy seeking to confine its meaning, and so contain the difficulty created. In *The Concept of Law*, first published in 1961, Hart argued that, although there was a 'penumbra of doubt', there nonetheless

54 *Handbook on complexity and public policy*

remained a 'core of certainty' (Hart, 1997: 123). Thus, while a judge would almost always have a choice to make, they could be sure of making the right decision in the vast majority of cases because the textual meaning was normally narrow. Any residual doubt was the price to be paid for the production of a system which applied general rules to specific contexts (Hasnas, 1995: 94).

Subsequent theorists went further. According to Dworkin, Hart left too much uncertainty in legal decision-making (Hasnas, 1995: 94–5). In the highly influential text, *Taking Rights Seriously* (1977), Dworkin argued that Hart's view of law as comprising rules alone was incorrect; there was also principle (Dworkin, 1977: 71–80). Dworkin reasoned that because judges were able to identify 'the "best" interpretation' of the materials before them (Hasnas, 1995: 94–5), they could reach correct decisions, eliminating indeterminacy by filling textual legal gaps with legal principles (Dworkin, 1977: 81; 1998: 337–47, 379–92; see also Hutchinson, 1987: 642, 647, 649; Mauthe and Webb, 2013b: 30). These interventions allowed lawyers to *appear* to internalize the Realist critique, that outcomes could not be deduced from the rules alone, while maintaining the belief that objectively correct outcomes were possible (Hasnas, 1995: 95).

Having suppressed the challenge from the Realists by containing their indeterminacy critique, lawyers also thought that they had avoided the implicit conclusion to be drawn from the central role given to the decision-maker's own preferences and context by the Realists: that all law is politics. It was at this point that the CLS emerged to challenge the suppression of indeterminacy and argue in support of the Realists' claim. The promise of the Realist critique was that 'legal theory could never be the same again' (Hunt, 1986: 5), that with Formalism seemingly dead, legal scholarship would be more honest about its explanatory capacity, at least in relation to adjudication. However, lawyers ignored the essence of the indeterminacy critique.

Although there were certainly shared motivating factors, the CLS was not a unified movement (Tushnet, 1984: 239). Nonetheless, there were three features common to many of the arguments made: the indeterminacy critique; the importance of the presence of destabilizing influences in social systems; and the perpetual nature of critique. Each of these shows a degree of alignment with the complexity approach and should also be seen as more than merely 'New' Legal Realism (Hunt, 1986: 43–4). Realism sought to deal with a growing dissatisfaction with the inability of the Formalist account to satisfactorily resolve legal questions for practitioners and policy-makers; it was pragmatic and focused on legal rules and their application in concrete cases. Conversely, the CLS viewed the law *as an institution* as not being capable of dealing with the problems it claimed to resolve. The movement was not just concerned with pragmatic legal issues of problem-solving in concrete cases, but also with the underlying assumptions which supported the presumed problem-solving capacity of the law and legal theory (ibid.: 43); this wider interest sets the CLS apart from the Realists (Tushnet, 1984: 239).

Against this background I now propose to discuss indeterminacy alongside the strong CLS position taken against the power-imposed stability of structures, and the connections between this account and the complexity concepts of emergence, contingency and critique. The anti-stability argument can be related to self-organization and emergence, while indeterminacy aligns with contingency and the challenge to claims of objectivity and final answers found in the complexity approach. The third element of CLS, perpetual critique, is closely connected with the two preceding elements and recognizes the emergent, contingent nature of assumptions, and the need to continually examine such claims, which I deal with in the last part of the following section.

CRITICAL LEGAL STUDIES AND COMPLEXITY

Destabilization and Emergence

The desirability of destabilization in CLS is best understood from an examination of the movement's political position. A common perception among detractors and certain proponents of CLS was that it was a left-leaning movement. It is likely that this view arose from the claim by some in CLS that the legal system perpetuated a capitalist narrative, having 'decisively implanted itself in both the academic, the political and the popular discourse of contemporary capitalist democracies' (Hunt, 1986: 4). Thus, for that strand of the CLS the critical project presented 'an identifiable alternative', one which could speak 'about the shape and character of a future alternative society' (ibid.: 2). Given the period during which the CLS was most active, it is not surprising that their critics quickly established a link between their position and that of Marxism (Tushnet, 1986: 506). The validity of that connection appeared to be cemented by the CLS attack on the Law and Economics movement (see Posner, 1977), which sought to explain how the law supported equilibrium economic theory, a school of thought intimately aligned with the market and theoretical trends in economics (Kennedy, 1998: 465–7; in complexity see Arthur, 1994; Arthur et al., 1997; Waldrop, 1994: 47–50, 141–2).

Superficially it appeared that the assertion that the CLS was left-leaning was accurate, but a more nuanced position can be adopted. It has been argued that the issue was not with capitalism, conservatism, market ideology or other similar notions per se, but with the assumptions behind theories that had allowed many of their characteristics to become accepted as natural facts of life. For example Tushnet, a leading light in CLS, argued that the attraction of small-scale decentralized socialism to some proponents of CLS was not founded on an innate leftist ideology, but was the 'embodiment of a critique of large-scale centralised capitalism' which sacrificed the importance of community and society to promote individual achievement (1986: 515). Yet the same critical scholar that argued against capitalism would challenge a socialist society on the basis that it denied 'the importance of individual achievement', and decentralization because it acted 'as an impediment to material and spiritual achievement' (1986: 515). On this reading the preference of the Crits was neither left- nor right-leaning, but was more accurately opposed to ingrained and cloistered power structures (Gordon, 1987: 198; but see White, 1986: 837–8). Their aim was to ensure that 'to every crucial feature of the social order there should correspond some form and arena of potentially destabilizing and broadly based conflict over the uses of state power' (Unger, 1983: 592; see also pp. 584, 592–3, 600, 612).

The difficulty was that instability had been defined by a system which tended to view it as a '*dysfunctional* disturbance of equilibrium' (Gordon, 1984: 70, emphasis added). Thus, the Crits had to demonstrate that a degree of instability was not an undesirable condition for the ordinary citizen, but was instead transformative and capable of overcoming a stability which 'cuts off access to genuine possibilities of transformation' (Freeman, 1981: 1235). For the Crits the stability of the status quo was upheld because

56 *Handbook on complexity and public policy*

only superficial elements of the system were open to question (Kairys, 1984: 248; see also Gordon, 1987: 214–16; Tushnet, 1986: 507–508), the more general organization of the social was artificially closed down only because it endangered the prevailing paradigm of a standardized, inflexible form of rationality and objectivity – the way things should be. The Crits believed in the power of the individual to change the world, to engage in asking new questions that might lead to actions that had far-reaching consequences out of all proportion to the original point of origin (see further, Gordon, 1984: 70–71; Kennedy, 1981: 1281–3). The difficulty appears to be that this was expressed, almost pleadingly, in aspirational rather than practical terms. Although the events that would signify the destabilization of settled power structures were clear, to be recognized through the shattering of assumptions and the empowerment of the individual to ask questions, what was lacking was an explanation as to why it was bad for elites to control the frames of reference. There was little discussion of how the processes of instability could be beneficial to the functioning of society. Here, the complexity approach can offer some answers.

The suppression of interaction, or interaction according to anything other than a controlled set of parameters, is the enemy of emergent, non-linear interaction (Waldrop, 1994: 142). The system organizes out of interactions to form a constantly shifting arrangement of concepts, rules, individuals, standards and structures. Or at least it should. If the parameters of the system are pre-determined, or if the range of questions to be asked limited, then the system is denied access to the emergent outcomes which feed off the continual input of energy in the form of new entropy-avoiding interactions (Flood and Carson, 1988: 14; Geyer and Rihani, 2010: 17; Harrison, 2006: 4; Waldrop, 1994: 147). CLS also viewed enforced stability negatively, equating it with stagnation. The complexity approach develops this position by explaining why it is that equilibrium, particularly when the stability it generates has been artificially constructed, amounts to 'a death state . . . where change does not occur' (Geyer and Rihani, 2010: 17). The limited presence or total absence of resources or energy, such as information and communication, which arises as a consequence of narrowly controlled interactive possibilities between system and environment, starves the system and leads to isolation and irrelevance. As such, it is dangerous to argue that interaction should only take place on the basis of foundations which are closely controlled in time and scope.

The complexity account shares and develops the view of CLS that instability can be a transformative force. However, the CLS position on instability rested on a limited, somewhat reductionist analysis. In CLS and other legal theory, such as Formalism, Realism, and Law and Economics, there was a tendency to reduce the essential complexity of the social system to particular parts with delineated functions or characteristics (Ruhl, 1996b: n. 4, p. 852). The nature of the stagnation observed by CLS originated with the assumptions and practices of particular institutions operating within what were viewed as relatively defined roles, but a number of CLS proponents did not leave these foundational premises behind. For many Crits, the remedy to the problems CLS perceived in society flowed from changing the parts in the hope that this would directly affect the whole: change law, change politics, fix society (see Hunt, 1986: 43). This approach oversimplified matters by ignoring the intricacy of the relationships and practices within and between the organizations under scrutiny. However, the reductionist sentiment of CLS can be overstated, as there remained a clear recognition of the challenge presented by the contingency of values in democratic society (Kairys, 1984: 244), and an explicit recognition of the indeterminacy of the application and meaning of law (Gordon, 1984: 114).

Complexity builds on the lesson developed by CLS that the creation and implementation of law and policy always operates under the particular preconceptions of the decision-maker. If it is constructed and applied on a narrow reductionist basis, which relies on a set of unquestioned assumptions and shuns self-reflexive critique, it is unlikely that all the relevant factors and practices will be considered during its formulation. The possibility that these processes do not fully understand the nature of the *perceived* problem, do not question the tools used to construct their perception, and so do not recognize the loaded nature of their view, casts doubt on the potential effectiveness of those measures.

The CLS position also observed that such a state of affairs, of wilful ignorance, is desirable for certain powerful groups. Although it may be possible to reconcile the reductionism of CLS with complexity theory, this principled CLS position against power-imposed stability does not explain with sufficient clarity the reasons why such a scenario is dangerous over the longer term. Indeed, a strong argument can be made that the centralization of power by those entrusted with decision-making creates clear lines of accountability and thereby supports democratic ideals. Yet, while on this reading there may be a rational argument for centralizing power or controlling criteria, such as those just noted, this could be detrimental to local or concrete experiences. On this view, consolidation ensures central accountability only because the requisite processes are located at the centre, and it does not improve the effectiveness of the rules or accountability *per se*. Similarly, while certainty might appear to be improved by more detailed regulation, this may render the gaps exposed by specific unanticipated contexts unfillable. The complexity argument also shows that regardless of the benefits claimed of centralized control, allowing an argument to rest on unexamined 'conceptual fallacies' can have significant negative consequences (Ruhl, 1996b: 860). For example, it is likely to lead to the misidentification of problems capable of solution through law, wasting of resources, a failure of aims, real difficulties with the enforcement of rules, and a lack of clarity about lines of responsibility, all of which stem directly from the suppression of emergent interaction and a failure to account for the contingency of the enterprise.

The essence of the argument is that if the view supporting the status quo on behalf of those already in power is permitted to continue to suppress the range of possibilities open to society, then the inevitable consequence is further suppression and stagnation. The aim should be to foster genuine discourse, disagreement and engagement. Although it has been observed that, because values are emergent and the system tends to rely on values which continue the existence of the system, complexity can also be viewed as preferring democracy because it generates energy through discourse (Webb, 2005: 239), it will also be necessary to ensure that powerful interests and monopolistic tendencies are contained. This is not to say that powerful interests could not capture an investigatory or adjudicative process, adopting a distortion of the complexity approach as an analytical lens, but it should be possible for other actors involved in the process to recognize that interest and question the value of its contribution. Moreover, capture of processes by powerful interests seeking to merely simulate a complexity analysis are likely to encounter the same failures as traditional reductionist approaches, as they seek dictatorial control over the terms of reference, rather than democratic interaction. Stagnation

58 *Handbook on complexity and public policy*

and suppression of interaction are as much anathema to complexity as they are to CLS. The difference is that, whereas CLS opposed suppression in principle, the complexity approach additionally explains *why* such suppression cannot be beneficial to the system.

Indeterminacy and Contingency

Contingency centres around the idea that what is 'could have been otherwise' (Cilliers and Preiser, 2010: 290), and that objectivity and final claims cannot be made because of the influence of context (Cilliers, 2005: 257; 2008: 47). This can be set alongside the CLS view which argued for the existence of short- and medium-term stabilities and regularities within the legal system, spawned out of an array of relatively permanent structures orientated around power, which allowed lawyers to make predictions for clients based on their experience over time (Gordon, 1984: 125). CLS observed that although these structures exist, and are governed by a functioning, predictable set of rules, there could be no reason, other than power, that would prevent the same set of rules leading to the opposite outcome (ibid.: 125). Thus, the perpetual presence of 'the indeterminate content of abstract institutional categories like democracy or the market' (Unger, 1983: 570) went unquestioned by traditional views because it ran counter to their predictive efforts. By failing to question the pragmatic models which allowed lawyers to function, the Crits argued that 'the indefinite possibilities of human connection' were closed down (ibid.: 579), and that the indeterminate nature of adjudication was being ignored. This reduction of complexity was viewed as highly detrimental to the functioning of society in general. To combat this, CLS sought 'to expose as ideology what appears to be positive fact or ethical norm' (Freeman, 1981: 1236), and to overcome the danger that a pragmatic description to help lawyers function could become a concrete understanding of the reality of processes (Gordon, 1984: 71).

By challenging the tendency of traditional approaches to transform pragmatic models into reality, the Crits revealed foundational assumptions which constrained the ability of the system to recognize the limits of its problem-solving capacity over essentially indeterminate decisions. It was remarked that where the reductionist liberal legal worldview was active, the principle aim was always 'that of rationalizing the real' (Gordon, 1981: 1018). If there is a defensible set of legal processes and principles, then it is possible to say that the law is rational so as to cast a particular legal act or policy as necessary, and to claim that it will work (ibid.: 1018). The Crits argued that this caused the law to be portrayed as 'somehow inevitable, natural, neutral, objective, scientific, God-given,' beyond question (Kairys, 1984: 248–9). In their effort to 'filter out complexity, variety, irrationality, unpredictability, disorder, cruelty, coercion, violence, suffering, solidarity and self-sacrifice' (Gordon, 1987: 200), these processes artificially reduced the nature of the legal system, and ignored the Crits' understanding of legal decision-making as a contested practice.

This anxiety over distortion (Gordon, 1987: 200), and to an extent delusion (Freeman, 1981: 1231), brought about by the exclusion from consideration of contestability in decision-making, of choice as being more than merely formulaic, shares a parallel with the critique from contingency in relation to uncertainty, lack of control, and the absence of final objective answers (Cilliers, 1995: 124; see also Richardson, 2005: 644). For complexity, a degree of simplification will always be necessary to make sense of the

Critical Legal Studies and a complexity approach 59

world (Richardson, 2005: 648), but because simplification denies the contingent nature of understanding, and the full range of emergent possibilities, distortion is unavoidable (Cilliers, 2007: 161). As a consequence of this, different actors will reach varied conclusions based on a range of unique selections of evidence and determinations of meaning (Cilliers, 2005: 263). These arguments are 'neither relativist nor vague' (Preiser and Cilliers, 2010: 269; see also Cilliers, 2005: 262–4), they simply acknowledge the limitations of our explanatory capacity. However, this understanding does indicate the limited value of perspectives which deliberately suppress other interpretations, or which present their conclusions as objective, natural or final. It suggests that we should not confine investigation and explanation to one perspective in the attempt to analyse a given problem, nor should we rely on hyper-detailed instruments to resolve problems.

On this reading, despite the apparent alignment between the notion of indeterminacy and the 'subjugating' societal consequences which flow from suppressing it (Freeman, 1981: 1233–4; Gordon, 1984: 112) and the idea of contingency, the latter can be seen as developing the position further. The specificity of the CLS argument, and the self-belief in the position, could be said to have undermined the importance of self-critique to the movement. Thus, in this respect CLS limited its application by specifically targeting one problem, the neo-formalism which arose following the avoidance of the Realists' challenge. Although they explained the limitations of the prevailing legal worldview, and offered principled solutions, it was a challenge against a specifically delineated rival. Much of their critique was directed at the assumptions and countermeasures deployed by that approach. The specificity of the initial attack, demonstrating why a particular set of assumptions was detrimental to legal reasoning, may also explain the subsequent fragmentation of the movement in its efforts to critique other problems in law (Neacsu, 2000: 430–33). However, while this specificity is not fatal to the wider utility of the CLS to complexity theory or law, it must be borne in mind. Yet it is clear that the complexity approach is not constrained in this way. The systems-theoretical approach to analysis which relies on a set of tools that the critic is free to employ as they wish is more flexible, and can be deployed to test the security of any claim (Geyer and Rihani, 2010: ch. 3; see also Mitleton-Kelly, 2003: 26). Moreover, because the political position of the complexity approach is emergent rather than innate or transcendental, with a natural preference for democratic discourse over entropy-inducing diktat, it actively counters claims that it is merely promoting a particular political perspective, a problem which (wrongly) plagued the CLS.

There is also an element of self-reflexivity to the complexity approach about its explanatory capacity that is not clearly expressed in CLS. The complexity approach is a singular view of the social; it claims that a systems-theory approach to critique will yield a useful result. However, the integral nature of contingency to the complexity approach means that it is not capable of producing the only correct answer, that would be immodest (Ruhl, 1996a: 1452; Webb, 2005: n. 36, p. 236). Moreover, the varied contexts of those employing a complexity analysis will lead to multiple interpretations of what complexity is, and the contribution it can make. Thus, any claim to have completely explained a given legal problem, or to have satisfactorily resolved a question of legal meaning is flawed, 'such pretence is not only hubristic, it is also a violation of that which is being modelled' (Cilliers, 2010: 8), and, as both complexity and the CLS agree, may produce something which is 'wholly artificial' and of little use to anyone (Webb, 2005: n. 43,

60 *Handbook on complexity and public policy*

p. 237). A failure to engage in discourse with alternative approaches, such as the rich CLS literature, to take part in 'the agnostics of the network' (Cilliers, 1995: 129; 1998: 120, 123; Preiser and Cilliers, 2010: 270), will isolate complexity theory, deny it energy, and so diminish its relevance.

Given the importance of genuine unconstrained discourse to the avoidance of entropy, multiple paths of exploratory inquiry should be encouraged. Thus, law- and policy-makers, and those who work with, interpret and enforce those policies and rules should reflect on their claims and, where necessary, revise them over time. This reflexivity is integral to the understanding in the complexity approach because it encourages the refining of ideas, self-critique and the perpetuation of discourse: the input of energy. Thus, a key requirement in the investigation of the range of contingent perspectives is ongoing critique. This process continually injects new energy into the debate, breaks open traditional frames of reference, and so creates new transformative possibilities, making it central to the generation of strategies to manage complex social problems.

Ongoing Critique and Self-reflexivity

Both approaches imply that an ongoing effort will be required to ensure that decision-makers are not permitted to lapse into the refrain that 'the law made me do it' (Kairys, 1984: 249; Cilliers, 2004: 24), that a particular set of unexamined assumptions *demanded* one outcome that was taken on that basis to be just, fair and rational. Thus, the principal concept which motivates both approaches is that of ongoing critique. For the Crits, it was expected that no explanation or solution could be legitimately presented which envisaged 'something more enduring than interminable critique' as the outcome of implementation (Tushnet, 1986: 516). This was because the purpose of critique was to challenge the existing order, not to provide a final solution (ibid.: 516). Consequently, the CLS was challenged by its opponents on the grounds that it proposed perpetual deconstruction without ever offering a way out (MacCormick, 1993: 142; see also Fischl, 1992: 800–802; Kelman, 1986; Neacsu, 2000: 419–22). However, given that the CLS enterprise focused on undermining received assumptions about the way things are, and how things should be decided, the *is* and *ought* question, in defined circumstances, it is unsurprising that the idea of a grand theory was unattractive to many of their number (Hunt, 1987: 5, see also Fischl, 1987: 532); this also further suggests that the reductionism of CLS can be overstated (see p. 56). In reply to the argument that CLS failed to offer a transformative critical device capable of correcting the perceived problems of law and legal rationalization, it could be said that this was because of the lack of interest in formulating a programmatic response to satisfy their critique. Thus, it would be more accurate to say that a CLS critique of an approach only suggested reform by implication of the flaws it highlighted. It was not that solutions-for-the-moment could not be constructed, but that they were best thought of as transient and imperfect, because they were 'the result of intensely pragmatic judgments about what *appears* useful in the present circumstances' (Tushnet, 1984: 241 emphasis added).

An additional explanation for the difficulties encountered by CLS can also be given. The CLS confronted the foundational beliefs of legal scholarship, and persuading those whose work rests on those unexamined premises to change will always be difficult. A key obstacle to critique in general is thus '[generating] the psychological experience of

Critical Legal Studies and a complexity approach 61

a threat' in the scholarly community, and persuading them to 'articulate the theoretical structures they take for granted' (Gordon, 1981: 1024). This problem also lies in the way of the critical approach of complexity. The complexity approach argues that where legal processes rest on unexamined, reductionist assumptions, law lacks the degree of self-reflexivity necessary to permit the observer, law-maker, judge, or similar to consider whether the premises from which they began, and under which they sought to take action, make sense or require re-examination. Indeed, work continues to be published which resolutely avoids such a confrontation (however, see Husa and Van Hoecke, 2013, for possible solutions; also consider Feldman, 1989; Mauthe, 2005; 2006; 2014). There appears to be only a very limited appetite for moving towards the self-critical position articulated by a critical complexity approach. While current understandings in legal scholarship disclose *an* explanation for how things could work, they only function satisfactorily if premised on unacknowledged '*specific boundary conditions*' (Eve, 1997: 275; emphasis in the original). The foundational assumptions for the success of this form of reasoning are that it should lead to frameworks of explanation for structures and phenomena which are all 'predictable' and 'orderly' (Geyer and Rihani, 2010: 6). Both CLS and complexity observe that lawyers have assumed that their approach to constructing knowledge can be predictable and orderly, and therefore objective and realistic (see Gordon, 1984: 114; Ruhl, 2008: 907–908). In my own field of public law it has been observed that this mode of reasoning is 'actually quite fixed' and is unlikely to change (Mauthe, 2005: 64, 70; see also 2006; 2014).

CONCLUSION

In this chapter I have sought to demonstrate the alignment between the CLS literature and that of complexity theory, with a view to mapping some issues for consideration in decision-making processes. More generally, I have suggested that a critical complexity approach and CLS offer tools to challenge reductionist claims about the nature of law and policy, and its experience in practice. The message from complexity theory is powerful. If we take account of the preceding critical experiences in law in combination with the critical complexity position outlined above, it can be seen that first, the negative consequences of enforced stability or confined perspectives are likely to be detrimental to the transformative capacity of law and policy creation and implementation processes. Secondly, closing down interaction between the models used by actors to appraise complex problems in scope or time will prevent self-critique and impede administrators, adjudicators and others from coming into contact with those holding competing views of the most effective way to construct and deploy, for example, regulatory efforts. This is particularly detrimental where existing approaches are premised on unexamined reductionist assumptions, and is likely to result in inequitable and ineffective conclusions based on the same type of assertions. Finally, as examination of CLS also demonstrates, destabilization as an aspect of emergence is important to ensuring the ongoing interaction, adaptability and longevity of the system.

In relation to broad questions of law and policy it has been shown that, for CLS, it was always important to maintain a degree of instability within the system so as to prevent powerful interests from monopolizing the terms of the debate. Where debate

62   *Handbook on complexity and public policy*

was captured, the consequence tended to be the suppression of the inevitable indeterminacy of meaning. Complexity theory builds on this by arguing that the foundational assumptions of decision-makers and other stakeholders in legal, policy and other processes are of central importance to the outcomes generated. Where the initial premises underlying reasoning remain unexamined there is a risk that reductionist rationalization and the avoidance of contingency will undermine the conclusions reached. Thus, both approaches show the importance of engaging in a constant self-critical review of beliefs, structures and processes to manage the inclination towards oversimplification and encourage the perpetuation of meaningful discourse through the continual supply of energy. It is clear that there are more connections to be explored regarding how CLS can be of benefit to complexity theory, pointing to a fruitful future discourse.

ACKNOWLEDGEMENTS

The author wishes to thank Amanda Cahill-Ripley, Paul Cairney and Barbara Mauthe for their comments on earlier drafts. Any errors remain the author's own.

REFERENCES

Arthur, W.B. (1994), *Increasing Returns and Path Dependence in the Economy*, Ann Arbor, MI: University of Michigan Press.
Arthur, W.B., S.N. Durlauf and D.A. Lane (1997), 'Process and Emergence in the Economy', in W.B. Arthur, S.N. Durlauf and D.A. Lane (eds), *The Economy as an Evolving Complex System, vol. II*, Reading, MA: Perseus Books.
Cilliers, P. (1995), 'Postmodern Knowledge and Complexity (or why anything does not go)', *South African Journal of Philosophy*, **14**(3), 124–32.
Cilliers, P. (1998), *Complexity and Postmodernism: Understanding Complex Systems*, London: Routledge.
Cilliers, P. (2001), 'Boundaries, Hierarchies and Networks in Complex Systems', *International Journal of Innovation Management*, **5**(2), 135–47.
Cilliers, P. (2004), 'Complexity, Ethics and Justice', *Tijdschrift voor Humanistiek*, **5**, 19–26.
Cilliers, P. (2005), 'Complexity, Deconstruction and Relativism', *Theory, Culture & Society*, **22**, 255–67.
Cilliers, P. (2007), 'Knowledge, Complexity and Understanding', in P. Cilliers (ed.), *Thinking Complexity: Complexity and Philosophy, Volume 1*, Mansfield, MA: ISCE Publishing, pp. 159–64.
Cilliers, P. (2008), 'Knowing Complex Systems: The Limits of Understanding', in F. Darbellay, M. Cockell, J. Billotte and F. Waldvogel (eds), *A Vision of Transdisciplinarity: Laying the Foundations for a World Knowledge Dialogue*, Lausanne: EPFL Press; Boca Raton, FL: CRC Press, Taylor and Francis Group, pp. 43–50.
Cilliers, P. (2010), 'Difference, Identity and Complexity', in P. Cilliers and R. Preiser (eds), *Complexity, Difference and Identity*, London: Springer, pp. 3–18.
Cilliers, P. and R. Preiser (2010), 'Glossary', in P. Cilliers and R. Preiser (eds), *Complexity, Difference and Identity*, (Issues in Business Ethics, Volume 26), London: Springer.
Dworkin, R. (1977), *Taking Rights Seriously*, London: Duckworth.
Dworkin, R. (1998), *Law's Empire*, Oxford: Hart.
Eve, R.A. (1997), 'Afterword: So Where Are We Now? A Final Word', in R.A. Eve, S. Horsfall and M. Lee (eds), *Chaos, Complexity, and Sociology: Myths, Models and Theories*, London: Sage, pp. 269–80.
Feldman, D. (1989), 'The Nature of Legal Scholarship', *Modern Law Review*, **52**(4), 498–517.
Finchett-Maddock, L. (2012), 'Seeing Red: Entropy, Property and Resistance in the Summer Riots 2011', *Law and Critique*, **23**(3), 199–217.
Fischl, R.M. (1987), 'Some Realism About Critical Legal Studies', *University of Miami Law Review*, **41**, 505–32.
Fischl, R.M. (1992), 'The Question that Killed Critical Legal Studies', *Law and Social Inquiry*, **17**, 779–820.
Flood, R.L. and E.R. Carson (1988), *Dealing with Complexity: An Introduction to the Theory and Application of System Science*, London: Plenum Press.

Freeman, A.D. (1981), 'Truth and Mystification in Legal Scholarship', *Yale Law Journal*, **90**, 1229–37.
Geyer, R. and S. Rihani (2010), *Complexity and Public Policy: A New Approach to Twenty-First Century Policy and Society*, London: Routledge.
Goldstein, J. (1999), 'Emergence as a Construct: History and Issues', *Emergence*, **1**(1), 49–72.
Goldstein, J. (2000), 'Emergence: A Concept Amid a Thicket of Conceptual Snares', *Emergence*, **2**(1), 5–22.
Goodrich, P. (1992), 'Critical Legal Studies in England: Prospective Histories', *Oxford Journal of Legal Studies*, **12**(2), 195–236.
Gordon, R. (1981), 'Historicism in Legal Scholarship', *Yale Law Journal*, **90**, 1017–56.
Gordon, R. (1984), 'Critical Legal Histories', *Stanford Law Review*, **36**(1/2), 57–125.
Gordon, R. (1987), 'Unfreezing Legal Reality: Critical Approaches to Law', *Florida State University Law Review*, **15**(2), 195–220.
Harrison, N.E. (2006), 'Thinking About the World We Make', in N.E. Harrison (ed.), *Complexity in World Politics: Concepts and Methods of a New Paradigm*, New York: State University of New York Press.
Hart, H.L.A. (1997), *The Concept of Law*, 2nd edn, Oxford: Oxford University Press.
Hasnas, J. (1995), 'Back to the Future: From Critical Legal Studies Forward to Legal Realism, or How Not to Miss the Point of the Indeterminacy Argument', *Duke Law Journal*, **45**, 84–132.
Heylighen, F., P. Cilliers and C. Gershenson (2007), 'Philosophy and Complexity', in J. Bogg and R. Geyer (eds), *Complexity Science & Society*, Oxford: Radcliffe Publishing, pp. 117–34.
Holland, J.H. (1995), *Hidden Order: How Adaptation Builds Complexity*, Reading, MA: Helix Books.
Hunt, A. (1986), 'The Theory of Critical Legal Studies', *Oxford Journal of Legal Studies*, **6**(1), 1–45.
Hunt, A. (1987), 'The Critique of Law: What is "Critical" about Critical Legal Theory?', *Journal of Law and Society*, **14**(1), 5–19.
Husa, J. and M. Van Hoecke (eds) (2013), *Objectivity in Law and Legal Reasoning*, Oxford: Hart Publishing.
Hutchinson, A.C. (1987), 'Indiana Dworkin and Law's Empire', *Yale Law Journal*, **96**(3), 637–55.
Kairys, D. (1984), 'Law and Politics', *George Washington Law Review*, **52**(2), 243–62.
Kelman, M.G. (1986), 'Trashing', *Stanford Law Review*, **36**, 293–348.
Kennedy, D. (1981), 'Cost-Reduction Theory as Legitimation', *Yale Law Journal*, **90**, 1275–83.
Kennedy, D. (1998), 'Law-and-Economics from the Perspective of Critical Legal Studies', in P. Newman (ed.), *The New Palgrave Dictionary of Economics and the Law, Vol. 2*, 2nd edn, Basingstoke: Macmillan, pp. 465–74.
Llewellyn, K.N. (1930), 'A Realistic Jurisprudence – The Next Step', *Columbia Law Review*, **30**(4), 431–65.
Llewellyn, K.N. (1931), 'Some Realism About Realism – Responding to Dean Pound', *Harvard Law Review*, **44**(8), 1222–64.
MacCormick, N. (1993), 'Reconstruction after Deconstruction: Closing in on Critique', in A.W. Norrie (ed.), *Closure of Critique?: New Direction in Legal Theory*, Edinburgh: Edinburgh University Press, pp. 142–56.
Mauthe, B. (2005), 'The Notion of Sovereignty and its Presentation within Public Law: a Critique on the Use of Theory and Concepts', *Northern Ireland Legal Quarterly*, **56**(1), 63–82.
Mauthe, B. (2006), 'Public Law, Knowledge and Explanation: A Critique on the Facilitative Nature of Public Law Analysis', *International Journal of Law in Context*, **2**(4), 377–92.
Mauthe, B. (2014), 'Public Law and the Value of Conceptual Analysis', *International Journal of Law in Context*, **10**(1), 47–63.
Mauthe, B. and T.E. Webb (2013a), 'In the Multiverse What is Real? Luhmann, Complexity and ANT', in A. La Cour and A. Philippopoulos-Mihalopoulos (eds), *Luhmann Observed: Radical Theoretical Encounters*, London: Palgrave, pp. 243–62.
Mauthe, B. and T.E. Webb (2013b), 'Realism and Analysis Within Public Law', *Liverpool Law Review*, **34**, 27–46.
Mitleton-Kelly, E. (2003), 'Ten Principles of Complexity and Enabling Infrastructures', in E. Mitleton-Kelly (ed.), *Complex Systems and Evolutionary Perspectives on Organisations: The Application of Complexity Theory to Organisations*, London: Pergamon, pp. 23–50.
Murray, J. (2006), 'Nome Law: Deleuze and Guattari on the Emergence of Law', *International Journal for the Semiotics of Law*, **19**, 127–51.
Murray, J. (2008), 'Complexity Theory and Socio-Legal Studies, Coda: Liverpool Law', *Liverpool Law Review*, **29**, 227–46.
Murray, J. (2010), *Deleuze & Guattari: Emergent Law*, Oxford: Routledge.
Neacsu, E.D. (2000), 'CLS Stands for Critical Legal Studies, if Anyone Remembers', *Journal of Law and Policy*, **8**, 415–53.
Posner, R.A. (1977), *Economic Analysis of Law*, 2nd edn, Boston, MA: Little, Brown.
Preiser, R. and P. Cilliers (2010), 'Unpacking the Ethics of Complexity: Concluding Reflections', in P. Cilliers and R. Preiser (eds), *Complexity, Difference and Identity*, London: Springer, pp. 265–88.
Rescher, N. (1998), *Complexity: A Philosophical Overview*, New Brunswick: Transaction.

64 Handbook on complexity and public policy

Richardson, K.A. (2004), 'Systems Theory and Complexity: Part 1', *Emergence: Complexity & Organisation*, **6**(3), 75–9.

Richardson, K.A. (2005), 'The Hegemony of the Physical Sciences: An Exploration in Complexity Thinking', *Futures*, **37**, 615–53.

Richardson, K.A., P. Cilliers and M. Lissack (2001), 'Complexity Science: A "Grey" Science for the "Stuff in Between"', *Emergence*, **3**(2), 6–18.

Ruhl, J.B. (1996a), 'The Fitness of Law: Using Complexity Theory to Describe The Evolution of Law and Society and its Practical Meaning For Democracy', *Vanderbilt Law Review*, **49**, 1407–90.

Ruhl, J.B. (1996b), 'Complexity Theory as a Paradigm for the Dynamical Law-and-Society System: A Wake-up Call for Legal Reductionism and the Modern Administrative State', *Duke Law Journal*, **45**(March), 849–928.

Ruhl, J.B. (1997), 'Thinking of Environmental Law as a Complex Adaptive System: How to Clean Up the Environment by Making a Mess of Environmental Law', *Houston Law Review*, **34**(Winter), 933–1002.

Ruhl, J.B. (2008), 'Law's Complexity: A Primer', *Georgia State University Law Review*, **24**, 885–991.

Ruhl, J.B. (2014), 'Managing Systemic Risk in Legal Systems', *Indiana Law Journal*, **89**, 559–603.

Ruhl, J.B. and H.J. Ruhl, Jr (1997), 'The Arrow of the Law in Modern Administrative States: Using Complexity Theory to Reveal the Diminishing Returns and Increasing Risks the Burgeoning of Law Poses to Society', *University of California Davis Law Review*, **30**(Winter), 405–82.

Ruhl, J.B. and J. Salzman (2002), 'Regulatory Traffic Jams', *Wyoming Law Review*, **2**(2), 253–89.

Ruhl, J.B. and J. Salzman (2003), 'Mozart and the Red Queen: The Problem of Regulatory Accretion in the Administrative State', *Georgetown Law Journal*, **91**, 757–850.

Tushnet, M.V. (1984), 'Perspectives on Critical Legal Studies', *George Washington Law Review*, **52**(2), 239–42.

Tushnet, M.V. (1986), 'Critical Legal Studies: An Introduction to its Origins and Underpinnings', *Journal of Legal Education*, **36**, 505–17.

Unger, R.M. (1983), 'The Critical Legal Studies Movement', *Harvard Law Review*, **96**(3), 561–675.

Urry, J. (2003), *Global Complexity*, Oxford: Polity Press.

Vermeule, A. (2012), *The System of the Constitution*, New York: Oxford University Press.

Waldrop, M.M. (1994), *Complexity: The Emerging Science at the Edge of Order and Chaos*, 2nd edn, London: Penguin Books.

Walker, N. (2006), 'European Constitutionalism in the State Constitutional Tradition', *Current Legal Problems*, **59**, 51–89.

Walker, N. (2010), 'Constitutionalism and the Incompleteness of Democracy: An Iterative Relationship', *Rectsfilosofie & Rechtstheorie*, **39**(3), 206–33.

Webb, J. (2005), 'Law, Ethics, and Complexity: Complexity Theory and the Normative Reconstruction of Law', *Cleveland State Law Review*, **52**, 227–42.

Webb, T.E. (2013), 'Exploring System Boundaries', *Law and Critique*, **24**, 131–51.

Webb, T.E. (2014), 'Tracing an Outline of Legal Complexity', *Ratio Juris*, **27**(4), 477–95.

White, E.G. (1986), 'From Realism to Critical Legal Studies: A Truncated Intellectual History', *Southwestern Law Journal*, **40**, 819–43.

5. 'What's the big deal?': complexity versus traditional US policy approaches
Michael Givel

INTRODUCTION

Policy outputs and outcomes occurring in the course of typical government business are often difficult, if not impossible, to comprehend with total certainty due to their frequently complex and even unpredictable nature (Miller and Page, 2007; New England Complex Systems Institute, 2011). Public policy outputs are what governmental political institutions including the executive, legislative and judicial branches do or do not do. For example, in the legislative branch, governments enact laws. Policy outputs can result in policy outcomes with unintended consequences, unexpected outcomes and novel events (Johnson, 2010; Miller and Page, 2007; Pressman and Wildavsky, 1983). A primary purpose of modern policy studies is to effectively evaluate policy outputs and outcome patterns and trends.

The unpredictable nature of policy outputs and outcome trends are known as emergent phenomena in all policy niches such as, for instance, energy or health. Emergent phenomena are sensitive to the initial policy situation such as a new welfare policy in a policy niche and are influenced by numerous negative or positive feedback loops. Positive policy feedback substantially increases benefits and reduces costs in relation to current political, economic or social public policies to the advantage of the greater society or certain groups or individuals in a society. Negative policy feedback reduces the political, economic or social benefits and increases costs of a policy to society, groups or individuals. Negative feedback can be caused by such factors as problems with implementing the policy or segments of a population challenging the policy that does not meet their needs. For instance, a social movement of poor people may rise to challenge and change government policies that allow a significant inequality of wealth and income.

Modern complexity science has developed several powerful methodological approaches, for example agent-based-modeling, network analysis, data mining, and scenario modeling to assess complex system behavior. In order to understand the nature of complex policy systems, a variety of interdisciplinary sources and several complex methodological approaches such as complex qualitative research analyses like archival research approaches can be applied. Through the evaluation of complex policy tendencies, policy specialists can provide recommendations to decision makers on the changing nature and direction of new policies, regulations and legislation. When accurately evaluating or predicting a large number of different policy trends and programs like social welfare, energy or a health care program, what is also essential is understanding the degree of change and ongoing impact of the policy (Taleb, 2008).

In this chapter, linear theory and methodology to comprehend public policy trends is related to analyzing government decision-making occurring in a unilateral direction

66 *Handbook on complexity and public policy*

similar to a line or permutation of a line. For instance, a political problem or issue may progress in a linear direction to the policy formulation stage. This theoretical and methodological approach is analogous to nineteenth-century Newtonian scientific physics in which the universe is said to operate in a precise, linear, mechanistic and clocklike manner. Other features of the linear theoretical approach include a tendency toward a standardized policy process from the initial policy problem or issue to implementation that results in a simplification and reduction of complex policy trends based on one established policy process. Newton based this linear methodological view in his *Philosophiæ Naturalis Principia Mathematica* on the 'sensorium of God' in which God viewed the universe as never changing and absolute (Newton, 1687). On the other hand, complexity theory and the methodology to measure complex policy system behavior assumes that government decisions may occur in complex or bilateral directions including interacting simultaneously on multiple levels of government. This theoretical and methodological approach is analogous to twentieth- and twenty-first-century scientific complexity theories such as specific relativity, chaos theory, quantum mechanics, fuzzy logic, artificial intelligence, computational modeling and web science.

Positivism

Currently, of the three major scientific methodological approaches now in use, including positivism, post-postivism and critical realism, only positivism is oriented toward measuring linear policy behavior (Popper, 2002). Positivism is an empirical methodology analogous to nineteenth-century linear Newtonian physics. Positivism is grounded in the scientific method and is derived from a variety of Enlightenment thinkers including August Comte and Henri de Saint-Simon (Crotty, 1998). In a scientific methodological context, positivism requires the 'value free' testing of a hypothesis with linked variables in which their behavior and interactions are observable. The scientific method includes collecting data, analyzing the data, interpreting the results of the data analysis, determining whether the original hypothesis was false in whole or in part, and retesting hypotheses that have been previously confirmed (Hinkel et al., 2003). Additionally, the scientific method tests linear causality and behavior of variables. If a number of the same or similar scientific hypotheses are tested and upheld, then this rises to the level of a robust scientific theory such as the biological theory of evolution.

Methodologically, positivism requires that empirical reality be measured by hard evidence or data that is observed directly. Any social phenomenon that is not directly known is excluded in positivist research. Positivism also assumes that there are no particular differences between the hard and social sciences in terms of measuring empirically observable phenomena (Mancias, 2007). Consequently, society and public policy as a subset of the greater society are viewed 'objectively' by positivists in the sub-field of public policy as behavioral in orientation.

Critiques of positivism, particularly after the Second World War, have indicated it is a distorted reductionist assumption of reality in which complex social trends including policy are relegated to quantitative or empirical qualitative logical terms that do not fully explain social realities (Creswell, 2007; Lincoln et al., 2011). Others have argued that positivism implicitly and explicitly supports the political and economic status quo. This is due to a recurrent failure in positivism to fully explore the underlying dynamics and structures of political and economic dominance and influence (Wyly, 2008). Another criticism of positivism has centered on the purported objectivity of the scientific researcher when in fact all researchers contain biases that are reflected in the scientific topics that they research (Pierre, 2011). Finally, social realities, including public policies, are not based on totally knowable and predictable policy trends as is argued by positivists, due to emergent and probabilistic future trends in society including public policies (Popper, 2002). As a result of these significant critiques of positivism, commencing after the Second World War, post-positivism and in the 1970s critical realism grew in prominence.

Post-positivism

Post-positivism is based in complexity and non-linear system behavior because post-positivists have observed that social reality relies on nested systems and processes that are not always observable through surface behavioral observations (Alvesson and Skoldberg, 2010). Rather, social systems including policy systems are often complex and abstract. Accentuating this abstract orientation is the acknowledgement by post-positivists that all scientists possess biases when conducting their scientific studies (Lincoln et al., 2011). A primary post-positivist methodological approach is determining how norms, values, symbols, discourse and language are embedded in and are reflective of social and policy systems (Lincoln et al., 2011). Given the complex nature of social realities and policy trends, post-positivists believe that scientific trends may be comprehended only as an emergent tendency (Alvesson and Skoldberg, 2010). The approach is in sync with complex system theory in public policy in which all policy behavior is congruent with complexity theory (Denzin and Lincoln, 2011).

Critical Realism

The third modern scientific methodology is critical realism, which materialized in the 1970s. Critical realism is akin to various complexity-oriented scientific theories of the twentieth and twenty-first centuries (Archer, 1995; Bhaskar, 1975; 1993; 1998; Callinicos, 2007; Givel, 2010). Critical realists acknowledge that scientists independently evaluate scientific trends based on empirical measurements, but also argue that the primary objective of science is to analyze what is real (Archer, 1995; Bhaskar, 1975, 1993, 1998; Callinicos, 2007; Givel, 2010). The real includes underlying and non-observable social realities and trends including most importantly political and economic inequality and social stratification, which play a key role in social relations (Archer, 1995; Bhaskar, 1975; 1993; 1998; Callinicos, 2007; Givel, 2010). In contrast to critics of classic pluralism who argue this approach upholds the status quo, analyses of inequality by critical realists can and do lead to significant social critiques of the political and economic status quo.

EARLY PUBLIC POLICY THEORY

All of this raises the important question: are the current and leading theories of public policy now in line with complex systems theory and methodology? US political scientist,

68   *Handbook on complexity and public policy*

Harold Lasswell, developed the earliest modern policy theory also known as policy sciences. The policy sciences were oriented toward solving applied public problems, utilizing a variety of interdisciplinary methodological approaches, and were grounded in linear positivism methodology (Birkland, 2011; deLeon, 2006; Lasswell, 1948, 1971; Theodoulou, 2013). Moreover, policy sciences were value-oriented, integrated with an anti-communist ideology that posited that public policy properly occurs in a democratic system consisting of public, civic and market sectors. This was in contrast to an entirely public, or what Lasswell described as totalitarian system (deLeon, 2006; Lasswell, 1948, 1971).

The basis for empirical analyses occurring in Lasswell's policy sciences theory was evaluating how the public policy decision-making process occurred. Within this process, public policy developed in a linear fashion from stage to stage (Anderson, 1975; Jones, 1970; Lasswell, 1956; Lasswell and Kaplan, 1971; Sabatier, 2007). This conception of policymaking known as the stages heuristic theory includes the following steps: a problem or issue reaching the public agenda, policy formulation, policy enactment, policy implementation, and feedback through policy evaluations (Sabatier, 2007). The stages heuristic theory remained as a leading theory until the 1980s. By that time, a number of critics argued that this theory was faulty because it did not adequately describe underlying policy drivers such as structural economic inequality or measure complex system behavior. This lack of measurement of complex system behavior did not account for complex policy interactions between the top, middle and bottom levels of government and did not describe non-linear government decision-making (Hjern and Hull, 1982; Sabatier, 1986; 2007).

MODERN US PUBLIC POLICY THEORIES

Since the fading of the stages heuristic model in the late 1980s as the predominant policy theory, a new wave of policy theories have emerged. Are these theories in sync with complex system theory and methodology or linear theory and methodology like the stages heuristic model? In 2007, Sabatier provided a definitive assessment of the 'more promising theoretical frameworks' in American public policy circles that have been developed and proposed (Sabatier, 2007). Describing the general selection process for these, Sabatier wrote in 2007: 'The first edition (of Sabatier's book) was criticized for its narrow selection criteria, particularly for only including frameworks that followed scientific norms of clarity, hypothesis-testing, acknowledgement of uncertainty, etc. Since I am unequivocally a social scientist, this criticism fell on deaf ears' (Sabatier, 2007: 11).

Sabatier's more promising policy theories include: institutional rational choice, punctuated equilibrium, multiple streams, advocacy coalition framework, policy diffusion, and large N-comparative studies (Sabatier, 2007). The entire selection criteria used by Sabatier for these American-based policy theories were: empirically tested with falsifiable hypotheses, provided a broad explanation of the policy process, and assessed a broad range of factors that political scientists have traditionally researched (Sabatier, 2007).

The theoretical grounding of these modern American policy theories originates from pluralism, which posits that many groups freely compete in the political marketplace and no one group is dominant (Baumgartner and Jones, 2009; Berry and Berry, 2007;

Blomquist, 2007; Dahl, 1961; Eldredge and Gould, 1972; Givel and Johnson, 2013; Gould, 1984; 1989; Joskow, 1995; Kingdon, 2010; Ostrom, 1998; 2007; Repetto, 2006; Sabatier, 2007; Sabatier and Jenkins-Smith, 1993; Wood and Doan, 2003; Zahariadis, 2007). The current policy theories that are in large part congruent with pluralism include: multiple streams, policy diffusion and large N-comparative studies (Berry and Berry, 2007; Blomquist, 2007; Kingdon, 2010). Institutional rational choice theorists also argue, like pluralism, that there is unfettered competition in the political market-place, but through rationalistic individual and group competition based on a form of neo-classical economics theory (Ostrom, 1998; 2007). Punctuated equilibrium theory and advocacy coalition framework theory provide an updated version of pluralism in which some groups have more resources and influence in the policy process than others (Baumgartner and Jones, 2009; Sabatier and Jenkins-Smith, 1993; Sabatier and Weible, 2007). In respect to complexity theory, multiple streams, advocacy coalition framework, and policy diffusion theories all acknowledge complex system behavior as a central feature of the policymaking process (Berry and Berry, 2007; Givel, 2010; 2012; Kingdon, 2010; Sabatier and Jenkins-Smith, 1993; Sabatier and Weible, 2007). On the other hand, punctuated equilibrium theory, large N-comparative studies and institutional rational choice are primarily linear in orientation (Baumgartner and Jones, 2009; Blomquist, 2007; Ostrom, 1998; 2007). All of these current US public policy theories use positivist, or in the case of punctuated equilibrium, a combination of positivism and post-positivist methodological approaches (Baumgartner and Jones, 2009; Berry and Berry, 2007; Blomquist, 2007; Kingdon, 2010; Ostrom, 1998; 2007; Sabatier and Weible, 2007). None use critical realism.

MODERN POLICY THEORIES UTILIZING COMPLEXITY THEORY

In addition to these prominent American policy theories being both complex and linear in theoretical orientation while primarily using positivist methodology, in modern times other prominent policy theories have emerged in policy studies linked to post-positivist or critical realist methodology. In particular, post-positivist research using discourse and interpretive methodology has been used by policy scholars associated with social constructionist theory (Ingram et al., 2007; Schneider and Sidney, 2009; Stone, 2001). In social constructionism in public policy every individual constructs reality based on her or his own perceptions. Collective social constructions by a larger number of political actors is linked to often competing perceptions of preferred public policies, of laws, rules and norms that may be adopted and implemented. Due to competitive and complex interactions of socially constructed reality based on various interpretive meanings and metaphors associated with preferred public policy actions, complex public policy continues to emerge from initial policy conditions.

Similarly, critical realist theory has emerged from the merger by Bhaskar and others of the philosophical traditions of transcendental realism emanating from Immanuel Kant and critical naturalism (Allison, 2006; Bhaskar, 1998; Givel, 2010; 2012; Romanell, 1958). Transcendental realism posits that all social science phenomena including public policy is based in basic underlying factors that generate social interactions and trends,

70   *Handbook on complexity and public policy*

particularly deep seated structural political and economic inequalities. Critical naturalism posits that all social science interaction including public policy outputs and outcomes is not based in supernatural or spiritual realms but in the natural world. Thus critical realism is a theory that holds that all social phenomena including public policymaking are based in underlying causes that are derived from human and natural factors. Due to the myriad interactions of these natural and structural factors, critical realists view social reality and policy behavior as complex (Clark, 2008).

THE THEORETICAL ROAD FORWARD

While there is a lack in totality of theoretical or methodological focus on complexity in the preferred US policy theories, these theories, along with social constructionism and critical realism, clearly exemplify that a new path to comprehend public policymaking is happening and necessary based on complexity theory and methodology. The following will explore and compare the theoretical bases of all pertinent complexity concepts related to public policymaking. Through this analysis, a theoretical model of complex policymaking will be provided as a means to bolster policy theory making and applied policy analyses.

Nature of Policymaking

In linear policy theory, government decisions are uni-directional, mechanistic and clock-like (Table 5.1). Public policies are often processed through a unified stages heuristic policy system. In contrast, the general nature of complex public policymaking linked to varying policy outputs and outcomes is predicated on self-organization by governmental policy institutions of governmental functions and structures like public agencies or legislative bodies. Potential public policies are based on the authoritative development, enactment and implementation of policies. Government policymaking authority is defined as the power that governments use to create public policies. This authority is derived from written laws and rules, extra-legal military, paramilitary or police actions, support of political leadership, and informal social norms and traditions that support government legitimacy and actions. Once policies are created, another possible occurrence in the policy process is operational implementation of the enacted policies. The type of policies implemented include: regulatory, redistributive and distributive policies (Lowi, 1964). Policy self-organization also occurs, in part, due to conversion of natural resources such as forests, mines or agriculture into useful goods and services available in the maintenance and building of policy institutions. Additionally, public policymaking occurs through complex and dynamic interactions of organized policy system parts. For instance, in education policy it is normal to evaluate smaller policy parts such as student-learning outcomes or teacher performance.

Countering this sophisticated creation and implementation of policies is the principle of entropy (Carnot, 1824; Lavenda, 2009; Rifkin, 1981). In entropy, orderly policy structures and approaches such as legislatures and policy trends drift toward disorder and unpredictability without continual concerted effort toward political self-organization and the implementation of public policies (Williams, 2003). This interaction between

Complexity versus traditional US policy approaches   71

Table 5.1   Complexity and linear models of public policymaking

Characteristic description	Linear model	Complexity model
Nature of policymaking	Government decision-making is linear, mechanistic, clocklike and conducted in a unified policy system	Governmental institutions that self-organize to implement emerging and complex policies; countered by entropic tendencies toward policy disorder
Context of policy process	Ordinarily based in one realm: government decision-making	Integrated in the physical universe, natural processes on earth, and all political and non-political social interactions and relationships
Space–time	Time is common and universal for all governments	Government policy spaces are linked to policy events; other policy spaces and events may impinge upon each other to change policy outputs and outcomes of both
Initial conditions	Policy issues or problems are converted into policies through set and generic policy stages in policy process	Initial conditions for each policy space and event differ due to complex interaction in the world; almost always leads to differing policy outputs and outcomes that are sensitive to the initial conditions
Nesting	Not relevant as focuses on surface observations of policy behaviors	Policy systems are hierarchically nested in a 4-dimensional and complex layering of other complex policy systems that exist at levels above and below normal government policymaking practices
Key strange attractors	Primarily include governments, interest groups and individuals that convert policy issues and problems into policies	Numerous with each having varying weight and influence on policy process, outputs and outcomes; economic and political inequality and dominance are a key strange attractor that may be influenced by other key attractors such as religion or culture depending on the individual society
Policy change over time	Consistent, predictable, linear or curvilinear policy outputs	Chaotic or stable and immensely varied in terms of policy output and outcome patterns and trends

developing and allocating through policy processes and system self-organization useful societal policies and the natural tendency toward disorder creates positive and negative feedback loops. If societal policies and resources become suboptimal, for instance, then a positive feedback loop may develop based on human free will to change the policy situation through governmental action. A negative feedback loop may develop due to policy institution inertia or a class of people that may freely block a policy change. For example, if severe economic inequality exists then the poor might demand that government policies redistribute income and wealth in a more egalitarian fashion. The rich who

72   Handbook on complexity and public policy

see this as a threat to their power and privilege may block this. If new government policy redistributes income and wealth more fairly, then a positive feedback loop has developed. If government policies, through the influence of the rich, maintain the status quo then a negative feedback loop has developed. Either way, there is a dynamic and ongoing interaction between policy and political institutions.

Context of Policy Process

Unlike what is postulated in linear and mechanistic policy models, that policymaking occurs through a governmental process, public policy occurs through multiple levels of interrelated and complex levels of influence (Figure 5.1 and Table 5.1). At the highest level of the physical universe are the vastly complex, often turbulent and changing, and interrelated physical forces of the universe that may directly or indirectly influence all biological life forms and social interactions and non-biological actions on earth. For instance, these forces of the universe, which are probabilistic in tendency, are manifest at the sub-atomic level in quantum mechanics. Quantum mechanics is directly related to such important modern technological advances as computer or electronic technology, both of which assist in the functioning of various sectors of society including governmental and economic (Al-Khalili, 2003).

Space–Time

In Albert Einstein's 1905 special theory of relativity and 1916 general theory of relativity, time was no longer an absolute common time as postulated by Newton (Einstein, 1961; Newton, 1687). Instead, space–time were interchangeable, with space–time being relative given particular observers who are moving at different velocities (Einstein, 1961). One recent possible scenario suggests that in any physical interaction space appears first followed by time (Ananthaswamy, 2013).

Like linear Newtonian physics, a primary assumption of linear policy theory is that time is absolute. This is reflected in the notion that all governmental decisions occur in common time. In complex public policy, space is defined as government venue and jurisdiction over a policy. Policy space is linked to time in the form of a policy event. However, policy events may occur relative to each other. For example, the 9/11 attack on the World Trade Center and the Pentagon occurred outside the usual timeframe of normal government operations. The 9/11 attack caused a robust positive feedback loop with short-term governmental legislation, criminal justice, and military and intelligence responses. Thus, public policy is based on policy space linked to a variety of policy events, with policy events sometimes operating relative to each other.

Initial Conditions and Nesting

Linear theoretical models do not account for unique emergent conditions (Table 5.1). Instead, linear models assume that policymaking occurs due to policy problems or issues in a generic linear fashion occurring in a non-complex manner usually in policy stages. Emergent public policy output and outcome trends stem from unique and complex initial conditions when a particular policy event in a policy space occurs (Williams,

Complexity versus traditional US policy approaches 73

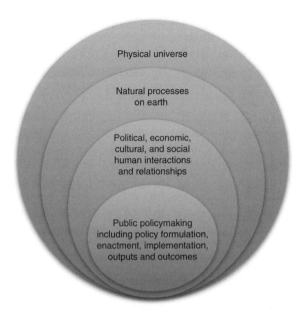

Figure 5.1 The context of public policymaking

2003). Because each policy, such as a healthcare program, is emergent, the probabilities are extremely high that no two policy events emanate from the same initial conditions.

Greatly compounding the complexity of future policy trends based on initial conditions is the notion that complex systems including policy systems are nested (see Figure 5.1). That is to say, complex and interrelated policy behavior is based on actions and policies that are conducted above and below where government policymaking normally takes place. As has been described earlier, public policymaking is situated within the laws of the physical universe, natural processes on earth, and a complex web of non-political and political human interactions. These policy influences are intertwined with actual policymaking processes. However, influences from below can also influence policy outputs and outcomes. For instance, a government may adopt policies caused by non-governmental actions like gay relationships that are then codified by law as gay marriage. Initial conditions are therefore not just horizontal but at the same time nested and vertical. This four-dimensional state of broad and in-depth horizontal and vertical policy interactions over time greatly increases the possibilities that all initial conditions of a policy are different. This leads to differing emergent policy output and outcome trends. By contrast, linear policy models assume that policymaking is observed by surface trends measured by policy behavior. Linear models do not ordinarily incorporate policy nesting.

Key Strange Attractors

Even though policy outputs and outcomes are not usually predictable, over time, there are universal characteristics in policymaking that emerge and are a crucial part

74 *Handbook on complexity and public policy*

of policymaking also equivalent to fixed points or in chaos theory's strange attractors (Table 5.1) (Williams, 2003). Primary fixed-point factors *tend to* move toward order, which can and often does influence policy outputs and outcomes (Givel, 2008). In developed Western nations, for instance, business interests wielding substantial resources and applying potential organizational pressure on governments can also play an important role (Givel, 2008). Policy scholars have noted that other prominent strange attractors or fixed points of order that may, at times, significantly shape policy outputs or mitigate primary factors include: religion, political culture, political leadership, crises, social movements, judicial review, ideology, and security concerns (Baumgartner and Jones, 2009; Givel, 2010). These fixed points can have varying weights of influence individually and as they interact with other fixed points of policy influence. In linear policy models, the assumption is that governmental actions, in tandem usually with group or individual behavior, are the primary driver of policymaking. This assumption does not incorporate the notion that policy drivers that can influence policymaking are often diverse.

Policy System Change over Time

Ordinarily, complex policy change occurs within a stable policy system (Table 5.1). However, a policy system crisis or even catastrophe may sometimes occur if a policy system is radically altered through a significant event such as a Great Depression or political revolution. At the crux of possible policy crises or catastrophes is the elimination or introduction of a new fixed point or strange attractor such as the radical restructuring of a government. Such crises can create a transitional period of time where unstable policy behavior occurs. For instance, from 5 September 1793 to 28 July 1794 during the transitional 'Reign of Terror' after the French Revolution, substantial policy instability occurred. Another substantial source of policy instability may occur when two or more policy systems merge. This may occur through a diplomatic agreement or by a hostile takeover during an armed invasion.

Normally, however, policy change occurs in a stable policy system instead of significant crises or catastrophes. Even then, policy output changes are often varied and complex. For example, as recent research in some policy niches including US Pacific Northwest forest policy and US auto efficiency policy have shown, attempts to induce a major alteration to a policy were not successful (Cashore and Howlett, 2007; Givel, 2008; Perl and Dunn, 2007). Another study in 2008 on US state tobacco policy found, despite a vigorous attempt by health advocates to change nine major state anti-tobacco policy areas, significant change from 1990 to 2006 did not occur (Givel, 2008). Instead, state anti-tobacco policy output patterns based on the number of states passing new legislation were varied and complex. Finally, from 1991 to 2009 in Bhutan, sweeping policy change occurred over five years that was neither dramatic, sharp, nor punctuated (Givel, 2012). The picture that is continuing to emerge regarding policy output and outcomes pattern analyses in these studies and many more, is that business-as-usual in policymaking is often complex and messy (Givel, 2012).

Complexity Model of Public Policy

The usual and normal nature and practice of public policymaking is one of fluctuating and complex contrasts. Despite order in policy systems through specialized political self-organization and the implementation of policies, uncertainty exists as to the nature of public policies and the emergent patterns of future policy outputs and outcomes. At the same time, while policy institutions actively move toward greater order, there is also a tendency for policy systems to move toward an entropic state. Although public policies are usually stable, they can occasionally be unstable due to policy system crises and catastrophes. While public policymaking occurs at the level of governmental decision-making, at the same time this decision-making is powerfully influenced by interconnected policy influences at higher and lower levels. Though policy events measured by time are linked to policy space measured by government venue and jurisdiction, other policy events linked to their own policy events and policy space may and can alter the functioning of policymaking. While initial conditions of policymaking are based on intricate policy interactions, each policy event that emerges is unique due to a varied complexity of policy factors.

This model of interacting, complex and often differing characteristics of policy-making provides a robust framework to assess and analyze normally complex policy behaviors (Table 5.1). Directly linked to this tendency toward order is disorder that challenges the policy order. Policy institutions and those with political influence in all societies attempt to maintain as much order as possible to maintain political power and economic and social privilege. But the struggle to maintain stability and control is never-ending as policy outputs and outcomes continue to emerge that challenge order and replace it through policy reform or even radical policy change due to crises and catastrophes.

Building more robust theoretical understandings and effectively analysing complex and normally emerging policy trends can now move us past the use of outworn linear policy theories and methodologies. These outworn theories provide, at best, an incomplete picture of how policies are developed, enacted, and their impact. The continuing advance of methodologies such as complex qualitative research approaches and complexity computer applications now significantly assist our understanding of complex policy trends. This continuation of robust complex theory building and applied concepts and knowledge offer a new and viable approach in understanding the nature and context of public policy trends as we progress into the twenty-first century.

REFERENCES

Al-Khalili, J. (2003), *Quantum: A Guide for the Perplexed*, London: Weidenfeld & Nicolson.
Allison, H.E. (2006), 'Transcendental Realism, Empirical Realism and Transcendental Idealism', *Kantian Review*, **11**, 1–28.
Alvesson, M. and K. Skoldberg (2010), *Reflexive Methodology*, London: Sage.
Ananthaswamy, A. (2013), 'Space Against Time: Is Space the Warp and Weft of Reality or Time – or Both, or Neither?', *NewScientist*, **218**, 35–7.
Anderson, J. (1975), *Public Policy-Making*, New York: Praeger.
Archer, M. (1995), *Realist Social Theory: The Morphogenetic Approach*, Cambridge: Press Syndicate of the University of Cambridge.

76 *Handbook on complexity and public policy*

Baumgartner, F. and B. Jones (2009), *Agendas and Instability in American Politics* (2nd edn), Chicago, IL: University of Chicago Press.
Berry, F.S. and W.D. Berry (2007), 'Innovation and Diffusion Models in Policy Research', in P. Sabatier (ed.), *Theories of the Policy Process* (2nd edn), Boulder, CO: Westview Press.
Bhaskar, R. (1975), *A Realist Theory of Science*, London: Verso.
Bhaskar, R. (1993), *Dialectic: The Pulse of Freedom*, London: Verso.
Bhaskar, R. (1998), *The Possibility of Naturalism: A Philosophical Critique of Contemporary Human Sciences* (3rd edn), London: Routledge.
Birkland, T. (2011), *An Introduction to the Policy Process: Theories, Concepts, and Models of Public Policy Making*, Armonk, NY: M.E. Sharpe.
Blomquist, W. (2007), 'The Policy Process and Large-N Comparative Studies', in P. Sabatier (ed.), *Theories of the Policy Process* (2nd edn), Boulder, CO: Westview Press.
Callinicos, A. (2007), *The Resources of Critique*, Malden, MA: Polity Press.
Carnot, S. (1824), *Reflections on the Motive Power of Fire and on Machines Fitted to Develop that Power*, Paris: Chez Bachelier, Libraire.
Cashore, B. and M. Howlett (2007), 'Punctuating Which Equilibrium? Understanding Thermostatic Policy Dynamics in Pacific Northwest Forestry', *American Journal of Political Science*, **51**(3), 532–51.
Clark, A. (2008), 'Critical Realism', in L. Given (ed.), *The SAGE Encyclopedia of Qualitative Research Methods*, Thousand Oaks, CA: Sage Publications, pp. 168–71.
Creswell, J. (2007), *Qualitative Research Design* (2nd edn), Thousand Oaks, CA: Sage Publications.
Crotty, M. (1998), *The Foundations of Social Science Research: Meaning and Perspective in the Research Process*, Thousand Oaks, CA: Sage Publications.
Dahl, R. (1961), *Who Governs? Democacy and Power in an American City*, New Haven, CT: Yale University Press.
deLeon, P. (2006), 'The Historical Roots of the Field', in M. Moran, M. Rein and R.E. Goodin (eds), *The Oxford Handbook of Public Policy*, Oxford: Oxford University Press.
Denzin, N.K. and Y.S. Lincoln (2011), *The Sage Handbook of Qualitative Research*, Thousand Oaks, CA: Sage Publications.
Einstein, A. (1961), *Relativity: The Special and the General Theory*, New York: Random House.
Eldredge, N. and S. Gould (1972), *Punctuated Equilibria: An Alternative to Phyletic Gradualism*, San Francisco, CA: Cooper and Co.
Givel, M. (2008), 'Assessing Material and Symbolic Variations in Punctuated Equilibrium and Public Policy Output Patterns', *Review of Policy Research*, **25**(6), 547–61.
Givel, M. (2010), 'The Evolution of the Theoretical Foundations of Punctuated Equilibrium Theory in Public Policy', *Review of Policy Research*, **27**(2), 187–98.
Givel, M. (2012), 'Nonpunctuated and Sweeping Policy Change: Bhutan Tobacco Policy Making from 1991 to 2009', *Review of Policy Research*, **29**(5), 645–60.
Givel, M. and E. Johnson (2013), 'Scientific Paradigms in US Policy: Is it Time for Complexity Science?', in P. Youngman and M. Hadzikad (eds), *Complexity and the Human Experience – Modeling Complexity in the Humanities and Social Sciences*, Singapore: Pan Stanford Press.
Gould, S. (1984), 'Toward the Vindication of Punctuational Change', in W.A. Berggren and J.A. Van Couvering (eds), *Catastrophes and Earth History*, Princeton, NJ: Princeton University Press.
Gould, S. (1989), 'Punctuated Equilibrium in Fact and Theory', in A. Solnit and S. Peterson (eds), *The Dynamics of Evolution*, Ithaca, NY: Cornell University Press.
Hinkel, D.E., W. Wiersma and S.G. Jurs (2003), *Applied Statistics for the Behavioral Sciences* (5th edn), Belmont, CA: Wadsworth.
Hjern, B. and C. Hull (1982), 'Implementation Research as Empirical Constitutionalism', *European Journal of Political Research*, **10**, 105–15.
Ingram, H., A. Schneider and P. deLeon (2007), 'Social Construction and Policy Design', in P. Sabatier (ed.), *Theories of the Policy Process* (2nd edn), Boulder, CO: Westview Press.
Johnson, L. (2010), *Science & Technology Innovation as a Complex Adaptive System: Applying the Natural Processes of Complexity to Policymaking*, paper presented at the American Political Science Association, Washington, DC available at http:papers.ssrn.com/sol3/papers.cfm?abstract_id=1657193.
Jones, C. (1970), *An Introduction to the Study of Public Policy*, Belmont, CA: Wadsworth Publishing Company.
Joskow, P. (1995), 'The New Institutional Economics: Alternative Approaches: Concluding Comment', *Journal of Institutional and Theoretical Economics*, **151**(1), 248–59.
Kingdon, J. (2010), *Agendas, Alternatives, and Public Policies*, update edn, with an Epilogue on Health Care (2nd edn), Upper Saddle River, NJ: Pearson.
Lasswell, H. (1948), *Power and Personality*, New York: W.W. Norton & Company.
Lasswell, H. (1956), *The Decision Process*, College Park, MD: University of Maryland Press.
Lasswell, H. (1971), *A Pre-View of Policy Sciences*, New York: American Elsevier.

Lasswell, H. and A. Kaplan (1971), *Power and Society*, New Haven, CT: Yale University Press.

Lavenda, B.H. (2009), *A New Perspective on Thermodynamics*, New York: Springer.

Lincoln, Y.S., E.G. Guba and S.A. Lynham (2011), 'Paradigmatic Controversies, Contradictions and Emerging Confluences, Revisited', in N.K. Denzin and Y.S. Lincoln (eds), *Sage Handbook of Qualitative Research* (4th edn), Thousand Oaks, CA: Sage Publications.

Lowi, T. (1964), 'American Business, Public Policy, Case Studies and Political Theory', *World Politics*, **16**, 677–715.

Mancias, P. (2007), 'The Social Sciences Since World War II: The Rise and Fall of Scientism', in W. Outhwaite and S. Turner (eds), *The Sage Handbook of Social Science Methodology*, Thousand Oaks, CA: Sage Publications.

Miller, J. and S. Page (2007), *Complex Adaptive Systems: An Introduction to Computational Models of Social Life*, Princeton, NJ: Princeton University Press.

New England Complex Systems Institute (2011), Paper presented at the International Conference on Complex Systems and the Association for the Advancement of Artificial Intelligence, Cambridge, MA.

Newton, I. (1687), *Philosophiæ Naturalis Principia Mathematica*, London: Jussu Societatis Regiæ ac Typis Josephi Streater.

Ostrom, E. (1998), 'A Behavioral Approach to the Rational Choice Theory of Collective Action, Presidential Address, American Political Science Association, 1997', *American Political Science Review*, **92**(1), 1–22.

Ostrom, E. (2007), 'Institutional Rational Choice: An Assessment of the Institutional Analysis and Development Framework', in P. Sabatier (ed.), *Theories of the Policy Process* (2nd edn), Boulder, CO: Westview Press.

Perl, A. and J. Dunn (2007), 'Reframing Auto Fuel Efficiency Policy: Punctuating a North American Policy Equilibrium', *Transport Reviews*, **27**, 1–35.

Pierre, E.A. St (2011), 'Post Qualitative Research: The Critique and the Coming After', in N.K. Denzin and Y.S. Lincoln (eds), *The Sage Handbook of Qualitative Research*, Thousand Oaks, CA: Sage Publications, pp. 611–25.

Popper, K. (2002), *The Logic of Scientific Discovery* (2nd edn), New York: Routledge.

Pressman, J. and A. Wildavsky (1983), *Implementation*, Berkeley, CA: University of California Press.

Repetto, R. (2006), 'Introduction', in R. Repetto (ed.), *Punctuated Equilibrium and the Dynamics of US Environmental Policy*, New Haven, CT: Yale University Press.

Rifkin, J. (1981), *Entropy: A New World View*, New York: Bantam Books.

Romanell, P. (1958), *Toward a Critical Naturalism*, New York: Macmillan Company.

Sabatier, P. (1986), 'Top-Down and Bottom-Up Models of Policy Implementation: A Critical and Suggested Synthesis', *Journal of Public Policy*, **6**, 21–48.

Sabatier, P. (2007), 'The Need for Better Theories', in P. Sabatier (ed.), *Theories of the Policy Process* (2nd edn), Boulder, CO: Westview Press.

Sabatier, P. and H. Jenkins-Smith (1993), *Policy Change and Learning: An Advocacy Coalition Approach*, Boulder, CO: Westview Press.

Sabatier, P and C. Weible (2007), 'The Advocacy Coalition Framework: Innovations and Clarifications', in P. Sabatier (ed.), *Theories of the Policy Process* (2nd edn), Boulder, CO: Westview Press.

Schneider, A. and M. Sidney (2009), 'What is Next for Policy Design and Social Construction Theory?', *Policy Studies Journal*, **37**(1), 103–19.

Stone, D. (2001), *Policy Paradox: The Art of Political Decision Making* (revised 3rd edn), New York: W.W. Norton & Company.

Taleb, N.N. (2008), *The Fourth Quadrant: A Map of the Limits of Statistics, The Third Culture*, available at http://www.edge.org/3rd_culture/taleb08/taleb08_index.html.

Theodoulou, S. (2013), 'The Contemporary Language of Public Policy: Starting to Understand', in S. Theodoulou and M. Cahn (eds), *Public Policy: The Essential Readings*, Upper Saddle River, NJ: Pearson.

Williams, G.P. (2003), *Chaos Theory Tamed*, Washington, DC: Joseph Henry Press, an imprint of National Academy Press.

Wood, D. and A. Doan (2003), 'The Politics of Problem Definition: Applying and Testing Threshhold Models', *American Journal of Political Science*, **47**(4), 640–53.

Wyly, E. (2008), *City Critical Urban Studies: New Directions*, Albany, NY: SUNY Press.

Zahariadis, N. (2007), 'The Multiple Streams Framework: Structures, Limitations, Prospects', in P. Sabatier (ed.), *Theories of the Policy Process* (2nd edn), Boulder, CO: Westview Press.

6. Can we discover the Higgs boson of public policy or public administration theory? A complexity theory answer
Göktuğ Morçöl

INTRODUCTION

On 4 July 2012, the scientists at the European Organization for Nuclear Research (CERN) declared that they had 'nearly discovered' the Higgs boson. This discovery generated excitement among particle physicists and the general public. After the centuries-old scientific quest in search of the most fundamental laws of the universe, have the scientists finally discovered the most fundamental particle of entire existence, what Lederman and Teresi (1993) call the 'God Particle'? Was this one of those rare events in history when physicists precisely and definitively discovered a universal truth?

I cannot answer these questions from a physicist's perspective; I will take The CERN's declaration that they had 'nearly discovered' the Higgs boson as the statement of truly what happened. Instead, in this chapter I will use the search for and the (near) discovery of Higgs boson as a metaphor for the aspirations of many scientists – including social scientists – to achieve the goal of creating a *universal, deductive and quantitative/precise science*. I will ask this question: is it possible to discover the God Particle of public administration or policy – the fundamental explanation of what public administration or policy is all about? I will answer the question from a complexity theory perspective.

The Higgs boson case illustrates that even within the most advanced fields of the natural sciences, such as particle physics, the results of scientific inquiries contain some degree of uncertainty. The level of uncertainty is much higher in the social science inquiries. It has been a longstanding debate whether the social sciences should emulate the methods of the natural sciences or use entirely different methods of inquiry because their subject matter is completely different from that of the natural sciences. Complexity theory contributes to this debate by demonstrating that uncertainty is inevitable in both the natural and social science investigations and that this is because of the complexity of both the natural and social phenomena.

In the natural sciences, the complexity is reducible to simple and linear explanations to some degree. This is why the CERN particle physicists 'nearly discovered' the Higgs boson, at a certain level of precision. Social phenomena are more complex. Even if one thinks that social scientists should emulate the methods of the natural sciences (for example, experiments), the results will be much less certain and subject to multiple interpretations. Complexity theory does not only suggest that the knowledge of social reality is irreducible to simple explanations, however. Complexity researchers use conceptual and methodological tools to understand the complexity of social phenomena: social network analyses, agent-based models, and comparative case studies. These tools are better than the reductionist and linear methods, such as experiments and linear regression models, in understanding social pheomena. I will briefly discuss complexity theory's concepts and methods and their relevancy for inquiries in public policy and administration in this chapter.

THE 'NEAR DISCOVERY' OF THE HIGGS BOSON AND ITS RELEVANCY

Higgs boson is a construct that signifies a subatomic particle that is theorized to be responsible for the existence of all the mass in the universe (Atteberry, n.d.). This construct is part of the so-called 'standard model' of particle physics. The model posits that the entire universe is made up of a set of basic building blocks: protons, neutrons and electrons, which are in turn made up of quarks and leptons. Quarks and leptons are indivisible, but these particles are not at the 'bottom' of existence because they do not have inherent mass. To gain mass, all particles must pass through a field known as the 'Higgs field'. As particles pass through the field, they interact with it in different ways and gain different masses. According to the standard model, the Higgs field needs a carrier particle to affect other particles: the Higgs boson.

The Higgs boson eluded discovery by physicists for decades. Finally, in July 2012, their efforts paid off – with a 'near discovery.' To understand why this was a *near* discovery, one should remember that the discovery of a subatomic particle always requires deductive inference. Nobody can see these particles with their naked eyes. Instead, scientists record the traces of particles they observe in experiments and compare them with the predictions of their models (for example, the standard model). A major problem in doing so is that many of these particles exist only for very short periods of time and therefore it takes well-designed experiments and very precise measurement tools to observe them. The inferential problem in these experiments is this: what is the probability of truly detecting a particle as predicted by the model (that is, that the detection is not a 'false positive')?

Stenger (2012) tells the story of how meticulously the CERN scientists designed and executed their experiments to achieve a very high level of precision in their measurements. He demonstrates that despite all their efforts, there still was some uncertainty in their findings. This uncertainly was a result of the deductive inference the scientists used. They looked for a particle that the standard model had predicted, but nobody had observed it before. Stenger notes that the CERN scientists detected a previously unknown particle 'beyond reasonable doubt' in their experiment. The probability of accidentally detecting a previously known particle in this experiment (that is, a false positive) was one in over three million. This level of precision the CERN scientists achieved in their measurements was as refined as it had ever been achieved in any scientific study. A comparison of this probability of 1/3 000 000 to the probability level that is typically used in most social science studies, 1/20 (that is, $p < .05$), would give some sense of the phenomenal accomplishment of the CERN scientists. However, whether the scientists discovered the expected Higgs boson, or a 'reasonable facsimile', in this experiment is still questionable.[1]

The Higgs boson is a good metaphor for the search for a universal, deductive and quantitative/precise science. The aspiration to develop such a science has a long history.[2] Briefly, this aspiration was codified into a workable set of scientific principles primarily

80   *Handbook on complexity and public policy*

by Descartes and Newton in the seventeenth century. Since then the name 'Newtonian science' has been used to represent this aspiration.[3] In the eighteenth and nineteenth centuries, social philosophers like Condorcet and Comte argued that social scientists should emulate the more 'mature sciences', like physics, and use mathematical/statistical methods. Hermeneutic and phenomenological philosophers countered them by arguing that social phenomena are inherently different from natural phenomena and therefore should be inquired differently. They argued that the natural and social sciences are fundamentally different in the sense that the former is explanatory, cumulative and predictive and the latter is interpretive and historically conditioned (Flyvbjerg, 2001: 29).[4]

The Newtonian science was further codified by the logical positivists of the early twentieth century as the 'scientific method'. The overall aspiration of the logical positivists was to formulate the codes of a universal science – codes that would be applicable to both natural and social sciences. The ultimate goal of science would be to make universal generalizations (that is, context-free laws, theories). In this code, deduction was heavily favored, but induction was tolerated. Quantitative methods were favored, and deemed superior to qualitative methods, because only quantitative empirical observations (measurements) could be precise.

THE RELEVANCY OF THE DEBATE FOR PUBLIC ADMINISTRATION POLICY

Many social scientists have followed the tenets of the Newtonian science and tried to emulate the natural sciences, while others opposed them. In public administration, the positions on the epistemological and methodological issues were crystallized in the seminal books by Herbert Simon and Dwight Waldo and their 1952 debate (Simon, 1947; 1952; Waldo, 1948; 1952a; 1952b). Simon promoted the ideals of logical positivism by arguing that facts have to be separated from values and that public administration research should be based on empirical facts, whereas Waldo argued that public administration is suffused with questions of value, democracy being its core value, and therefore public administration inquiry has to be normative in nature.

The studies on the methods used in the public administration and policy studies in the following decades show that Simon's argument prevailed in the literature, at least in one important sense: quantitative studies dominated the literature for decades. Radin (2000) and Yang (2007) observe that policy analysts used quantitative methods predominantly in the 1960s and 1970s. Yang notes that quantitative methods lost their predominance in the 1980s, but they began to re-emerge in the 1990s. In the 2000s, quantitative/statistical analyses were prevalent in the papers published in the top journals in public administration (Raadschelders and Lee, 2011). Several studies show that quantitative methods have been prevalent in the curricula of the educational programs in public administration and policy and political science since the 1980s (Hy et al., 1981; 1987; LaPlante, 1989; Waugh et al., 1994; Jenkins-Smith et al., 1999; Schwartz-Shea and Yanow, 2002; Bennett et al., 2003; Morçöl and Ivanova, 2010). The prevalence of the use of quantitative methods in the literature and educational programs is indicative of the influence of the Newtonian/positivist science in public administration and policy and political science, as quantitative analyses depend largely on realist ontologies (that reality exists independently and is measurable) and objectivist epistemologies (that researchers are independent of realities they study and that they can separate facts from values) (Raadschelders and Lee, 2011).

The predominance of quantitative methods has not ended the debates on what is the appropriate epistemology and methodology for public administration and policy and political science. Waldo and his followers formulated their vision for public administration at the 1968 Minnowbrook conference. The members of the New Public Administration movement, which emerged from the conference, promoted and used post-positivist and post-modernist/post-structuralist methodologies in their studies (Riccucci, 2010: 13–14). These methodological applications have been represented in the articles published in the journal *Administrative Theory & Praxis* and some others.

King et al.'s (1994) and Flyvbjerg's (2001) books and the discussions that ensured their publications exemplify the ongoing epistemological and methodological debates in the broader social science literature.[5] King et al. argue that the ultimate goal of both natural and social sciences is to draw valid inferences (King et al., 1994: 46) and that the essence of the unity of science is in scientific methods (King et al., 1994: 7–8). They also argue that the rules of scientific inference are applicable to both quantitative and qualitative methodologies and extend the applications of the quantitative rules of inference (that is, deductive and mathematical rules) to qualitative methodologies.[6] Flyvbjerg questions whether a unified science (inclusive of both natural and social sciences) is possible and reminds the reader that despite all their efforts for more than a century, social scientists have not been able to emulate the natural sciences in developing a unified theoretical or methodological framework (Flyvbjerg, 2001: 30–32). He also critiques those who think that the social sciences should emulate the methods of the natural sciences and argues that it is not possible to develop cumulative and predictive social science theories (Flyvbjerg, 2001: 25). He also cites Thomas Kuhn's argument that even the natural sciences cannot have common universal principles or methodology of inquiry; instead scientific inquiry progresses through shifts from one paradigm to another (Flyvbjerg, 2001: 27). Flyvbjerg argues that even such a paradigmatic unity is not possible in the social sciences. Riccucci (2010) observes that there is no unifying paradigm in public administration and this is because the public administration research is driven by multiple values (effectiveness, efficiency, equity, legitimacy, and accountability in service delivery) and it is multidisciplinary.

COMPLEXITY THEORY

Complexity theory has important implications for the Newtonian search for a universal, deductive and quantitative/precise science. I summarize these implications in this section. More extensive discussions can be found in Morçöl (2002; 2012). Any discussion on the implications of complexity theory should begin with a clarification of the notions of complexity and simplicity.

Complexity versus Simplicity

One of the foundational principles of the Newtonian science is the principle of 'Occam's razor': the simplest scientific explanation is the best explanation (for example, a model that uses the least number of independent variables, or the least number of elements in

82 *Handbook on complexity and public policy*

its mathematical formulation, is the best one). This simplicity allows scientists to make universal generalizations. If there are too many variables or elements in an equation, this means that there are too many contingencies, which would prevent the researcher from making generalizations to large populations or a large range of phenomena. Einstein's formula of $E = mc^2$ is one of the best examples of the Occam's razor principle. It has only three elements that are related to each other in a relatively simple manner: energy (E), mass (m), and speed of light (c). This simple formula applies to large realms in the known universe; there are no terms in it that apply to the contingencies in a particular galaxy or a planet.

Can the entire universe be explained with a simple formula? Einstein thought so. Complexity theorists do not have a unified answer to this question, but they would all agree that the complexity of physical/chemical/biological realities is not (at least easily) reducible to simple explanations. Prigogine and Kauffman demonstrated that complexities abound in the physical, chemical and biological worlds (Prigogine, 1996; Prigogine and Stengers, 1984; Kauffman, 1993; 1995). The notion of emergence, one of the key notions in complexity theory, is that the properties of emergent structures are irreducible to those of their components.[7] For instance, once a water molecule is formed, its properties are qualitatively different from (irreducible to) those of the hydrogen and oxygen atoms.

Social scientists simplify as well. Some argue that this is the way it should be. King et al. (1994) argue, for example, that the aim of any (social) scientific investigation should be to explain as much as possible with as little as possible. They acknowledge that reality is complex, but we have to make theoretical generalizations (that is, simplifications) and use research designs that simplify (pp. 9–10). Policy theorists Stokey and Zeckhauser (1978) and Sharkansky (2002) acknowledge that public policy processes are complex and they observe that policy analysts and policymakers typically ignore complexities and attempt to devise simple solutions to simply defined policy problems. Complexity theorists Axelrod and Schelling not only demonstrated that societies are complex, but they also showed how social complexities can be conceptualized and investigated (Axelrod, 1997; 2006; Axelrod and Cohen, 2000; Schelling, 2006). Others did the same for the public policy and administration processes (Geyer and Rihani, 2010; Gerrits, 2012; Rhodes et al., 2011; Teisman et al., 2009).

An important question is: is it possible for the human mind not to simplify and grasp complex realities as they are? As a reviewer of this paper noted, we all seek to simplify to understand realities we experience and none of us can perceive the whole world all at once. Indeed the human mind has evolved to select segments of reality, and thus simplify; this is a survival mechanism our species used since its beginning (Ornstein and Ehrlich, 2000). However, as Ornstein and Ehrlich note, this mechanism created a mismatch with today's complex world, which human beings helped create. For instance, our simplifying minds cannot easily grasp the complexities of global warming and this mismatch is an impediment in taking actions to reverse the warming trend.

If simplification is a mechanism of the human mind, then is there an appropriate level or form of simplification? Complexity theorists do not have a unified answer to this question, and they make various forms of simplifications as they conceptualize the phenomena they study. However, a significant contribution of complexity theory is to counter the Occam's razor principle (that the simplest scientific explanation is the best

The Higgs boson of public policy or public administration theory?   83

explanation). More importantly, complexity theorists have offered conceptual and methodological tools of understanding complexities. They demonstrated that non-linear, self-organizational, emergent and co-evolutionary processes generate inherently complex systems and conceptualized the mechanisms of these processes (Morçöl, 2012: 21–138). Also complexity researchers either developed or adopted methods of inquiry that are based on recognizing complexities to certain degrees, such as social network analyses, agent-based simulations, and comparative case studies.

A proposition I made earlier in this chapter should be clarified here: social phenomena are more complex than natural phenomena. It is beyond the scope of this chapter to fully explain and support this proposition. However, I can briefly say that social systems are more complex than natural systems in general because the former are composed of the actions and relationships of human beings, each of whom is a complex biological and psychological system. Many of the concepts of complexity theory were generated in the natural sciences (for example, Kauffman, 1993; 1995; Prigogine, 1996; Prigogine and Stengers, 1984). These complexity theorists demonstrate that thermodynamic and biological systems are inherently complex. What makes social systems more complex – on top of the fact that humans are subject to the complexity of the natural thermodynamic and biological phenomena – is that each human being is an interpretive and purposeful actor and the structural properties of social systems emerge through the interactions of these actors. There are no interpretive or purposeful actors in thermodynamic or biological systems. The increased level of complexity in social system requires additional, or different, conceptualizations to investigate them.

The inherent complexity of natural and social systems has important implications for the universality of scientific knowledge, deductive methodologies, and quantification and precision in measurement. In the following sections, I will focus on the implications for studying social systems, particularly policy/governance systems.

Universal Generalizations

To better understand the implications of complexity theory for scientific inquiries in general, and social scientific inquiries in particular, we need to consider the problems of making universal generalizations and the importance of context. Many philosophers and scientists strived to formulate universal theories, at least universal generalizations of some aspects of physical, biological or social reality. The most recent and ambitious example of such efforts is the 'theory of everything'. This putative theory would bring together all the known theories of the physical universe to explain them in one simple and comprehensive formula.[8]

The roots of the aspiration to develop a universal theory can be traced back to the definitions of an 'ideal theory' by Socrates, Descartes and Kant (Flyvbjerg, 2001). According to these philosophers, an ideal theory must be *explicit* (it should be well-reasoned, clearly stated and detailed), *universal* (it should apply to all places at all times), *abstract* (it should not require references to concrete examples), *context-independent, systematic* (it must constitute a whole), and *complete and predictive* (it must cover all variations in the phenomenon under study) (Flyvbjerg, 2001: 38–9).

Is such an ideal theory possible in the social sciences? Marx, Freud, Chomsky and Levi-Strauss are among those who do think that it is possible in principle (Flyvbjerg,

84 *Handbook on complexity and public policy*

2001). Giddens, Foucault and Bourdieu are skeptical or critical of the proposition that an ideal social theory is possible. Common among these skeptics and critics is the understanding that each human actor is unique and his/her actions take place in specific contexts.

Context becomes particularly important in making generalizations about social phenomena, because, unlike natural objects (for example, molecules, atoms and subatomic particles), each individual human being has unique characteristics, as well as characteristics that are common with other individuals. This combination of uniqueness and commonality is because context is important. If it were possible to categorize all human beings into clearly defined groups (like hydrogen atoms versus oxygen atoms), it would be possible to make context-free generalizations for each of the categories (for example, all the hydrogen atoms have one electron and one proton). The 'Dreyfus paradox' succinctly captures the dilemma that context creates for social theorists. According to Dreyfus, an ideal theory has to exclude the context of everyday human activity (an ideal theory must be universal, abstract and context-free) in order to make predictions, but by doing so it makes predictions impossible because every human action takes place in, and its direction is influenced by, its context (Flyvbjerg, 2001: 40–42).

Giddens's (1984) concept of 'double hermeneutic' can help us understand the problem of context better. In Giddens's conceptualization, human beings are both the objects and subjects of an inquiry (they are both the observed and the observers) and the ways human beings interpret themselves and their relationships with their interpreters (observers) determine the ways they act. A social scientist cannot understand the behavior of a human being without understanding the interpretation of the individual of his/her own behavior (that is, the meaning that the individual attributes to his/her own behavior). In other words, every human being imputes a particular meaning to his/her behavior and this meaning is derived from his/her personal experiences and acculturation. Because no two personal experiences are alike, all human beings are different and they interpret their behaviors differently. But because human beings also receive common codes of interpreting themselves and their worlds through acculturation processes, there is the tendency in human societies to homogenize interpretations. Understanding the context means understanding the meanings human beings impute to their own behaviors and the behaviors of others based on their unique personal experiences and homogenizing acculturation processes. The observer (scientist, researcher) is also a human being who is in a relationship with the observed human being while interpreting their behaviors and interpretations. The interpretations of the observer are influenced by the former's personal experiences and acculturation as well. The observer must understand the contexts of the behaviors and interpretations of the observed, as well as the context of his/her own interpretations. This dual nature of interpretations makes it impossible to meet the requirements of the 'ideal theory' in the social sciences.

Many social researchers recognize the importance of understanding the context of individual actions, but they argue that the difficulties it poses can be overcome to some extent. King et al. (1994), for example, note that even the thickest (most contextual descriptions) include abstractions; qualitative researchers make simplifications and generalize as well (pp. 42–3). Flyvbjerg (2001) counters this argument by saying that because of the inherent uniqueness of individual human beings and the complexity of

The Higgs boson of public policy or public administration theory? 85

their contexts, social scientist should prefer using case methods, which generate concrete, practical and context-dependent knowledge (pp. 70, 84–6).

Complexity theorists recognize the importance of context in understanding not only social phenomena, but also natural phenomena. Rössler (1986) and Prigogine and Stengers (1984) highlight the importance of the *context of the observers* of natural phenomena. According to Rössler, a researcher's knowledge of the universe cannot be 'exophysical' (detached from it); it should be 'endophysical' (situated within the universe) (Rössler, 1986: 320). Prigogine and Stengers (1984) concur and argue that we (observers) are 'subject to intrinsic constraints that identify us as part of the physical world we are describing. It is a physics that presupposes an observer situated within the observed world' (Prigogine and Stengers, 1984: 218). Casti (1994) stresses that knowledge is a participatory process (Casti, 1994: 98) and that complexity or simplicity of a system is defined both by the nature of the systems studied and the relationship of the observer with a system (Casti, 1994: 269).

Once the inherent contextuality of the knowledge of an observer is recognized, is it still possible to make universal generalizations in the natural sciences? Complexity theorists are split in their answers to this question. On one side, Kauffman (1995) thinks that the '[universal] laws of complexity' can be discovered and Barabási and Albert (1999) demonstrate the universal applicability of mathematical power laws. On the other side, Prigogine and Stengers (1984) argue that it is not possible to make universal generalizations about systems' behaviors. Because natural systems self-organize, each system has a unique history and unique behavioral patterns, which nullify determinism in the form of predictability of future behavior, except in some pockets of reality in the universe. Therefore, the knowledge of each system must include an understanding of its unique history and behavior and the contexts that surround them. This precludes the possibility of universal generalizations.

Leaving aside the questions regarding whether and to what extent there are universally generalizable patterns in the natural world, I argue that in the social sciences universal generalizations are not possible and generalizability is more limited than it is in the natural sciences. This is because social systems are more complex than natural systems, because they are constituted by the activities of and relationships among interpretive and purposeful actors. However, I do not agree with Flyvbjerg's (2001, ch. 3) assertion that theory is not possible at all in social sciences. Although universal theories are not possible, context-sensitive generalizations are possible and useful in social inquiries.

Deductive Inferences

In the deductive approach, a researcher draws his/her hypotheses from a theoretical generalization and tests them empirically. This strict deductive approach is not compatible with the methods complexity researchers typically use: social network analyses, agent-based simulations, or comparative case studies. Even when a researcher begins with a theoretical generalization, the applications of these methods will generate results that will not definitively confirm or falsify the generalization. Instead, researchers discover patterns that do not necessarily fit any prior explanations.

A detailed discussion of these three types of methods is beyond the scope of this chapter. A brief discussion on agent-based simulations (ABS) can be illustrative, instead.

86   *Handbook on complexity and public policy*

Theoretical generalizations are simplifications. ABS researchers simplify too. ABS models are dynamic, unlike the theories and models used in the deductive approach, such as the models neoclassical economists use, which are based on the assumption that static equilibria exist (Miller and Page, 2007: 71). In ABS, systems only occasionally settle into equilibria; more often they evolve from one state to another. ABS models are constructed to capture both these static states and dynamic change, which does not allow testing theories definitively.

ABS modelers do not seek to build or test theories; instead they explore how collective patterns emerge. Epstein and Axtell (1996) point out that this exploration of emergent patterns can be studied in a 'generative social science', whose purpose is not to 'explain' social phenomena with nomothetic generalizations (universal laws, theories), but to understand the generative processes that ensue from agent interactions in complex systems. Similarly, Axelrod (2007) argues that ABS provides a 'third way of doing science', different from, but also inclusive of, deduction and induction:

> Like deduction, it starts with a set of explicit assumptions. But unlike deduction, it does not prove theorems. Instead, a simulation generates data that can be analyzed inductively. Unlike typical induction, however, the simulated data comes from a rigorously specified set of rules rather than direct measurement of the real world. While induction can be used to find patterns in data, and deduction can be used to find consequences of assumptions, simulation modeling can be used as an aid intuition (Axelrod, 2007: 92–3).

Quantitative versus Qualitative Research

Are quantitative or qualitative methods more compatible with complexity theory? Complexity researchers use both. For example, agent-based simulations are quantitative (for example, Epstein and Axtell, 1996; Axelrod, 2007), while the case studies some complexity researchers use are qualitative (for example, Teisman et al. 2009). Before assessing the merits of quantitative and qualitative methods, a closer look into their assumptions and implications is needed.

Quantitative methods require that researchers make two implicit assumptions: measurement units are uniform and there are some stable patterns in the relations among the units. Qualitative methods, on the other hand, are used to investigate unique individual cases, with no assumptions of uniformity or stability. Quantitative methods enable researchers to make generalizations about the stable relations among uniform units and verify precise predictions driven from them. Qualitative methods, on the other hand, are not suitable to make generalizations or verify precise predictions; instead, they allow researchers to understand the contexts of individual behaviors/events.

In order to detect a Higgs boson, as the CERN researchers did, one has to assume that all Higgs bosons are alike and that their relations to other particles in the universe are stable so that the traces of Higgs particles can be detected reliably in experiments. The CERN researchers used a universal theory, the standard model of particle physics, which was formulated mathematically, drew hypotheses (predictions) from it, and devised the experiment to detect the Higgs with a high level of precision.

As effective, and impressive, as the accomplishments of quantitative methods are in the Higgs case, they would not be as effective in understanding the properties and evolutions of complex systems. As Prigogine and Stengers (1984) observe, the properties and evolutions of complex systems have some commonalities, as well as some uniqueness; they are also dynamic and self-organizational and the relations among their elements are non-linear. Quantitative methods, like agent-based simulations and social network analyses, help researchers discern the patterns of the behaviors of complex systems. To better understand their properties and behaviors, a researcher would need to develop a qualitative understanding of the relationships among its elements and the relationships of the system with other systems.

Smith and Jenks (2006) argue that complex systems can only be understood qualitatively, because their borders cannot be delineated clearly, nor can their future behaviors be predicted precisely (Smith and Jenks, 2006: 23). I do not think that these characteristics exclude the possibility of using quantitative methods in studying complex systems. Quantitative methods, like agent-based simulations, simplify complexities and, as such, yield incomplete knowledge. This is true for qualitative methods as well; they cannot yield complete knowledge either. Both quantitative and qualitative methods should be used in a complementary fashion to understand complex systems, as I argued elsewhere (Morçöl, 2012: 248). Even then, our understanding of complex systems will be incomplete.

IMPLICATIONS FOR PUBLIC POLICY AND ADMINISTRATION

Will it ever be possible to discover the Higgs boson of public policy and administration, the fundamental explanation of what public policy and administration is all about? Should social scientists in general, and public policy and administration researchers in particular, even try to do that? In the story of the Higgs boson a universal theory was tested deductively, by testing its predictions quantitatively and with precision. Complexity theory suggests that social scientists in general, and public policy and administration researchers in particular, do not even need to try to follow the storyline of the Higgs boson. It suggests that scientific inquiry could, and should, seek contextual information, not context-free, universal generalizations. Complexity researchers do use quantitative methods such as agent-based simulations, but they do so 'generatively', not deductively.

Many public policy and administration researchers followed the storyline of the Higgs boson. Examples include the 'administrative/management science' conceptualizations of the early twentieth century, the rational long-term planning models of the 1960s, and the rational choice models of the late twentieth and early twenty-first centuries. These efforts have been criticized widely. For example, Flyvbjerg (2001) argues that the social sciences have never been, and will never be, able to develop predictive and explanatory theories and that they should not even try to emulate the natural sciences (Flyvbjerg, 2001: 4). Similarly, Raadschelders (2011) argues that the efforts of some public administration scholars 'to establish a science that is replicable, objective, and generalizable' that is based on quantitative empirical studies have not been successful; there is not even a consensus about what this science should be like (Raadschelders, 2011: 916).

The implications of complexity theory resonate somewhat with Flyvbjerg's and Raadschelders' arguments, but these implications should not lead us to conclude on a pessimistic note. Complexity theory projects a vision of scientific inquiry whose goal

88   *Handbook on complexity and public policy*

is to understand the non-linear dynamics of policy/administrative systems/networks. Recent developments in the areas of governance and network studies are highly compatible with complexity theory's implications. Network theorists point to the complex and self-organizational nature of governance/policy networks and highlight the uncertainties in their behaviors (for example, Agranoff and McGuire, 1999; Koppenjan and Klijn, 2004). Social network analyses are being used at an increasing frequency in the studies of complex governance networks (for example, Graddy and Chen, 2006; Kapucu, 2006; McGuirk, 2000).

Complexity theory contributes to these studies in two main ways. First, it provides an ontological and epistemological framework for them. This framework recognizes the incompleteness and uncertain nature in the knowledge of complex governance systems. It shows that a science based on universal generalizations and deductive methodologies is misguided. It shows that quantitative and qualitative methods can be used in a different way: generatively, as Epstein and Axtell (1996) put it. Second, the methods complexity researchers use, particularly social network analyses and agent-based simulations, help policymakers, analysts and administrators to identify the key actors and factors in policy/governance systems. With this knowledge, they can find best ways to influence these systems to generate desired outcomes. Social network analyses have been used to identify the key actors in emergency management systems (Kapucu, 2006), for example. Agent-based simulations have been used in understanding the processes in which macro-level policies emerge from the interactions of agents under different conditions (Maroulis et al., 2010). These simulations can help researchers understand where to find 'policy leverages' (Axelrod, 1997). The dynamic and interpretive tools that agent-based simulations researchers use are suitable for a continual improvement approach in policy analysis and evaluation studies.

It is highly unlikely that there will be a discovery, or near discovery, of the Higgs boson of public policy or administration any time in the future. Instead, we should look at the emerging studies on governance and network studies, combined with the insights of complexity theory, to help us better understand policy and administrative processes.

ACKNOWLEDGMENTS

An earlier version of this chapter was presented at the 74th National Conference of the American Society for Public Administration, New Orleans, LA, 15–19 March 2013. I want to thank Paul M. Chalekian of the University of Nevada, Reno, and the editors of this book for their highly valuable comments on the earlier versions of this chapter.

NOTES

1. In Stenger's (2012) own words: 'I have looked at the results just reported by the two experiments at CERN, Atlas and CMS. Both show 5-sigma signals for a range of secondary particles at a mass of 125–126 GeV. The standard model of elementary particles and forces predicted a 4.6-sigma effect at that mass, although the value of mass itself was not predicted. Not only is each individual result significant, rejecting the null hypothesis at a probability of one in over three million, the fact that two independent experiments agree surely makes the case for a previously unknown particle at 125 GeV proven beyond a reasonable doubt.

The Higgs boson of public policy or public administration theory? 89

Whether it is the long sought-after Higgs boson or a composite of known particles is yet to be definitively established.'

2. A more detailed version of the historical account presented in this section can be found in Morçöl (2002).

3. Arguably the term 'Cartesian' could also be used to name this set of principles. The term 'Cartesian' refers to the principles developed by Descartes, particularly in logic and mathematics. His contributions to the later developments of scientific principles are undeniable. However, Newton, who came a generation after Descartes, deserves more credit because of his extensive contributions to a wide range of scientific areas of study (mechanics and optics) and mathematics.

4. More specifically; according to Wilhelm Dilthey, one of the founders of hermeneutics, 'the natural sciences developed causal explanations of 'outer' events, whereas the social sciences were concerned about an 'inner' understanding of meaningful conduct. The social scientific understanding was an understanding of the meaning of what the other person did and grasping of his or her experience of the world. . ... According to Dilthey and other hermeneutic theorists, empathic understanding was itself an explanation' (Morçöl, 2002: 98).

5. These books were at the centers of two streams of debates in the 1990s and 2000s. King et al.'s book (1994) was critically assessed in a special issue of the *American Political Science Review* (Caporaso, 1995; Collier, 1995; King et al., 1995; Laitin, 1995; Rogowski, 1995; Tarrow, 1995). The publication of Flyvbjerg's book (2001) coincided with the emergence of the Perestroika movement in political science (Perestroika, 2000; Laitin, 2003; Monroe, 2005). Although the book and the movement were not directly related, the proponents of Flybjerg's arguments and the Perestroika movement coalesced because of the commonalities in their core arguments. The proponents and critics of Flybjerg's book and the Perestroika movement are debated in the chapters of Schram and Caterino's edited book (2006).

6. King et al.'s (1994) method of demonstration is an extension of deductive and quantitative reasoning and their definition of 'qualitative' is controversial. They define 'qualitative research' as small-n research, as opposed to large-n research, for example. Then they apply the principles of deductive inference typically used in large-n research to small-n research, using mathematical formulae. Their definitions and demonstration method are obviously not compatible with how hermeneutic or phenomenological scholars define the terms 'qualitative' and 'quantitative' and the methods they would use to compare the two.

7. There is an ongoing discussion among complexity theorists on the notion of emergence. They address issues such as whether the properties of emergent structures are indeed irreducible and what the mechanisms of emergence are. For extensive discussions on emergence, see Epstein (2006), Holland (1998), Morçöl (2012), and Sawyer (2005).

8. The Wikipedia entry on the theory of everything notes: 'The term mainly refers to the desire to reconcile the two main successful physical frameworks, general relativity which describes gravity and the large-scale structure of spacetime, and quantum field theory, particularly as implemented in the standard model, which describes the small-scale structure of matter while incorporating the other three non-gravitational forces, the weak, strong and electromagnetic interactions' (http://en.wikipedia.org/wiki/Theory_of_everything; accessed on 6 March 2013).

REFERENCES

Agranoff, R. and M. McGuire (1999), 'Managing in network settings', *Policy Studies Review*, **16**(1), 18–38.

Atteberry, J. (n.d.), 'What exactly is the Higgs boson?', accessed 11 December 2014 at http://science.howstuffworks.com/higgs-boson1.htm.

Axelrod, R. (1997), *The Complexity of Cooperation: Agent-based Models of Competition and Collaboration*, Princeton, NJ: Princeton University Press.

Axelrod, R. (2006), *The Evolution of Cooperation*, (rev. edn), Cambridge, MA: Basic Books.

Axelrod, R. (2007), 'Simulation in the social sciences', in J-P. Rennard (ed.), *Handbook of Research on Nature-inspired Computing for Economics and Management*, Hershey, PA: Idea Group Reference, pp. 90–99.

Axelrod, R. and M.D. Cohen (2000), *Harnessing Complexity: Organizational Implications of a Scientific Frontier*, New York: Basic Books.

Barabási, A-L. and R. Albert (1999), 'Emergence of scaling in random networks', *Science*, **286**(5439), 509–12.

Bennett, A., A. Barth and K.R. Rutherford (2003), 'Do we preach what we practice? A survey of methods in political science journals and curricula', *PS: Political Science and Politics*, **36**(3), 373–8.

Caporaso, J.A. (1995), 'Research design, falsification, and the qualitative–quantitative divide', *American Political Science Review*, **89**(2), 457–60.

Casti, J.L. (1994), *Complexification: Explaining a Paradoxical World Through the Science of Surprise*, New York: Harper Perennial.

90 *Handbook on complexity and public policy*

Collier, D. (1995), 'Translating quantitative methods for qualitative researchers: The case of selection bias', *American Political Science Review*, **89**(2), 461–6.

Epstein, J.M. (2006), *Generative Social Science: Studies in Agent-based Computational Modeling*, Princeton, NJ: Princeton University Press.

Epstein, J.M. and R. Axtell (1996), *Growing Artificial Societies: Social Science from the Bottom up*, Washington, DC: Brookings Institution Press.

Flyvbjerg, B. (2001), *Making Social Science Matter: Why Social Inquiry Fails and How it can Succeed Again*, Cambridge: Cambridge University Press.

Gerrits, L. (2012), *Punching Clouds: An Introduction to the Complexity of Public Decision-making*, Litchfield Park, AZ: Emergent Publications.

Geyer, R. and S. Rihani (2010), *Complexity and Public Policy: A New Approach to 21st Century Politics, Policy and Society*, London: Routledge.

Giddens, A. (1984), *The Constitution of Society: Outline of the Theory of Structuration*, Berkeley, CA: University of California Press.

Graddy, E.A. and B. Chen (2006), 'Influences on the size and scope of networks for social service delivery', *Journal of Public Administration Research and Theory*, **16**(4), 533–52.

Holland, J.H. (1998), *Emergence: From Chaos to Order*, New York: Basic Books.

Hy, R.J., P.B. Nelson and W.L. Waugh (1981), 'Statistical backgrounds and computing needs of graduate students in political science and public administration', *Teaching Political Science*, **8**, 201–12.

Hy, R.J., W.L. Waugh and P.B. Nelson (1987), 'The future of public administration and quantitative skills', *Public Administration Quarterly*, **11**(2), 134–49.

Jenkins-Smith, H., N. Mitchell and C. Silva (1999), 'The state of public policy graduate education in leading American political science departments', *Policy Currents (Newsletter of the Public Policy Section of the American Political Science Association)*, **19**(3–4), 7–10.

Kapucu, N. (2006), 'Interagency communication networks during emergencies: boundary spanners in multiagency coordination', *American Review of Public Administration*, **36**(2), 207–25.

Kauffman, S. (1993), *The Origins of Order: Self-organization and Selection in Evolution*, New York: Oxford University Press.

Kauffman, S. (1995), *At Home in the Universe: The Search for Laws of Self-organization and Complexity*, New York: Oxford University Press.

King, G., R.O. Keohane and S. Verba (1994), *Designing Social Inquiry: Scientific Inference in Qualitative Research*, Princeton, NJ: Princeton University Press.

King, G., R.O. Keohane and S. Verba (1995), 'The importance of research design in political science', *American Political Science Review*, **89**(2), 475–81.

Koppenjan, J. and E-H. Klijn (2004), *Managing Uncertainties in Networks: A Network Approach to Problem Solving and Decision Making*, London: Routledge.

Laitin, D.D. (1995), 'Disciplining political science', *American Political Science Review*, **89**(2), 454–6.

Laitin, D.D. (2003), 'The Perestroikan challenge to social science', *Politics & Society*, **31**(1), 163–84.

LaPlante, J.M. (1989), 'Research methods education for public sector careers: the challenge of utilization', *Policy Studies Review*, **8**(4), 845–51.

Lederman, L. and D. Teresi (1993), *The God Particle: If the Universe is the Answer, what is the Question?*, New York: Bantam Doubleday Dell Publishing.

Maroulis, S., R. Guimerà, H. Petry, M.J. Stringer, L.M. Gomez, L.A.N. Amaral and U. Wilensky (2010), 'Complex systems view of educational policy research', *Science*, **330**(6000), 38–9.

McGuirk, P.M. (2000), 'Power and policy networks in urban governance: local government and property-led regeneration in Dublin', *Urban Studies*, **37**(4), 651–72.

Miller, J.H. and S.E. Page (2007), *Complex Adaptive Systems: An Introduction to Computational Models of Social Life*, Princeton, NJ: Princeton University Press.

Monroe, K.R. (ed.) (2005), *Perestroika!: The Raucous Rebellion in Political Science*, New Haven: Yale University Press.

Morçöl, G. (2002), *A New Mind for Policy Analysis: Toward a Post-Newtonian and Postpositivist Epistemology and Methodology*, Westport, CT: Praeger.

Morçöl, G. (2012), *A Complexity Theory for Public Policy*, New York: Routledge.

Morçöl, G. and N. Ivanova (2010), 'Methods taught in public policy programs: are quantitative methods still prevalent?', *Journal of Public Administration Education*, **16**(2), 255–77.

Ornstein, R. and P. Ehrlich (2000), *New World New Mind: Moving Toward Conscious Evolution*, Cambridge, MA: Malor Books.

Perestroika (2000), 'On the irrelevance of APSA and APSR to the study of political science', accessed 4 March 2013 at http://archive.org/details/OnTheIrrelevanceOfApsaAndApsrToTheStudyOfPoliticalScience.

Prigogine, I. (1996), *The End of Certainty: Time, Chaos, and the New Laws of Nature*, New York: The Free Press.

The Higgs boson of public policy or public administration theory? 91

Prigogine, I. and I. Stengers (1984), *Order out of Chaos: Man's New Dialogue with Nature*, New York: Bantam Books.

Raadschelders, J.C. (2011), 'The future of the study of public administration: embedding research object and methodology in epistemology and ontology', *Public Administration Review*, **71**(6), 916–24.

Raadschelders, J.C. and K-H. Lee (2011), 'Trends in the study of public administration: empirical and qualitative observations from Public Administration Review, 2000–2009', *Public Administration Review*, **71**(1), 19–33.

Radin, B. (2000), *Beyond Machiavelli: Policy Analysis Comes of Age*, Washington, DC: Georgetown University Press.

Rhodes, M.L., J. Murphy, J. Muir and J.A. Murray (2011), *Public Management and Complexity Theory: Richer Decision-making in Public Services*, London: Routledge.

Riccucci, N.M. (2010), *Public Administration: Traditions of Inquiry and Philosophies of Knowledge*, Washington, DC: Georgetown University Press.

Rogowski, R. (1995), 'The role of theory and anomaly in social-scientific inference', *American Political Science Review*, **89**(2), 467–70.

Rössler, O.E. (1986), 'How chaotic is the universe?', in A.V. Holden (ed.), *Chaos*, Princeton, NJ: Princeton University Press, pp. 315–20.

Sawyer, R.K. (2005), *Social Emergence: Societies as Complex Systems*, Cambridge: Cambridge University Press.

Schelling, T.C. (2006), *Micromotives and Macrobehavior*, New York: W.W. Norton.

Schram, S.F. and B. Caterino (eds) (2006), *Making Political Science Matter: Debating Knowledge, Research, and Method*, New York: New York University Press.

Schwartz-Shea, P. and D. Yanow (2002), '"Reading" "methods" "texts": how research methods texts construct political science', *Political Science Quarterly*, **55**(2), 457–86.

Sharkansky, I. (2002), *Politics and Policymaking: In Search of Simplicity*, London: Lynne Rienner Publishers.

Simon, H.A. (1947), *Administrative Behavior*, New York: Macmillan.

Simon, H.A. (1952), '"Development of theory of democratic administration": Replies and comments', *American Political Science Review*, **46**(June), 494–6.

Smith, J. and C. Jenks (2006), *Qualitative Complexity: Ecology, Cognitive Processes and the Re-emergence of Structures in Post-humanist Social Theory*, London: Routledge.

Stenger, V. (2012), 'Higgs and significance', *Huffington Post*, 5 July, accessed 6 July 2012 at http://www.huffing tonpost.com/victor-stenger/higgs-and-significance_b_1649808.html.

Stokey, E. and R. Zeckhauser (1978), *A Primer for Policy Analysis*, New York: W.W. Norton.

Tarrow, S. (1995), 'Bridging the quantitative–qualitative divide in political science', *American Political Science Review*, **89**(2), 471–4.

Teisman, G.R., A. van Buuren and L. Gerrits (eds) (2009), *Managing Complex Governance Systems: Dynamics, Self-organization and Coevolution in Public Investments*, London: Routledge.

Waldo, D. (1948), *The Administrative State: A Study of the Political Theory of American Public Administration*, New York: Ronald Press.

Waldo, D. (1952a), 'Development of theory of democratic administration', *American Political Science Review*, **46**(March), 81–103.

Waldo, D. (1952b), '"Development of theory of democratic administration": Replies and comments', *American Political Science Review*, **46**(June), 500–503.

Waugh, W.L., R.J. Hy and J. Brudney (1994), 'Quantitative analysis and skill building in public administration graduate education', *Public Administration Quarterly*, **18**(2), 204–22.

Yang, K. (2007), 'Quantitative methods for policy analysis', in F. Fischer, G.J. Miller and M.S. Sidney (eds), *Handbook of Public Policy Analysis: Theory, Politics, and Methods*, Boca Raton, FL: CRC Press, pp. 349–68.

7. The policymaker's complexity toolkit
Jim Price and Philip Haynes with Mary Darking,
Julia Stroud, Chris Warren-Adamson and Carla Ricaurte

> Evolution is cleverer than you are.
> Orgel's Second Rule

INTRODUCTION

As discussed in the introductory chapter, one of complexity theory's most useful contributions to policymaking and implementation may be the provision of a unifying language between different disciplines and professions, between 'theorists' and 'practitioners'. Whilst not quite yet the 'Esperanto' of scientific and social enquiry (Lissack, 1999), the framing of concepts with similar phraseology, models and conceptual metaphors might help both policymakers and practitioners communicate more effectively and begin to promote a common understanding of approaches to enquiry and policy implementation across disciplinary and cultural boundaries (Klein, 2004; Cooper et al., 2004; Price, 2005). Indeed contributors to this chapter come from different disciplines within one academic institution and have all found 'complexity' to be a commonly understood framework (if not completely uniformly interpreted), through which productive conversations regarding policy and practice can occur. We introduce the Brighton Complex Systems Toolkit (CSTK, 2012) to represent an example of successful collaborative practice, and frame a modified version as a suggested 'Complex Practice Toolkit', advocating a complexity approach, informed by an 'evolutionary' discourse, for both policymakers and those implementing policy (practitioners).

This chapter first discusses our interpretation of what might be called 'hard and soft' complexity, provides a concrete example of using complexity in practice to develop a cross-sector 'toolkit' for practitioners, and then goes on to suggest and explore seven tools for a 'Complex Practice Toolkit' for policymakers, based on our experiences and informed by theoretical perspectives.

Our intention is to help the policymaker and practitioner appreciate rule and value-based interactions between different agents (with different strategies) in their defined systems, to enable communication, information capture and recall, leading to pattern recognition, sense making and the evaluation of outcomes. The ultimate goal is to select and amplify successful work-streams and damp down of those less successful in an adaptive or evolutionary sense. We acknowledge that evolution and adaptation will occur in most complex systems (that is those involving humans), but in the area of policy-making, we hope to guide those in positions of influence to facilitate what has been termed an 'ecology of innovation' (Goldstein et al., 2010), and hope that the 'Complex Practice Toolkit' might act as a catalyst for action and change.

HARD AND SOFT COMPLEXITY

The Toolkit assumes that 'natural complexity' (that is, that complexity all around us in the natural world) and its scientific study have generated an inter-disciplinary 'complex systems science' where fundamental theory has been both generated and tested. Described by some as 'hard' complexity, or perhaps more usefully by Morin (2007) as 'restricted' complexity (in so far that there is a tendency to look for essential characteristics of different but related systems in a neo-reductionist manner), this area remains important to the policymaker since it embraces empiricism and data collection. 'Big data' relating to any defined system, can now be mined, and trends showing successful 'adaptations' to the changing environments identified. More difficult to apply and explore in recent years has been 'soft', 'metaphorical' or 'general' complexity (Morin, 2007), and we discuss the influence these concepts might have on culture and communication within a given system.

THE BRIGHTON KNOWLEDGE EXCHANGE PROJECT

The Brighton Knowledge Exchange project (http://www.brighton.ac.uk/sass/complex-systems/) brought academic and scientific expertise together with policymakers and practitioners in order to develop an approach which we consider to be significantly different from current advice, and which incorporates a language which is meaningful to both sides.

The Brighton project developed as a dialogue between public and voluntary service organizations which took place over the course of 12 months in 2010–11 as part of an ESRC funded knowledge exchange grant. The principal aim of the dialogue was to explore how complexity models and metaphors could be used to develop applied methods to support the organization, management and professional delivery of public services. Participant organizations ranged from small community organizations to statutory service providers working with the local authority. The scope of public service activities represented was equally diverse, including substance misuse, adult learning and social care commissioning. Time pressures and resource constraints faced all 12 project partners, and facilitated a sense of common purpose and a sharp focus on the practical application of ideas.

A combination of methods were used to support partner dialogue at both practical and conceptual levels, but in particular an online environment was created through which partners could contribute discussion points, videos, podcasts, references and documents. This environment helped to generate a set of resources which were drawn upon and further developed via face-to-face workshop sessions. These sessions included system mapping workshops and 'critical reading' groups in which ideas concerning a specific article were shared and examined.

Whilst methods of engagement varied to suit a range of partner needs, the 'back and forth', iterative examination of ideas and practice remained central to the project. In this way, partners began to use complexity concepts to articulate patterns of activity that were fundamental yet that had been frequently unacknowledged. For example, one partner from a community adult learning centre found complexity concepts offered a

94 *Handbook on complexity and public policy*

counterpoint to conventional funder and commissioning conceptualizations of their work (each individual learning programme they offered had previously been treated as a separate entity). Complexity models allowed them to re-ground their strategic planning activities around a holistic view of their service provision, creating a space for them to articulate the means by which they adapted their curriculum in response to feedback from the community.

Since the end of the project, community partners have continued to work with academic partners, forming a 'Community Research and Evaluation Gateway'. This forum has been successful in supporting community organizations in their articulation of *core values* (cf. 'rules') and their need to orientate organizational, management and professional practices around these values. The holistic, complex and frequently radical thinking that underlies the interventions these organizations design is frequently at odds with the policy and funding environment. Empowering community organizations to stay close to their values through engaging them in academic research partnerships has created the possibility for triggering 'upstream' policy change. Their attentiveness to community feedback, professionalism with respect to data and information use and radical thinking in the face of complex community issues have now offered a capacity to effect change within local policy systems. Not having previously had the time and resource to articulate these elements of their professional practice has been recognized as the principal factor which left them in a disempowered 'response' mode of policy engagement. The Community and Research Evaluation Gateway now aims to address and mitigate this by lending voice to community organizations and enabling system change.

The experience of the Brighton Project led to the development of an initial toolkit (CSTK, 2012). Implementation has led to some modifications and it has been developed into what we term the 'Complex Practice Toolkit', comprising seven 'tools' for policymakers aligned with an evolutionary or bottom-up approach to management. Although similar to approaches recommended by other management strategists (Beinhocker, 2007; Grisogono 2006; Axelrod and Cohen 2000), our new and radical 'Complex Practice Toolkit' is based on the pragmatic outcomes of a grounded knowledge exchange project, involving both 'theorists' and 'practitioners' engaged in policymaking and its implementation. The proposed toolkit can be summarized in Table 7.1.

Table 7.1   Complex Practice Toolkit: seven tools for policymakers

1.	Define your policy system
2.	Lead through facilitation, distribution, direction and rules
3.	Give up the illusion of prediction and control – encourage self-organization
4.	Take a realistic view of human behaviour (heuristics and rules of thumb)
5.	Use information scanning and resource management to enable economic evolution
6.	Evolve a portfolio of 'policy experiments'
7.	Empiricism, data and evaluation: do more of what works and less of what doesn't

SEVEN 'TOOLS' FOR POLICYMAKERS

1. Define your Policy System

Regarding policy-making in practice, one important thing to do initially is to define the complex adaptive system(s) in which policy is both formulated and implemented. Defining a system implies that it might be 'identifiable', that is, possess an identity, and that identity should necessarily be 'different' from that of another. In complex systems, this identity is a function of the dynamic nature of differences between agents and groups of agents (the important philosophical and ethical consequences of the relationship between difference, identity and complexity are well explicated elsewhere (Cilliers and Preiser, 2010)). Lichtenstein et al. (2006: 6) define the 'system' as being comprised of 'agents, individuals as well as groups of individuals, who "resonate" through sharing common interests, knowledge and/or goals due to their history of interaction and sharing of worldviews'. Considering public agencies and communities of individuals as systems such as this offers the prospect of planned and facilitated interactions in what Conn (2010) terms 'spaces of possibilities' which can be recognized and exploited, offering a way to acknowledge the asymmetry which often exists between the top-down and bottom-up in terms of relationships and organizational dynamics. Individuals in the community have strong horizontal peer relationships at a tangent to the vertical hierarchy of the external public agencies, thus requiring adequate support in this 'space of possibilities' which exists between communities and other structures. It is suggested that practitioners and other intermediaries need to work together in this space for mutual benefit.

Defining the system may be aligned with the, often temporary, 'space of possibility' in an organizational interaction, with support and facilitation of the 'space', together with planning for the measurement and evaluation of outcomes becoming an important task for the policymaker / leader (Sanderson, 2006). An example might be the implementation of a series of stakeholders' or policy teams' workshops, using methods such as 'Open Space' or 'World Café' (Malby and Fischer, 2006; Olson and Eoyang, 2001), in which the attendees first 'define the complex system' from their own perspectives, connect and share information and plans for potential outcomes, and begin to develop a system culture based on shared values and norms. This was indeed the process undertaken in the development of the initial Brighton Toolkit, and the 'Community Research and Evaluation Gateway' may be seen as the reality of a 'space of possibility'.

Determining whether the system has changed or evolved to a new state, such that it becomes necessary to redefine it, can be checked with participants intermittently, yet might also be informed by 'horizon scanning' by the facilitators of the planned interactive activities.

Intervening in the system

In the Brighton project, we used Meadows' short paper (1997) as the focus for a workshop on how to intervene in a system. Participants became interested in how Meadows had identified a range of possible interventions and how this highlighted the limited range of practices endorsed by current policy and public service management models. For example, Meadows' model positioned practices concerning resources and information use as low impact compared to those associated with empowerment, purpose and

96 *Handbook on complexity and public policy*

paradigm change. Community organization partners in particular felt that dominant, funder-driven organization and management practices frequently so stifled articulation of the more radical elements of their strategic vision that they felt unable to properly empower the client groups and communities with whom they worked. A number of intervention points were identified and incorporated into the toolkit for use by practitioners.

The original Brighton Toolkit developed Meadows' framework with prompt questions for use in workshops or educational sessions, and important elements of the framework have been developed in the CPT presented here. Readers might wish to try out the Brighton Toolkit in its own right instead of or as an adjunct to the CPT (http://www.brighton.ac.uk/sass/complex-systems/ToolkitFramework.pdf). The authors would welcome feedback on success or otherwise with the use of either.

2.   Lead through Facilitation, Distribution, Direction and Rules

Policymakers wishing to facilitate change or a review of strategy within their system may see themselves as educators or facilitators, but the overlap with *leadership* may be larger than imagined. Leadership in a complex environment presents a challenge, both conceptually and practically. Accepting that groups of humans interacting as a system make the system complex and adaptive, how should one best see 'leadership'?

In the last decade it has been recognized that 'decentralization' is a driving force for many organizations (Castellani and Hafferty, 2009; Lichtenstein et al., 2006; Osborn et al., 2002), a good example in the UK being the NHS (Department of Health, 1997; 2000). As organizations cope with these decentralizing forces, there is a growing recognition that traditional 'top-down' theories of leadership may not be helpful, and may be, at best, overly simplistic (Osborn et al., 2002). Social processes are seen as too complex and 'messy' (Klein, 2004) to be attributed to one individual or a planned series of events (Finkelstein, 2003; Marion and Uhl Bien, 2001). This leads to increasing doubt about the notion of leadership resting within the character or characteristics of the leader (Seers, 2004), and whether the leader can indeed 'act on' an organization exogenously (Meyer et al., 2005).

A novel way to approach leadership from a complexity perspective is to see leadership as an emergent event, what Lichtenstein et al. (2006) see as 'the outcome of relational interaction among agents' in the system. These authors have expounded the notion of 'complexity leadership theory', seeing leadership as transcending the individual and being a phenomenon associated with the 'system' (Hazy, 2006; Marion and Uhl-Bien, 2001). Viewing the organization as a 'complex adaptive system' defines leadership by relationships and interactions rather than by hierarchical structures. Leadership may then be seen as a 'distributed' process (Brown and Gioia, 2002; Gronn, 2002), and may provide an integrated theoretical framework for explaining interactive dynamics that have in recent years been acknowledged by a variety of emergent leadership theories, for example shared leadership (Pearce and Conger, 2003), collective leadership (Weick and Roberts, 1993) distributed leadership (Gronn, 2002), relational leadership (Drath, 2001) adaptive leadership (Linsky and Heifetz, 2002), and leadership as an emergent organizational meta-capability (Hazy, 2006). This concept of 'distributed leadership' has major implications for how leadership is constructed, retaining the notion of a 'system', not dissimilar from Senge's (1990) 'learning organization', that is, reifying the 'system' in which the agents interact. However, at the other extreme lies another perspective which goes further and touches on the importance of 'self' in the relational interactions that occur in the policymaker's 'system'.

Drawing on the work of Mead (1934) and Elias (1939), Stacey has developed the notion of the 'complex responsive process' of human relating, referring to the *actions of human bodies* as they interact with each other, so constituting the *social*, and as each interacts, *at the same time*, with himself or herself, so constituting *mind/self*' (Stacey and Griffin, 2006: 14; italics in the original). Griffin (2002) built on Stacey's work on complex responsive processes in proposing a model of leadership which moves beyond the systematic approach to complex leadership. His ideas involve the patterning of one-to-one, group-to-individual and group-to-group interactions described by Mead and Elias, as a model for both developing leadership and explaining ethics in any system. The notion of 'participative self-organization' is seen as a way in which these issues are played out, and theories of ethics are seen as theories of leadership, both being action into the future and therefore about the identity of persons who are both changeable and stable. The role of leader emerges in the interaction in a group, and those participating are continuously creating and recreating the meaning of leadership themes in the local interaction with which they are involved (ibid.: 203). Groups then 'tend to recognise the leadership role in those who have acquired a greater spontaneity, a greater ability to deal with the unknown as it emerges from the known context' (ibid: 204).

So what does this mean for 'leadership' in the policymaking world? The 'leader' may perhaps be best seen though this lens as 'facilitator', bringing agents together, establishing interactive frameworks and catalysing interactions in the 'space of possibilities'. The leader can offer the 'distributed leadership' model to agents within the system, allowing dynamic shifts in leadership concepts and roles as the system evolves. The leader can help agents define their system from within, both temporally and spatially. S/he can also influence the direction of evolution of the system, not through being 'directive' but through offering a temporal framework and expectations (that is, likely lifetime of the 'system') and creating a vision of one or several possible futures (that is, end-states of the system). Expectations of ways of behaving (ethics and values) will also need explication, through the exploration and agreement of 'simple rules' (explicit and implicit), at a high enough level to avoid over-specification (Zimmerman et al., 1998; Plsek and Wilson, 2001; Lichtenstein et al., 2006; Goldstein et al., 2010). In summary, complexity theory informs the 'policymaker leader' as shown in Table 7.2.

Table 7.2 Complexity and policymaker leadership

Leader as facilitator	Convene and catalyse interactions of agents in the system
	Offer 'distributed leadership' as a model for system dynamics
Define the system	Offer temporal and spatial visions of the system
	Allow agents to define the system from within
Direction but not directive	Outline temporal and pragmatic outcomes
	Explore expectations and create a vision of future states as a range of possibilities
Explore and explicate 'rules'	Explicate ethics and values using exploration and agreement of implicit and explicit rules

98   *Handbook on complexity and public policy*

3.   Give up the Illusion of Prediction and Control -- Encourage Self-organization

Most recent complexity theorists in the management world realize that 'traditional' ways of managing and leading are severely limited in the domain of the 'complex adaptive system' (for example Wheatley, 1992; Zimmerman et al., 1998; Stacey et al., 2000; Stacey, 2005). The hierarchical managerial system lies at the opposite end of a managerial spectrum to that which is evolutionary and ecological in its perspective, and diffusion of ideas and innovation occurs in markedly different ways between these two extremes (Greenhalgh et al., 2005). Relaxing one's grip on the need for control and prediction is difficult for the traditional hierarchical manager and is also difficult for the political policymaker who may be required to have an overall strategy defined by ideological priorities. A government minister or local mayor needs to translate this with the help of policy managers and administrators to core values and then priority tasks and related output objectives. But political leadership is mistaken if it attempts too much intervention at the micro management level. The need for using the power of 'self-organization' soon becomes apparent, and the role of the manager or leader changes to facilitating the self-organization process to occur (Olson and Eoyang, 2001). So what exactly do we mean by 'self-organization' and how can we promote it?

Self-organization refers to the emergence of new orders brought about by the interaction of social actors within the system without the intervention of a single central or external controller. There are resonances with the 'bottom-up' rather than the 'top-down' approach to policy implementation recorded in the public administration literature (Hill, 2012). Bottom-up policy making acknowledges the realism of professional discretion, as highly trained public professionals like doctors, teachers and social workers craft the detail of policy in their working lives, and exercise their own judgements and priorities into the policy frameworks they are given. Better therefore that they really are able to feel they can participate in and share the value base of their government if real long-term progress is to be made. The alternative is mistrust, organizational conflict and various manifestations of local resistance to change. New orders emerging from self-organization can include the creation of new rules and social practices as well as working groups or management bodies within the system, but these can be defensive and protective, or creative and innovative (Ricaurte-Quijano, 2013). Self organization does not have a built-in public benefit.

For Boons et al. (2009: 235–6), self-organization has four key characteristics:

1. Self-organization is a driving force that emerges from the interactions and actions of individuals in the system, often amidst unstable and constraining conditions.
2. Self-organization causes processes to follow unexpected trajectories since there is no unique agent controlling actions.
3. Self-organization is closely related to values, perceptions and interests of actors within the system. These understandings are changing constantly since actors feed back to each other through communication.
4. Self-organization is driven by the individual need to survive (self-interest, competition, or conservative self-organization), but also by the ambition to contribute to changing the state of things and having an impact on the system (public interest, cooperation, or dissipative self-organization).

The policymaker needs to be aware of the unpredictability of outcome, but also that outcome more likely to be positive and within certain parameters is more likely to occur if the values, perceptions and interests of the actors are: (a) respected and (b) in line with those of the 'system' itself, that is, the organizational culture. The wise policymaker will embrace the notion of self-organization, and put energy into creating the conditions for it to occur within the defined system, as much in line with a positive vision of the future as possible. The following practical tips may help:

1. Identify processes in the system that could be dealt with through interaction, such as the identification of perturbations or conflicts, discussion of alternatives, decision-making, implementation and the monitoring of action.
2. Set up mechanisms and encourage practices of communication and information sharing. Communication should allow individuals to identify interdependences in roles, shared and conflicting points of view, interests and values. Communication and dialogue should also allow the identification of what actors within the system would like to maintain and what they would like to change.
3. Encourage the emergence of agreed bottom-up rules and practices, which should facilitate rather than constrain voluntary action, leadership, negotiated decision-making, cooperative action, transparency and accountability.
4. Allow horizontal communication based on dialogue and negotiations between actors. Decisions should be distributed, that is, they should not be identifiable as the choice of a single actor, but the result of dialogue and multiple perspectives.
5. Encourage implementation mechanisms that rely on the alignment of individual interests and perceptions, as well as their roles and competences, that is, that rely on interdependencies, cooperative action and mutual benefit.

4. Take a Realistic View of Human Behaviour (Heuristics and Rules of Thumb)

In recent years much light has been shed on reasons behind the fallacy that humans are rational beings who make logical decisions based on cognitive processing of all the available facts (Tversky and Kahneman, 1974; Kahneman and Tversky, 1984, Kahneman, 2011; Thaler and Sunstein, 2008). The importance of subjective heuristics and biases, environmental influences and context / history are much more important than we would perhaps like to believe. Even if the normal rules of engagement and institutional values and culture might be expected to cause an individual or group of individuals to act in a certain way, other factors, often hidden, but related to other aspects of system dynamics (such as those listed by Meadows, 1997), might cause quite unexpected actions to occur, sometimes quite contrary to the expected values of the system.

The public sector is full of operational examples where difficult decisions have to be made in the midst of complex environments, and sometimes the stakes are very high, with decisions resulting in life or death. A key aim for policymakers becomes the creation of a policy environment that supports difficult decision making, rather than undermining it. The limited rationality of even one experienced practitioner is a given, but they need a supportive and open environment where they can check their own assessment with that of others. The culture should be for shared responsibility, rather than scapegoating. The

100 *Handbook on complexity and public policy*

tragedy of child deaths, due to abuse, when the case is known to public services, such as the UK case of Victoria Climbié, provides an example.

The cultural features of the organization responsible for Climbié's care might be analysed through the lens of complexity using the notion of the 'strange attractor'.[1] Like isobarometric troughs on a weather map, these cultural features (*inter alia*, leadership, funding, values, anxiety) may be seen to self-organize, endure, deepen, become shallow, engulf and be engulfed over time. Thus an organization delivering a vital area of public policy can move from a stable attractor pattern to a non-linear chaotic attractor pattern, and then 'flip' to another stable, yet very different 'attractor state'. This change may relate to external and/or local political pressures, requiring managers and practitioners with experience, confidence and system sensitivity to recognize high-risk priorities and contain fragmentation. The practice dynamic between the organization's attractor thrust towards scientific managerialism (based on the premise of some certainty linked to performance targets and risk management checklists) and the adaptive 'edge of chaos'[2] activity that recognizes inherent instability and high levels of risk must be negotiated (Czerwinski, 1998). The latter acknowledges the need for highly developed expert and qualitative practices, with supportive shared, collective working and decision-making. Classical managerial shortcuts will be unable to refine a positive culture of decision-making based on practice experience and an open sharing of multidisciplinary perspectives. Supporting this view, Houchin and MacLean's important study (2005) of the building of an organization on complexity lines identified the lack of management of workers' anxiety (effectively an all-powerful attractor) as a critical factor in the eventual undermining of the project, and complements Menzies Lyth's psychodynamic account of organizational anxiety in a health context in an earlier decade (Menzies Lyth, 1989).

In the case of Climbié, where there was, *inter alia*, a breakdown of the formal child protection system (local authority, health, police in particular), Lord Laming's subsequent report highlighted the responsibility of senior management and recommended the introduction of new structures and protocols (Laming, 2003). This rigid and reductionist response notwithstanding, there has been a turnaround in the effectiveness and processes of Hackney Children Services since the event. Perhaps of more relevance to the change is the adoption of a systemic approach to the organization, recognizing the embeddedness of denial, fear and anxiety as key 'attractor' features of the child protection system (Ferguson, 2005; Smith, 2005), and also the introduction of a 'Meadows approach' to staffing that allows for building 'stocks' and 'buffers'. Knowledgeable and experienced staff have been appointed, and a managerial style has evolved which has recognized the dynamic attractor processes between systems, coupled with the maturity to manage the uncertainty implicit in such work (Goodman and Trowler, 2011).

In this case study, we have identified and highlighted anxiety in particular as a powerful, dominant attractor in the complex systems of child protection and child placement. The public management facilitation of good local decision-making requires policy to be implemented in a manner that optimizes the use of collective professional experience and participation.

5. Use Information Scanning and Resource Management to Enable Economic Evolution

The use of information is one key instrument available to managers and policymakers identified in Meadows' (2008) system model and in the recent Brighton toolkit development. Policymakers in a complex system need to be aware of the importance of wide information scanning and its management to facilitate policy evolution 'pilots'. Narrow and erroneous information processing is cited as one of the causes of the 2008 Financial Crisis (Haynes, 2012).

Information comes in different forms. A distinction recognized by social scientists is *quantitative* and *qualitative*. The former puts the emphasis on data, or bits of information that are stored as numerical counts, or at least attempts to numerically code bits of information into mutually exclusive categories. While the benefits of such types of information are their potential to be precise about definition and to make comparisons of quantity, there will be issues of reliability of measurement over time and its validity (validity is whether the measure is useful in the context of its application). For example, in the managerial practice of performance management, there has been much debate about the use of quantitative performance indicators (PIs) (Haynes, 2003). Do PIs measure what they are supposed to measure (does the school mathematics test really measure mathematics skills?), and are these skills useful in an external context (will those mathematical skills taught successfully, really be the ones future industry requires of its graduates?)?

Qualitative information moves the managerial focus away from individual data items and categories represented by numbers and digits towards a synthesis of items that might depend much more on situational context. For example, a personal or household address in a database is several bits of information that nevertheless come together to provide a specific piece of geographical information: a house number, a road, and a neighbourhood, a town, combines to give the information of a business or private address. Most policy-related managerial decisions and judgements are essentially qualitative, even though they may be composed of different quantitative bits of information. They involve a synthesis of information. Complex policy systems require a synthesis of judgement.

Policymakers should begin by assessing what quantitative information they already have available at their disposal and what it tells them about the operation of their policy system. Most modern public sector institutions and agencies have a very large amount of information collected within their routine processes and operations and this is useful in describing and understanding system dynamics and behaviour. Policymaking requires managers to understand the interface of their own system with other systems. Other influencing systems will be overlapping or in close proximity to their own.

A first task for the policy manager and their support staff is to assess what information is available and what is of an acceptable quality to be used. Existing information can be repackaged and redefined to make it fit for answering questions of analysis. A simple example is that two variables like height and weight can be combined together to become the Body Mass Indicator (BMI) and this can begin to provide understanding about the prevalence of obesity in the population.

Is there still additional information that needs to be collected for better decisions to be made? If this is possible and can be prioritized, such new data collection costs will need to weighted against the likely advantage of having yet more information available and

102 *Handbook on complexity and public policy*

its ability to cover existing gaps of knowledge. Another important activity is the periodic review of the collection of routine data and information, and asking whether there are currently any key changes in the core statistical information required. Given the nature of change in dynamic systems and their interaction with the environment, data quickly becomes irrelevant, and new questions need to be answered.

Having made a judgement about the range, diversity and quality of information available and what to do about it, it is futile to wait for the perfection of information collection and data sources (as this will never be achieved in a complex evolving system) but instead one should continue with the analysis and synthesis of the information that is available. Quantitative indicators can certainly give insights into the health of an overall system and how it is evolving, but this requires a dynamic qualitative judgement and active decisions rather than automated static modelling and decision tools.

Regular trend analysis is clearly important in this regard. Meadows (2008: 20) argues that system thinkers need to understand change over time rather than just examining one-off events. This allows a judgement about whether a system is approaching a major limit, or if it is about to attain a key goal. History seems to be full of examples where policymakers and managers became fixated with some particular indicators and chose to ignore or deny the importance of others. The build-up to the recent Financial Crisis was an example of this. Economists were obsessed with consumer inflation targeting and the corresponding setting of central bank interest rates, but ignored asset price inflation, in particular property price inflation fed by cheap debt and global financial flows. Similarly, policymakers thought that balance of trade figures were no longer important to scrutinize as imbalances in exports and imports would be self-corrected by inward and outward investment. This ignored the different social and financial impacts that flows of money in and out of countries have when compared to flows of goods, and the fact that balance of payments trends were related to – if not determining – particular labour market trends and consumer patterns. Before the Financial Crisis broke there were early data signs of the trouble brewing, but few spotted them in advance.

At the more micro level policy managers can use information to try and change organizational behaviour. The timing and placement of information and its analysis into organizations can certainly contribute to changes in interaction and behaviour as information influences debate and decisions. Public health information, such as the use of hospital infection rates, has contributed to major changes in UK hospital practices, with some success in controlling and reducing dangerous infections.

A growth area in the application of information in recent years has been risk management and planning. Here information is sought to assess risk in advance of an organization undertaking a particular activity. While this can provide a useful exercise in scenario planning where hard information is combined with managerial qualitative judgements, risk analysis can become dysfunctional if it tries to rely too much on historical statistical models, as complexity theory demonstrates that historical analysis is not consistently reliable as a good guide for future events. Sudden and unexpected growth in a policy system's instability makes historical data analysis patterns particularly unreliable as an aid to future decision making.

Signs of a policy system moving into chaos (high instability) require rapid and precise interventions and resource targeting. An example of this was the failure of the Japanese Government and central Bank of Japan to prevent an escalation of its own financial

The policymaker's complexity toolkit 103

crisis in the 1990s when its banking system had imploded with bad debt and loss of confidence. It failed to act quickly with low interest rates, banking reform and restructuring, something that had become an urgent priority, and also did not provide enough policy clarity to drive an economic stimulus on the back of low interest rates (Turner, 2008). Contradictory messages and actions were communicated by policymakers, including a since much criticized upwards consumer tax adjustment. Instability needs strong leadership and judgement with focused and targeted use of resources to deal with system behaviours and interactions that are no longer responding to the normal policy and management processes.

There is much debate about whether information systems can be used to identify edge of chaos movements in data that are likely to move into tipping points and increased instability, transformational changes or major crisis events. Several authors have argued that this was possible in the build-up to the global financial crisis, with house prices and household debt levels rising faster than wages, and credit bubbles exceeding real growth. Such trends could be observed in routine data that was available. Discerning such qualitative system change in quantitative data is a matter of judgement and cannot easily be programmed automatically into an information system. This is why writers like Taleb (2005; 2007) and Mandelbrot and Hudson (2008) have warned against automated 'quant' models that make decisions to spend money. Instead qualitative assessments of quantitative data are essential to any management of risk. Qualitative judgement of patterns is likely to be more successful than deterministic modelling of fixed algorithms. As one commentator put it, 'looking in the rear view mirror is not always the best guide to the road ahead' (Ellsworth, 2002: 164).

Seddon (2008) has put a stress on entry into a policy service system, and avoiding what he calls 'demand failures'. He observes that dysfunctional processing of work, as demands enter a public policy system, are often the source of much inefficiency and waste, with inappropriate outputs then provided and disappointing outcomes resulting. Examples would be bureaucratic and standardized forms of assessing need that do not allow flexibility and adaption to individual differences and circumstances. Where services and products are complex this leads to a reductionist approach rather than a holistic understanding of the service process and end output. For example, components might be ordered in advance for a housing repair, on the basis of a tenant phone call report and address location, but these components ordered from different specialist sources are not found to fit together to meet the needs of a specific property repair. Better for a crafts person to assess the job at the local scene and then to assess the relevant parts directly, thus allowing individual circumstances to be taken into account early on in the process (Seddon, 2008). Delays and waste are reduced. With complex medical cases, circumstances can change while the patient is awaiting specialist referral and so an over-rigid referral from a first general assessment to a later secondary hospital appointment may be inappropriate by the time the person gets to see the relevant surgeon and they then join a queue for a different specialist. The alternative is a much more adaptive and evolving primary care system that interconnects more easily with secondary resources and allows for changing circumstances.

The growing focus on strategic, outward-looking management in the public sector in recent decades has meant policymakers and policy managers have become better at seeing how public agencies can adapt to external economic and social change. An

104 *Handbook on complexity and public policy*

important example is the need for the personal tax system to adapt to the increased numbers of women working and of single households and cohabitations and the growth in part-time employment. One might also point to the need for public health and health care systems to move from managing chronic infectious diseases to lifestyle-related conditions such as cardiorespiratory disease, cancer, mental ill-health and obesity. Such social changes require public policy and public services to constantly evolve and adapt and to appreciate when current static services become dysfunctional. The creative and dynamic use of information to make decisions about the deployment of resources is at the core of the successful management of a complex public policy system.

6.   Evolve a Portfolio of 'Policy Experiments'

The generation of novelty in an 'ecology of innovation' (Goldstein et al., 2010) is a crucial aspect of a 'selection' take on policy development. Often the novel aspects might be quite noticeable, positive in their outcome, and fit into the bracket of 'positive deviance' (Dorsey, 2000); however, many innovations might be much smaller 'fluctuations', and easily overlooked. Policymakers need mechanisms to both detect these fluctuations and to reinforce and amplify them if they have the potential to be more broadly beneficial. Using information, especially regarding inputs and outputs, in order to obtain a more holistic view of the system (Hübler, 2005), as well as identifying patterns and trends and distinguishing meaningful 'signals' from background 'noise', becomes an important part of the policymaker's role.

In order to set up these policy experiments to search for novelty, there must be a degree of heterogeneity, and it has been shown that the right combination of difference (in backgrounds, perspectives, heuristics and mental models) is much more important in homogeneous teams for improving performance than the possession of high-level skills (Page, 2007). In fact, the degree of difference among members of the social system has been shown to be proportional to the degree of novelty that the system will generate (ibid.). This essentially underlines the importance of avoiding so-called 'groupthink', when new information is discounted in favour of a single, prevailing mental map (Janis, 1972). Axelrod and Cohen (2000) remind us of the important tension within a Complex Adaptive System (CAS) between 'exploration' (creating new, untested agents or strategies) and 'exploitation' (copying tested types that have proved to be the best to date), an important trade-off for the policymaker to recognize. At one extreme 'eternal boiling' might occur, when the level of novelty, temperature or noise is so high that the system remains permanently disordered, and any potentially valuable structures are destroyed before they can be usefully employed (one might argue that the recent changes in the UK National Health Service (NHS) organizational structure have pushed the system towards this extreme). At the other end of the scale 'premature convergence' occurs when variability is lost too quickly, and speedy imitation of initially successful models prevents future system improvements (ibid.: 44). The wise policymaker will remain alert to the potential pitfalls of this trade-off at either end of the spectrum. Managerial judgements about this dynamic have become critical in the development and practice of multidisciplinary teams to solve so-called 'wicked' policy problems: examples being substance misuse, long-term unemployment, and how best to provide integrated long-term health and social care. Increasingly governments look to non-government organization (NGO) providers and perhaps private bodies for cost-effective solutions.

There is also the need to know that with every project there is the possibility of failure. In fact the reality is that the project is very likely to fail, because, as Ormerod points out to us candidly: 'most things fail' (2005). Indeed recent thinking in the field of safety systems offers a view of failure as an emergent property of a system, with an attempt to move beyond the Newtonian analysis of failure in complex systems (Dekker et al., 2011). The policymaker's real dilemma is to design programmes of work that allow for failure, and do not assume, for instance, that regional pilot schemes should always proceed to national 'roll out' and wider adoption in due course, whatever the outcome (Gov.uk, 2012). Success in such challenging policy circumstances depends on good communication and negotiation skills in a world that adapts best via relational collaborations rather than adversarial contacting. Many of the failures of privatization seem to be linked to an early and uncompromising split between purchaser and provider where purchasers have imperfect information, but are tied quickly to long-term and expensive solutions. The provider inevitably makes bold promises to win the contract that in time they find are impossible to deliver. Examples in the NHS are poor innovation in hospital design and poorer quality buildings on completion (House of Commons Treasury Select Committee, 2011). Such a contract can quickly become adversarial where the energies go into the payment of lawyers and arguments that are not in the public interest and waste precious resources. Relational contracting allows for more constructive communication about what is working and what is failing, and how resourcing needs to change to reflect new circumstances. Given a world that has become more innovative with public policy, with ever more experimental partnerships with NGOs and private bodies, good and open systems of evaluation are an imperative. A strengthening of the relationship between research and policymaking is needed to increase the learning in an evolutionary and adapting complex policy system.

7. Empiricism, Data and Evaluation: do More of What Works and Less of What doesn't

The policymaker's experiments in novelty and innovation will generate data, and this will need evaluation. As some have cogently argued, this 'evaluation' should be closely aligned to the notions of complexity and emergence, and the modernist privileging of objective, value-free knowledge, derived through quantitative social science methodologies, should be tempered with a more hermeneutic and qualitative approach to the nature of information used in evaluation (Sanderson, 2000; 2006). In the past, policy evaluation has been seen as rational and 'goal-orientated' following a 'goal-driven' interpretation of policy itself. But this instrumental, Newtonian 'social physics' view of policy evaluation (Marion, 1999) is now more widely challenged, and since the late 1990s, the *critical realist* stance, more in tune with complexity principles, has become more central to policy evaluation (Pawson and Tilley, 1997; Byrne, 1998; Haynes, 2003). Indeed the notion of critical realism can be further aligned with the notion of dissipative social systems by adopting Bhaskar's (1998) idea of 'social naturalism', which presents a 'transformational model of social action'. Similar to Conn's 'space of possibilities' model (Conn, 2010) described above, this model posits three levels of social reality:

1. society as a 'structural entity';
2. the individual, subject to socialization in particular social contexts but nevertheless with significant powers of agency;
3. an intermediary level comprising a 'position-practice system' of 'rules, roles and relations' that regulate the interactions between the individual and social levels.

This intermediary level equates to Conn's 'space of possibilities' and links our proposed model to the early thinking of complex systems related to social policy. Sanderson (2000) also points out the importance of seeing evaluation as 'craft' or 'practice', and the development of 'communicative competence', at institutional and individual level, key in the success of an evaluation strategy.

Seeing evaluation as equivalent to 'selection' in evolutionary reams, Axelrod and Cohen (2000) suggest that social activity is used to support the growth and spread of valued criteria. An example of this might be prize competitions, for example the Nobel Prize, whereby the processes of refining prize criteria, selecting judges, including nominees in publicising winners will serve to disseminate the underlying goals that motivated the creation of the prize. Such activity increases the use of the criteria imported in the prize, which may be far more effective than direct advocacy of those criteria (ibid.: 158).

Another factor important to bear in mind is that of time. Looking for shorter-term, more granular measures of success as a proxy for longer-term broader goals may be the only practicable solution. Sabatier (1986) identified the problem of 'premature assessment' of the effects of policies and programmes, a problem which besets evaluation in many policy fields. It has been argued that a timeframe of 10–15 years is needed in order to identify how policies and programmes are working as a basis for learning. In particular, such a timeframe is needed to be able to distinguish, on the one hand, the effects of the implementation structure and, on the other hand, the validity of the relevant policy theory, the latter being crucially dependent in complex systems upon longer-term impacts. In reality, one may need to make an evaluative decision in the much shorter term, and proxy measures are often essential. An example in a game of chess might be the number of pieces taken, or the dominance of the central board area, or for the military or emergency services, conducting simulation exercises in a context as near to reality as possible without genuine war disaster will also provide valuable, quick experience. Although the nature of complex systems means that any outcome or evaluation of lower-level factors involved might be at risk of misattribution, since no complex system can ever be truly modelled (Cilliers, 1998; 2001), the evaluation of shorter-term simulations can be valuable for selection, even with acknowledged reduction in validity (Axelrod and Cohen, 2000).

Complexity theory therefore has important implications for evaluation in the context of policy initiatives to address key economic and social problems, in that it requires us to challenge some basic assumptions that underpin traditional approaches; for instance how we can gain access to and analyse social processes, how policies to change such processes are formulated and implemented, and how such policies 'work' in promoting change. It requires us to recognize that evaluation is necessarily itself a highly complex endeavour, a craft or 'practice' (Sanderson, 2000), which, in common with other crafts and expert practices, may require tools to facilitate the process, and to become expert in the craft is likely to require extensive, repetitive and deliberate practice (Ericsson et al., 1993; Simon and Chase, 1973).

CONCLUSION

We have presented seven 'tools' in a 'Complex Practice Toolkit' for policymakers, based on principles collated from other experts in the complexity and policy field over many years. We present this model in a contextual basis, with grounded examples from practice, acknowledging the diverse disciplinary backgrounds which complexity enjoys. The insights and principles emerging from fundamental theory are presented as recommendations for application to practice.

Our seven tools approach indicates that one should: (1) define the system one is working in (as nearly as possible) to make things manageable; (2) lead in one's sphere of influence through facilitation, distribution, direction and by using some 'simple rules' based on values; (3) openly acknowledge the limitations and 'illusion' of prediction and control and encourage self-organization in an appropriate way; (4) take a realistic view of one's colleagues' behaviours and seek to understand the heuristics and 'simple rules' which set the culture in the context in which they operate; (5) use information and resource scanning to maximize the intelligence one has to hand to make economically sensible decisions; (6) try things out via a portfolio of policy experiments or pilots (which are always evaluated); and (7) using the evaluation of these experiments, do more of what works and stop that which does not (acknowledging that these outcomes may well be context-specific, and generalization may be risky).

In this chapter have also shown how Meadows' (1997) ideas for intervening in a complex system, which influenced the development of the Brighton Complex Systems Toolkit (CSTK, 2012), have also shaped the development of the 'Complex Practice Toolkit', described here. We hope that for both models, further iterations and 'improvements' will occur, using the knowledge co-created to date to construct a 'fitter' model for future use. We encourage the reader to try these models in practice and would welcome feedback on the 'Complex Practice Toolkit', to improve its applicability and utility. We will of course respond in the spirit of collaborative creativity.

ACKNOWLEDGEMENTS

The original Brighton Toolkit was output from the Economic and Social Research Council (ESRC) funded project: *Systems and Complex Systems. Policy and Practice. Approaches in Public Policy and Practice*. RES-192-22-0083.

NOTES

1. In using this term we adapt the physics definition minimally to mean 'a state toward which a system tends to evolve'.
2. In using the term 'edge of chaos' here, we mean a critical phase of evolution of a system when rapid evolution occurs, and the system could either become chaotic (and potentially cease to exist), or revert to a more stable pattern (attractor), which may or may not resemble the previous stable state or 'attractor'. It is therefore the state of maximal creativity, yet also maximal anxiety.

108 *Handbook on complexity and public policy*

REFERENCES

Axelrod, R. and D. Cohen (2000), *Harnessing Complexity: Organizational Implications of a Scientific Frontier*, New York: Basic Books.

Beinhocker, E. (2007), 'McKinsey HEEDNet seminar', DEFRA.

Bhaskar, R. (1998), 'Societies', in M. Archer, R. Bhaskar, A. Collier, T. Lawson and A. Norrie (eds), *Critical Realism: Essential Readings*, London: Routledge.

Boons, F., A. van Buuren, L. Gerrits and G. Teisman (2009), 'Towards an approach of evolutionary public management', in G. Teisman, A. van Buuren and L. Gerrits (eds), *Managing Complex Governance Systems: Dynamics, Self-Organization and Coevolution in Public Investments*, New York and London: Routledge.

Brown, M.W. and D.A. Gioia (2002), 'Making things click: distributive leadership in an online division of an offline organization', *Leadership Quarterly*, **13**, 397–419.

Byrne, D. (1998), *Complexity Theory and the Social Sciences: An Introduction*, London: Routledge.

Castellani, B. and F. Hafferty (2009), *Sociology and Complexity Science*, Berlin: Springer.

Cilliers, P. (1998), *Complexity and Postmodernism*, London: Routledge.

Cilliers, P. (2001), 'Boundaries, hierarchies and networks in complex systems', *International Journal of Innovation Management*, **5**(2), 135–47.

Cilliers, P. and R. Preiser (2010), *Complexity, Difference and Identity: An Ethical Perspective*, New York: Springer.

Conn, E. (2010), 'Community engagement in the social eco-system dance: tools for practitioners', paper prepared for the International Workshop on Complexity and Real World Applications, Southampton, 21–27 July 2010.

Cooper, H., S. Braye and R. Geyer (2004), 'Complexity and interprofessional education learning', *Health and Social Care*, **3**(4), 179–89.

CSTK (2012), 'Brighton Complex Systems Toolkit', available at http://www.brighton.ac.uk/ sass/ complex-systems/ToolkitFramework.pdf.

Czerwinski, T. (1998), *Coping with the Bounds: Speculations on Nonlinearity in Military Affairs*, Washington, DC: National Defense University Press.

Dekker, S., P. Cilliers and J.H. Hofmeyr (2011), 'The complexity of failure: implications of complexity theory for safety investigations', *Safety Science*, **49**(6), 939–45.

Department of Health (1997), *The New NHS: Modern, Dependable*, London: HMSO.

Department of Health (2000), *'The NHS Plan: A Plan for Investment, a Plan for Reform'*, London: HMSO.

Dorsey, D. (2000), 'Positive deviant', Fast Company Publishing.

Drath, W.H. (2001), *The Deep Blue Sea: Rethinking the Source of Leadership*, San Francisco, CA: Jossey Bass.

Elias, N. (1939), *The Civilising Process*, Oxford: Blackwell.

Ellsworth, R. (2002), *Leading with Purpose*, Stanford, CA: Stanford University Press.

Ericsson, K.A., R.Th. Krampe and C. Tesch-Romer (1993), 'The role of deliberate practice in the acquisition of expert performance', *Psychological Review*, **100**(3), 363–406.

Ferguson, H. (2005), 'Working with violence, the emotions and the psycho-social dimensions of child protection: reflections on the Victoria Climbié Case', *Social Work Education*, **24**(7), 781–95.

Finkelstein, S. (2003), *Why Smart Executives Fail*, New York: Portfolio.

Goldstein, J., J.K. Hazy and B.B. Lichtenstein (2010), *Complexity and the Nexus of Leadership*, New York: Palgrave Macmillan.

Goodman, S. and I. Trowler (2011), *Social Work Reclaimed: Innovative Frameworks for Child and Family Social Work Practice*, London: Jessica Kingsley.

Gov.uk (2012), 'Iain Duncan Smith: Early roll out of Universal Credit to go live in Manchester and Cheshire', available at https://www.gov.uk/government/news/iain-duncan-smith-early-roll-out-of-universal-credit-to-go-live-in-manchester-and-cheshire.

Greenhalgh, T., G. Robert, P. Bate, F. MacFarlane and O. Kyriakidou (2005), *Diffusion of Innovations in Health Service Organisations: a Systematic Literature Review*, London: BMJ Books /Blackwell.

Griffin, D. (2002), *The Emergence of Leadership: Linking Self-Organization and Ethics*, London: Routledge.

Grisogono, A.M. (2006), 'DST Node Report', CMS net Forum, 27–28 November, available at www.complex systems.net.au/documents/grisogono_dsto.ppt.

Gronn, P. (2002), 'Distributed leadership as a unit of analysis', *Leadership Quarterly*, **13**, 423–51.

Haynes, P. (2003), *Managing Complexity in the Public Services*, Maidenhead: Open University Press.

Haynes, P. (2012), *Public Policy Beyond the Financial Crisis: An International Study*, London: Routledge.

Hazy, J.K. (2006), 'Measuring leadership effectiveness in complex socio-technical systems', *Emergence Complexity & Organization*, **8**(3), 58–77.

Hill, M. (2012), *The Public Policy Process*, 6th edn, London: Pearson.

Houchin, K. and D. MacLean (2005), 'Complexity theory and strategic change: an empirically informed critique', *British Journal of Management*, **16**(2), 149–66.

House of Commons Treasury Select Committee (2011), *Seventeenth Report: The Private Finance Initiative*, London: Houses of Parliament, available at http://www.publications.parliament.uk/pa/cm201012/cmselect/cmtreasy/1146/114602.htm.

Hübler, A. (2005), 'Predicting complex systems with a holistic approach', *Complexity*, **10**(3), 11–16.

Janis, I. (1972), *Victims of Groupthink*, Boston: Houghton Mifflin.

Kahneman, D. (2011), *Thinking, Fast and Slow*, London: Allen Lane.

Kahneman, D. and A. Tversky (1984), 'Choices, values and frames', *American Psychologist*, **39**(4), 341–50.

Klein, J.T. (2004), 'Interdisciplinarity and complexity: an evolving relationship', *Emergence Complexity and Organization (E:CO)*, **6**(1–2), 2–10.

Laming, H. (2003), *Report of the Inquiry into the Death of Victoria Climbié*, HMSO, Cmnd 5730.

Lichtenstein, B.B., M. Uhl-Bien, R. Marion, A. Seers, J. Douglas Orton and C. Schreiber (2006), 'Complexity leadership theory: an interactive perspective on leading in complex adaptive systems emergence', *Complexity and Organization*, **8**(4), 2–12.

Linsky, M. and R.A. Heifetz (2002), *Leadership on the Line: Staying Alive Through the Dangers of Leading*, Boston: Harvard Business School Press.

Lissack, M. (1999), 'Complexity: the science, its vocabulary, and its relation to organizations', *Emergence*, **1**(1), 110–26.

Malby, B. and M. Fischer (2006), *Tools for Change: An Invitation to Dance*, Chichester: Kingsham Press.

Mandelbrot, B. and R. Hudson (2008), *The Misbehaviour of Markets: A Fractal View of Risk, Ruin and Reward*, London: Profile Books.

Marion, R. (1999), *The Edge of Organization: Chaos and Complexity Theories of Formal Social Systems*, Thousand Oaks, CA: Sage.

Marion, R. and M. Uhl-Bien (2001), 'Leadership in complex organizations', *Leadership Quarterly*, **12**, 389–418.

Mead, G.H. (1934), *Mind, Self, and Society: From the Standpoint of the Social Behaviorist*, Chicago: Chicago University Press.

Meadows, D.H. (1997), 'Places to intervene in a system', *Whole Earth Review*, Winter.

Meadows, D.H. (2008), *Thinking in Systems: a Primer*, White River Junction, VT: Chelsea Green Publishing.

Menzies Lyth, I. (1989), *The Dynamics of the Social*, London: Free Association Books.

Meyer, A., V. Gaba and K. Colwell (2005), 'Organizing far from equilibrium: nonlinear change in organizational fields', *Organization Science*, **16**, 456–73.

Morin, E. (2007), 'Restricted complexity, general complexity', in C. Gershenson, D. Aerts and B. Edmonds (eds), *Worldviews, Science and Us: Philosophy and Complexity*, Singapore: World Scientific, pp. 5–29.

Olson, E.E. and G.H. Eoyang (2001), *Facilitating Organizational Change: Lessons from Complexity Science*, San Francisco, CA: Jossey Bass.

Ormerod, P. (2005), *Why Most Things Fail . . . and How to Avoid it*, London: Faber & Faber.

Osborn, R.N., J.G. Hunt and L.R. Jauch (2002), 'Toward a contextual theory of leadership', *Leadership Quarterly*, **13**, 797–837.

Page, S. (2007), *The Difference: How the Power of Diversity Creates Better Groups, Firms, Schools and Societies*, Princeton: Princeton University Press.

Pawson, R. and N. Tilley (1997), *Realistic Evaluation*, London: Sage.

Pearce, C. and J. Conger (2003), 'All those years ago: the historical underpinnings of shared leadership', in C. Pearce and J. Conger (eds), *Shared Leadership*, London: Sage, pp. 1–18.

Plsek, P. and T. Wilson (2001), 'Complexity, leadership, and management in healthcare organisations', *BMJ*, **323**: 746.1.

Price, J.M. (2005), 'Complexity theory and interprofessional education', in *The Theory–Practice Relationship in Interprofessional Education*, Occasional Paper No. 7, Higher Education Academy: Health Sciences and Practice.

Ricaurte-Quijano, C. (2013), *Self-organisation in Tourism Planning: Complex Dynamics of Planning, Policy-making, and Tourism Governance in Santa Elena, Ecuador*, PhD Thesis, Brighton: University of Brighton, available at http://ethos.bl.uk.

Sabatier, P. (1986), 'What can we learn from implementation studies?', in F.X. Kaufmann, G. Majone and V. Ostrom (eds), *Guidance, Control, and Evaluation in the Public Sector*, Berlin: de Gruyter, pp. 313–26.

Sanderson, I. (2000), 'Evaluation in complex policy', *Systems Evaluation*, **6**(4), 433–54.

Sanderson, I. (2006), 'Complexity, "practical rationality" and evidence-based policy making', *Policy & Politics*, **34**(1), 115–32.

Seddon, J. (2008), *Systems Thinking in the Public Sector*, Axminster: Triarchy Press.

Seers, A. (2004), 'Leadership and flexible organisational structures', in G.B. Green (ed.), *New Frontiers of Leadership*, LMX Leadership: The series, Greenwich, CT: Information Age Publishing, pp. 1–31.

110   *Handbook on complexity and public policy*

Senge, P. (1990), *The Fifth Discipline: The Art and Practice of the Learning Organization*, New York: Doubleday.
Simon, H.A. and W.G. Chase (1973), 'Skill in chess', *American Scientist*, **61**, 393–403.
Smith, M. (2005), *Surviving Fears in Health and Social Care: The Terrors of Night and the Arrows of Day*, London: Jessica Kingsley.
Stacey, R. (ed.) (2005), *Experiencing Emergence in Organizations*, London: Routledge.
Stacey, R. and D. Griffin (eds) (2006), *Complexity and the Experience of Managing in Public Sector Organisations*, London: Routledge.
Stacey, R., D. Griffin and P. Shaw (2000), *Complexity and Management: Fad or Radical Challenge to Systems Thinking?*, London: Routledge.
Taleb, N. (2005), *Fooled by Randomness: The Hidden Role of Chance in the Markets and Life*, London: Penguin.
Taleb, N. (2007), *The Black Swan: The Impact of the Highly Improbable*, London: Penguin.
Thaler, R.H. and C.R. Sunstein (2008), *Nudge: Improving Decisions about Health, Wealth, and Happiness*, New Haven: Yale University Press.
Turner, G. (2008), *The Credit Crunch: Housing Bubbles, Globalisation and the Worldwide Economic Crisis*, London: Pluto Press.
Tversky, A. and D. Kanheman (1974), 'Judgement under uncertainty: heuristics and biases', *Science*, New Series, **185**(4157), 1124–31.
Weick, K.E. and K.H. Roberts (1993), 'Collective mind in organizations: heedful interrelating on flight decks', *Administrative Science Quarterly*, **38**(3), 357–81.
Wheatley, M. (1992), *Leadership and the New Science*, New York: Berrett Koehler.
Zimmerman, B., C. Lindberg and P. Plsek (1998), *Edgeware: Insights from Complexity Science for Health Care Leaders*, Irving, TX: Edgeware VHA inc.

8. Effective policy making: addressing apparently intractable problems
Eve Mitleton-Kelly

INTRODUCTION

Policy and decision makers are often faced with complex problems, which appear very difficult to address or even unsolvable and which challenge existing methods and approaches. The problems are complex and multidimensional, yet they are addressed in a relatively simplistic way, usually addressing a single dimension. For example the emphasis may be on culture, or finance, or new technology, when all those as well as many other dimensions may be contributing to the problem space.

This chapter will describe an approach based on complexity theory, which has been developed and tested over two decades in 30 different research projects. The approach, which is part of the EMK methodology, identifies the multiple dimensions of the 'problem space' which may include social, cultural, technical, physical, political, economic, leadership and other dimensions.

Organizations often feel overwhelmed by what they perceive as the enormity of such a task because it is seen as an endless list of issues to be addressed. However, by identifying a set of critical co-evolving clusters, policy makers are able to set up enabling environments that can effectively address complex challenges.

Company EnF, for example, was convinced that the problem they were facing was lack of organizational integration following an acquisition, due primarily to different national cultures (Mitleton-Kelly, 2005). The parent was based in the UK while the acquired company (a set of small firms that had already gone through a series of mergers and acquisitions) was based in Norway, Sweden and Finland. The parent spent two years and a great deal of time and money trying to address the problem of 'different national cultures'. The difference in national cultures certainly was one of the contributing factors, to the lack of integration, but only one. When the multiple dimensions of the problem space were explored, several others were found to be more significant. By identifying the key dimensions and the critical co-evolving clusters of issues within those dimensions, the company was able to set up an enabling environment that effectively addressed the problem.

That particular case used two tools to identify the problem space; one was based on the psychological profiles of the top managers in all four countries and provided quantitative data, which was an excellent starting point when working with engineers, who relate more readily to numbers. The core tool, however, was qualitative and was based on a set of interviews in all four countries and an in-depth method of individual and group analysis. Although the analysis involved 14 volunteers from the organization, who were trained to do the interviews and participated in the analysis, these 'researchers' were not the problem owners.

111

112 *Handbook on complexity and public policy*

Although the interviewing and analysis processes are a necessary starting point to understand the key issues involved, it is one step removed, as it does not directly involve the problem owners. This chapter will focus and describe a more powerful approach, which works directly with the problem owners, to identify the multi-dimensional problem space. This was used recently (March 2013) when working with a government agency in Indonesia, to help them change their organization to enable them, in turn, to address the significant challenge of deforestation. It is with the addition of this process that very difficult and complex problems can now be addressed. Some highly complex problems, such as climate change, cannot be solved, but they can be addressed more effectively. The methodology has been tested repeatedly in many different organizational contexts in the private, public and voluntary sectors, both with individual organizations and with multiple agencies, to address global challenges such as the spread of pandemics.

The approach that will be discussed has two key aims: (a) to identify the multi-dimensional problem space and the critical co-evolving clusters; and (b) to address those clusters, by setting up an enabling environment that addresses all the key interacting dimensions at the same time, while also supporting the organization at multiple scales: at individual, group and organizational level.

THE INDONESIAN CASE STUDY

The researchers worked with a Government Agency (GA) responsible for spatial planning and forests, which is facing a major challenge in terms of deforestation. Indonesia has the third largest forest in the world and retaining that forest is therefore very important to climate change. The approach taken was to help the GA change its own organization first, to enable it to address the problem of deforestation. During the preliminary exploratory work which included a set of interviews, it became clear that the issues being faced internally, such as corruption, nepotism and ethnic groupings, were very similar to those faced in the broader environment. By addressing those issues internally the Government Agency would be better able to address them externally.

Part of the preliminary work done involved a visit by a team of 15 to London in December 2012, who were introduced to ten principles of complexity theory (Mitleton-Kelly, 2003a) which help policy makers to understand organizations as complex social systems. This set the ground for understanding the need for a change in policy, management style, culture and organizational structure. During a workshop in Jakarta in March 2013, a larger group of 30 was again introduced to the ten principles, before going onto the exercise to identify the multiple dimensions of the problem space.

TEN PRINCIPLES OF COMPLEXITY THEORY

The ten principles identified and used by Mitleton-Kelly (2003a) in 30 research projects over a 20-year period, are:

1. Interconnectivity
2. Interdependence

3. Feedback
4. Emergence
5. Self-organization
6. Exploration of the space of possibilities
7. Co-evolution
8. Historicity
9. Far-from-equilibrium
10. Creation of New Order.

Complexity theory looks at complex behaviour and focuses on the relationships between entities, not just the entities themselves. In a human context, social interaction and the relationships of *interconnectivity* and *interdependence* give rise to complex behaviour; these relationships are intricate, intertwined and non-linear. This means that an intervention in one part of the system may have unpredictable consequences in another part of the system, which may appear distant or apparently unrelated. IT (Information Technology) professionals are familiar with this phenomenon. Changing one part of an intricate, interconnected and interdependent computer system will affect other parts of that system, and that impact is not predictable.

Policy makers should be, but are not always, aware that the same phenomenon applies to complex social systems. For example, making significant financial changes (financial dimension) in an organization or an economy will have an impact on the culture (cultural dimension), the ways of working (organizational dimension), relationships (social and political dimensions), what technology can be used and how (technical dimension) and even how space is used (physical dimension). The recent financial restrictions in several European countries provide a good example, with Greece offering the most dramatic illustration. The consequences of financial constraints had a significant impact on businesses (some ceased operating, others changed significantly in order to survive and some new ones came into being); and on individuals through unemployment with consequent poverty and loss of homes. The pattern of these consequences could have been predicted, but not the specific incidences or some of the other impacts including some positive outcomes.

The most recent reports indicate that both Greece and the UK are turning their economies around (at the time of writing). The questions are why and how; understanding the characteristics of organizations and economies as complex social systems will both provide an explanation, as the theory provides an explanatory framework, and help policy makers work with those characteristics to achieve desirable goals.

When the survival of a complex system is threatened, usually when an event external to the system has such a significant impact that it pushes the system far-from-equilibrium, in the sense that it can no longer continue operating under its existing regime, the system will either find a 'new order', that is, find a new way of being and operating, or it will die. These are the two extremes of the bifurcation at the critical point. However, at that critical point the system will become highly innovative; it will *self-organize* and *explore its space of possibilities*. In other words it will try to find viable alternatives and may also become more self-reliant. Self-organization is defined as being spontaneous and not directed by an external agent. For example when a group comes together spontaneously, because it recognizes that a task needs to be undertaken, and among its members it has

114   *Handbook on complexity and public policy*

the skills to do it, then it goes ahead to do that task without any external direction: that is *self-organization*. This is to be distinguished from self-management, when a senior manager chooses the team and the objectives, but allows the team freedom on how that objective is to be achieved.

Exploration of the space of possibilities is when the system explores different options to find one or more that are viable. The search will inevitably involve failure of some of the experiments. If that failure is not recognized as an essential part of the process of exploration and blame is assigned, then the culture which develops is said to be a 'blame culture', and one of the consequences is that it restricts innovation. On the other hand when an experiment fails, but is used as the basis for learning, then innovation and learning are enabled. The idea is to ask 'what could have been done differently?' and to try again until the experiment succeeds. There are, however, two provisos to this approach. One is that the experiment has been undertaken responsibly and with serious intent and the failure is not due to ineptness or carelessness. The second proviso is that a dead-end needs to be recognized. There is no point in wasting resources when the initial idea or objective is unsound.

When a system is pushed far-from-equilibrium it will therefore try different options to ensure its survival. Self-organization and exploration of the space of possibilities are two of the characteristics of complex systems that it will employ. But others come into play at the same time. For example the outcome of the experiments will not be predictable, partly because of *emergence* and partly because of the *co-evolutionary* dynamics active within the system. Emergence is the process of transition from micro-agent interaction to macro-structures. The outcome is systemic and is more than the sum of the parts. A creative meeting is an example: the ideas flow between the participants and trigger new ideas which would not have been initiated without the interaction. By contrast, if the same group of individuals are put in separate rooms and given exactly the same question/problem to address, the resulting separate answers will never be the same as the outcome from the team while interacting; nor can the separate answers be added together to produce the insights gained through interaction. These interactions are an integral part of the team working together as a 'system' and are therefore systemic. They cannot be pulled apart and the individual answers cannot be added together to produce the same outcome: the outcome from the interaction is therefore more than the sum of the parts.

Emergence is seen by most complexity scholars as a bottom-up process, but that is only half the story. Once the emergent comes into being, it affects the interacting entities in two ways. It both constrains certain behaviours while at the same time opening up new possibilities. Organizational culture is a good example of such an emergent which both constrains certain behaviours of its members, while at the same time opening up possibilities for those members by virtue of belonging to that organization, which the individual would not enjoy by him/herself.

Emergence therefore needs to be considered as a dual bottom-up and a top-down process, which is systemic.

Co-evolution takes place within a biological or social ecosystem. It does not happen in isolation. When entity A ('entity' may refer to an individual, a group, an organization, an economy, a society, and so on) takes a decision or action which affects B to such an extent that B has to change its behaviour, this is known as adaptation. B has adapted to

Effective policy making: addressing intractable problems 115

changes in its environment. However, if B's changed behaviour, in turn, influences A to such an extent that A has to change its behaviour, then that is known as co-evolution. *Co-evolution* is therefore *reciprocal influence that changes the behaviour of the interacting entities*. It is a very powerful dynamic and one that most policy makers do not take into account when developing policy.

The processes of co-evolution, emergence, self-organization, exploration of the space of possibilities, interaction and interdependence will all involve feedback, both positive and negative. *Positive feedback* is active when, for example, house prices keep on rising because buyers want to buy before prices rise any higher, but to ensure a successful sale they will often offer a higher price than they originally intended. The higher the prices go, the more buyers panic and rush to buy at the highest price they can afford. Counterintuitively, positive feedback destabilizes the system.

Negative feedback on the other hand tends to bring the system back into equilibrium. These systems are usually machine type systems such as central heating thermostats. When the temperature drops below the desired level, the heating will be switched on to raise the temperature to the designated level. The behaviour of machine type or complicated systems can be predicted and controlled. Unfortunately this does not apply to complex systems. Policy makers often confuse the two and assume that what applies to a complicated system will also apply to a complex one. This is particularly evident with economic policy when two fundamental errors are made. One is to assume that a complex system like the economy has a single equilibrium point, when it may have *multiple equilibria points*. The other fundamental error is to assume that the *right amount of correction* can be applied in the *most timely* manner. But such dynamics are not linear or predictable. They are subject to emergence and co-evolution and their outcome can neither be predicted nor controlled.

Finally, another mistake is to assume *historicity* in the sense that the past or an individual's/country's history will determine the future. The past may *influence* the future, but it cannot determine it, because of all the complex dynamics described above.

The ten principles of complex social systems have been used over the past two decades to understand the dynamics of organizations and social institutions. The principles are *scale invariant* in that they apply to interactions between two individuals as well as to interactions between several entities, which may be groups or entire societies.

The principles and logic of complexity not only help us to understand how organizations and social institutions work, but also provide a sound basis for addressing complex multidimensional problems, which appear to be extremely difficult to address or even unsolvable. These dimensions may be social, cultural, political, economic, technical, physical, financial and other. They appear intractable because only a single dimension is addressed at any one time. Complex problems have clusters of interconnected and interdependent dimensions, which have to be understood as a whole and addressed at the same time and at multiple scales. The Indonesian Government Agency applied these principles to understand their multi-dimensional problem space and to agree on actions to address these problems by setting up an enabling learning environment.

116 *Handbook on complexity and public policy*

INTERVIEW FINDINGS WITH THE INDONESIAN TEAM

Before the workshop in Jakarta with the team of 30, to identify the multi-dimensional problem space, a set of interviews were conducted by one of the local facilitators, who also acted as the translator. The data from the interviews was analysed into common themes, dilemmas or tensions and underlying assumptions. Dilemmas are defined as equally desirable objectives that appear not to be achievable at the same time. They are used to raise the question 'what could be done differently to achieve both objectives at the same time?' Many dilemmas only *appear* as exclusive either/or, when a change in approach can accommodate the inclusive both/and, and achieve both objectives at the same time. Underlying assumptions are defined as the underlying beliefs, usually unspoken, that guide decision making.

The interviews and the analysis provide an outline of the problem space, which was used as a working framework for the workshop. The fundamental difference between the interview analysis and the workshop is that the former is undertaken by an outsider, a researcher or consultant outside the organization, while the work done during the workshop is a direct outcome by the problem owners.

The following findings also provide some of the historical background to the case. (The interviews and analysis were undertaken by one of the local facilitators/researchers and were conducted in Indonesian. The EMK methodology was used both for conducting the interviews and for the analysis.)

Common Themes

The crucial importance of a strong leader. Interviewees saw the 1990s as a 'golden era' for the organization; it was acknowledged as competitive, but also as a highly collegiate atmosphere. They described how the organization then broke down under a new Head in 2001 who tended to over-promote individuals on the basis of ethnic affiliation, and how this legacy continues to plague the Agency, even after the current Head took over the organization in 2007.

Nepotism and collusion in the appointment of staff to key positions, often on ethnic lines, leading to factionalism and loss of trust across the organization. One interviewee described the culture of over-promotion in terms of 'picking a fruit too early means that it will rot rather than ripen'. Individuals often lack the skills needed for the posts they are assigned to ('wrong man, wrong place'), leaving a sub-set of technically competent staff to bear the workload. This in itself is a source of jealousy and resentment. The interviews suggested two possible reasons:

1. The desire for both money (cash for positions) and control in a context where, traditionally, individuals often compete for influence as 'big men', in this case by acquiring the power to farm out positions in the bureaucracy.
2. Lack of clear criteria or SOPs (Standard Operating Procedures) to guide recruitment by the Staffing Division, leaving the system open to abuse.

Confusing and uneven *allocation of roles and responsibilities.* Interviewees traced this to a Ministry of Home Affairs Decree aimed at rationalizing local government. With the

Effective policy making: addressing intractable problems   117

possibility that this might result in redundancies, the Provincial Governor established new structures inside and outside the Government Agency (GA) to shield his bureaucratic cadre from cuts. These, however, also served to duplicate the work of the GA's existing divisions.

Loss of respect and status as the pre-eminent development planning agency at provincial level. This meant that line agencies and district agencies no longer felt the need to report to or seek the GA's advice.

A culture of 'survival' leading to a focus on individual rather than organizational resilience. The organizational system did not provide any formal support or induction procedures for new recruits. As a consequence, staff developed their own survival mechanisms, for example by establishing oneself as a good technician that can be relied upon to do the job; allying oneself with a 'big man' within the organization; mobilizing one's own charismatic power to command the personal respect of others; or simply deciding to go with the flow.

Tensions/Dilemmas

The current Head of the GA was respected for his integrity and openness. But staff members were frustrated by his hands-off leadership style. He was perceived as avoiding difficult decisions (perhaps his own survival mechanism). This may be a genuine cause for concern or it may be that they are unused to being granted such high levels of trust and confidence.

The current Head's open leadership style was not reflected down through the organization. Interviewees highlighted:

- Cultural barriers to communication. Staff members wait to be called in by their line managers rather than approach them with a problem. They also see little need to take the initiative in calling for change in the expectation that change can only be mandated from above.
- The prevalence of 'single fighters', especially amongst those more competent staff members who others rely on to deliver the bulk of the organization's work. These individuals tend to be more competitive, and to perceive others as incapable.

Individuals see informal networks as a positive means of self-organization in the interests of the organization, and where they might otherwise feel unsupported, lost and lonely (a survival strategy). But those outside these networks perceive them as factions, the workings of aspirant 'big men' seeking to construct their own alliances. This is a major barrier to trust and communication across the organization.

Underlying Assumptions

The interviews highlighted a commonly held assumption that the GA succeeds or fails depending on the individual in charge of the organization. This comes from an understanding that the biggest changes in organizational culture over the past 15 years have come with changes in leadership.

So attention is now focused on *who* to secure as a replacement Head after the

118   *Handbook on complexity and public policy*

current Head retires in 2013 (and who might be in a stronger position to determine that outcome); rather than on the *systems* that the GA needs to put in place to improve staffing and reduce political interference, irrespective of who the Head might be.

It would appear that the group does not yet fully believe in the 'organizational self' in their collective capacity to initiate change without a mandate or firm guidance from above. Their understanding of self-organization and taking the initiative is still more about survival than about change.

This is very likely a product of the 'authoritarian mindset' that became engrained in Indonesian society under the Soeharto regime – and this may explain a tendency among some in the group to view that period of the GA's history as a 'golden age'.

MARCH 2013 WORKSHOP IN JAKARTA

The researchers requested that a complete cross-section of the organization be chosen to participate in the 8-day workshop, which included leadership exercises in addition to the complexity sessions. A group of 30 was chosen from across the hierarchy from the most junior to the Head of the organization. They had never been in the same room together before. The Indonesian culture is very hierarchical and as a rule senior officials do not normally talk to junior ones. Yet they found themselves in the same room doing the same exercise.

The complexity sessions started on a Wednesday, with an introduction to complexity theory, as less than half were present at the London workshop. Over the following two days the team was taken through a process of identifying the multiple dimensions in their problem space. They started in small groups, discussing their perception of the problem/ challenge facing them and then worked individually noting what they, as individuals, had identified as the different dimensions of the problem space. There were flip charts around the walls with headings of the different dimensions (social, cultural, physical, political, financial, technical, and so on). The team then went around the room, as individuals, without further discussion, writing on the flip charts.

This is a counter-intuitive exercise in a complexity workshop as the participants are being asked to separate a complex problem into individual categories. The participants experience a great tension in isolating and categorizing the complex problem into simple categories. This is intentional and is designed to help participants experience that difficulty. When the individual work is done, the facilitators ask them to find connections between the different dimensions. This process becomes very active and dynamic and needs to be 'orchestrated' appropriately. The idea is to identify the key clusters of connected dimensions across issues. In academic language we are beginning to identify *interaction, connectivity* and *interdependence* between the dimensions, on the way to understanding some of the *co-evolutionary* dynamics which underlie the problem space.

The approach is based on that understanding; it is not enough to list the dimensions and the issues within those dimensions, the essential requirement is to understand how the multiple dimensions of each key issue interact and change each other, that is how they *co-evolve (that is, reciprocal influence which changes the behaviour of the interacting entities)*. In addition it is essential to see how the different issues also co-evolve and change the organizational landscape.

Effective policy making: addressing intractable problems 119

By the end of Friday, there was a great deal of tension and discomfort in the room; issues that were never talked about were being articulated and listed in full view. One of these issues was ethnic and religious groupings, or 'gangs' as they were called. It became clear that it was the junior members of staff who had identified that issue. It was not a moral criticism, it was because they felt excluded from the 'gangs', in the sense that when a position became vacant or a job needed doing, the senior manager would award it to members of his ethnic or religious group irrespective of whether that individual had the right qualifications for the job. This of course was also linked to corruption and nepotism, which were the other two sensitive issues.

The discussion was not easy and one had to understand the local culture to avoid offending the participants. We could not apply Western values to an Indonesian culture. The local culture expects that the senior male family member will support his own family and give preference to a family member if a position becomes vacant and he is able to influence such an appointment. This had to be acknowledged and honoured. However, a distinction was made between the honourable and accepted support of family with extending the same help to an acquaintance, especially when money changes hands. It was also important to focus on the underlying issue and to use the language and concepts that allowed the team to talk about a very difficult topic. The term 'transparency' was found acceptable and captured the related issues of corruption, nepotism and ethnic groupings. It also provided a way to agree on practical actions the following week when the team was focusing on the enabling environment.

As mentioned above, the tension in the room was tangible and that evening the two facilitators, who also acted as translators, were inundated with telephone calls from different groups within the team who felt very uncomfortable with the process. To defuse the tension it was decided to give up our Sunday break and to meet with the different groups individually. These meetings and the fact of being listened to, plus the break between Friday and Monday, and the thought that on Monday we were going to be much more positive and develop the conditions for an enabling environment, all helped to change the atmosphere significantly. The other aspect was social. We were in a hotel in the middle of nowhere, two and a half hours' drive from Jakarta, and were constantly in each other's company; we also had all our meals together. We therefore had to talk to each other and get to know one another. This informal social interaction helped enormously with the overall process. Researchers wishing to use this approach should, however, be warned that they need to be experienced enough to handle the conflict and tension which will inevitably arise from the process. If these tensions are not addressed appropriately and effectively, a great deal of damage could be done to the organization, as deep seated issues have been brought to the surface which need resolution.

ADDRESSING A COMPLEX PROBLEM

A complex problem can be addressed effectively only by identifying the critical co-evolving clusters of issues within the multiple dimensions. It is then possible to set up an *enabling environment* that addresses all the dimensions at the same time and at multiple scales (individual, group, organizational, societal). Identifying and addressing one issue in isolation from all its related issues is based on a mistaken assumption. No issue within

120   *Handbook on complexity and public policy*

a social ecosystem can exist in isolation; it is related to and is interdependent with other issues. Furthermore, assuming that a single solution will effectively address a complex problem over time is also a mistaken assumption. The problem and its environment are co-evolving and changing each other in the process. At the same time the processes of emergence, self-organization and exploration of the space of possibilities are also in operation; they may be slowed down through particular interventions, but cannot be stopped altogether.

A solution that may have been effective at time T1 is unlikely to still be effective over time, as the context changes. An enabling environment on the other hand assumes that the context will change and that policies will need to be adjusted accordingly.

PREPARATION FOR THE ENABLING ENVIRONMENT

By Monday morning everyone was highly enthusiastic and ready to start on the positive part of the process. We spent two more days on the Enabling Environment, with a great deal of discussion in small groups and in plenary sessions, and some very important agreements were made. The findings from the interviews and from the workshop were brought together to identify the key clusters of issues that had to be addressed. These clusters are the related issues listed under the different dimensions. This identifies inter-connectivity, interdependence and potential co-evolution and demonstrates those relationships. Once rich clusters are identified the next step is to ask the question 'why'. It is easy to answer by explaining 'what' happened and 'how' it happened, but extremely difficult to think through 'why' it happened and what would have stopped it happening. This is when very good facilitation skills are essential. If the 'why' question is not answered satisfactorily then the deep underlying causalities will not be brought to the surface and understood and the process will remain at a superficial level and the conditions being set up for the enabling environment will be inadequate or even incorrect.

If the deep underlying and multiple causalities are brought to the surface, then they can be addressed. That does not mean that addressing them will be easy or straightforward, but at least the real challenge will have been identified. Part of that challenge will be setting up the appropriate conditions to address the inter-related causalities. Again, asking 'why' these conditions cannot be set up, will provide further insights and a deeper understanding of the problem space and is likely to lead to a more effective enabling environment.

To provide an indication of the practical application of the methodology and an understanding of how the enabling environment begins to take shape, the key theme clusters and related actions agreed during the exercise are summarized below. It should be noted, however, that this is just a first step that needs further discussion and refinement. It also needs to be applied at multiple scales, and this exercise took place at a later time.

A   Leadership

One key outcome was the understanding that not everything depended on the leader of the organization. The idea of distributed leadership was introduced and constantly reinforced during the workshop and was captured by the following quotation: 'the ability to

Effective policy making: addressing intractable problems 121

change and improve the GA lies in the hands of each of us, and does NOT depend on a single leader'.

B Building a Sense of Ownership and Responsibility Towards the GA as an Organization

B1 Increase understanding of the GA's roles and functions
As discussed above, the individuals within the organization were focusing on individual survival and placed little, if any, emphasis on organizational cohesion and identity. This affected their sense of ownership and responsibility. Another contributing factor was lack of understanding of roles and functions within the GA and how each contributed to, and was part of, the whole organization. One action agreed therefore was to raise awareness of the different roles and functions by organizing monthly seminars where individuals would present their work to their colleagues. Another action was to present the results of the March workshop to others in the organization.

B2 Develop personal integrity as the basis for organizational integrity
A key aspect, however, was to develop their sense of personal and organizational integrity by being open to learn and to receive feedback; by having faith in themselves and the courage to speak the truth; by respecting each other and by looking to the future rather than to the past.

B3 Increase transparency
In practical terms, it was agreed to increase transparency within the GA in three key areas: (1) increased budget transparency by having open discussions of the budget between Divisions and Sub-Divisions, to enable them to understand each other's priorities and needs. This was fundamental as lack of transparency gave rise to suspicion. (2) Transparency over recruitment decisions, to overcome the fear of nepotism. One important aspect was agreement to have presentations by new staff on the skills that they were bringing to the GA to ensure that they were suitably qualified. Another related decision was to develop the capacity of existing (permanent) staff rather than to continue to recruit new people on short-term contracts. (3) Transparency over procurement decisions, especially office facilities, as the discrepancy between the allocation of these facilities was causing a great deal of tension. It was agreed to actively communicate about the allocation of these facilities, as well as to introduce procedures to formally request such facilities; also to hold meetings to review the procurement chain for goods and services, and together to seek solutions to the problems being faced.

B4 Multi-disciplinary pilot activities that bridge different Divisions
Lack of awareness and knowledge about the work of different Divisions was the underlying issue and it was agreed to hold a Discussion Forum to share knowledge about Sustainable Development, and to discuss problems related to Monitoring and Evaluation, and other related topics. One Division worked with local communities and it was agreed to initiate a pilot study to develop tourism potential in a local community; to have joint studies to assess the root causes of environmental degradation; and to support three specific villages undertaking local initiatives. The key here was the establishment

122 *Handbook on complexity and public policy*

of a learning environment to learn and support each other's activities that fitted into 'C', the next major heading below.

C   Develop a Learning Environment to Facilitate Co-evolution

C1   Increase capacity to communicate

Developing a learning environment would help the GA change, by individuals learning from each other and thus accelerating the rate of co-evolution (McKelvey, 2002; Volberda and Lewin, 2003; Benbya and McKelvey, 2006). This is dependent on good communications, and one of the key agreements was to 'increase the capacity to communicate' by improving personal communication skills and actively interacting with other colleagues. In practical terms they agreed to set up a Cross-GA Discussion Forum and to start with a discussion on Vision 2100, the Blueprints for a Sustainable Economy and Land Use, and the Provincial Spatial Plan. If these discussions lead to an improved understanding of what is needed to achieve the vision and objectives and how individuals and teams can contribute, it will also help with instilling a sense of ownership and responsibility.

The GA culture is highly bureaucratic with the strongly held belief that formal procedures are the correct way to bring about organizational change. It was therefore agreed that an administrative order by the Head of the GA was needed, which will mandate a regular cycle of meetings between staff and Sub-Division Heads, Sub-Division Heads and Division Heads, and Division Heads and the Head of GA, to develop a culture of two-way communication. The intention was to improve the flow of information to more junior staff and to facilitate the hearing of concerns from below.

To balance the formal mandates, the team was encouraged to think about more informal methods of communication, such as monthly coffee mornings for managers to update staff on new developments, and developing social media (Facebook) to exchange ideas on organizational change. In addition they were encouraged to actually take time each day to talk to someone else in the organization to get to know what they are doing or to raise a question. In a Western context this kind of 'networking' may come easily, but in an Indonesian organizational culture it will take effort to do it. The team was also encouraged to make a short note for themselves articulating the benefits of such interactions to help them 'see' the point of these informal exchanges.

C2   Develop a culture of induction and mentoring

One of the key challenges that had developed over a number of years was lack of technical knowledge. When senior staff were appointed who were not suitably qualified, the more junior qualified staff carried the burden. At the same time new staff had no induction or mentoring and the knowledge that was available was not passed on. It was therefore important to break that cycle and to set up processes for both induction of new staff and mentoring of junior staff. Furthermore, it was agreed that younger staff would be encouraged to join in activities and meetings outside the GA and that all staff would be encouraged to come together routinely to share, learn and reflect.

The induction process would include presentations by new staff on the skills they were bringing to the GA; regular staff 'rolling' between Divisions and Sub-Divisions to learn each other's jobs, both of new and longer-serving staff; that new staff would be involved in all activities and would be encouraged to interact with their colleagues, both horizontally and vertically.

C3 The planning process should make appropriate use of staff
It was important to ensure that the planning process made proper use of staff, in accordance with their assigned roles. It was therefore agreed that individuals would actively take the initiative to join in planning meetings, to improve their understanding of the process; also that responsibility for the development of planning documents would be shared across staff, so that each has a role and a contribution to make, as well as an opportunity to learn; to regularly invite external experts to provide advice on planning documents; and finally to consult on the GA's work plans at Sub-Division level, to provide opportunities for all staff to offer input and feedback.

C4 Improve technical and management skills
One of the underlying problems was the lack of technical and management skills throughout the GA and it was agreed that it was necessary to improve those skills by providing technical training to improve planning skills, as well as training to improve management skills. The latter would be compulsory for all GA managers (Heads of Divisions and Sub-Divisions).

APPLYING THE PRINCIPLES OF COMPLEXITY

The exercise ended by going back and linking the agreements that will provide the basis for setting up the enabling environment, to complexity theory (Mitleton-Kelly, 2003a).

Develop connectivity and feedback in order to build a sense of ownership and a learning environment. The GA has developed a sense of individual survival and also finds it difficult to think systemically. The idea of inter-connectivity and interdependence within an organizational system is not well understood and during implementation of the workshop agreements, this is going to be one of the key points of emphasis. During the presentation on complexity theory the principles of connectivity (Kauffman, 1993) and feedback (Arthur, 1990; 1995; 2002) attracted a great deal of discussion as the concepts appeared difficult to understand. Connectivity and feedback are based on interactions and communication, both formal and informal, and both were emphasized in the actions agreed above, especially the informal activities, which were not well developed within the GA.

Enable self-organization. The reliance on authority and on the organization's leader as the source of all meaningful actions and initiatives is going to be difficult to overcome, yet there are signs that this is happening. The principle of self-organization (Kauffman, 1993) was discussed at length and individuals were encouraged to initiate actions and take responsibility. Again it will take time, for this to work through, but the willingness to try was evident. The team was encouraged to give informal actions some emphasis and to take time to meet with colleagues informally and to initiate those meetings.

Facilitate distributed leadership at all levels, not just formal leadership. This again was to overcome the embedded idea that only appointed individuals can be leaders and that only they can initiate actions. Actively encouraging distributed leadership (Bennett et al., 2003), will make a significant difference to the culture of the GA.

124   *Handbook on complexity and public policy*

Expect the emergent (on 'emergence' see Checkland, 1981; Kauffman, 1993; Maturana and Varela, 1992; Varela, 1995, Mitleton-Kelly, 2007) and the unexpected and treat it as an opportunity to innovate. Use it *to explore the space of possibilities*. This was not going to be easy within the GA culture. Both self-organization and exploration of the space of possibilities (Mitleton-Kelly, 2003a) were seen as needing prior authorization and permission. The emphasis on enabling informal actions was a first step in encouraging the staff to take initiatives.

Develop multiple micro-strategies. In an organizational culture that constrains self-organization and distributed leadership it will be difficult to try out new and different ways of doing things. Yet the team saw quite clearly that they cannot continue to repeat old methods. To develop a new and different organization they will have to do things differently. In addition, they were encouraged to develop multiple micro-strategies in the sense that different ideas could be explored in parallel in different contexts. For example local communities could try different approaches to addressing the same problem or achieving the same objective. The key, however, was to learn from the exploration and the different micro-strategies; some will not work, while others will succeed, and the GA will need to understand 'why' these micro-strategies succeeded or failed in that context. If this understanding then leads to improved approaches or in applying successful strategies in different contexts, then the organization would have created a learning environment.

Facilitate and accelerate co-evolution. The setting up of a learning environment to improve communication and learning from each other would be fundamental in accelerating change through an active co-evolutionary process. Co-evolution takes place whether we intend it or not, but the *rate of co-evolution* (McKelvey, 2002; Volberda and Lewin, 2003; Benbya and McKelvey, 2006) can be affected by deliberate action within the appropriate enabling environment.

Create new order, that is, develop a new GA. The ability of a complex system to create new order (for example a new structure, new relationships, or a new way of working) is one of the key distinguishing characteristics of complex systems (Prigogine and Stengers, 1985; Nicolis and Prigogine, 1989). Complicated machine-type systems are not able to create new order.

The GA has been pushed far-from-equilibrium by acquiring a new Head and a new Governor. Both these men are imposing new strategies that mean the GA cannot continue to operate as it did in the past and has to change. The workshop and follow-up support will help it to explore new options and find a new way of operating or a 'new order'. Complexity theory has provided the GA with an explanatory framework to help it understand itself as a complex social system with particular characteristics. By using the ten principles of complexity they will be able to work in a different way to overcome past weaknesses and develop new strengths to enable them to cope within an increasingly difficult and hostile environment.

SOME INSIGHTS GAINED BY THE GA IN A COMPLEX OPERATING ENVIRONMENT

During the problem identification process, the workshop brought to the surface some very sensitive issues that were not normally discussed, such as 'gangs', but also

Effective policy making: addressing intractable problems 125

corruption and nepotism. Articulating these issues and discussing them in plenary with a team, which crossed hierarchical boundaries, was a big step forward for the GA. One of the key insights was that the organization was interdependent and that the ability to change and improve was in their hands and not entirely dependent on a single leader. The concept of co-evolution was another key insight: that no individual or team is powerless, and that they do not exist in isolation; they are part of a bigger social eco-system that is constantly changing and co-evolving internally as well as with its broader external environment.

Another insight was that the behaviour of human complex systems cannot be predicted or controlled (although it can be enabled and facilitated through an enabling environment), and that it is emergent (more than the sum of the parts). Other insights were associated with other complexity principles such as self-organization and exploration of the space of possibilities, and these ideas can be applied to some of the initiatives they are trying to encourage with local communities.

It was the combination of an understanding of complexity principles in practice and the workshop process of identifying the problem space and the enabling environment, which provided the deep insights. The theory alone would not have been enough, and the workshop process without the theoretical underpinning would also not have worked as well. The theory explained the process and underpinned the new insights.

THE OPPORTUNITIES AND CHALLENGES THE GA TEAM FACES

During the enabling environment (EE) process, these extremely sensitive issues were addressed in terms of 'increasing transparency' inside the GA and by developing personal integrity as the basis for organizational integrity. The issues also involved a better understanding of the GA's roles and functions. The transparency issue included budget transparency, recruitment decisions/processes, and procurement decisions.

The EE process identified some key areas that need to be addressed and these are the opportunities and challenges facing the GA in future. If they cannot address them effectively, they will not be able to go forward as a new and different organization. One of the key insights was the 'creation of new order', which in the GA context meant a new type of organization with different procedures and a different culture. It will be necessary during implementation to ensure that this new organization emerges, and this will include difficult issues such as taking practical steps to increase transparency in all three areas (budget, recruitment, procurement) and to encourage distributed leadership. Most of all the GA will need to learn how to live with the unexpected emergent, how to self-organize and how to explore the space of possibilities without seeking prior approval from senior management.

One of the main outcomes from the workshop was a willingness to *co-create a learning environment* and this will be key in setting up the enabling environment to address all the key issues identified when exploring the multi-dimensional problem space.

The learning environment included a focus on improved communication; on better preparation of new staff such as induction and mentoring; on actively participating in the planning process; and on improving technical and management skills. All of these

126 *Handbook on complexity and public policy*

topics have a strong practical component as well as a cultural element; the two influencing each other and changing in the process (co-evolving). The implementation stage will need to actively set up the practical aspects and to nurture the culture as it is emerging.

CONCLUSIONS

The chapter described an approach that is part of the EMK Methodology, which identifies the multiple dimensions of the problem space and the critical clusters of issues that need to be addressed. This enables policy makers to set up an enabling environment that effectively addresses a complex problem.

Ten principles of complexity which underpin the methodology were outlined and the approach was described using a recent case study with an Indonesian Government Agency to address some very sensitive issues including corruption, nepotism, ethnic and religious groupings.

What this chapter demonstrates is that complex problems that appear intractable may often be the result of inadequate or inappropriate approaches. Complex problems can be addressed effectively by understanding the problem space; this understanding is based on identifying the multiple dimensions and the co-evolving dynamics of critical clusters. A list of dimensions and issues is simply not enough. Such a list may also be counterproductive as it often leads to inaction through the enormity of what needs to be addressed. Identifying the co-evolving clusters on the other hand provides an excellent starting point, which captures the complexity of the problem space and highlights the different complexity principles such as connectivity, interdependence, feedback, emergence, self-organization, exploration of the space of possibilities and co-evolution.

By addressing all the inter-related dimensions in the key clusters, at multiple scales, it is possible to set up an enabling environment that is sustainable and is able to continue to be effective as the environment changes.

Much detail is provided in the above example, not because other policy makers can use those details, but to illustrate the depth and effectiveness of the process, but also the practical nature of the approach. It is the methodology as a whole that needs to be understood and adapted to different contexts, with the strong proviso that it should only be used by experienced researchers and facilitators able to handle the conflicts and tensions that will inevitably arise. Another proviso is that the policy makers need to be involved throughout the process, as they will be responsible for implementing the enabling environment and need to understand *why* the different decisions were taken.

Policy makers themselves will benefit by understanding the characteristics of their organizations as complex social systems and by being aware of the powerful dynamics that can both aid them and seriously constrain their efforts, if such understanding is missing.

An understanding of complexity theory can therefore (a) provide an explanatory framework to help policy makers work with the characteristics of their organization as a complex social system; and (b) provide a methodology that is able to address complex problems effectively.

The approach has been used with complex problems within single organizations (Mitleton-Kelly, 2003b) as well as multiple organizations (Mitleton-Kelly, 2011); and at

Effective policy making: addressing intractable problems 127

a global scale with several nations. It has been used to address lack of integration following a merger or acquisition; lack of alignment between the IT department and the rest of the business; project governance for a national project delivered locally by multiple partners for the Royal British Legion; to develop a framework of governance for government with six Government Administrations; leadership in the NHS (UK National Health Service), DEFRA (Department for the Environment, Food and Rural Affairs), and many other organizations in the private and public sectors; sustainable development in communities; to develop learning organizations; the 'design' of organizations and the emergence of new organizational forms; co-evolutionary sustainability; innovation in the private and public sector; disaster risk reduction in West African States; the relationship between policy and outcomes; uncertainty and risk in decision making for Local Government; contingency planning related to evacuation following a major disaster. It is currently being used to identify the multi-dimensional problem space of pandemics with the World Economic Forum, Global Agenda Councils on Complex Systems and Catastrophic Risks.

Appropriate parts of the methodology have been used to address problems within family foundations, community trusts, SMEs, European Commission agencies; global companies, government departments/agencies and government administrations; and multiples agencies addressing global issues.

REFERENCES

Arthur, W.B. (1990), 'Positive Feedbacks in the Economy', *Scientific American*, 262, February, 92–9.
Arthur, W.B. (1995), *Increasing Returns and Path Dependence in the Economy*, Ann Arbor, MI: The University of Michigan Press.
Arthur, W.B. (2002), 'Is the Information Revolution Over? If History is a Guide, it is not', *Business 2.0*, 3(3), March, 65–72.
Benbya, H. and B. McKelvey (2006), 'Using Co-evolutionary and Complexity Theories to Improve IS alignment: A Multi-level Approach', *Journal of Information Technology*, 21, 284–98.
Bennett, N., C. Wise, P.A. Woods and J. Harvey (2003), *Distributed Leadership: A Review of Literature*, National College for School Leadership.
Checkland, P. (1981), *Systems Thinking, Systems Practice*, Chichester: Wiley.
Kauffman, S. (1993), *The Origins of Order: Self-Organisation and Selection in Evolution*, Oxford: Oxford University Press.
Maturana, H.R. and F. Varela (1992), *The Tree of Knowledge: The Biological Roots of Human Understanding*, Boston, MA: Shambhala Publications.
McKelvey, B. (2002) 'Managing Co-evolutionary Dynamics', paper presented at the 18th EGOS Conference, Barcelona, Spain, 4–6 July.
Mitleton-Kelly, E. (2003a), 'Ten Principles of Complexity and Enabling Infrastructures', in E. Mitleton-Kelly (ed.), *Complex Systems and Evolutionary Perspectives of Organisations: The Application of Complexity Theory to Organisations*, Bingley: Emerald Group Publishing Limited, pp. 23–51.
Mitleton-Kelly, E. (2003b), 'Complexity Research: Approaches and Methods: The LSE Complexity Group Integrated Methodology', in A. Keskinen, M. Aaltonen and E. Mitleton-Kelly (eds) and 'Foreword' by S. Kauffman, *Organisational Complexity*, Helsinki: Finland Futures Research Centre Publication Series, 6, TUTU Publications, 56–78.
Mitleton-Kelly, E. (2005), 'Co-evolutionary Integration: The Co-creation of a New Organisational Form Following a Merger or Acquisition', *E:CO (Emergence: Complexity & Organization)*, 8(2), 36–47.
Mitleton-Kelly, E. (2007), 'The Emergence of Final Cause', in M. Aaltonen (ed.), *The Third Lens. Multiontology Sense-making and Strategic Decision-making*, Aldershot: Ashgate Publishing, pp.111–25.
Mitleton-Kelly, E. (2011), 'Identifying the Multi-Dimensional Problem-space and Co-creating an Enabling Environment', Chapter 2 in A. Tait and K.A. Richardson (eds), *Moving Forward with Complexity*, Litchfield Park, AZ: Emergent Publications.

128 *Handbook on complexity and public policy*

Nicolis, G. and I. Prigogine (1989), *Exploring Complexity*, New York: W.H. Freeman.
Prigogine, I. and I. Stengers (1985), *Order Out of Chaos*, London: Flamingo.
Varela, F. (1995), paper given at the 'Complexity and Strategy Conference', London School of Economics, London, May.
Volberda, H.W. and A.Y. Lewin (2003), 'Co-evolutionary Dynamics Within and Between Firms: From Evolution to Co-evolution', *Journal of Management Studies*, **40**(8), 2111–36.

PART II

METHODS AND MODELLING FOR POLICY RESEARCH AND ACTION

9. Complexity theory and political science: do new theories require new methods?
Stuart Astill and Paul Cairney

INTRODUCTION

A key argument in the complexity theory literature is that it represents a new scientific paradigm providing new ways to understand, and study, the natural and social worlds (Mitchell, 2009: x; Mitleton-Kelly, 2003: 26; Sanderson, 2006: 117). Broadly speaking, its opponent is 'reductionism', or the attempt to break down an object of study into its component parts. The broad insight from complexity theory is that reductionism is doomed to failure because complex systems are greater than the sum of their parts. Elements interact with each other to produce outcomes that are not solely attributable to individual parts of a system.

In political science and policy studies this argument is used, in a similarly broad way, to challenge particular brands of 'positivism' associated with the attempt to generate 'general laws' about the social world. The generation of laws is problematic because complex systems are associated with often volatile arrangements and unpredictable outcomes even if long-term, regular patterns of behaviour can be identified in many areas. Policy issues may be subject to significant levels of uncertainty and ambiguity, be impervious to control by policymakers and, in some cases, appear to be intractable or 'wicked' (Geyer and Rihani, 2010: 5; 74–5; Room, 2011: 6–7; Klijn, 2008: 314). Further, they may be better studied and explained if we develop methods better equipped to analyse complexity rather than simply seeking analytical simplicity. This point is expressed most strongly by Lewis and Steinmo (2008), who argue that the identification of a new ontological perspective (this is how the world is) has a knock-on effect for epistemological (this is how we should gather knowledge of it) and methodological (this is how we should conduct the research) perspectives.

Yet this is easier said than done. We know more about how we should *not* study complexity than how to do it in a meaningful way; to recognize complexity but to achieve enough analytical clarity to make research manageable. For example, does the study of complexity require new methods derived from other disciplines, or do we simply incorporate an understanding of complexity in well-established social science methods? Is there potential to combine methods from a range of natural and social science disciplines? The potential benefit of new methods may be considerable but also problematic. In particular, if we adopt a new way of thinking and new methods, and reject other approaches, it may be difficult to compare the information from new and old approaches. It may be difficult to accumulate knowledge in policy studies if we reject the insights of the past (Cairney, 2012; 2013). Another approach is to be pragmatic and focus on complexity as a new source for modelling behaviour; to pursue the latest best hypothesis as a working model of complex system dynamics and strive to improve further. The choice may be

132   *Handbook on complexity and public policy*

between an attempt to completely reinterpret the world and its past, or to provide a new methodological perspective to compare with a range of others.

In this context, the aim of the chapter is to outline a wide range of methods that are potentially useful in the study of complex systems. First, it outlines established methods in political science – broadly described as quantitative, qualitative and deductive – and explores our ability to combine their approaches and insights to understand complex systems. Second, it identifies the advantages of methods and approaches that could be used more in political science, including mathematical approaches (such as pattern-based thinking), modelling and simulation, network analysis and experimental research designs.

ESTABLISHED METHODS AND COMPLEXITY: ELEMENTARY STATISTICS AND QUANTITATIVE METHODS

The use of descriptive statistics is a first simple step. Identifying mean and median values, and breaking units of observation into two different groups to show that they have different characteristics, are simple and effective ways of exploring data and tackling puzzles in political science. However, the potential drawback is that simple measures do not help us to fully understand the social world, particularly when it does not produce simple and predictable patterns. The nature of complex systems means that they display non-linear features, such as when the same input produces either a dampened or amplified (and hence seemingly 'exaggerated') output, as negative or positive feedback (a good example is when people ignore or pay disproportionate attention to the same issue – Jones and Baumgartner, 2005). So, a small change in the explanatory variable (such as a sudden shift in attention to an issue) can cause an enormous change in the thing that we are trying to understand (the dependant variable, such as a change in policy or policy networks).

In other examples, we may be dealing with events that produce a high impact but have a very low probability of occurring. So, looking simply at the distributions of how often these events happen over time will be limited. In a system with complex feedback mechanisms, the distributions of events, average measures, correlation analyses and basic cause and effect assumptions become potentially misleading. In such cases, we need different approaches to describe and explain behaviour.

In some cases, simple analysis may produce problematic policy advice when, for example, we consider how to manage risk effectively or allocate funds. Consider the extreme example of counter terrorism analysis – what if interventions and policies to prevent many low-level incidents increase the probability of catastrophic large-scale incidents (as discussed by Ryan, 2011)? In other cases, the routine use of 'cost–benefit analyses' may be highly misleading if based on typical averages or linear assumptions rather than more sophisticated techniques such as simulations that take non-linearity and complexity into account. This is a feature of, for example, climate change issues with uncertain effects – flood defences, investment in green technology or biodiversity payback – and areas in which it is unwise to assume that human behaviour will be predictable and based on obeying simple rules. Examples include public transport versus road policies (where feedback loops of better roads can increase congestion), public health

Complexity theory and political science 133

investment against traditional medical spending, economic interventions such as interest rate setting to control inflation (where expectations of the effectiveness of the measures in some part determine the effectiveness of the measure) and changes to social security benefits (the uptake of a benefit change may rise or fall in significant disproportion to its changing value). In other areas it may be wise to anticipate the risk associated with small-probability–high-cost problems. A classic example is the £0.5bn cost of the failure of the contractor Metronet, responsible for London Underground maintenance (NAO, 2009). We cover further issues around cost–benefit analysis in the section on modelling.

Linearity and Complexity

Simple quantitative methods reveal their fragility when we look at the basic theory behind the most commonly used approaches. To understand this problem, we have to understand what is meant by linear, non-linear and complex or chaotic – because basic quantitative methods depend greatly on the assumption of linearity and, frequently, the assumption of 'normality' (Osborne and Waters, 2002).

The easiest way to grasp understanding of linearity is when it refers to two sets of observations; one set of observations is, for example, something that we are able to change in a system (or something that we know changes in a system) and that we can measure. The other set of observations is of something that changes as a consequence of the known changes in the first set of observations. Simply illustrated, we could increase the temperature and see how high the column of mercury climbs in a thermometer. The first observation contains the known changes (of the 'independent variable') and we hypothesize that the second observations will change as a result (the 'dependent variable').

Now that we have two sets of observations we can understand linearity, empirically, in this environment: linearity exists if we put the two sets of observations on a graph and the 'best fit' relationship between them is a straight line (using 'ordinary least squares', where the sum of the squares of all the distance between observed points and the best-fit line is minimized). This means that the change in the dependent variable is always the same amount for any given amount of change in the independent variable, and continues to be so, no matter where we start. So, we should find that in a thermometer the mercury climbs 2 mm for every 1 degree of increase in temperature. This is the same whether we go up from −2 degrees to +8 degrees or whether we go from +28 degrees to +38. By extension, multiples of this relationship hold absolutely; the mercury climbs 4 mm for every 2 degrees increase and 1 mm for every 0.5 degrees.

We can, with some modifications, treat 'transformed linear relationships' in the same way. These are the relationships that are not linear but not 'complex' in the sense we are concerned about. For example, we see that a country's Gross Domestic Product (GDP, a measure of a country's output, or goods and services produced, per year) per capita varies in a consistent relationship with life expectancy at birth (see Figure 9.1).

This relationship is not linear, but with a simple mathematical equation it can be made so. Or, at least, it can be treated in statistics and modelling in almost the same way as a true linear relationship. Other consistently predictable relationships, with different shaped curves, can often be 'transformed' even if they have a 'squared' or greater level of complication; each part of the function that defines the shape of the curve can be handled as if there are a small number of key variables that affect the outcome. Statistical

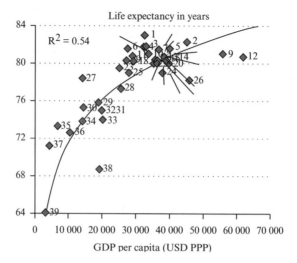

Source: OECD (2011).

Figure 9.1 Life expectancy at birth as a transformed linear relationship

methods that link changes in several independent variables with one dependent variable outcome – multivariate regression – are based on the same theory of linear relationships. Consequently, some research problems may be even 'bigger' (more variables, more dimensions) but not inherently more 'complex', since we can produce representations of linear relationships and hence call upon our host of familiar methods.

The Underlying Philosophy of Linearity and Complexity

What happens when, instead of taking these relationships – and our ability to represent them using statistical programmes and visual aids – for granted, we consider the underlying philosophy? The basic idea is that we take a number of observations to represent the totality of the observations. This sampling process is founded upon mathematical assumptions about the probability that the selected observations will fit certain statistical parameters. The 'mean average' (the total of observations summed and then divided by the number of observations) depends on the assumption that the data fits the 'normal' ('Gaussian') distribution, often known as the bell curve; or, at least, that the means of repeated sets of observations fit this pattern. More specifically, the assumption should be that any errors from sampling or measurement are normally distributed – just as many errors are positive as negative, and big errors are less likely than small in the classic 'Gaussian' pattern. If they suspect that this is not true, researchers can at least be comforted that taking the median average (or using quartiles, quintiles, deciles, and so on) is taking a 'non-parametric' approach (using the order of values, not the values themselves) and is much more robust when using non-linear data in complex systems.

Usually researchers will not worry about this; they will base their readiness to use a given method on having seen it applied, successfully, in many other situations. If they are

Complexity theory and political science   135

particularly worried, they may run some basic tests on the data to assess its suitability for being subjected to some given sampling and analysis techniques. Often these tests can be as simple as plotting the data against a time scale to see if there is anything irregular or 'non-Gaussian'; making an x–y plot of bivariate data or a frequency plot (histogram) to see if the data put into bins is regularly or irregularly distributed. If one has done a linear regression and the differences between the real data and the fitted line predictions ('the errors') are plotted, one can test to make sure that the correct number of errors (roughly) fall within one, two and three standard deviations of the mean. Interesting research questions may arise when the tests are not 'normal'.

When we study complexity, or complex systems, it is much more common that the conditions under which standard statistical methods are valid will not exist – and it will not always be apparent from standard tests that this is the case. Some data depend on human feedback and decision making over time, producing something that, within limits, looks like a classic time series that can be regressed. However, once a particular point is passed, the relationships break down. If we don't understand the problem fully, we may take this regular looking series and generalize the findings – producing seriously false conclusions.

Standard tests are not designed to identify these rare-but-important patterns or risks. For example, the problem with very simple statistics such as 'an average' is twofold. First, 'improbable events' of a very high magnitude, such as earthquakes higher than 5 on the Richter scale, do not conform to the 'normal distribution' of events – or the related 'binomial distribution' which describes the probability of an event's magnitude falling between two numbers. Second, estimates of future events 'formed by observations collected over short time periods provide an incorrect picture of large-scale fluctuations' (Herbert, 2006). Our observations may be best seen as data regarding a relatively short snapshot in time, subject to major change. This problem requires us to match quantitative observation with qualitative and theoretical work that helps us define terms, understand systems, consider the context and our assumptions about why certain relationships hold at a certain time, and reflect on the limitations of our studies and the conclusions we can draw.

The most common illustrative problem relates to the impossibility of relying on correlation in a non-linear system. The very essence of the 'butterfly effect' of complex systems and non-linearity is that, as we vary some independent variable(s) consistently, we cannot hope to find a recognizable, or even necessarily repeatable, pattern emerging in the dependent variable that we can express in any reductive way. In many cases, independent variables interact with each other, or cause changes in dependent variables, which then feed back as input effects on the independent. This results in a process where they either dampen or reinforce each other, amplifying or sending to undetectable levels the observable results. It produces what is effectively an unpredictable situation: variation, so small as to be unmeasurable in the real world, creating large differences in outcomes, and undermining the ability of equations to represent this effect by the usual means. The outcome is a degree of variation or change that is difficult or impossible to plot as a linear or curvilinear relationship. This is a particular definition of 'chaos': a deterministic system (we know its rules), having no randomness, that is still unfamiliar, and its next state is unpredictable.

This set of warning signals should be heeded but not be taken as problematic for the political scientist wishing to make progress in understanding the world. Instead, we see this as a great opportunity for thoughtful research that prioritizes a need to understand

136   *Handbook on complexity and public policy*

the system and the puzzle before embarking on choosing a method, collecting evidence and diving into analysis. The thoughtful use of the multi-methods and 'solving the puzzle' approach that we advocate will add something to classic 'linear' situations (if they can be identified) as well as increasing the chances of sensibly handling complexity when it is present.

Qualitative Methods and Case Study Analysis

Patterns of complex systems may defy prediction, but qualitative approaches help explain how and why a system produces such results. Qualitative methods are too frequently dismissed by those who promote the so-called 'hard' quantitative methods. Yet to generate reproducible analysis, insight and conclusions, we must be able to describe systems, generate hypotheses and understand the huge range of potential variables. Assumptions, which always exist, must be explicable and explored. When complex systems involve humans it is particularly vital that the skills of qualitative analysis are understood and effectively used.

There are two main approaches worthy of particular mention. First, we may use interviews to examine how policymakers and other actors understand the policymaking world in which they operate. A significant part of the literature refers in some way to 'complexity thinking', or the need to change one's view of the world to better operate within it (see, for example, the chapter by Price et al. in this volume, Sanderson, 2009 on 'intelligent policymaking', and Cairney, 2015). In-depth interviews help us compare recommendations on complexity thinking to the types of understandings and methods adopted by policy participants. The results may be used, for example, to explain how and why governments seek simple solutions to complex problems.

Second, we may use interviews alongside documentary analysis to build up a detailed picture of a case study – sacrificing (perhaps temporarily) breadth for depth. In such research, we seek to explain the outcomes of one case with reference to a wide range of explanatory variables and detailed analysis of how (rule-bound or rule-influenced) people interact with each other. We may draw upon complexity themes to examine, for example, how people interpret the rules in which they operate and what effect a change in those rules would have on the operation of the policymaking system. We may also explore broad concepts such as 'emergence' in the absence of central control, which may have a particular meaning in policymaking systems with central governments. Qualitative analysis may be necessary to turn a set of very broad concepts, derived largely from studies of the natural and physical world, into something more meaningful when applied to social interaction.

The full richness of qualitative, semi-structured insight can be highly successful in untangling puzzles about political systems that involve complexity. However, even the 'thickest' qualitative case study extracts the pertinent elements and leaves others behind. Case study research is a 'modelling' approach whereby a simplified version of the world is being used to draw general conclusions. We sift carefully and select a small number of elements on which to focus in our simplified model of the world. Therefore, as with any method, we must be extremely aware of human cognitive biases, such as the bias that leads human minds to see patterns and, hence compelling solutions where none exist.

When we look at deductive methods below we will see the crucial part that qualitative

Complexity theory and political science 137

and case study methods play in helping us to understand complex systems. Qualitative methods work well alongside new toolbox quantitative methods that focus on exploring, modelling, 'playing' with, and describing systems through simulation, networks, taxonomy and structuring – bringing qualitative and quantitative into a single methodological space.

Modelling and Deduction

The aim of modelling is to build a set of assumptions that define a model and then manipulate this restricted world logically to arrive at conclusions that would have not been accessible in any other way (see below). This approach is used widely in economics and political science (rational choice theory, see Hindmoor, 2006). Simple models have been used to explore the outcomes of collective action based on the actions of individuals seeking to fulfil their preferences and following simple rules. Classic examples of thought experiments include the 'paradox of non-voting', in which we consider why someone would vote if they knew they could not influence the outcome of an election, or take part in pressure group activity if they could 'free ride' and enjoy the benefits without engaging. Or, more applied institutional rational choice considers what rules could be devised to encourage particular kinds of cooperative behaviour, or what level of state intervention is appropriate (Ostrom et al., 2014).

These models are built on assumptions about how individuals think and act, including those that make unrealistic assumptions regarding their ability to consider information and rank their preferences, and more applied behavioural/ psychological methods, often using experimental approaches, which introduce more realistic or more detailed assumptions about human behaviour (Kahneman, 2012). The latter are used to reveal the 'biased' nature of the human mind, helping us to consider which parts of a complex system can be modelled and which have to be 'off model' adjusted or even interpreted through the lens of qualitative knowledge.

The Challenge of Complexity: How can we Combine the Insights of Multiple Methods?

Complexity research may not be about definitive explanation or maximizing 'explanatory power'. We are trying to understand how the system operates; to identify the *sufficient* set of key features to describe the system and its interaction with the world in a more complete way. In the case of both quantitative and qualitative methods, we run the risk of oversimplifying the issue, by focusing on a system's essential features, and identifying trends or levels of stability that do not exist, by generalizing from a snapshot in time (see Little's chapter in this volume on this point).

When we have looked at specific methods we have not wholly endorsed or dismissed them – it is largely a question of understanding the methodological assumptions of each approach so that when we apply them in the new world of complexity we do not fall into a trap of drawing false conclusions. We should understand the methods a little more deeply, so that we can understand how to question and examine the results that they throw up. The last step is to consider what we may or may not conclude about the puzzle, based on a retrospective assessment of our whole investigation.

As an illustrative example, consider a finding that was observed initially from a case

138   *Handbook on complexity and public policy*

study – say, comparing UK and French policymaking networks – but perhaps only verifiable through another method such as quantitative analysis. In a comparative study on policy formation (Astill, 2005), one of the interviewees talks about the policymaking implications of 'the network being much larger' in the French case. This person has not any in-depth knowledge of the UK case and only knows a little of what they have been told by the interviewer and what they know about the generality of the UK and French political systems. We can think about how this idea might typically be handled in a case study analysis and how it is handled using mixed methods. How can we understand the number of actors in a policy process to be much bigger or smaller than a comparator? At first, this idea would seem a fairly simple observation that could be drawn from a case study; several interviewees will mention that the policy process involved a larger number of actors than the comparator – so, it certainly *appears* that there are more actors being taken into account. However, the good political scientist is careful, and will start to construct in her head a set of criteria by which she could defend this inference if she wished to present it as evidence.

Reduction is always a powerful tool in these circumstances. Perhaps the political scientist would imagine a very simple decision-making process that involved two people, and another exactly the same that involved four. How would she establish to her satisfaction, from case study evidence, that the one was bigger than the other? It cannot be simply that more people are mentioned. It must be that more people are mentioned in the context of having a relationship with others that meets some criteria of having an impact on the emerging policy. This is the first lesson: creating the question through inquiry. We have arrived at what amounts to a formulation of the question, 'people involved in the policy process'. Although this is important and simple to understand, even the best political scientists working on case studies do not necessarily do this in a fully conscious way; it is done in a way more akin to how we subconsciously coordinate hand and eye to catch a ball.

In the complex world we realize very soon that, within the world of dynamics, interdependency and feedback, the interaction between a quantitative method, the assumptions and definitions we make about the question, and the flexibility of the quantitative methods, are crucial. What the quantitative analysis does for us, in this relatively easy to compare case, is to give us a way of holding exactly the same assumptions constant when examining the two policy processes, and thereby facilitating a convincing assessment of the number of actors involved. If we change the assumptions, the answer will change and, by working in a structured way with both the assumptions and the data, we can learn about the robustness of our conclusion and start to discuss the sensitivity of the answer. With the traditional case study approach, even if the assumptions were stated, the basis of aggregating the subjective evidence leaves us with a mostly subjective result in contrast to the aggregation in the quantitative approach which still uses the subjective views of the interviewees (about who they were connected to) yet leads us through rigorous analysis towards a more objective conclusion.

We can also use this illustrative case to make an important point about complexity and networks that is generalizable to many of our methods in political science. If any node in a network is only aware of the local conditions surrounding itself, then any inference that is made by the node about the overall network, from these local conditions, has no guarantee of being accurate. The case study process relies on accumulating local views,

Complexity theory and political science 139

which may or may not be accurate, while the quantitative method relies on using that information to produce a representation of the whole network and then examining that to discover its properties. The combination of methods is important, and may be supplemented by others.

COMPLEXITY AND THE FOUNDATIONS OF A NEW METHODS TOOLBOX

Thinking in Patterns, Moving beyond the Lines and Solving Puzzles

We advocate 'creative pattern-based thinking', combining models and a degree of formalization, to open up new perspectives for enquiry. This is a valuable mind-set for the complex world where curiosity and discovery may need to be promoted above proof and definitive conclusions. The aim is that a researcher will use logically-structured thinking about patterns, systems and structures to add to the understanding of a subject, rather than simply trying to force numbers onto problems. We should not pursue dry statistical methods that leave no room for expressive thinking in political analysis. The pattern-based and puzzle-solving aspect of mathematical thinking, as a distinct branch of thought, is characterized through its contrast with the usual 'scientific approach' that political science tends to adopt when using numbers. Our preferred approach is to construct insights by beginning with axioms and definitions which are then explored, reconfigured and logically manipulated – a process that has particular relevance and value when handling complexity in political science.

The key to the approach is the combination of insights from qualitative methods with the use of logic (that is, solving puzzles) in an abstracted world (a model) – the result of which is then brought back to the real world for assessment. It differs from a common approach in quantitative methods – reconciling a priori hypotheses with experimental or observational data. In political science the latter is particularly problematic because it is virtually impossible to replicate conditions to test our findings. This impossibility of replication is even more acute in the complex world due to the non-linear effects of tiny changes producing massively different outcomes. The alternative is to use logical and mathematical angles as our quantitative approach, allowing us to find the patterns in complex phenomena that are not amenable to ordinary methods.

We can see a classic complex scenario in Figure 9.2. The smaller charts show how outcomes change their nature considerably, as one small parameter changes in the real world or in a model. The bottom left graph shows that, over time, the outcome we are interested in stays entirely constant. The single curved line on the large chart represents the finding that, if we use a different, higher, parameter r, we will see the outcome increase – but it remains one single outcome and stays steady over time. The middle bottom chart shows how, with a tiny change to one part of the environment represented by a small increase to the r parameter, the thing we are interested in fluctuates between two observable states over time, very regularly. This effect is seen on the main chart: as we move along, increasing the r parameter, we see the observed levels of x go from just 1 to 2 at around an r value of 3.2. As the r parameter gets bigger, these two outcomes – that are cycled between – get further and further apart. Next we see, from the main chart,

140 *Handbook on complexity and public policy*

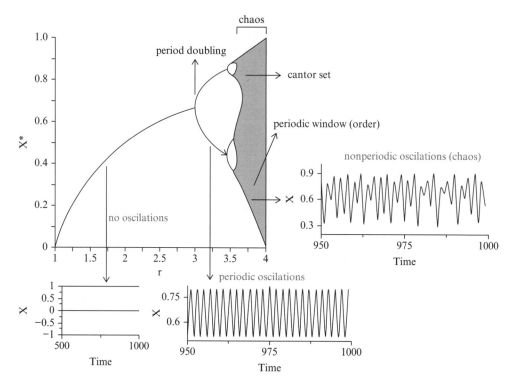

Source: Sardanyés (n.d.).

Figure 9.2 *'Chaos'*

that after a little movement more in *r*, four outcomes are being cycled around. Last of all, the parameter changes a little more and then suddenly, without any warning, the system moves into chaos – the outcome can be at any level in the next time period without our ability to predict when (or if) it will return back to a previously seen value. This is seen in the right-hand time series and the 'cantor set' that shows this situation on the main chart. We have found structure *under certain conditions*; we can discuss what those conditions are and how parts of the system relate to each other in predictable and unpredictable ways.

The reader could ask themselves if they had observed a reasonable sample of the outcome variable, over time, with *r* parameters reasonably varying between 3 to 3.6, what conclusion may have been drawn using traditional analysis? The answer is probably 'almost any'.

Figure 9.2 illustrates what is known as 'period doubling' (using this type of 'logistic map'; see Strogatz, 2001), a very common equation for modelling for competitive systems and many other human and natural situations. In political science it could be used to model political participation, with natural intergroup communication creating a growing group, feeding demand, but the *r* parameter representing a crowding effect as too many people restrict the efficacy of strategic group action.

Complexity theory and political science   141

What the mathematics of complexity, fractals (never-ending and ever more detailed patterns) and chaos tells us is that, if we know about complex phenomena, we can start to have a greater insight into when systems are stable or not and which variables are sensitive or not.

Another useful way of expressing such a situation is to define areas of stability or instability as illustrated in Figure 9.3, where for two parameters (in linked logistic maps) the researcher has shown where the outcome is stable and where it is unstable. For the two different parameters l1 and l2 we see multiple outcomes that are stable (period 1, 2, 4, 8, etc.), the light grey area from the top right corner indicates chaos, and white the quasiperiodicity (where there is periodicity – i.e. a tendency to recur at intervals – but irregular in its repetition). These images help us see patterns, in what looks like 'randomness', arising from even basic, simple, rules-based complex models where standard techniques fail us.

The Mandelbrot set (Figure 9.4) is another famous image that shows where outcomes of a phenomenon are stable or unstable. It represents, for two parameters in a simple equation (where the result is fed back in to the same equation repeatedly), where the process gives us an actual answer, represented by the black part, while the grey/white part shows no proper solution (the result, in fact, zooms off to infinity). The different shades only show how long it takes the calculation to reach a given value. Illustrating our focus on exploration, rather than description or prediction, it shows that in this simply-defined yet complex phenomenon, where it is impossible to predict where the next point will fall, the pattern in which the points actually do fall is certainly and very deeply patterned.

A New Toolbox

These patterns, expressed as pictures, demonstrate the limitations to presenting data in a linear way, as stable outcomes. They help warn against simply carrying out standard statistical tests. They help show why we should not apply methods without appreciating the data and the assumptions about how our methods analyse data. They emphasize the need to know more about our system from qualitative inquiry. They may also prompt us to add some more methodological 'tools', consisting of:

- visualization and presentation of data to see patterns and understand variability, irregularity and stability;
- understand data not as dry observations but as emergent from systems of agents interacting with each other and the rules they follow;
- describing systems and data and trying to understand sub-phenomena through taxonomy, distance, relations and allied concepts;
- learning about the systems that give rise to such data and thereby being able to abstract key points, and model using simple mathematics, networks analysis, or other simulation-based models;
- building models or simulations and then altering key parameters and assumptions to see when results either vary little or become unrecognizable.

The revisiting of conclusions in the light of the original assumptions is a key step in this analytical process.

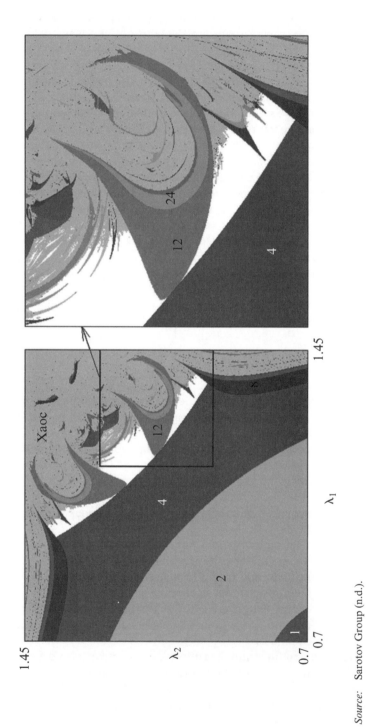

Source: Sarotov Group (n.d.).

Figure 9.3 Non-identical coupled logistic maps

Complexity theory and political science 143

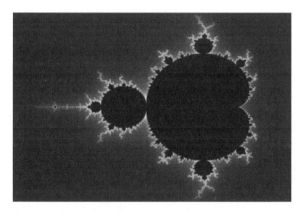

Source: Beyer (n.d.).

Figure 9.4 Mandelbrot set

Modelling and Simulation

Models are created to represent a simplified version of reality to answer certain questions. The question is sometimes obvious, such as in the case of an aeroplane wing tested in a wind tunnel. The question is also limited: an aeroplane wing constructed to test wind resistance may not be as useful if set on fire to test safety. It could be designed for both purposes, or we could accept its strength as a simple model for one purpose and recognize its limitations in others. One model cannot represent all aspects of reality, so being a good indicator of one aspect is a strength rather than a weakness. In the study of political systems, we also need to think about the purpose of the model, but its primary purpose (and its limitations) may not be as obvious.

In some disciplines (such as economics) the role of modelling is often far more complicated, including the pursuit of mathematical completeness, lots of equations and extensions. Yet to handle complexity (and other phenomena), a model can be both simple and valuable. Modelling is about the promotion of learning by identifying insights and seeing divergences between reality and the model to highlight mistakes and put into context the limitations of our models.

Even in the realm of complexity, we have to simplify our analysis to seek an explanation of outcomes that do not arise from large numbers of objects or from complicated rules. Rather, we have to attribute explanation to a reasonable number of definable (through taxonomy) sub-elements that relate to each other according to some easy-to-simplify rules. The interaction of these 'few enough to be defined' and 'simple enough to be described' elements produces, through feedback and interaction, outcomes in our complex system. In that context, modelling embeds an emphasis on understanding and simplifying agents and processes under complexity. The art of building a model is useful to study complex systems if it is done with an appropriate level of self-awareness: identifying, and being able to link in a systematic way, the elements of the system; starting from a very coarse verbal description onwards; refining description to give typologies and names to those elements, and then explaining how they work together.

144 *Handbook on complexity and public policy*

Take the simple example of a public service that is open freely to those that want to use it – such as a free training course at a library. If we take, as a basic assumption, that this is a drop-in service, then we can start to think of the elements of our system: there are individuals who may or may not use the service, there are those who run the service and there are other resources available. There will be trainers and a room with desks and, eventually, some limit on capacity. If we know the course runs once a week, we may want to know how big the room should be and how many trainers. We assume that, if there are too few resources, then the experience will be such that many of the attendees may not come back the next week as they will get no value out of it. The trainers and the room are the resources, the service users are those who are benefiting from these resources. The resources are limited. So, we have something that we know (the relationships between the different elements of the system), we have a taxonomy (resources and users) and some simple rules that relate to users' decisions based on resources and a time dimension.

This approach, and level of simplicity, can be found in a classic model that is very popular with, for example, biologists and ecologists – where a population of a rapidly reproducing organism has only a limited source of food (or other resource) to survive on. Such models often introduce another kind of individual that is the predator to the main population (political science scenarios, where two kinds of agent are competing for resources and one has the ability to eliminate or exclude the other, may be less obvious but not uncommon). Our point is that a model may be simple but still effective and widely used in scientific study. There are several ways of treating such models. However, when we want to handle complexity, we must never forget the first way – of learning from this type of simplification and reflecting back on the real world to see what we have learnt about the system just by conceptualizing the model. We can also beg, borrow, steal or invent mathematical ways of representing our system from biologists, physicists, economists and many more. With such equations we can then insert initial conditions and scenarios – then the result, under the model's conditions, can be produced and compared to real life. This is the simplest form of simulation. Multiple runs of scenarios, subtly or grossly differentiated, can easily and quickly be multiplied for simple models (even using accessible and easily learnt tools such as Microsoft Excel). Further, into the world of simulation we can build what is known as an 'agent-based model' (ABM), where we set up a computer program that will create agents that, as the program runs, obey the rules we have decided to include in the model.

One of the greatest benefits of simulation (and especially agent-based simulation, ABS) is that it is a highly intuitive (and often visual) method that can shock with counter-intuitive results. However knowledgeable researchers think they are, no one is capable of intuiting the results of multiple interacting agents or rules even in the simplest of environments or with small numbers. A striking example, in agent-based simulation, is the flow of people through an emergency exit that can be increased by putting a barrier at the right distance in front of the door (Figure 9.5). This is a counter-intuitive result which translates back into the real world and has been used in the design of large spaces in buildings.

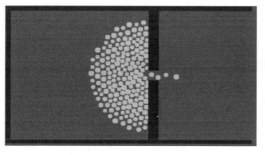 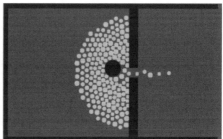

Source: Open ABM (2012).

Figure 9.5 Escape dynamics

RELATIONS AND THE BASIS OF NETWORKS

Network (or Social Network) analysis satisfies many of the features we were keen to find in our 'new toolbox': they are visually valuable, they emphasize interactions and lead thinking towards emergence and rules. They prompt thinking about systems in a taxonomical way – identifying nodes/agents (of differing or similar kinds) and links between those nodes. They are data rich and can take quantitative data (arising from qualitative enquiry) into a sound modelling and analysis environment. They can also be the basis of engaging and illuminating simulation. In most cases, we are talking about the communication ability of the network, from simple analyses like the time it takes a train to get from one place to another, to more complicated measures of power and influence.

The simplest types of network focus on structure and interdependency within a system; we identify the nodes (which can be anything from metro stations to brain cells, but we would most likely use a human or organizational entity) and a binary (one or zero) indicator as to whether a link exists or not between any two nodes. Even at this level we can say a great deal about systems and start to introduce quite sophisticated measurements that start to show complex properties. We can also see the different kinds of measures that we have. The first kind of measure describes the network overall; this could be the number of nodes in the network, the average number of links per node, or more complicated measures that assess the clustered-ness of the network or the average of all the shortest paths from every node to every other node (a connectedness measure). The second kind of measure is of an individual node in a network; this could be the number of immediate neighbours it has, the average path length to every other node, or a centrality measure of the node in the network.

With these measures we can start to see how complexity manifests itself in a network. The removal of, say, just one link in a network can have two extremely different results dependent on only a very small difference in starting conditions. We could remove one link from a network of thousands of nodes and links and it could have a minimal effect or split the network into two. In practical analysis we can construct hypotheses about, for example, when networks would be robust to changes such as removing only one link (or some given number or percentage of links) and when they would be sensitive.

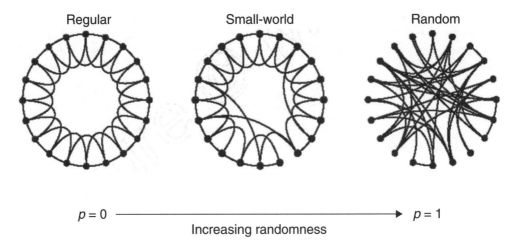

Source: Watts and Strogatz (1998).

Figure 9.6 Small world networks

One way to explore complexity is to examine 'small world networks'. Watts and Strogatz (1998) show how a regularly arranged network, where every actor is linked only to their nearest neighbours, contrasts with a random network where there is no 'locality' at all and network links can occur to anyone, anywhere (Figure 9.6).

The movement from the highly clustered regular network, where everyone only knows immediate neighbours, to the random, where anyone knows anyone with equal probability is conceived of through 'rewiring' the network. This rewiring effectively introduces 'shortcuts' which can be taken to occur, replacing a link, with probability p. When this probability is zero, no links are rewired and the graph stays regular, with p set to 1, all the links are rewired and a random network results. Somewhere in between, curious things happen.

In the regular clustered network, an individual at any point of the graph only knows people who already know several of her contacts. Even her contacts' contacts that she does not know are still well connected in her locality. However, this regular arrangement means that in order to discover a connection to the vast majority of the 'non-local' actors in the network, the number of steps needed is generally very high. In fact, most of the network is only reachable through multiple step connections. This gives the network a 'long average path length'.

The random graph is, not surprisingly, opposite in these properties; there is very little 'cliquishness' and it happens rarely that one actor's contact will know many of her other contacts. However, average path length for the whole network is short because there are so many cross-cutting links to every part of the structure. What Watts and Strogatz (1998) show is that, as the rewiring factor, p, increases, the change in these two characteristics is not as we might expect through intuition. In fact, the average path length for the whole network drops rapidly as just a few rewirings are made while the cliquishness remains high for much longer before falling away.

What we have in a small world is a relatively clustered network where, surprisingly, anyone in the network can be reached through relatively few steps. This analysis produces three vital points for complexity in political science. First, one cannot guess at the communication ability of a network by looking at a drawing of it, forming an impression by qualitative appreciation, or by 'doing some of the work'. We must be cautious about drawing generalities when we think we can see patterns. One needs to have complete information on a network to be able to calculate the parameters. This is generally lacking in political science which generally describes networks metaphorically without detailed measurement (Dowding, 1995).

Second, we should recognize the limits to using qualitative interviews with actors in the networks to assess the nature of the network. Actors/nodes in small world networks (and by extension other networks) may have no idea about which sort of network they are in. They can only observe their local conditions and this is no guide to the real nature of the whole network.

Finally, we should recognize the upfront assumptions that we make when doing political science investigation. If the entire nature of the network is susceptible to a link added or taken away, or a couple of links being reported as strength '4' rather than strength '9', then consider how sensitive our analysis is to the definitions we provide, the link we didn't count, or the missed node in the sample. Our discussion and appreciation of the method we adopt, and the rigour involved, becomes more critical than ever.

CONCLUSION: COMPARING METHODS

If complexity theory involves new ways of thinking about the world, we may need new methods to aid that thinking. For example, we cannot rely on quantitative methods based on the assumption that the relationship between key independent and dependent variables is linear and enduring. Rather, complex systems may produce a combination of long periods of stability punctuated by extreme events, irregularity or cycles of behaviour less amenable to prediction. Similarly, case studies involving in-depth qualitative analysis only take us so far. For example, if nodes within networks vary markedly across the system as a whole, we cannot rely on methods that focus on one small part of the system. In some cases, a combination of methods may aid analysis but also struggle to capture and explain complex system behaviour. In each case, our ability to explain what is going on may be hampered by our initial inability to describe it adequately.

It is in this context that we should consider the value of other methods, such as modelling and simulation, formal network analysis and experimental methods. In each case, the initial aim is to understand how complex systems behave – by simplifying them enough to allow us to model the interaction between agents and nodes and observing the effects. For example, ABS is useful as a way to monitor the effects of small changes in rules or in the networks bringing people together (see the chapters by Edmonds and Gershenson, Johnson, Hadzikadic, Whitmeyer and Carmichael, Morçöl, and Bilge in this volume). Experimental methods allow us to monitor changes according to the ways in which people understand or adapt to complex systems. Such methods aid analysis partly by allowing us to visualize these interactions and to consider why some changes produce rather counter-intuitive effects. Indeed, this exploration is the key to 'complexity

148   *Handbook on complexity and public policy*

thinking', which focuses on the various ways in which elements of the system can interact with each other, and the effect of changing the rules or types of interaction.

These methods may differ from mainstream methods in political science, but our approach to how to organize, conduct and analyse research should not. The process involves: identifying a research problem and linking it to established theories; thinking about how to address the research problem and which methods are most appropriate (or if a mix of methods can be valuable); explicitly identifying the assumptions we make about what to study and how to study it; conducting the research and considering the results; and, reflecting, self-critically, about how our theories, methods and assumptions influence the research process and results – and how we might improve the research design. In that context, these methods are only as useful as the people using them.

REFERENCES

Astill, S. (2005), *Networks that Form Policy: The Case of Pension Reform*, PhD thesis, London School of Economics, available at http://lse.academia.edu/StuartAstill.

Beyer, W. (n.d.), 'Mandel zoom 00 mandelbrot set.jpg', accessed 17 March 2014 at http://commons.wikimedia.org/wiki/File:Mandel_zoom_00_mandelbrot_set.jpg.

Cairney, P. (2012), 'Complexity Theory in Political Science and Public Policy', *Political Studies Review*, **10**(3), 346–58.

Cairney, P. (2013), 'Standing on the Shoulders of Giants: How Do We Combine the Insights of Multiple Theories in Public Policy Studies?', *Policy Studies Journal*, **41**(1), 1–21.

Cairney, P. (2015), 'How Can Policy Theory Have an Impact on Policy Making?', *Teaching Public Administration*, **33**(1), 22–39.

Dowding, K. (1995), 'Model or Metaphor? A Critical Review of the Policy Network Approach', *Political Studies*, **43**(2), 136–58.

Geyer, R. and S. Rihani (2010), *Complexity and Public Policy*, London: Routledge.

Herbert, B. (2006), 'Student Understanding of Complex Earth Systems', in C. Manduca and D. Mogk (eds), *Earth and Mind: How Geologists Think and Learn about the Earth*, Geological Society of America Special Paper 413, pp. 95–104.

Hindmoor, A. (2006), *Rational Choice*, Basingstoke: Palgrave Macmillan.

Jones, B. and F. Baumgartner (2005), *The Politics of Attention*, Chicago, IL: University of Chicago Press.

Kahneman, D. (2012), *Thinking Fast and Slow* (UK edn), London: Penguin.

Klijn, E. (2008), 'Complexity Theory and Public Administration: What's New?', *Public Management Review*, **10**(3), 299–317.

Lewis, O. and S. Steinmo (2008), 'Taking Evolution Seriously', European University Institute, Florence, available at http://spot.colorado.edu/~steinmo/TakingEvolution.pdf.

Mitchell, M. (2009), *Complexity*, Oxford: Oxford University Press.

Mitleton-Kelly, E. (2003), 'Ten Principles of Complexity and Enabling Infrastructures', in E. Mitleton-Kelly (ed.), *Complex Systems and Evolutionary Perspectives of Organisations*, Amsterdam: Elsevier.

National Audit Office (NAO) (2009), *The Failure of Metronet*, available at http://www.nao.org.uk/wp-content/uploads/2009/06/0809512.pdf.

OECD (2011), 'Life Expectancy at Birth', in *Health at a Glance 2011: OECD Indicators*, Paris: OECD Publishing.

Open ABM (2012), 'Escape Dynamics', accessed 20 May 2014 at http://www.openabm.org/book/3138/64-escape-dynamics.

Osborne, J.E. and E. Waters (2002), 'Four Assumptions of Multiple Regression that Researchers should Always Test', *Practical Assessment, Research & Evaluation*, 8(2), accessed 20 August 2013 at http://pareonline.net/getvn.asp?n=2&v=8.

Ostrom, E., M. Cox and E. Schlager (2014), 'Institutional Rational Choice', in P. Sabatier and C. Weible (eds), *Theories of the Policy Process*, 3rd edn, Chicago, IL: Westview Press.

Room, G. (2011), *Complexity, Institutions and Public Policy*, Cheltenham, UK and Northampton, MA, USA: Edward Elgar Publishing.

Ryan, A. (2011), 'Military Applications of Complex Systems', in C. Hooker (ed.), *Philosophy of Complex Systems*, Oxford: North Holland.

Sanderson, I. (2006), 'Complexity, "Practical Rationality" and Evidence-based Policy Making', *Policy & Politics*, **34**(1), 115–32.

Sanderson, I. (2009), 'Intelligent Policy Making for a Complex World: Pragmatism, Evidence and Learning', *Political Studies*, **57**(4), 699–719.

Sardanyés, J. (n.d.), 'Chaos', accessed 17 March 2014 at http://complex.upf.es/~josep/Chaos.html.

Sarotov Group (n.d.), 'Non-identical Coupled Logistic Maps', accessed 17 March 2014 at http://www.sgtnd.narod.ru/science/atlas/eng/charts/nonind.htm.

Strogatz, S. (2001), *Nonlinear Dynamics and Chaos*, Chicago, IL: Westview Press.

Watts, D.J. and S.H. Strogatz (1998), 'Collective Dynamics of "Small-world" Networks', *Nature*, **393**(6684), 440–42.

10. Complexity modelling and application to policy research
Liz Johnson

INTRODUCTION TO KEEPING COMPLEXITY SIMPLE

The goal of this chapter is to keep complexity simple and to provide the basic theories, vocabulary, and applications of complexity science methodologies for extending and advancing policy research beyond the limitations of traditional research methods. Yet there are currently no agreed upon or rigorous definitions of complexity and what it entails (L. Johnson, 2009; Rescher, 1998). Complexity means unique things to various people within and across disciplines (L. Johnson, 2009; N. Johnson, 2009). What complexity means to the discipline of policy studies will be reviewed, as well as what it could mean in the future. To start, simple definitions of the types of systems are helpful for clarity and for developing models.

DEFINITIONS

Simple system: All systems are comprised of at least two or more constituent parts or elements. The characteristics and interacting behavior of the elements depend on the type of systems. Behavior in simple systems can be accurately predicted and controlled due to the limited number of inert parts. Simple systems exhibit cause and effect relationships, limited interactions, limited feedback loops, few variables, centralized decision-making, and are decomposable, in contrast to 'irreducible' complex systems (Casti, 1994). Modeling complex adaptive policy phenomena as simple systems has been the traditional approach in policy research, yet few true simple systems exist in policy.

Complicated system: A complicated system is made up of many elaborately interconnected parts where the parts make up the whole. Prediction in complicated systems is challenging at the agent level, yet macro behaviors or characteristics can be adequately described in statistical or probabilistic terms, as well as through qualitative methods. A majority of policy research in the twentieth century was conducted based on complicated systems and linearity assumptions.

Complex system: A complicated system differs from a complex system in critical ways. A complex system is made of diverse, interrelating and interdependent parts or agents, that can be challenging to understand, describe, manage, design, and/or change, where the parts are greater than the whole (Magee and deWeck, 2004). Complex systems arise from independent agent interactions where there is not adaptation, but emergence is still possible (Wolf-Branigin, 2013). What makes complex systems worthy of exploration is that they 'pulse with life', do not remain in equilibrium, and are capable of producing patterns and structures (Page, 2009). Life itself is an emergent complex phenomenon, of

Complexity modelling and application to policy research 151

complex systems composed of patterns and structures, which emerge from the bottom-up, without being inherently built into the system, as is policy (Page, 2009). Complex systems do not fall into simple, discernible patterns, and outcomes in complex systems do not fit onto statistics' standard normal bell curve distribution (Page, 2009).

Complex adaptive system (CAS): At times the term 'complex systems' is used interchangeably with complex adaptive system (CAS). Both complex systems and CAS arise through dynamic, non-linear interactions and embody emergent potential, which exceeds the sum of its parts. Both types of systems are characterized by positive and negative feedback, which can amplify or dampen change (Ramalingam et al., 2008). In a complex system, there is no adaptation to the environment in response to learning and evolving feedback, like in a CAS (L. Johnson, 2009; Page, 2009). Agents may adapt in complex systems but the system does not adapt (Wolf-Branigin, 2013). A CAS is a complex system or systems made of diverse, interrelating and interdependent parts or agents, where the parts are greater than the whole, while agents and the whole system can adapt (L. Johnson, 2009; Page, 2009).

Complexity science: Complexity science is comprised of complexity theory. Complexity theory is often presented metaphorically, but theory-wise deals with functions and processes of changing, interrelated systems (Cronbach, 1988; Holland, 1998; Wolf-Branigin, 2013). Wolf-Branigin (2013) specifies complexity theory as 'this emerging paradigm in the physical and social sciences [that] seeks to understand how agents self-organize and then how they continually use feedback to produce emergent behavior' (p. 175). Complexity theory can be confused with chaos theory, which is important to differentiate when operationalizing policy research. Both theories are based on non-linear dynamics, how phenomena change over time, and can be sensitive to initial conditions (Warren, 2008). Chaos is characterized by unstable, complex behavior that does not repeat itself (Kellert, 1993). Also in chaotic systems, external influences can have simple causes and are the deterministic result of their interrelating parts and external causes (Kellert, 1993; Page, 2009; Wolf-Branigin, 2013). Complexity is comprised of underlying structures that upon initial examination appear random, but in effect exhibit underlying patterns (Wolf-Branigin, 2013).

We know from complexity science that actors in the whole of the complex system can influence almost everything but control almost nothing (Fullan, 2003; Page, 2009). Complexity is built on the interconnections between parts that comprise a complex system (Gribbin, 2004). Complexity science can be thought of as a loosely organized set of principles and ideas that have emerged resulting in a more effective means to describe and understand the processes and dynamics of change (Ramalingam et al., 2008). The science of complexity provides the means to internally represent the experience of change by describing our collective reality as processes (Casti, 1994: 273). Complexity is a way to understand life as living systems comprised of relationships, patterns, processes and context (Capra, 2005). Non-linear phenomena can be understood through complexity as a mathematical language with the possibility of unifying life through the integration of biological, cognitive and social dimensions (Capra, 2002). Casti (1994) suggested that complexity science is a means to eventually assess the limits of reductionism as the accepted problem-solving approach to research. A complexity approach allows for clarifying philosophy of science debates of order versus chaos, random versus determinism, equilibrium versus non-equilibrium, and analysis versus syntheses in policy research (Wolf-Branigin, 2013).

152 *Handbook on complexity and public policy*

Complexity is a subjective concept because we are not capable of observing all interactions of social phenomena bias-free. Many disciplines in the natural sciences use experimental design to remove subjectivity from theories, methodologies and analysis. Policy research is limited in applying experimental design and usually settles for less rigorous approaches in theory development. Researchers, as observers of complex social phenomena, abstract the characteristics, mechanisms and interactions for formalization in policy modeling. Complexity modeling in simulation form is just one tool to facilitate formalization in order to gain better understanding, insight and prediction of some aspect of the connected social world (Gilbert and Troitzsch, 2005).

Social, economic and ecological systems are the major global systems that policy sub-systems are nested within. Furthermore, 'no systems live in isolation' (Casti, 1994: 278). The systems and their parts dynamically interact with other parts and with the environment to influence their own futures. The combination of the parts interacting at the individual- or micro-level give rise to system-wide, global or macro behaviors. The micro- and macro-level systems can influence each other, while interacting in a dynamic environment. System-wide patterns can emerge from the interdependent interactions of adaptation from autonomous agents at the individual level. People can be represented symbolically as agents in simulations. Examining the behavior of dynamic social systems can provide information on system trends in the form of patterns that would otherwise be unpredictable (Ogula, 2008). Applying complexity science to policy studies allows for increased understanding and anticipation of critical patterns in social systems in order to facilitate more effective strategies and decisions for policy interventions. A research goal in complexity science is to identify consistent patterns, trends and tendencies in the reproduction of system behaviors, so appropriate strategies can be developed for system enhancement, prevention of cascading system failures and sustainability.

Complex systems, as open systems, exchange energy or information through numerous direct and indirect feedback loops causing 'rich' nonlinear interactions (Cilliers, 2000). The systems adapt, possess memory and have history, which impacts the systems' behavior. Consequently, these factors have implications for how policy research phenomenon are conceptualized, operationalized, analyzed and interpreted. The search of literature on complexity in public policy reveals an abundance of studies that mention that research phenomena are complex. Yet there are surprisingly few examples of valid public policy research that incorporates a complexity science framework throughout the entire research process.

Complexity concepts can help researchers investigate when and why productive systems emerge and how they can be sustained (Miller and Page, 2007). In a complex system or CAS, everything is connected to everything else, whether directly or indirectly (Richardson et al., 2001: 8). Consequently, there is the problematic nature of boundaries in researching these living systems. When researchers attempt modeling complex phenomena like policy, traditional reductionist tools fall short (Miller and Page, 2007). A reductionist tool can be thought as one that reduces a system to its smallest scaled elements for experimentation. Further, it is impossible to reduce a system to its constituent parts without killing it, according to Miller and Page (2007). Casti (1994) argued that ignoring any part of systems' processes, or cutting connected parts, usually destroys and alters essential elements of the structure or behaviors. This argument is significant because of the complex connections between a policy's purpose, intent, systems, interdependencies and agents. Moreover, when you slice up policy systems into subsystems, as with other complex systems, you lose critical information and links that make that system a system (Casti, 1994). An alternative way to model systems is to think in terms of the inherent and innate adaptive processes, as well as the enmeshed connections of agents inherent in social systems (Miller and Page, 2007).

A complexity science approach to complex phenomena, as in policy research, allows for investigating the social landscape as dynamic and co-adaptive (Miller and Page, 2007). Complexity science serves as a supplemental, complementary and alternative approach to traditional research (L. Johnson, 2009; Wolf-Branigin, 2013). Still it is critical to recognize the nature of complexity, whereby agents compete for limited resources – like energy, power, money and space – in the environment. Also, complexity, as a framework, provides the means for anticipatory proactive thinking and preparedness instead of exclusive retrospective prediction from traditional methods (Lissack, 2013). Traditional research methods are not yet able to account for emergence. Yet emergence serves as an integral component in complexity research and in complex phenomena. Policy research can be enhanced by what if, what could be, and what should be models from complexity, in addition to what was modeled in traditional policy research (L. Johnson, 2009; Lissack, 2013).

RATIONALE FOR WHY POLICY RESEARCHERS NEED NEW RESEARCH TOOLS

Policy represents some of the most pressing real-world challenges in research today. Many of the advances in complex systems research have come from studying phenomena in physics, physical systems, networks, evolutionary biology and chaos theory. These disciplines have shown that while phenomena can behave complexly in time and space, their underlying laws that govern behavior are known. In the case of complex human systems, such as societies and global economies, the behaviors in time and space are complex, yet the systems are unpredictable because the underlying laws are unknown (Strogatz, 2008). With the world becoming more complex, the science of complexity science offers a means to tackle some of the most challenging problems that we face in policy research. Researchers and policy makers need to adapt their thinking and approaches accordingly. As Einstein stated, 'We can't solve problems using the same kind of thinking we used when we created them' (BrainyQuote, 2013).

The time has come to examine complexity in the world and its impact on policy under a new lens of opportunity. Everything that happens in the world is new, and the world does not repeat itself (Bar-Yam, 2010). Complexity modeling represents and describes unique ways to think about mimicking and learning about the complexities and systems in the world. Bar-Yam (2010) stated, 'We want to generalize ideas that can be used to take the past to the future.' A major goal in complexity research is to capture details of interactions and processes. Furthermore, complexity allows for a shift of research questions from comparing past group average behaviors and linear relationships, to questions that ask about patterns, trends, spatial relationships, group membership, structure, time course of events, and interactions at the system-level of social groups (Tabachnick and Fidell, 2001; Wolf-Branigin, 2013).

154 *Handbook on complexity and public policy*

Importantly, complex policy problems cannot be solved assuming that entities are discrete and separated, based on simple relationships that aggregate (Lissack, 2013). Traditional scientific thinking and research has been successful but can be improved upon to create new opportunities by bridging with complexity. In order to find greater meaning in information and research, context is critical; and complexity as a context in research practice and policymaking is essential (Lissack, 2013). Also, the actions and behaviors of reflexive, anticipatory actors, which are prevalent in policy, are not well served by traditional research rules, usually established for non-thinking, non-reflective, and non-anticipatory actors. Simply, complicated and complexity are not the same (Lissack, 2013). Complexity is about interdependency and interrelatedness, whereby relationships and networks are important (L. Johnson, 2009; Lissack, 2013). If a research phenomenon is complex and adaptive, then thinking and research should match, and complexity science should be used.

Policy research can benefit by capturing more complexity in research and by focusing on what is important instead of what is easy to measure (Coburn, 2003). Agent-level interactions and agent-level data remain key to understanding aggregate system behavior. Applied to policy research, complexity can contribute novel insights, novel discoveries, and uncover connections between phenomena that were previously thought to be unrelated (Wolf-Branigin, 2013). For example, Wolf-Branigin (2013) discovered that researchers' tacit acceptance of residual errors (the observable estimate of the unobservable statistical error) as random noise or error, misses an opportunity to possibly discern patterns and meaning in the unobservable. Additionally, effective complexity simulation models can demonstrate visual interactions thought to be unrelated, as well as contribute meaning to the unobservable from traditional research tools.

Mathematical equation-based models can adequately address some policy problems. Yet traditional science has limited tools to learn about the world and can take us only so far with linear approximations. Also, there are complex aspects of the world that just could not be worked out until now (Bar-Yam, 2010). Social phenomena that include change, interrelationships and agency require more than equations. Powerful complexity tools make the work of discovering the intricacies of complexity in policy more tractable and understandable. Furthermore, complexity's new tools create shifting perspectives so new questions about the world can be asked. With these new exciting tools, a powerful new scientific door can be opened that expands the boundaries of what we can know about complex adaptive systems in policy (Bar-Yam, 2010).

Additionally, complexity science offers novel ways to think about the policy process that go beyond traditional qualitative and quantitative methods and tools. There are a multitude of interactions that give rise to a social system behavior that consequently generates complexity, equilibrium or chaos (Page, 2005). Simulations can provide the means for modeling a social system's transition into and through the phases of complexity, equilibrium and chaos throughout the policy process. Also, the mechanism of control in social systems, what directs this still mysterious process, can be examined more deeply and on different system levels (Miller and Page, 2007).

The key to translating complexity mechanisms into useful policy is learning how parts of systems interact and give rise to patterns, which contrasts to the traditional hypothesis testing or confirmatory approach (Wolf-Branigin, 2013). The focus is on examining the dynamic relationships of the whole and interdependent sub-systems. Social systems differ from physical systems in that actors learn, communicate, have history and adapt. Actors are represented as agents in simulations. In simulation topologies, agents can move in free continuous space, interact with local neighbors in cellular automata, can be connected in networks, move over geographical information system (GIS) designated space, or be modeled without the importance of spatial interactions (Macal and North, 2006). Agents have volition and can choose their level of connectedness, interdependence and responsiveness. Social actors actively seek connections and adjust them in response to environmental and social cues. In research, an emphasis should be focused on evolutionary learning and dynamic connections of the actual policy and policy actors, in addition to representation of static institutions, laws, regulations and other traditional policy instruments (OECD, 2008).

Incorporating a variety of tools can make theory better (Miller and Page, 2007). We have ample room in policy studies to improve our theories. A triangulation approach that adds complexity methodologies to quantitative and qualitative research strategies can help address the gaps in understanding interconnected parts that is necessary for improved understanding of social phenomena. The use of multiple approaches and strategies can facilitate strengthening the research process and the analysis of results. It is important to note that some social systems are complicated, predictable, and linear in nature and are best served by traditional research approaches. However, complex social systems with their non-linear interactions, adaptations and emergent characteristics require additional modeling that simulation can provide.

COMPARISON OF COMPLEXITY METHODOLOGICAL TOOLS TO TRADITIONAL RESEARCH

Public policy, like other social science disciplines, has been predominantly driven by empirical statistical research methods, attempting to describe and predict *what is* through the probabilistic average of means, and more sophisticated methods of pseudo-control groups versus treatment groups. This approach has offered a wealth of knowledge, yet does not usually account for complexity, dynamic consequences or emergence in complex adapting social systems found in policy. Traditional research methods offer prediction and justification for decision-making in policy. Whereas that is an accepted approach, it can fall short as an isolated systems approach and accounting for processes, complexity, emergence and dynamism in policy research. Inferences can be made from traditional research, but outcomes do not always match reality in the policy realm. As the classic example goes, no matter how many times we observe white swans, it does not justify researchers claiming that all swans are white (Popper, 2002). Taleb (2008) argued that statistical modelers are unable to prove that their models work. More important to policy makers, if models don't work, are the outcomes neutral? For example, statistics are not gathered until after the policy event (Taleb, 2008). Consequently, novel occurrences or emergent system behavior cannot be adequately predicted with traditional tools. Complexity can account for novel occurrences and emergence in policymaking.

Effective policymaking and implementation for complex systems is challenging. In relation to policy impacts, you can be very wrong prediction-wise and there be little impact on policy. At other times you can be slightly wrong, which results in major

156 *Handbook on complexity and public policy*

impacts (Taleb, 2008). Complexity tools can identify tipping points and identify system vulnerabilities and strengths in order to better guide and sustain policy actions. Statistics are limited, providing a binary output of whether an event is true or false (Taleb, 2008). However, statistics are often an effective predictive tool for non-rare and non-novel events in policy. Yet, in policymaking, degrees of true and false can be critical. Binary outputs alone are not adequate in policymaking to create policy that is efficient, equitable and feasible. Accurately predicting a policy event like an election or war is helpful, but what is critical is nuanced information about the specific direction and degree of impacts on the agent- and system-level. Furthermore, the rarer the event, the more theory and observations you need, and the greater the estimations error, according to Taleb (2008). Complex systems exhibit outputs whereby unlikely events are more likely, in contrast to normal bell-shaped distribution predictions. Consequently, bell curve estimates and predictions of future system behavior provide an incorrect depiction of large-scale changes in complex systems (Herbert, 2006). Given the static nature of traditional research, with limited explanatory power, supplemental tools like complexity methodologies can address these weaknesses and limitations.

Also, traditional quantitative research does not allow for co-variance of interactions between the dependent variable and independent variables and those variables interacting at the same time. For example, in traditional research, there can be uncertainty about interaction terms and impacting variables. Complexity methodologies permit variables to interact both ways with each other, at the same time, and even interact with the environment. Additionally, simulation methodologies can provide variations or sliders of variable measures, instead of single snapshots or a time series of snapshots from specified times. Ideally, all slider variations or variable parameter ranges can be simulated into a model to display a more detailed picture of complexity interactions. Due to the advances in computing technology, complexity models can capture simultaneous interactions of dependent and independent variables, as well as qualitative variable interactions. Research can be conducted on multiple scales at the same time, like including federal, state and local systems. Further, population samples can include diverse, heterogeneous agents with flexible, individual cognitive decision-making capabilities. Moreover, agents do not have to be optimizing and can be simply goal oriented, so models can have multiple interdependent sources of influence on policy outputs and outcomes (see Table 10.1).

Table 10.1 Traditional research versus complexity research

Traditional research	Complexity
Reduction	Whole system
Repeatable on same system	Rerun computer simulations
Precise	Flexible, versatile
Static	Dynamic, process, networked
Optimizing	Adapting, emergence
Homogeneous agents	Heterogeneous agents

RESEARCH MODELING PERSPECTIVES FOR COMPLEXITY

Human life, as we perceive it, occurs in relation to time and physical, geographic space. Yet complex systems of diverse scale, as in policy, can be evaluated in terms of both physical description and simulation, in relation to time and space. First off, data from both quantitative and qualitative data can be used in simulations, and combined to strengthen models. Qualitative data offer information and a snapshot on *what is* from a narrative perspective. Complexity modeling is about representing subtle and deep structures from within and across system dynamics, which can be represented in simulation movie-like format. This is done by means of identifying networks, connections, strength of connections, and interactions of social phenomena at the individual- and system-wide-level, in order to translate them into dynamic computer simulations, with corresponding output graphs of selected variable interactions.

The advances in complexity science and computing allow policy researchers to ask not only mechanism-based questions of *what* works, but also effect-based questions like *why* and *how* policy works. Complexity can simulate micro-level agent behavior interactions also with macro-level social behavior (Diepold et al., 2010). Through modeling, policy researchers can capture the properties of locality of interactions on a micro-, meso- and macro-level. Applying this research tool to challenging problems found in policy studies can provide researchers with models, whereby complex systems are directly represented. The methodology includes modeling the behavior and interactions of social systems in the form of policy simulation labs that are capable of creating artificial societies and communities (Epstein, 2006). Simulation experiments enable the creation of generative, dynamically interacting social structures, instead of relying on the prevailing approach of static equilibrium according to Epstein (2006). Simulation results can be tested by comparing how well observed structures of traditional research, validated by research statistics, match structures generated in simulation (Epstein, 2006). Further, policy research in complex systems that synthesize computer simulations of virtual policy scenarios can reveal the implications of system-wide and individual-level agent trends, leading ideally to more effective policymaking and implementation.

EFFECTIVELY MODELING COMPLEXITY STARTS WITH CONTEXT AND ASKING THE RIGHT QUESTIONS

1. What policy system or systems would you like to model?
2. What ontological objectives, like what is, what could be, what was, and what should be, best fit your research goals?
3. What type of policy do you want to model: past, current, future, futuristic or hypothetical?
4. Is the model's conceptualization based on a problem, event, intervention, hypotheses, theory, question, experimental design, narrative, data or combination?
5. What is the model structure type or how can behavior best be characterized?
6. Are the research phenomena of interest bottom-up self-organizing; top-down with centralized authority; fixed number of agents and fixed environment; non-fixed

number of agents in interactive environment; or evolutionary and adaptive agents and environment?

Complexity modeling requires taking into account nonlinear interactions, self-organization, networks made of many parts, emergence, and structure or behavior that is neither random nor regular (Braha, 2010). It is a creative process skill that takes time to develop, but can be achieved with practice. Simulation software for modeling complexity research ranges from simple to very advanced. Developing models, as part of an interdisciplinary team, is ideal to garner varying perspectives, feedback, analysis and programming support.

BENEFITS OF APPLYING COMPLEXITY MODELS IN POLICY RESEARCH

Applying complexity can benefit policy research in important and significant ways. It provides the means to generate complexity in simulation policy labs, so researchers can observe the interacting dynamics of agents, the collective and environment. The idiom 'seeing is believing' is a strength. Simulations may display visual evidence not discernible or visible from traditional research, though interpretation may be difficult. The visual simulation process is similar to an action-packed infomercial, in contrast to a static snapshot found in the predominant forms of policy research. Simulations can make policy research become dynamically alive, with conceptually-simulated experiments that are not constrained or limited by the boundaries of physical reality. This allows for creativity and generative systems thinking, beyond the limits policy researchers have been able to explore. Researchers can get mired by the ease and comfort of traditional research approaches, failing to contribute significantly to their area of policy expertise. Simulations can serve as a spark plug of innovation to positively augment traditional research tools. Since complexity methodologies are rule-based, different disciplines can understand the rules and participate on an interdisciplinary level (L. Johnson, 2009).

Simulations can use available quantitative and qualitative data to model assumptions, reducing time on the data-gathering process and costs. Case studies tend to be fairly easy to translate, given the narrative aspect of the research designs. Additionally, software programs can process huge amounts of data very quickly. In fact, some software programs can process entire samples of populations, such as 30 000 000 agents to represent the population of Afghanistan. There are a variety of software programs available to researchers, with some free to the public. Software like NetLogo even provides sample social science models to guide research (L. Johnson, 2009).

Using simulation as policy labs offers great flexibility for use in policy research. Researchers can systematically repeat experiments with single variable values varied alone or in combination with other variables and the environment. The complex interactions can then be systematically explored and consequently greater variation in policy interventions can be simulated. Complex system interactions can then be simulated with competing or varied policy interventions and can potentially reveal new alternatives from emergence. Also, at the agent level, simulations have the potential to capture the impact of complex interactions by varying policy interventions (L. Johnson, 2009).

Complexity modelling and application to policy research   159

Furthermore, many events in policy happen simultaneously. Complexity can account for simultaneous interactions between agents, type of organization of agents, between the organizations, and the environment. For example, agents can interact complexly at the intrasystem level by collectively arranging tasks for self-organization. At the inter-system level agents from different systems can interact complexly across their system boundaries, potentially causing adaptation. At the hierarchical level, even though there are constraints of vertical control, agents can interact complexly and potentially cause adaptation. Each level of agent self-organization can influence the collective whole with their corresponding behaviors. Consequently, parts of the collective whole have modes of representing organization structures of complexity. A change, no matter how small, has the potential to change the multi-level collective whole. A change can follow all paths of influence to impact internal-, horizontal- and vertical-levels at the same time (L. Johnson, 2009; Lewis, 2006).

Many in policy studies approach research like classical physics, by assuming that it is possible to isolate and concentrate on a well-defined system (Peat, 2010). An isolated system-approach implies that a social system can be studied disconnected from its inter-acting environment, separating agent behaviors from environmental fluctuations. Social systems usually exhibit non-linearity, which calls into question the use of an isolated systems-approach for complex policy systems. Peat (2010) claimed in relation to policy that it is wrongly assumed that a well-defined intervention will result in predictable out-comes. There are multiple interactions and interdependencies with multiple causes and impacts. According to Peat (2010), it is wrongly assumed, if deviations occur, that exer-cising control can fix the situation. Whereas this may work for simple systems, complex systems are hard to control and predict. Multiple forces and externalities shape the future of policy and do not aggregate simply (Hammond, 2009). Additionally, researchers cannot realistically collect all the needed data to predict and control policy. Complexity can provide the means to discover nuanced behavior patterns in systems that are deemed unpredictable and provide insight on the direction in which the system is moving (Ogula, 2008). Also, system tipping points and thresholds can be discerned in order to better adapt policy interventions and appropriate system-enhancing and sustaining actions.

Another advantage to complexity methodologies is that they can address non-equilibrium system dynamics since policy systems do not typically operate in a state of equilibrium. Further, policy is often designed and implemented incrementally, which repeats patterns of protectionism and small attempts to solve big, system-wide problems. Policy actors, political systems and institutions rarely aspire to fully solve problems from a systems approach. Complexity allows for political and administrative system con-straints to be modeled. Measures of policy effectiveness, efficiency, equity and feasibility can be included in models and further analysed. Also simulations can reveal and visually display interactions at the agent- and system-level, to better understand connectivity, adaptation and interdependent actions. The goal should be system improvement, sus-tainability, and how to avoid system deterioration, which is not usually how traditional policy research is approached. Knowledge of critical connections, interactions and interdependencies could add to a base of research knowledge, in order for more optimal policy recommendations. Complexity provides a framework to analyse policy from a lens of dynamism, processes, mechanisms and systems (L. Johnson, 2009).

Finally, policy research usually deals with patterns of behavior, such as social justice,

160   *Handbook on complexity and public policy*

feasibility, effectiveness and efficiency, but not in the context of complex adaptive systems. Pender (2004) argued that complexity in policy studies is often treated as a 'hidden free good' or even ignored. There needs to be a connection mechanism for researchers and society at large, in order to see how complexity and policy could best mesh. Simulations are tools that can bridge the gap displaying complexity in a simpler form for mutual understanding. Policy makers should take responsibility for keeping the public updated on the costs, benefits and potential externalities of complex systems policy. So, once policy simulation models are generated, they can be made fully accessible to the general public and posted on a neutral Internet site. There is a benefit to providing the public with visual representations of what could be in policy, so they can better assess their public investment. This approach not only provides greater transparency and understanding, but also offers a viable means for the public to be directly involved in manipulating simulation variables and analysing various possible policy outcomes. Simulation modeling can serve as a bridge to the future, facilitating public feedback and engaging society in order to strengthen understanding of complexity and policy interactions (L. Johnson, 2009).

CHALLENGES AND LIMITATIONS OF APPLYING COMPLEXITY MODELS IN POLICY RESEARCH

Complexity methodologies are not perfect tools, nor the answer to solving all policy problems. However, if properly designed, implemented, verified and validated, the tools can add significantly to policy research. Yet according to Macal (2005), no computational model will ever be fully verifiable and validated. The methodologies provide innovative ways to approach policy research in order to capture processes, interdependencies and interactions, not just probabilistic and output means of data. Still, operationalizing underlying micro-scale rules that direct and explain macro-scale outcomes is challenging (Braha, 2010). As additional research tools, they are intended to complement and enhance traditional research, not replace them, according to Achorn (2004). Complexity modelling requires systems thinking, knowledge of complexity science, and often an interdisciplinary approach, to ensure that model assumptions are valid. Knowledge of programming and software capabilities is also beneficial in order to develop good models or to be an adequate contributor to an interdisciplinary research team. Translating theories, system rules, assumptions and qualitative data into reducible programming language can be challenging. Computers do not naturally possess a programmer's subtle, refined knowledge and insight into human behavior, which can skew intended research objectives. There is also the risk of creating an effective model and not having the computer code accurately represent the model assumptions, system rules, data, theory or goals of research.

Some systems are too complex to model. Researchers do not have the capacity to model everything. Researchers do not have perfected knowledge of complex systems, which makes perfect modeling impossible. There is always the risk in any modeling of omitting critical variables, interactions, assumptions and mechanisms. Additionally, the inner workings of the systems may not be apparent yet. Complexity research in policy can result in non-significant outcomes and no emergence. There is continued debate on the value and limitations of complexity methodologies. Policy studies lag behind other disciplines, like economics and sociology, in the acceptance and application of complexity methodologies. Consequently, there is opportunity to cultivate knowledge in complexity science and its methodologies in order to capture the nuances needed for policy studies and more effective decision-making in policy.

METHODS AND TOOLS OF COMPLEXITY SCIENCE APPLIED TO POLICY RESEARCH

Advances in computing have facilitated progress into research of the unanticipated consequences of policymaking and the recognition of the unrealized opportunities of capturing policy system dynamics. Powerful computers and user-friendly software make possible the exploration of policy system dynamics which is out of reach by pure mathematical means. A variety of complexity tools are being applied to develop policy research through virtual system-building simulations. System parameters, or specific research variable values, can be easily adjusted with tools that are represented as visual, calibrated sliding bars in simulations. Researchers can change one or all variable values in a single simulation run, then change parameters for the next one. Impacts can be measured by graphed measurement tools like those used in traditional research (L. Johnson, 2009).

Agent-based modeling (ABM) and Network analysis are among the most common tools because of the flexibility in designing research and the availability of free community software. Agent-based models have developed from combining features from complexity system dynamics, emergence, genetic evolutionary algorithms, and game theory (Wolf-Branigin, 2013). Network analysis is a collection of nodes, hubs and connections between links that follow specific dynamic laws where the objective is to identify stable and weak network configurations. Data mining searches allow for pattern discovery in large sets of data that are complexly related. Scenario modeling techniques use hypothetical models as a basis to create varied artificial policy futures. Sensitivity analysis uses techniques that vary scenarios by determining the actual policy impact from the projected output through evaluating how changes in research variables impact the targeted variable. Dynamical system modeling incorporates feedback loops along with rate and flows or rates over intervals of time, which capture inflows and outflows in policy, like those used in population studies. Then the growth rate can be modeled exponentially or logistically (L. Johnson, 2009).

Developing complexity models can range from simple to advanced, requiring a range of programming skills. Programming languages include C, C++, Python, Java and others. Free simulation software packages include NetLogo, StarLogo, Repast, and SoNIA Social Network Image Animator, MATLAB, Mathematica, MS Excel and others are commercially available applications.

Network Analysis

Networks are the glue that hold the universe and its social systems together, whether in a physical state, abstract, visible or invisible form, and have defined structural characteristics. They follow dynamic laws and can take form in a visual of a map of relationships

162　*Handbook on complexity and public policy*

or links. Networks are ubiquitous, found in social, policy, biological, environmental and technological systems. Since everything is connected to everything else, Network analysis can be used to explore the critical thresholds in policy and how robustness in policy can be sustained (Barabasi, 2003). Networks can be referred to as graphs made up of nodes (vertices) and links (edges), where nodes may have states and links may have weights and direction (Sayama, 2010). Networks have properties hidden in their construction that enhance or limit our ability to empower or influence them (Barabasi, 2003). Visual and simulation representations of networks can divulge the mechanisms and offer insights in order to identify stable or weak configurations.

Research variables can be thought of as relationship types instead of categorical variables. Consequently, they provide the communication, infrastructure, production and transportation for complex adaptive systems. Networks in the real world operate in a constant state of dynamism and flux. They are not fixed, and continually evolve, adding new parts. Their structures are emergent phenomena, an outcome of how the parts follow governing rules. However, the visual representations of Network analysis can be shown in a single iteration, series of iterations, final iteration, or even in movie form from large amounts of streamed data (Social Network Animator, n.d.). In fact, systematic relationships and network patterns between entities in policy systems may prove more important to system performance than just studying the attributes of entities alone.

Networks have properties that can be measured. 'Size' can be thought of as the total number of nodes. 'Degree' is the number of links one node has to others in the network. 'Average degree' is the number of links per one node. 'Connectivity measures' include the average path length, which is the shortest path or number of steps it takes to get from one node to another in order to determine efficiency of movement or information in a network. 'Clustering coefficients of connectivity' is the degree or measure of how nodes cluster together. Connectivity also includes 'degree distribution', which provides an estimation of how connectivity is distributed within the network. 'Degree centrality' is simply how many connections a node has. 'Closeness centrality' is how close the node is to other nodes in the network, while 'betweenness' is how many shortest paths intersect the node (Sayama, 2010).

Types of Networks

The laws of growth and preferential attachment processes govern networks (Barabasi, 2003). Preferential attachment in networks dictates that a new node will connect with a node that has more links to other nodes. In fact, the probability a new node will link to another node is proportional to the number of links the favored node has (Barabasi, 2003). A few well-connected centers attract most new nodes, transforming into hubs of power and influence.

Improving knowledge of networks can better inform the practice of policymaking and research. Within networks there can be a few highly connected centers (essentially scale-free) in relation to lots of poorly connected nodes. For example, a leader of a democratic government would be a highly connected node or hub. The longer a hub has been in place, the greater the number of links will attach to it, creating a cumulative advantage. When the environment is filled with too much information, preferential linking makes certain hubs easier to find. Also, the greater the capacity of a hub, the faster it grows (Robb, 2004). These networks are very tolerant to random failures since they can absorb 80 percent of their nodes before they collapse. In contrast, they are very vulnerable to attacks on the hubs, allowing for 5–15 percent hub loss before collapse (Robb, 2004). This type of network is efficient and fast-communicating. Networks are not purposely designed. They evolve adaptive properties of robustness. 'Network typology provides direct information about the characteristics of network dynamics' (Braha, 2010).

The characteristics of varying network typologies can guide us into more effective use of Network theory application to research. An ideal in building knowledge in policy would be to have the capacity of every agent to connect with other agents, whether directly or indirectly. In other words 'almost every node is reachable from almost every other' (Wilensky, 2005). An example of this network typology is a giant component network that responds to random events. This approach does not apply well to policy networks since they are not structured with every agent connected to every other agent. This network would require too much energy, time and complexity for efficiency.

In small-world networks, most nodes are not neighbors of each other but can be reached from every other node by a small number of hops. In application to policy, the characteristic structure is limiting. An example of a small-world network in policy is a political caucus. Agents or entities can be equally represented in the network. There is the potential for some direct links to be dropped. These dropped links could be critical. As the name 'small-world' implies, this typology is for small worlds of information connections.

Conceptualization of Networks in Policy Research

Dynamical state changes are important to consider in complex policy network research. There are dynamics on networks, which is a state transition on dynamical nodes, and dynamics of networks, which results in changes like self-organization and growth (Sayama, 2010). Many complex policy networks show both dynamics at the same time. What is important to consider is the actions of diffusion, regulatory processes, culture propagation, information propagation, and population dynamics on the policy network. For example, if a small fraction of nodes diffuse policy, will other nodes accept, reject or ignore this? And will the transition be smooth or sharp? Networks have wide application and have been applied to a range of spheres such as policy process, terrorism, policy diffusion, nation building, contagions, voting, wars, civic engagement, governance, innovation, and more in policy.

There are a variety of network approaches that are used in policy research. For example, policy networks, characterized by informal decentralization and horizontal relationships, can account for a wide range of actors, government institutions, non-government institutions, spatial governance, political behaviours, and power differentials in policy sectors and decision-making (Kenis and Schneider, 1991). A policy network can be defined as 'a set of relatively stable relationships which are of non-hierarchical and interdependent nature linking a variety of actors, who share common interests with regard to a policy and who exchange resources to pursue these shared interests acknowledging that co-operation is the best way to achieve common goals' (Börzel, 1997; see also Peterson and Bomberg, 1999: 8).

164   *Handbook on complexity and public policy*

Social Network Analysis (SNA) is based on the assumption that social units have multiple levels of relationship properties and seeks an understanding of the interdependence of who knows who knows whom (Wolf-Branigin, 2013). Social networks are characterized by assumptions that social networks shape us, we shape networks, our friends influence us, we influence our friends, and networks have a life of their own, displaying emergence (Christakis and Fowler, 2009). The category of the whole (complete or saturated) social network allows for the examination of connections of all network members. The category of the ego-centered social network allows for the examination of individuals, groups, organizations or societies, with the ego as the focal node. The neighborhood is made up of the ego and all the connections to the ego that are directly adjacent. This type of approach is helpful for providing information on how one ego differs from another's, and the demography of an ego network's cohesion and differentiation from a micro perspective (Hanneman and Riddle, 2005).

Designing and Developing Network Analysis for Research

When designing and developing network analysis, it is important to start with a concept or rough visual draft of what entity relates or connects to what in the policy systems. It is also important to identify nodes and hubs, as well as rationale and theory (if available) for designation. Also, it is beneficial to think in terms of iterative steps of simplifying, emphasizing and explaining research phenomena while modeling, and evaluating after each step (Kornhauser et al., 2009). The goal in simplifying is to model relationship connections and the strength of links, which represent the phenomena of interest in the policy systems. The next stages are as follows: evaluate the model to check if it is as simple as possible. Check to see if the important relationships are emphasized visually and if the connections or links are easily observable, then evaluate for possible improvement. Check to see if the model conveys a clear focal point and the visual representation of the main idea conveys a meaningful point (Kornhauser et al., 2009).

EFFECTIVE MODEL BUILDING CONSIDERATIONS FOR NETWORK ANALYSIS

1. Why do the entities relate or connect and does this change during the research period being studied?
2. How do the entities relate or connect and does this change during the research period being studied?
3. What is the direction of relationships or connections: one-way, two-ways, all ways, multi-ways, alternating, combination, or no direction?
4. To what degree do entities relate or connect: totally, partially, once then drop the connection, intermittently, randomly, or combination?
5. If appropriate, how will connections be weighted? Is there data to substantiate weighting? And does weighting change?
6. What is the time period or amount of time that nodes and/or hubs connect to other nodes or hubs?
7. Are network interactions fixed or do they randomly change and adapt over time?

AGENT-BASED MODELS (ABMs)

ABMs are useful in policy research when there are numerous interrelated factors, complex interactions between agents, heterogeneous populations, and when agents learn and adapt (Achorn, 2004). The approach is considered mainstream in economics and sociology but is still in the early stages in policy research. Also, ABMs are one of the most generalized frameworks for simulating CAS, whereby the researcher can create a fixed or a varying number of virtual individuals or agents and simulate their behaviors and actions explicitly in a computer (Sayama, 2010). The steps for creating an ABM include abstraction, conceptualization, design, operationalization, programming, implementation, verification, validation and evaluation.

The goal of ABM research should first include empirical understanding, by inquiring into why large-scale regularities evolve and persist; given there is little top-down control. Second, the approach allows for evaluating if the policy design is effective. Third, ABM can direct researchers toward increased exploration into the causal mechanism in the policy systems (Axelrod and Tesfatsion, 2013). Finally, the goal of ABM research should not be exclusively predication since simulations do not offer a high level of accuracy for specified policy event outcomes.

Conceptualization of ABMs in Policy Research

ABM is a characterization and depiction of social reality that allows for the visualization of agent movements and interactions (Gilbert and Terna, 1999). Planning to conceptualize, create, implement and analyze an ABM should include data (if available), randomness, and solid logic to match the model to actual reality. Models are created on rules, based on theories, assumptions, equations and principles. Developing an ABM starts with assumptions about agents and corresponding interactions and then uses computer simulations to create 'histories' with the capacity to reveal the dynamic consequences of the model's assumptions (Axelrod and Tesfatsion, 2013). Designing ABMs requires asking and re-asking questions in order to refine research designs. First off, what policy type do you want to model? Options include past, current, future or hypothetical policy. Varying operationalizing approaches and a combination of conceptualization strategies can work for ABMs. Like traditional research, researchers can start with a policy problem, questions, theory and hypotheses. Policy process, interventions, impacts and externalities can be modeled. Also, as with traditional research, ABMs can be directed by data, events and/or narratives. ABMs have the capacity to simulate counterfactuals and experimental simulation with the treatment and control groups of agents.

Designing and Developing ABMs for Research

Creating an ABM requires abstraction and identification of system structures, as well as essential elements and connections, based on the conceptualizations and model assumptions. System structures in models include:

1. Bottom-up system regulation with self-organization, which is preferable to top-down modeling since systems in the real world cannot be totally controlled.

166   *Handbook on complexity and public policy*

2. Top-down system regulation where agent behaviors are directly influenced by central authority or a combination of bottom-up and top-down (for example: government agency, dictator).
3. Fixed number of agents and fixed environment (for example: swarm).
4. Interaction between agents and environment, which can be evolutionary with agent replacement like birth and death (for example: predator–prey).

To design the model, researchers specify agent attributes and actions at the individual level and system-wide level, as well as dynamics of agents interacting within the environment and social collective. Agents have internal properties (for example: can learn, make decisions, histories, and purpose), perceive the environment, have autonomy, and follow predefined rules. In order to model this, researchers identify the agent types (for example: individuals, groups, institution, cities) and substantiate with a theory of agent behavior (Macal and North, 2006). The steps to create an ABM include describing the possible interactions and outcome states for each part. In order to do this, it is necessary to identify agent relationships and substantiate with a theory for agent interactions (Macal and North, 2006). Next, describe how each state of each part changes over time or from event-driven simulations through interactions with other parts (Sayama, 2010). It is helpful to find similar simulations and research to help guide model creation. Start simple and add layers of interactions. The complexity should be in the model outcomes, not the design. Agents and the environment can be sketched or created in PowerPoint to show, through the use of arrows, how parts are connected and the directions of interactions.

Implementing ABMs requires the translation of the model specifications into computer programming language. There are a variety of software tools that can model ABMs. Some are offered for free on the Internet, like NetLogo, and offer user-group service, providing answers to programming questions. Computer code requires 'if', 'then' and 'else' statements sequenced in logic format. The implementation stage should be a straightforward and unambiguous transformation of model specifications that are logically sequenced into computer code.

However, researchers are constrained by an inability to imagine all the interaction possibilities that real complex systems demonstrate in relation to policy research. Verification and validation serve as critical elements for further model development. The purpose of verification and validation is to determine if the model is complete, consistent and unambiguous. Verification ensures that the model is programmed without errors and algorithms have been incorporated correctly (Macal, 2005). Additionally, validation is an issue since there is no controlled experiment. The objective of model validation is to ensure that the model represents and correctly simulates behaviors of the real-world system, addressing the right problem with reasonable arguments. Validation is where researchers can check to see if simulations produce outputs that quantitatively and qualitatively match those observed in the real social world (Gilbert, 2008) (see Figure 10.1).

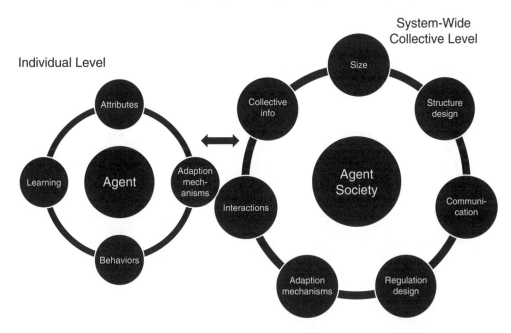

Figure 10.1   Design levels of ABMs

EFFECTIVE MODEL-BUILDING CONSIDERATIONS FOR ABMs

1. Are the model assumptions based on valid theory, rules, descriptions and logic?
2. Are model rules simple, fixed, sophisticated, adaptive, evolutionary, or a combination?
3. Are the model elements simple, and is the complexity in the simulation outcomes, not in the design?
4. What kind of rules do the agents and the environment follow: simple, fixed, adaptive, random, sophisticated-layering, threshold-based and/or combination?
5. Energy is not unlimited in policy systems, so how is energy or resources (for example time, energy, money and information) represented? How is energy allocated? How does energy increase or deplete, and at what rate?
6. How is the fitness function – a measure of an agent's competency and success with goals and meeting demands of the environment – represented?
7. What is the time scale of the model (for example year, day, or hour for each iteration)? How do you translate agent behavior time into a process that matches actual reality? Does the simulation start at a specified time or at zero?
8. How are networks, feedback, communication, information, adaptation and knowledge represented?

168 Handbook on complexity and public policy

CONCLUSION

There has been significant progress in the understanding of complex adaptive systems in policy, but the goal should not be to control complexity in policy but to harness it (Page, 2009). Moreover, policy theory does not adequately take into account how policymaking interrelates and intersects with complexity in the real world (Givel and Johnson, 2013). If research cannot account for components, dynamics and relationships among subsystems and levels in complex policy systems, there is opportunity. Simply, if a research phenomenon is complex, then research should be able to systematically address its complexity by incorporating and integrating complexity science throughout the entire research process (Johnson, 2013). Complexity provides the means to move from research that describes *what is* to further inquiry into *what could be*. Additionally, incorporating social values into the research even allows for modeling *what should be*. Complexity methodologies can be tools for improving perspective and the practice of policymaking and policy research. Some, like Pagels (1988), say that the future of the world lies with people who master the new paradigm of complexity. I say that the future lies in investigating even more deeply, the frontier of complexity and policy research.

REFERENCES

Achorn, E. (2004), 'Integrating agent-based models with quantitative and qualitative research methods', paper presented at the Association for Active Educational Researchers, accessed at http://www.aare.edu.au/04pap/ach04769.pdf.

Axelrod, R.M. and L. Tesfatsion (2013), 'A guide for newcomers to agent-based modeling in the social sciences', in L. Tesfatsion and K.L. Judd (eds), *Handbook of Computational Economics, Vol. 2: Agent-Based Computational Economics*, Amsterdam: Elsevier/North-Holland, pp. 1647–59.

Barabasi, A.L. (2003), *Linked: How Everything is Connected to Everything Else and What it Means for Business, Science, and Everyday Life*, New York: Penguin Group.

Bar-Yam, Y. (2010), 'Complexity lectures', personal collection of Y. Bar-Yam of New England Complex Systems Institute, Cambridge, MA.

Börzel, T.A. (1997), 'What's so special about policy networks? An exploration of the concept and its usefulness in studying European governance', *European Integration Online Papers*, available at http://eiop.or.at/eiop/texte/1997-016.htm.

Braha, D. (2010), *Introduction to Complex Networks*, Cambridge, MA, New England Complex Systems Institute.

BrainyQuote (2013), accessed at http://www.brainyquote.com/quotes/authors/a/albert_einstein.html.

Capra, F. (2002), *The Hidden Connections*, London: HarperCollins.

Capra, F. (2005), 'Complexity and life', *Theory, Culture & Society*, **22**(33), 33–44.

Casti, J.L. (1994), *Complexification: Explaining a Paradoxical World Through the Science of Surprise*, New York: HarperCollins.

Christakis, N.A. and J.H. Fowler (2009), *Connected: The Surprising Power of our Social Networks and How they Shape our Lives*, New York: Little, Brown and Company.

Cilliers, P. (2000), 'What can we learn from a theory of complexity?', *Emergence*, **2**(1), 23–33.

Coburn, C.E. (2003), 'Rethinking scale: Moving beyond numbers to deep and lasting change', *Educational Researcher*, **32**(6), 3–12.

Cronbach, L. (1988), 'Playing chess with chaos', *Educational Researcher*, **17**(6), 46–9.

Diepold, K., F.J. Winkler and B. Lohmann (2010), 'Systematic hybrid state modeling of complex dynamical systems: The Quad-I/HS framework', *Mathematical and Computer Modeling of Dynamical Systems*, **16**(4), 347–71.

Epstein, J.M. (2006), *Generative Social Science: Studies in Agent-based Computational Modeling*, Princeton, NJ: Princeton University Press.

Fullan, M. (2003), *Change Force with a Vengeance*, New York: RoutledgeFalmer.

Gilbert, N. (2008), *Agent-Based Models*, Los Angeles, CA: Sage Publications.

Gilbert, N. and P. Terna (1999), 'How to build and use agent-based models in social science', *Mind and Society*, **1**(1), 57–72.

Gilbert, N. and K.G. Troitzsch (2005), *Simulation for the Social Scientist* (2nd edn), New York: Open University Press.

Givel, M. and L. Johnson (2013), 'Scientific paradigms in the United States: Are we ready for complexity science?', in P. Youngman and M. Hadzikadic (eds), *Complexity and the Human Experience: Modeling Complexity in the Humanities and Social Sciences*, New York: Pan Standard Publishing.

Gribbin, J. (2004), *Deep Simplicity: Chaos, Complexity and the Emergence of Life*, London: Allen Lane.

Hammond, R.A. (2009), 'Complex systems modeling for obesity research', *Preventing Chronic Disease Public Health Research, Practice and Policy*, **6**(3), 1–10.

Hanneman, R.A. and M. Riddle (2005), 'Introduction to social network methods', available at http://faculty. ucr.edu/~hanneman/.

Herbert, B.E. (2006), 'Student understanding of complex earth systems', in C. Manduca and D. Mogk (eds), *Earth and Mind: How Geologists Think and Learn about the Earth*, Boulder, CO: Geological Society of America.

Holland, J.H. (1998), *Emergence: From Chaos to Order*, Cambridge, MA: MIT Press.

Johnson, L. (2009), *Agent-based Model Overview: A Guide for Public Policy Practitioners*, Charlotte, NC: Complex Systems Institute.

Johnson, L. (2013), 'Applying complexity to qualitative policy research: An exploratory case study', *Journal of Social Science for Policy Development*, **1**(1), 1–14.

Johnson, N. (2009), *Simply Complexity: A Clear Guide to Complexity Theory*, Oxford: Oneworld Publications.

Kellert, S. (1993), *In the Wake of Chaos: Unpredictable Order in Dynamical Systems*, Chicago, IL: University of Chicago Press.

Kenis, P. and V. Schneider (1991), 'Policy networks and policy analysis: Scrutinizing a new analytical toolbox', in B. Marin and R. Mayntz (eds), *Policy Networks: Empirical Evidence and Theoretical Considerations* (9th edn), New York: European Centre for Social Welfare and Policy Research.

Kornhauser, D., U. Wilensky and W. Rand (2009), 'Design guidelines for agent based model visualization', *Journal of Artificial Societies and Social Simulation*, **12**(2).

Lewis, C. (2006), *Complex Systems – Webs of Delight*, available at http//:www.calresco.org/lucas.cas.htm.

Lissack, M. (2013), 'Epi-thinking', *Modes of Explanations*, available at http://www.modesofexplanation.org/.

Macal, C.M. (2005), *Threat Anticipation: Social Science Methods and Models*, Chicago, IL, available at http://jtac.uchicago.edu/conferences/05/.

Macal, C.M. and M.J. North (2006), 'Introduction to agent-based modeling and simulation', *MCS LANS Informal Seminar*, Argonne, IL: Argonne National Laboratory.

Magee, C. and O. deWeck (2004), 'Complex system classification', paper presented at the International Council On Systems Engineering (INCOSE), Cambridge, MA, available at http://dspace.mit.edu/handle/1721.1/6753#files-area.

Miller, J.H. and S.E. Page (2007), *Complex Adaptive Systems an Introduction to Computational Models of Social Life*, Princeton, NJ: Princeton University Press.

OECD (2008), 'Applications of complexity science for public policy: New tools for finding unanticipated consequences and unrealized opportunities', based on a workshop at the Ettore Majorana International Centre for Scientific Culture, Erice, Sicily, 5–7 October 2008.

Ogula, D. (2008), 'Attractors, strange attractors and fractals', available at http://www.informaworld.com/10.1081/E-EPAP2-12001068.

Page, S.E. (2005), 'Agent based models', available at http://cscs.umich.edu/~spage/palgrave.pdf.

Page, S.E. (2009), *Understanding Complexity*, Chantilly, VA: The Teaching Company.

Pagels, H.R. (1988), *The Dreams of Reason*, New York: Simon and Schuster.

Peat, F.D. (2010), 'Non-linear dynamics (chaos theory) and its implications for policy planning', available at http://www.fdavidpeat.com/bibliography/essays/chaos.htm.

Pender, H. (2004), 'Public policy, complexity and rulebase technology', The Australia Institute working paper.

Peterson, J. and E.E. Bomberg (1999), *Decision-making in the European Union*, New York: St Martin's Press.

Popper, K. (2002), *The Logic of Scientific Discovery*, (2nd edn), New York: Routledge Classics.

Ramalingam, B., H. Jones, T. Reba and J. Young (2008), 'Exploring the science of complexity: Ideas and implications for development and humanitarian efforts', London: ODI Working Papers.

Rescher, N. (1998), *Complexity: A Philosophical Overview*, New Brunswick, NJ: Transaction Publishers.

Richardson, K.A., P. Cilliers and M. Lissack (2001), 'Complexity science: A "gray" science for the "stuff in between"', *Emergence*, **3**(2), 6–8.

Richardson, K.A., G. Mathieson and P. Cilliers (2000), 'The theory and practice of complexity science: Epistemological considerations for military operational analysis', *SysteMexico*, **1**(1), 25–66.

Robb, J. (2004), 'Scale-free networks', available at http://globalguerrillas.typepad.com/globalguerrillas/2004/05/scalefree_terro.html.

170 *Handbook on complexity and public policy*

Sayama, H. (2010), *Fundamentals of Modeling. Agent Based Models*, Cambridge, MA: New England Complex Systems Institute.

Social Network Animator (n.d.), available at http://www.stanford.edu/group/sonia/.

Strogatz, S. (2008), *Chaos* (Vol. 1), Chantilly, VA: The Teaching Company.

Tabachnick, B.G. and L.S. Fidell (2001), *Using Multivariate Statistics*, (4th edn), Needham Heights, MA: Allyn & Bacon.

Taleb, N.N. (2008), 'The fourth quadrant: A map of the limits of statistics', available at http://www.edge.org/3rd_culture/taleb08/taleb08_index.html.

Warren, K. (2008), 'Chaos theory and complexity theory', in T. Mizrahi and L.E. Davis (eds), *Encyclopedia of Social Work* (20th edn), Washington, DC: NASW Press, pp. 227–33.

Wilensky, U. (2005), 'NetLogo giant component model', accessed 15 September 2011 at http://ccl.northwestern.edu/netlogo/models/GiantComponent.

Wolf-Branigin, M. (2013), *Using Complexity Theory for Research and Program Evaluation*, New York: Oxford University Press.

11. Policymaking as complex cartography? Mapping and achieving probable futures using complex concepts and tools
Kasey Treadwell Shine

INTRODUCTION

While recognizing that there is a vast body of literature on complexity and chaos in the natural sciences, as well as their potential applications to the social sciences, this chapter uses the term 'complexity theory' as a shorthand reference to the general principles of complex phenomena. It reflects an ontological and epistemological perspective on phenomena that exhibit unexpected, unpredictable properties and behaviours that collectively differentiate them from other phenomena (and hence require a different mode of scientific inquiry). In the natural sciences, these 'complex' phenomena are emergent (greater than the sum of their parts) and non-reproducible (non-linear; contingent on other phenomena and systems; extremely sensitive to small changes in conditions), yet also maintain observable, non-conscious patterns throughout time and space – a kind of dynamic equilibrium, or what I call 'stability-through-change'. This chapter argues that socially complex phenomena can be viewed in the same light.

This chapter explores how working from a complexity theory perspective can allow for a very different perspective on developing and implementing new strategies to address persistent social problems. There are many debates concerning the degree to which complexity theory can and should be applied to socially complex phenomena, metaphorically or otherwise. This chapter draws together some 'first principles', to argue that socially complex phenomena can be expected to behave in the same way as theoretically complex phenomena. This in turn allows for a view of policymaking as a guided but emergent and complex process of social change towards desired, unknown but not unknowable future(s).

The chapter then draws out complex tools and concepts, which examine and seek understanding of socially complex phenomena through the lens of theoretical complexity. These can assist in achieving social change through adaptive and reflexive complex policymaking. This section also explores how social complexity (unlike natural complexity) is affected by conscious agency, ideology and power, requiring considerations of values and 'complex ethics'.

Next, the chapter applies these insights to a 'work in progress', that of trying to adopt a multi-dimensional approach to address child poverty and promote child well-being in Ireland. Two examples of 'key features' of child poverty are used to illustrate how taking a complex view of these issues recasts policymakers as compass-bearers for desired future(s), driving from the policy-swamp of the present an emergent and complex process of social change, with other stakeholders as key players both in the process of mapping a path to the future and in acting as lookouts for way-markers along the way.

172 *Handbook on complexity and public policy*

Policymaking in this view becomes a kind of complex cartography, guided by complex ethics and utilizing complex tools to create the map for convergence to a desired future.

The chapter concludes with a few thoughts of what such a view might mean for evidence-based policymaking. If such an 'evidence base' is always/already changing, perhaps it is better to think of complex evidence as 'informed thinking'. This might seek out from within the system positive emergent patterns that should be maintained, rather than looking to reproduce conditions that in complex systems are unlikely to result in consistently similar outcomes.

STARTING THE JOURNEY: EXAMINING SOCIALLY COMPLEX PHENOMENA THROUGH A COMPLEX LENS

The appeal of complexity theory for understanding and analysing social complexity has generated a wealth of literature applying complex ideas to a vast range of subjects. It can offer insights and tools, for example, in terms of thinking about non-linear causality, systemic effects, dynamism and seemingly self-organizing patterns. In some cases, such complexity is 'observable' through modelling and mathematical techniques. But the literature is also full of debates about whether complexity theory in the social sciences can be applied qua theory, or is better applied as a loose metaphor that adopts the language but perhaps not the fundamental logic or principles of complexity sciences. Often, complexity concepts and theories are conflated with chaos theory.[1] This makes arguments for a rigorous application of complexity to socially complex phenomena even more difficult: 'Stating the themes [of complex systems] is simple. It is their interleaving that is so terribly uncertain' (Kauffman, 1995: 186). This chapter takes the view that social complexity can be treated as theoretically complex primarily as applied in the complexity sciences (and especially drawing from ideas of co-evolution, emergence and stability through change), without delving too deeply into broader debates (see Byrne, 1998; Eve et al., 1997; Kiel and Elliot, 1996; Manson, 2001; Morçöl, 2001 for a good summary of these debates).

However, a substantial body of literature examining social complexity does extrapolate and apply concepts from complexity or chaos theories. This is especially prevalent in disciplines such as political science and public policy, where there is a strong impetus to produce practical, workable solutions to complex problems (see for example Cairney, 2012; Fuchs and Colliers, 2007; Geyer, 2012; Hausner (ed.), 1997; Kooiman (ed.), 1993; Loorbach, 2010; Sanderson, 2009). This diverse range of literature collectively establishes a solid foundation from which to derive a 'complex toolkit' and a well-developed lens for looking at the social world in a complex way, which I will draw upon in the rest of the chapter.

EXPECTING THE UNEXPECTED: COMPLEXITY AS A DRIVER OF SOCIAL CHANGE

It is also important to start from a few first principles. Like many people (I suspect), what most attracted me to the complexity sciences was the way in which it sought to discover 'rules' or 'patterns' underlying complex phenomena. Not in a traditional, Newtonian and reductionist view, but as something that was inherent in and intrinsic to complex systems themselves (see for example Cohen and Stewart, 1994; Gell-Mann, 1994; Kauffman, 1995; Bak, 1996; Prigogine and Stengers, 1984). These patterns were emergent and ordered even as context, chance and contingency ensured that they would never be repeated in quite the same way again. Such intrinsic properties seemed to help to make sense of both expected and unexpected outcomes that I observed in the course of doctoral research, where well-intentioned policies and programmes to address poverty and social exclusion rarely led to what was intended, and even seemed to sometimes become part of the problem. These issues also arose when examining housing management in Ireland (Shine, 2006) and in experiences when attempting to engage in evidence-informed policymaking activities (Shine and Bartley, 2011).

Opening the Spaces of the Possible

This sense of emergent ordering, of stability-through-change, that complex phenomena exist in a state of flux even as order emerges and is observable and is always/already changing (rather like Schrödinger's cat), gives a dynamism and a force to that complexity. It is an engine which *if* – and this is a big if – it were used well could create significant and positive social change. The open systemic view where both order and change are necessary is one of the first principles of complex systems, and it does not always get sufficient attention as the fundamental process by which potentialities can become probabilities.

Cohen and Stewart (1994) identify this 'deeper theme' of emergence: 'of high-level patterns arising from the indescribably complex interaction of lower-level subsystems. We distinguish two kinds of emergence, "regular" and "super". We call regular emergence "simplexity" and the super version "complicity". The archetypal examples of complicity are evolution and consciousness' (1994: 397). Simplexity 'is the emergence of large-scale simplicities as direct consequences of rules', a prime example of which is Newton's laws of physics (1994: 411). Simplex rules govern very many observable phenomena. Cohen and Stewart collectively call these 'features': 'simple general concepts related to the mathematical or verbal world in which the laws are formulated' (1994: 408). 'Conservation of energy' is a feature of the 'rule' of Newtonian physics.

Complicity is the opposite, an emergence that opens up the spaces of the possible, 'in which totally different rules *converge* to produce similar features, and so exhibit the same large-scale structural patterns' (1994: 414; emphasis added). Change and emergence that results in large-scale ordering (features) can and does come from many different 'rules' – unknown, but not unknowable in so far as emergent patterns can be observed. This has important implications in later arguments that policymaking can actively work towards a desired future(s) not through control and command rules (or, to use the language of Cohen and Stewart, simplex rules) but through complicit 'rules' which are diverse and convergent.

Another 'first principle' of complexity in the natural sciences is that of a description of observable patterns, behaviours and outcomes that can be identified as complex even if the underlying drivers for those patterns cannot be deduced. Evolution, for example, is a robust description for the outcomes of natural selection, but 'natural selection' does

174 *Handbook on complexity and public policy*

not have a physical basis in reality, even as it acts to shape reality. Similarly in the social arena, the idea of socially transmitted 'memes' and the field of memetics has wide acceptance, as an analogy to the perpetuation of genes and genomic expression (see Dawkins, 1976[2006]). However, while genes have a physical structure that can be examined and modelled, memes are not 'real' in the same way.

Social Complexity as Theoretical Complexity

The implications for social complexity are twofold. One, if it is possible to describe natural complexity in terms of its emergent properties even if those properties may not have physical structures, the same logic would seem to apply to observable emergent properties in social complexity. This allows us to expect socially complex phenomena will act as theoretically complex phenomena, without trying necessarily to connect the two empirically. This moves the debate away from whether social complexity is 'metaphorical' or 'real'. Instead, it seems to be an accurate description of what is happening for many phenomena, although we are not entirely sure why and we are struggling to name it. In turn, expecting complexity allows a consideration of leveraging social complexity (as dynamic flows of potentiality and change) to create better (more optimal, more open, convergent) outcomes.

The second implication of expecting socially complex phenomena to behave as if they are complex is to consider its dynamic effects as a 'force' shaping the social fabric. Here the analogy used is that of gravitational wells, which can warp and distort the space–time continuum; and of strange attractors, whereby phenomena 'loop' in complex temporal dynamic patterns (recognizing that strange attractors are a central tenet of chaos theory). Without pushing the analogy too far, such a perspective poses interesting questions about the nature of advantage and disadvantage. I suggest these might be 'features' of social complexity, an expression of what might be called 'negative' and 'positive' attractor states.[2] Like strange attractors, these might be pooled by dynamic forces that exert a gravitational pull on the social fabric.

CREATING AND SHAPING THE MAP: INTRODUCING COMPLEX POLICYMAKING

Using the first principles set out above leads to a consideration of the role of policymaking and of policymakers in a socially complex world, and looks at how complexity might offer practical tools and ideas to develop new approaches to persistent social problems. There are two aspects to this approach: one, how to recognize and understand the patterns and behaviours of socially complex phenomena, to create a (dynamic, always/already changing) map of what is happening so that policymaking can be more adaptive and responsive. The second is how to use complex tools in policymaking so that it can be proactive in guiding the way forward, looking toward desired futures and working to re-shape the map as needed.

Taking a complex view of policymaking threatens to destabilize the notion that it can have any kind of productive effect; that at best, policymakers can only seek to imperfectly manage complexity, recognizing that they 'sit' in a dynamic, contingent, adaptive

Policymaking as complex cartography? 175

open system (or set of systems), highly context-dependent for its trajectory (or stasis). Schön's description of the 'policy swamp' and Stacey's contested and dynamic policy-making interactions are apt descriptions of policymaking and policymakers hanging on at the edge of chaos (Schön, 1979; 1983; Stacey, 1995). Unexpectedly, however, things can turn out all right. Small changes can have large effects; suddenly, optimum equilibria can be established.

The importance of expecting the unexpected is a critical element of making policymaking more adaptive and responsive. It also can make it more difficult for policymakers to act. They may become paralysed by uncertainty and be more likely to fall back into more comfortable (if less successful) strategies to manage complexities, for example by applying rigid notions of new public management techniques (Parsons, 2001; 2002; Marston and Watts, 2003; see also Whitt, 2008). Institutional structures, organizational cultures and power dynamics also have a role to play in how well (if at all) policymakers are able to fully engage with the uncertain (but not unknowable) in socially complex systems (Shine and Bartley, 2011). Yet complexity is not simply about change, chance and contingency; it is also about stability, potentiality and emergent order. So perhaps it is better to speak of also expecting that patterns will and do emerge. Complex policymaking can therefore play a role in convergence to emergent order (i.e. stability), and in mitigating negative unexpected outcomes (i.e. 'negative' change).

Guiding the Way: Adaptive and Reflexive, Complex Policymaking

How can adaptive and reflexive complex policymaking produce 'convergence to emergent order(ing)'? Drawing upon experiences of efforts to promote collaborative evidence-informed policymaking (Shine and Bartley, 2011), it appears that policymaking seems to work best when policymakers and other stakeholders are receptive to collaboration and where knowledge is co-produced, shared and collectively 'owned'. From a complex point of view, knowledge seems to be an essential component of the dynamic forces driving social complexity. A first principle in generating a collective, convergent emergent order(ing), therefore, is to have the most open, diverse and universally 'pooled' knowledge(s) possible. A second principle of more responsive and adaptive complex policymaking is to recognize that social complexity can be consciously affected, and that utilizing competing values for 'creative tensions' may be transformative in opening up the spaces of the possible (drawing from the 'competing-values framework' of Cameron and Quinn, 1999[2006]; Quinn, 1988; Talbot, 2003; 2005; 2008a; 2008b). This leads into a third principle: that why, how, by whom and in what way ordering is achieved is not valueless and has real effects on people's lives, their communities and the social fabric generally.

A final principle is that 'many hands make light work'. Convergent ordering in different parts of the system can produce the 'tipping points' for larger-scale emergent patterns, which have ripple effects throughout a complex system or interlocking systems (similar to Bak's concept of self-organized criticality; see Bak, 1996; Bak and Creutz, 1994; Moss, 2002). Drawing on the idea of complicit 'rules' for generating change means that its impetus can come through non-local interactions, the actions of many different actors, or chance and contingency that may not be generated consciously (drawing from ideas in actor network theory; see, for example, Latour, 2000; Law, 1992). Multiple and

176   *Handbook on complexity and public policy*

diverse complicit rules are convergent rather than reductive; this differentiates them from what otherwise might be seen as a reintroduction of a positivist perspective in the process of convergent ordering (see, for example, Smith and Graetz, 2006; Stacey, 2003).

The concepts of 'regulated autonomy' and 'patches' are two tools for working with complicit rules in order to generate convergent ordering. Church (1999) argues that organizations work best if they are allowed to be self-organizing, relational and emergent networks at multiple levels of management. This, he suggests, can be achieved by 'regulated autonomy' by which local actors can act independently but within the overall parameters of the organization and its goals. Applied to policymaking, both across policy areas and from international to local spatial levels, regulated autonomy implies that the 'best' convergent ordering emerges from allowing actors to work independently and collectively at various different levels, but with some commonly defined goals or outcomes. Fuchs and Collier (2007) make similar arguments to Church in their discussion of 'interactive autonomy'.

From a different but related perspective, Kauffman (1995) puts forward the idea of 'patches' in his discussion of co-evolution.[3] These are interdependent, interacting localizations within systems (or between systems) acting in their own 'self-interest', but there is an equilibrium that emerges that drives all patches to self-optimization, thus creating benefits to the system(s) as a whole. These ideas have sometimes been taken forward in the applications of 'fuzzy logic' to organizational and other problems (see, for example, Imam et al., 2004; Habib and Shokoohi, 2009). In theory, the idea of working in 'patches' (which, crucially, are interdependent and interactive) is appealing, as again the impetus for change in the system under policymaking scrutiny may come from 'inside out' (within the system) and 'outside in' (without the system), with the added value that this could lead to systemic improvements. The literature on patches does point out, however, that the size and number of patches does matter (Seel, 1999) and there is much to be developed before the concept of patches can be applied in practice.

Introducing North: the Role of Transformative Causality

Stacey (2003; see also Stacey 2000; 2001) raises several important questions in the use of complex tools for more adaptive and responsive policymaking. He argues that organizations do not exist of themselves outside the 'patterning of people's interactions' (2003: 326). In other words, organizations are often thought of as independent entities, that they have a reality and a structure that exists outside of the people who make up those organizations. Stacey disagrees, arguing that conceptualizing organizations as operating separately or existing independently of people creates tensions in thinking about causality. This risks a dichotomous (either/or) or dualist (both/and) implicit assumption that causality 'starts' or 'exists' somewhere independently of people (2003: 327–30). Rather, Stacey draws on Mead and Elias (as quoted in Stacey, 2003) to think of the one and the many as a 'paradox' whereby there is no 'choice' to be made in resolving the paradox. Stacey then draws upon Hegel to suggest that a different logic is required to think about these paradoxes, in which a 'transformative' causality drives stable instability and instable stability (2003: 329). Other authors have highlighted similar ideas relating to emergent, paradoxical or interdependent learning and knowledge (see, for example, Allen, 2002; Hearn and Rooney, 2002; Hearn et al., 2003; Nicolaides and Yorks, 2008).

Policymaking as complex cartography?   177

These ideas go to the heart of the idea that policymakers and others might both expect and use social complexity to create social change. Complex systems have no inherent 'direction'. By their nature, they seek homeostasis, equilibrium by stability through change. In natural complexity, dynamism arises from always/already changing disruptions within and without the system. In social complexity, such disruptions may also come from conscious choices, driven by ideology, power, or to use Hegel's phrase, from 'desire for desire of the other' or mutual recognition (as quoted in Stacey, 2003: 329). These intentional 'disruptions' create the conditions for convergent and emergent ordering (echoing instable stability) and unexpected patterns (echoing stable instabilities), and, through the idea of transformative causality, the tools by which to work together, to seek desired futures.

Where does this leave policymakers and others in the complex, emergent ordering process of policymaking? The above points suggest that complex policymaking can be productive, indeed transformative. However, the fact remains that people, and values, matter: by whom, for whom and for what purpose transformative causality is driven profoundly affects the 'direction' of social change and what future(s) it might achieve.

Guiding the Journey 'North': the Role of Complex Ethics

I suggest therefore that 'complex ethics' are needed to guide transformative change, facilitating a 'direction' to a desired future(s) and marking out the outer boundary beyond which change becomes systematically destructive, chaotic, destabilizing (. . . here be monsters . . .) (see also Shine, 2006). I draw in part from the growing body of literature that examines the role of agency, knowledge and relational aspects in driving more responsive and adaptive policymaking, and which implicitly or explicitly draws upon complexity thinking (see, for example, Craft and Howlett, 2012; Howlett, 2011; Stewart and Ayres, 2001; Tyfield, 2012; Wagenaar, 2007; Wilsford, 2010).

Embodying and acting within complex ethics has the potential to open out the spaces of the possible in terms of what policymaking can and should achieve. In this view, complex policymaking is a process of finding convergent, complicit 'rules' to where we might go (toward a desired future), which, in turn, helps to determine where we should go from the present, and is guided by complex ethics marking out where we should not go. The challenge and the opportunity of this complex policymaking process is that policymakers need not be the only ones discovering convergent 'rules' or using complex ethics to define the outer boundaries beyond which lie unacceptable conditions or consequences. Letting go of (the illusion of) control and 'diving in' to the policy swamp does not necessarily imply all then becomes chaos. Rather, it is an opportunity to work with others, to gain dynamic flows of potentialities and opportunities for wider systemic effects, opening up new ways of how change is leveraged and by whom.

Complex ethics rests upon the capacities of policymakers and others to recognize emergent patterns, drawing upon interdependent learning and diverse, distributed knowledge, in turn guiding reflexive and evaluative actions that seek intentionally to direct dynamic flows of potentialities and possibilities to achieve probable future(s). This involves identifying and supporting emergent ordering that is 'good' (that is, drawing dynamic flows toward desired futures, and/or opening up the spaces of the possible); ameliorating patterns that are 'bad' (that is, those that are diverting dynamic flows and

178   *Handbook on complexity and public policy*

lessening their transformative possibilities); and being alert to how those patterns change and can be changed (which may yet prove to be 'negative' or 'positive').

Again, some practical tools can help us identify and be sensitive to emergent and emerging patterns in mapping the way to desired future(s). The diverse field of 'futures studies' examines potential future(s), how likely any given future may be and how to help to move those future(s) into something that can be potentially 'knowable' even where they are presently 'unknown'. Drawing on both futures studies and the sustainable development literature, the idea of 'backcasting' sets out the methods and tools for first imagining the desired future and then works out the steps that might be necessary to achieve that future from the present, with reflexive checks and balances along the way, to take corrective action (see, for example, Meppem and Gill, 1998; Robinson, 1982; 2003). Wangel (2011) further suggests including social structures and agency (how and by whom change happens) in backcasting, echoing notions of complex ethics.

Seeking the solution to a problem from within the system is a feature of the literature linking the theory of autopoietic systems with complexity theory. Again, there is some work to develop ideas from this literature to use as policymaking tools. Nevertheless, they bring into consideration the self-(re)producing role of reflexivity and feedback loops, behaviour and language (which provide power to knowledge) in helping to drive and sustain emergent ordering (Abma, 2004; Eve et al., 1997; Goldspink and Kay, 2004; Stewart and Ayers, 2001).

Nowcasting is a final tool that can help identify emergent patterns. It is a relatively new application to predict short-term trends, primarily in economics (see for example, Banbura et al., 2010; Giannone et al., 2008) but also applied, for example, in analysing trends through social media (for example, see Lampos and Cristianini, 2012). At European level, the use of nowcasting for poverty risk is being developed through the EUROMOD tax and benefit micro-simulation model (Eurostat, 2013). Nowcasting provides a way for 'monitoring' or being sensitive to newly emerging patterns in the short- to medium-term. Data and indicator developments in the social sciences may provide analogous techniques, and through these provide 'way-markers' to guide policymakers and others as they create, shape and re-shape the map in the achievement of desired future(s).

A COMPLEX POLICYMAKING 'WORK IN PROGRESS'

I now explore the adoption of a multi-dimensional approach to child poverty as a 'work in progress'. It is useful first to provide some background to recent developments in this area. Ireland has had a long tradition of developing anti-poverty programmes and policies, especially since becoming a member of the European Economic Community in the 1970s. Since 1997, Ireland has adopted successive national anti-poverty strategies to address poverty and promote social inclusion. The current strategy, the National Action Plan for Social Inclusion 2007–2016 (NAPinclusion) (Government of Ireland, 2007), takes a life-cycle approach to set out both a national target[4] and a range of high-level goals with a number of associated actions, with responsibility for these falling across a range of government departments. Children are one of the life-cycle groups within the NAPinclusion, and four of its twelve high-level goals seek especially to address educational disadvantage and promote income adequacy through income supports for children and families.

Policymaking as complex cartography? 179

In 2011, the NAPinclusion national poverty target was reviewed and revised, in light of worsening poverty outcomes due to the economic recession as well as European Union and other international developments (including the adoption of a European poverty target under the EU 2020 Strategy).[5] Arising from the review of the national poverty target, the Irish government adopted a new National Social Target for Poverty Reduction[6] and agreed that new sub-targets would be set for child-poverty reduction, and for poverty reduction in jobless households.

Separately, the Department of Children and Youth Affairs (DCYA) was established in 2011 with a view to integrating a child-centred approach in policy and practice, coordinating services, policies and programmes that span multiple government departments. DCYA is responsible for the development of a Children and Young People's Policy Framework, which will be the overarching framework for integration and coordination across government, with the aim of improving children's outcomes and life chances.

In 2013, EU member states adopted the Social Investment Package, which recommends the adoption of a social investment approach across the life-cycle, in order to maximize all people's human, social and financial capital (EU Commission, 2013). The package includes a specific recommendation on investing in children to break the intergenerational cycle of disadvantage, which adopts a multi-dimensional approach to both address child poverty and promote well-being across three pillars: access to adequate resources, access to quality and affordable services and the right of the child to participate. The package also reiterated a commitment to a 2008 EU Commission recommendation on Active Inclusion, which has knock-on effects on child poverty, for example, through recommendations around parental employment and access to services to facilitate active inclusion in labour markets.

These developments, both at Irish and EU levels (and reinforced by international developments such as the publication of the OECD's Report: *Doing Better for Children* (Chapple and Richardson, 2009)), set the stage for the development of a multi-dimensional approach to address child poverty and promote well-being in Ireland. Such an approach recognizes that there is a continuum and a nested context to children's lives: addressing child poverty without also promoting broader well-being for all children will be likely to reproduce structural and other inequalities, limiting life chances and potentially destabilizing society as a whole.

Exploring a Complex Policymaking 'Work in Progress'

The rest of this chapter explores my experiences of coordinating one element of the Irish multi-dimensional approach, through developing the sub-target for child poverty reduction. As a complex policymaking 'work in progress', the following is an explorative illustration of working through the issues raised in the first half of this chapter: from recognizing, understanding and expecting social complexity as complex; to applying complex tools; to considering the role of policymakers and others, especially through complex ethics. Taken together, this exploration intends to illustrate the process by which a more responsive and adaptive policymaking approach might be taken as a kind of complex cartography, mapping its way from the policy swamp of the present toward desired future(s).

180 *Handbook on complexity and public policy*

Seeing 'Features' of Child Poverty in a Different Light Through a Complex Lens

There is not enough space to examine all issues relating to child poverty, so two aspects or enduring 'key features' are chosen as illustrative examples. These are educational disadvantage and the processes of social inclusion/exclusion. They are chosen for a number of reasons. They are manifestly socially complex, persistent and multi-dimensional. Both 'features' have implications over time (intergenerational impacts) and space (from international to local levels) and are a cause and a consequence of many interrelated factors. Considering these in the nested context of children's lives raises issues about the necessary interconnectedness and 'direction' of policymaking, the role of policymakers and others and valued choices.

EDUCATIONAL DISADVANTAGE AND EDUCATIONAL ATTAINMENT: TWO 'SIDES' OF THE SAME CONTINUUM?

Educational disadvantage covers a range of phenomena from child-specific factors (temperament, pro-social behaviours, aspects of school readiness); to parental and family factors (the quality of the home-learning environment, maternal educational attainment); school factors such as curricula, teaching methods and school culture; and neighbourhood factors, such as safety and security and prevalence of anti-social behaviour (for example, see Kellaghan et al., 1995; GUI, 2009; Smyth and McCoy, 2009; Smyth and Banks, 2012 for Irish studies).

The persistence of educational disadvantage, despite considerable efforts to address it, has generated much debate. Taking a complex perspective, a number of questions emerge, for example, does the persistence of educational disadvantage point to a self-organizing pattern that maintains stability through change? What drives that self-organization, especially given considerable efforts to disrupt 'stability' through a range of policies, programmes and practices? More broadly, what is the relationship between educational disadvantage and educational advantage (attainment), and can a complex perspective provide insight into why both educational disadvantage and educational attainment remain stable? Is there something in that relationship that perpetuates poor outcomes on the one hand, and better outcomes on the other?

Educational Disadvantage as a Necessary Counterbalance to Attainment

In working through these questions, it is useful to start with the idea that education can be seen as one of many possible iterations of a complex system, in that it appears to be dynamic, socially complex, systemic and yet also 'stable' or self-organizing. Educational advantage and disadvantage can be seen as emergent 'features' of the 'system' of education, converging through the interactions of many different factors across numerous spatial and temporal scales. Yet the observable persistence of the current system of education and of the features (and trajectories) of educational advantage/disadvantage, collectively, suggest that some kind of dynamic, complex ordering is at work. In other words, despite widespread recognition of the need for large-scale reforms in education including the need to tackle educational disadvantage as well as efforts to pursue those reforms in policy and practice, little seems to be changing.

By complex reasoning, what might in fact be happening is a kind of systemic homeostasis: not inertia, because there are quite clearly efforts to create change and in effect 'destabilize' the current system. This homeostasis is a continual, dynamic process of systemic re-balancing to maintain a given iteration where advantage is largely conferred by a relatively narrow definition of 'attainment'. Perhaps educational disadvantage persists because it must necessarily be a counterbalance to sustainable educational attainment (as it is currently constituted). To close the feedback loop and so establish homeostasis, the 'feature' of educational disadvantage acts to siphon off instable dynamic flows to establish an always/already changing stable instability, a negative attractor state as it were, continually twinned with a positive attractor state of educational advantage with its 'instable stability' of attainment.

The second point in positing educational advantage/disadvantage as attractor states is that these might exert a kind of gravitational 'pull' on the social fabric. My experiences with practitioners suggest an implicit recognition of this pull, of the cumulative negative spiral by which many experiencing poverty and disadvantage seem to be affected – a step forward, two steps back. This, in turn, provides further insights into educational disadvantage as warping the social fabric. Its effects therefore extend far beyond the 'system' of education, for example, on the capacities of parents and children to 'escape' the pull of that negative attractor state; in the resources required to 'flip' that attractor state into a more positive one (especially if that fundamentally destabilizes the positive attractor state of advantage/attainment); and in intergenerational and neighbourhood effects. This also closes off the possibilities for using other dynamic flows – such as might be generated from active inclusion in labour markets – for creating more positive outcomes.

Opening up the Spaces of the Possible: Finding Complicit and Valued 'Rules' for Educational Advantage

It is important to note that, as with all complex phenomena, this systemic view incorporates chance, contingency and non-linear dynamics, and thus does not necessarily imply conscious actions to perpetuate educational disadvantage (or, advantage at the expense of disadvantage). However, conscious actions do have a role to play in social complexity. The above insights suggest that policymaking efforts to treat educational disadvantage discontinuously from efforts to expand educational reforms will likely be ineffective. Below I explore some of the ways that complex policymaking can contribute to 'opening up the spaces of the possible', for example, by seeking to achieve 'equality of opportunities' as a desired future (an aim reflected in commonly cited policy principles, but often difficult to achieve in implementation or practice).

The first step in thinking about how complex policymaking might 'open up the spaces of the possible' is to consider the relationship between educational advantage, attainment (as it is currently constituted) and educational disadvantage. In light of the above discussion, the challenge of policymaking, here, is to disrupt the 'negative' attractor state of educational disadvantage, without destabilizing educational advantage as a 'positive' attractor state. In effect, complex-policymaking efforts should seek to change the current iteration of the 'system' of education through convergent 'rules' and dynamic flows

182 *Handbook on complexity and public policy*

in order to generate a better (more open, more productive) emergent order whilst still maintaining the trajectories of educational advantage, so as to allow the potentiality of 'equality of opportunity' to be realized as a desired future.

Such convergent 'rules' could start the process by changing (re-ordering through dynamic flows) the relatively narrow pathway of 'attainment' as a point-in-time achievement. They could seek to create transformative, diverse and multidimensional pathways of 'educational advantage'. These 'rules' could leverage the factors that lay 'behind' attainment and extend to all areas of children's lives. Thus broad policy interventions to address educational disadvantage could, for example, be aimed at building children's capacities and skills, to improve pro-social behaviours and self-regulation; fostering healthy eating and supporting nutritional needs; recognizing and connecting diverse learning environments (home, school, peer networks and so on); supporting parents' role and engagement in children's learning; and fostering safe and secure play (and play facilities) in the community. From a complex point of view, all of these can generate dynamic flows, creating synergies and opportunities for cancelling out the 'negative' pull of educational disadvantage while maintaining the 'positive' pull of educational advantage.

A second consideration here is to broaden the spaces of the possible through multiple pathways of 'attainment', whether in informal or formal settings. There is an increasing body of evidence to suggest that having broad-based competencies and life-skills especially in the area of self-regulation, and instilling a life-long 'love of learning', are strongly positively correlated to children's current well-being and long-term outcomes (Bradshaw et al., 2008; Freeney and O'Connell, 2010; Raver, 2012; Zimmerman, 2008; Zins et al., 2004). Here, the use of backcasting techniques and regulated autonomy could be helpful in identifying what, when and where 'attainment' is achieved, with way-markers of 'successful' attainment to guide the way toward realizing the potential of educational advantage in providing a desired future of equality of opportunities.

These complex tools can also be useful in considering when, and where, to intervene. For example, transitions in a child's educational/learning life are likely to be points where instability and change are at their highest. Intervening at these points could provide natural 'tipping points' through which pathways of educational advantage and attainment are re-adjusted and re-oriented, aligning them to the long-term goal of a desired future. Likewise, a focus on the early years may help to provide a tipping point to establishing a 'love of learning' and educational advantage, potentially avoiding the pull of educational disadvantage altogether.

From Possibilities to Probabilities: the Role of Complex Ethics

There are many challenges in achieving buy-in and commitment from a wide range of stakeholders for re-defining 'attainment' and adopting a wider focus on improving educational advantage for all. Deciding why, when, where, how and for whom to intervene is critical. Further, such decisions must be balanced against pragmatic realities (resource constraints, political will, ideological concerns and so on) which may determine how much is achievable in the short- to medium-term.

Recognizing and using interdependent learning and co-produced, distributed knowledge seems to be critical to whether these decisions can be transformative over the long term. From a complex perspective, such learning and knowledge might be seen as bringing together diverse and dynamic flows, a process of 'letting go' of control and letting many hands make light work (see also Dunlop, 2009; Sitterle and Kessler, 2012). Yet it is also here where complex ethics comes into play, as policymakers and others make reflexive choices to guide the transformation of potentialities and possibilities into probabilities. In this, policymakers and policies play a crucial role in establishing the trajectory of change, using tools such as regulated autonomy, nowcasting and responsive evaluation: they are, in effect, compass-bearers for the future (see also Craft and Howlett, 2012; Howlett, 2011).

A sense of complex ethics allows policymakers and others the ability to consider 'intentional disruptions', which are needed when emergent ordering is significantly diverging from the pathways toward a desired future. It also allows them to 'let go' of pathways that are significantly converging toward that future. In practical terms as it relates to educational advantage/disadvantage, this means recognizing that most children are progressing along 'typical' trajectories that lead to educational advantage already. Re-defining attainment and widening educational advantage is unlikely to disadvantage these children, and, in fact, there is considerable evidence to suggest such changes will be hugely beneficial. These children will not need interventions, so long as the spaces of the possible remain open enough that most, within the nested context of their lives and with the support of, for example, mainstream services and supports, will find their own way to a desired future. Of course, policymakers and others must also be sensitive to emergent patterns that signal divergence, and be particularly vigilant at transition points for all children, as these can have negative destabilizing effects (as well as potentially positive effects).

At the same time, educational disadvantage as a negative attractor state can be said to distort or make atypical pathways for some children, making it difficult (although not impossible) to access or divert onto 'typical' pathways toward educational advantage. Broadening the spaces of the possible is a step for accessing or diverting to 'typical' pathways; but to transform these possibilities into probabilities, policymakers and others must also make deliberate choices or intentional disruptions. Again, intervening at transition points and in the early years presents good opportunities to leverage significant change. Graduated change over time is another possibility, especially as chance and contingency may provide an unexpected 'tipping point' that transforms graduated change into significant systemic change. However, by its nature such change is unpredictable (although sensitivity to emergent patterns, for example by using nowcasting techniques, could signal some likelihood of these occurring). Although graduated change is likely to be resource intensive and may not be sufficient to disrupt atypical pathways, pragmatic realities may require such an approach in the short- to medium-term. However, there is a risk that non-directed graduated change will reinforce systemic homeostasis over the long term.

THE 'SOCIAL INVESTMENT' APPROACH AND SOCIAL INCLUSION

The second 'feature' of child poverty explored here is that of social exclusion, or its corollary in child well-being and social inclusion. Promoting social inclusion as a key element in addressing poverty has long been a central policy principle at European and

184 *Handbook on complexity and public policy*

Irish levels, as elsewhere. More recently, the focus has been on its fundamental role, not only in providing a better quality of life for all, but also in achieving social solidarity and cohesion as well as promoting resilience and capacities for individuals. Conversely, the damaging effects of social exclusion, particularly for children and families, have also received much attention in recent policy efforts. These dual-policy concerns are reflected, for example, in the EU Social Investment Package and in the Recommendation on *Investing in Children: Breaking the Cycle of Disadvantage*, in which one pillar for policymaking action is the participation of children in decisions that affect their lives and in their inclusion in all aspects of civil society.

Although always implicit in social inclusion discourses, the greater attention to issues of social cohesion and solidarity, resilience and capacities (and conversely, to disenfranchisement and vulnerability) suggests a different policy approach to social inclusion/exclusion is emerging: one that does not simply seek to bring 'them' (the socially excluded) into 'us' (the socially included), but one that recognizes mutual benefits for all as well as the collective damage that is done when people are not able to realize their full potential. Positively, this 'social investment' approach seeks to maximize all people's human and social capital in order that they might fully participate in civil society. Others are critical of this approach, arguing that it can reinforce the predominance of economic discourse/rationale in the social sphere, recasting citizens as productive units for a global economy and potentially re-creating the issue of 'blaming' the poor for their 'failure' to be 'productive' (Perkins et al., 2004; Morel et al., 2012; see also McCall, 2010).

Social Inclusion/Exclusion as Forms of Co-evolution

A complex view of this emerging shift in perspective can help to inform responsive and adaptive policy responses that are able to make the most of this 'social investment' approach while militating against potentially negative or unintended consequences. Drawing on ideas and insights already sketched out above, social inclusion appears to open up the spaces of the possible, generating convergent 'rules' that are guided by policy makers and others, which ultimately seeks the full participation of all in every aspect of civic society, with a high quality of life across all aspects of well-being. From a complex perspective, social inclusion might therefore be interpreted as a form of beneficial co-evolution: in driving forward a positive process of dynamic change that fundamentally alters the social fabric and levels the playing field. Conversely, 'social exclusion' could be interpreted as closing the spaces of the possible, dissipated by divergent 'rules' and the pull of negative attractor states, a form of antagonistic co-evolution that creates 'peaks' and 'valleys' in the social fabric, isolating individuals, families and communities (see also Byrne, 1999).

The advantage of the social-investment approach is that it is implicitly focused on investing in potentialities to realize probabilities of inclusion, equality and well-being for all. Complex policymaking responses seem well suited to implementing this approach. However, conceptualizing social inclusion and exclusion as push–pull forces on the social fabric suggests that the adoption and implementation of the social investment approach must both reinforce 'positive' change and mitigate 'negative' change. To be sustainable as a desired future, the social investment approach needs to both sustain social, economic and civic capital for all as well as provide additional investment for those who are excluded. Otherwise, it may run the risk of reproducing inequalities, of making the 'peaks' higher even as it tries to raise 'valleys' in the social fabric. It may also destabilize the 'peaks' of the 'included'.

This in turn suggests that for responsive and adaptive policymaking efforts to be effective here, both mainstreamed/universal and targeted policies, supports and services are needed because they fundamentally affect what is possible. And, from a complex view, taking this both/and approach has the potential to establish more equal, more dynamic and more open flows of change, thus increasing the possibilities for multiplier effects, self-sustainability (positive feedback loops, stability-through-change) and crucially, less resource-intensive efforts over the long term. Such considerations are important in light of recent debates in the United Kingdom and in the United States and to some degree in Ireland that have asked whether the balance of policy efforts is too heavily in favour of universalism (see, for example, Barnett et al., 2004, Government of Ireland, 2010; ISSA, 2013; Whitehurst, 2013).

CONCLUSIONS

This chapter has argued that is it possible to conceptualize a complex policymaking process as a kind of complex cartography, mapping the way to a desired future(s) from the policy swamp of the present. Such an approach starts from the 'first principles' that social complexity appears to exhibit similar features, properties and behaviours to complexity as it is conceptualized in the natural sciences.

It may not (yet) be possible to connect the two empirically, but expecting socially complex phenomena to behave as complex allows for a different perspective. One, it sensitizes policymakers and others to emergent patterns, to recognizing and understanding complexity in the issues they face. Two, the expectation of complexity leads, in turn, to questions about how policymaking can positively leverage dynamic flows of change, potentiality and possibility, generated by socially complex phenomena. Complex policymaking then becomes a process of finding convergent, complicit 'rules' that produce 'super' emergence (emergent order) that opens up the spaces of the possible. Three, there is a substantial body of literature that provides practical tools through which complex policymaking might implement adaptive and reflexive policy responses.

Finally, the idea of complex ethics brings into consideration a key difference between social and natural complexity, namely that the former can be affected by conscious decisions, power and ideology. Making deliberative and evaluative choices, working with others in shaping the map to the future, policymakers, here, act as 'compass bearers' to the desired future(s), correcting and re-orienting emergent ordering and pathways.

Exploring two 'features' of child poverty has illustrated how the ideas and insights generated in the first half of the chapter could be used to inform a policymaking 'work in progress', in order to develop a multi-dimensional approach to child poverty. A complex understanding of educational disadvantage/advantage suggests that the 'system' of education in its current iteration uses both to maintain systemic homeostasis, an instable/stable stability-through-change. Complex policymaking might, then, focus on addressing the broad continuum of advantage and disadvantage within the nested context of

186 *Handbook on complexity and public policy*

children's lives, and re-defining pathways of 'attainment', to generate convergent 'rules' towards a desired future. In discussing 'typical' and 'atypical' pathways, the role of complex ethics has also been explored. This helps policymakers and others to decide when to 'let go' and allow 'many hands to make light work'; when to be vigilant (for example at transition points); and when to 'intentionally disrupt' atypical pathways, to reconnect to pathways of educational advantage.

Finally, social inclusion/exclusion was examined in light of the emerging shift to a social-investment approach, which by its nature seeks to invest in 'potentiality'. However, conceptualizing inclusion/exclusion as different forms of co-evolution raises important considerations about how this approach should be implemented. This view strongly suggests that successful implementation needs to take both a mainstreaming and a targeted approach. Otherwise, it runs the risk of continuing to deform antagonistically the social fabric, reproducing inequalities and closing off the spaces of the possible for those 'excluded', and potentially destabilizing the 'peaks' of the 'included'.

The arguments for a complex policymaking as a kind of complex cartography, in turn suggest new directions for the role of evidence-informed policymaking (EIPM). In this, data, research and evidence play a vital role in sensitizing policymakers and others to emerging patterns. EIPM becomes part not only of understanding what was and what is, but also what might be (and, with complex ethics, what should not be). A complex view of EIPM might ask, not 'what works', but 'what are the best possible solutions for all': recognizing and embracing change, over time and space, as the potential to achieve desired future(s).

NOTES

1. The former might be described as surface simplicity underpinned by deep complexity, as in the social organization of ants or birds; the latter, as deep simplicity determining surface complexity, as in the proverbial 'butterfly effect' (drawing on Cohen and Stewart, 1994).
2. I use the term 'attractor states' as a generic term describing the general phenomenon of attractors, instead of using the terms 'strange attractors', 'hidden attractors', 'self-excited attractors' or other forms of attractors, all of which have precise definitions through mathematical formulae and modelling (see, for example, Leonov and Kuznetsov, 2013). There are also a class of attractors known as 'Julia sets', which effectively act as strange repellers. See http://en.wikipedia.org/wiki/Chaos_theory#Strange_attractors (accessed 15 September 2013).
3. Co-evolution refers to the process of how interacting species deform fitness landscapes such that they constantly have to evolve, and evolve again (see Kauffman, 1995).
4. The overall goal of the NAPinclusion is to reduce the number of those experiencing consistent poverty to between 2 per cent and 4 per cent by 2012, with the aim of eliminating consistent poverty by 2016.
5. The Europe 2020 Strategy contains as one of its five headline targets a target in relation to poverty: to lift at least 20 million people out of the risk of poverty and exclusion by 2020. Member states are required to set national poverty targets, using appropriate national indicators, in support of the EU target (European Commission, 2010).
6. To lift at least 200000 people out of the risk of poverty and exclusion between 2012 and 2020 by reducing the number experiencing consistent poverty to 4 per cent by 2016 (interim target) and to 2 per cent or less by 2020, from the 2010 baseline rate of 6.2 per cent (Department of Social Protection (2013), Annual Report 2012, Government of Ireland, Stationery Office, available at http://www.welfare.ie/en/downloads/ar2012.pdf, p. 8).

REFERENCES

Abma, T. (2004), 'Responsive evaluation: The meaning and special contribution to public administration', *Public Administration*, **82**(4), 993–1012.

Allen, P. (2002), 'Evolution, emergence and learning in complex systems', paper presented at 'Tackling Industrial Complexity: The ideas that make a difference', 9–10 April, Downing College, Cambridge, UK, accessed 24 September 2013 at http://www2.ifm.eng.cam.ac.uk/mcn/pdf_files/part5_1.pdf.

Bak, P. (1996), *How Nature Works: The Science of Self-Organised Criticality*, New York: Copernicus Press.

Bak, P. and M. Creutz (1994), 'Fractals and self-organised criticality', accessed 28th July 2013 at http://lattice guy.net/mypubs/pub123.pdf.

Banbura, M., D. Giannone and L. Reichlin (2010), 'Nowcasting', *ECARES Working Paper*, 2010-0212010, Brussels: ECARES.

Barnett, W.S., K. Brown and R. Shore (2004), 'The universal vs. targeted debate: Should the United States have preschool for all?', *Preschool Policy Matters*, (6), April, New Brunswick, NJ: National Institute for Early Education Research.

Bradshaw, C.P., L.M. O'Brennan and C.A. McNeely (2008), 'Core competencies and the prevention of school failure and early school leaving', in N.G. Guerra and C.P. Bradshaw (eds), 'Core competencies to prevent problem behaviours and promote positive youth development', *New Directions for Child and Adolescent Development*, (122), 19–32.

Byrne, D. (1998), *Complexity Theory and the Social Sciences: An Introduction*, London: Routledge.

Byrne, D. (1999), *Social Exclusion*, Buckingham: Open University Press.

Cairney, P. (2012), 'Complexity theory in political science and public policy', *Political Studies Review*, **10**(3), 346–58.

Cameron, K. and R. Quinn (1999) [1st edn], (2006) [latest edn], *Diagnosing and Changing Organizational Culture: Based on the Competing Values Framework*, San Francisco, CA: Jossey Bass/Wiley and Sons.

Chapple, S. and D. Richardson (2009), *Doing Better for Children*, Paris: OECD.

Church, M. (1999), 'Organizing simply for complexity: Beyond metaphor towards theory', *Long Range Planning*, **32**, 425–40.

Cohen, J. and I. Stewart (1994), *The Collapse of Chaos: Discovering Simplicity in a Complex World*, London: Penguin Books.

Craft, J. and M. Howlett (2012), 'Policy formulation, governance shifts and policy influence: Location and content in policy advisory systems', *Journal of Public Policy*, **32**(2), 79–98.

Dawkins, R. (1976), *The Selfish Gene*, (30th anniversary edn, 2006), Oxford: Oxford University Press.

Dunlop, C. (2009), 'Policy transfer as learning: Capturing variation in what decision-makers learn from epistemic communities', *Policy Studies*, **30**(3), 289–311.

European Commission (2010), *Europe 2020: A Strategy for Smart, Sustainable and Inclusive Growth: Communication from the Commission*, Brussels: EU Commission Publications Office.

European Commission (2013), 'Social investment package', EU Commission 20.2.2013 COM (2013) 83, EU Commission: Brussels, accessed 24 September 2013 at http://ec.europa.eu/social/main.jsp?catId=1044&lang Id=en&newsId=1807&moreDocuments=yes&tableName=news.

Eurostat (2013), 'Using EUROMOD to nowcast poverty risk in the European Union', *Eurostat Methodologies and Working Papers*, Luxembourg: Publications Office of the European Union.

Eve, R., A.S. Horsfall and M.E. Lee (eds) (1997), *Chaos, Complexity, and Sociology*, Thousand Oaks, CA: Sage.

Freeney, Y. and M. O'Connell (2010), 'Wait for it: Delay-discounting and academic performance among an Irish adolescent sample', *Learning and Individual Differences*, **20**(3), 231–6.

Fuchs, C. and J. Collier (2007), 'A dynamic systems view of economic and political theory', *Theoria: A Journal of Social and Political Theory*, (113), August, 23–52.

Gell-Mann, M. (1994), *The Quark and the Jaguar: Adventures in the Simple and the Complex*, London: Little, Brown.

Geyer, R. (2012), 'Can complexity move UK policy beyond "evidence-based policy making" and the "audit culture?" Applying a "complexity cascade" to education and health policy', *Political Studies*, **60**(1), 20–43.

Giannone, D., L. Reichlin and D. Small (2008), 'Nowcasting: The real-time informational content of macroeconomic data', *Journal of Monetary Economics*, **55**(4), 665–76.

Goldspink, C. and R. Kay (2004), 'Bridging the micro–macro divide: A new basis for social science', *Human Relations*, **57**(5), 597–618.

Government of Ireland (2007), *National Action Plan for Social Inclusion 2007–2016*, Dublin: Government Stationery Office.

Government of Ireland (2010), *A Policy and Value for Money Review of Child Income Support*, Dublin: Government of Ireland.

188 *Handbook on complexity and public policy*

Growing Up in Ireland (2009), *Key Findings: Child Cohort – The Education of 9-year-olds*, Dublin: Office of the Minister for Youth and Children.

Habib, F. and A. Shokoohi (2009), 'Classification and resolving urban problems by means of fuzzy approach', *World Academy of Science, Engineering and Technology*, **36**, 894–901.

Hausner, J. (ed.) (1997), *Beyond Market and Hierarchy: Interactive Governance and Social Complexity*, Cheltenham, UK and Northampton, MA, USA: Edward Elgar Publishing.

Hearn, G. and D. Rooney (2002), 'The future role of government in knowledge-based economies', *Foresight*, **4**(6), 23–33.

Hearn, G., D. Rooney and T. Mandeville (2003), 'Phenomenological turbulence and innovation in knowledge systems', *Prometheus*, **21**(2), 231–45.

Howlett, M. (2011), 'Public managers as the missing variable in policy studies: An empirical investigation using Canadian data', *Review of Policy Research*, **28**(3), 247–63.

Imam, S.A., A.C. Johnson and M.O. Askar (2004), 'Bringing the corporation to life: The implications of biological systems theory for business strategy', *Proceedings of ISOneWorld Conference*, Las Vegas, pp. 421–8.

International Social Security Association (ISSA) (2013), 'Child Benefits in the UK: The beginning of the end of universality', *Social Security Observer*, (22), September, accessed 20 September 2013 at http://www.issa.int/-/child-benefits-in-the-uk-the-beginning-of-the-end-of-universality-.

Kauffman, S. (1995), *At Home in the Universe*, London: Viking.

Kellaghan, T., S. Weir, S. Ó hUallacháin and M. Morgan (1995), *Educational Disadvantage in Ireland*, Dublin: Department of Education/Combat Poverty Agency and Educational Research Centre.

Kiel, L.D. and E. Elliott (eds) (1996), *Chaos Theory in the Social Sciences*, Ann Arbor: University of Michigan Press.

Kooiman, J. (ed.) (1993), *Modern Governance: New Government–Society Interactions*, London: Sage.

Lampos, V. and N. Cristianini (2012), 'Nowcasting events from the social web with statistical learning', *ACM Transactions on Intelligent Systems and Technology (TIST)*, **3**(4), 72.

Latour, B. (2000), 'When things strike back: a possible contribution of "science studies" to the social sciences', *British Journal of Sociology*, **51**(1), 107–24.

Law, J. (1992), 'Notes on the theory of the actor-network: Ordering, strategy and heterogeneity', *Systems Practice*, **5**(4), 379–93.

Leonov, G.A. and N.V. Kuznetsov (2013), 'Hidden attractors in dynamical systems. From hidden oscillations in Hilbert–Kolmogorov, Aizerman, and Kalman problems to hidden chaotic attractor in Chua circuits', *International Journal of Bifurcation and Chaos*, **23**(01).

Loorbach, D. (2010), 'Transition management for sustainable development: A prescriptive, complexity-based governance framework', *Governance*, **23**(1), 161–83.

Manson, S.M. (2001), 'Simplifying complexity: A review of complexity theory', *Geoforum*, **32**, 405–14.

Marston, G. and R. Watts (2003), 'Tampering with the evidence: A critical appraisal of evidence based policy-making', *The Drawing Board: An Australian Review of Public Affairs*, **3**(3), 143–63.

McCall, V. (2010), 'Cultural services and social policy: Exploring policy makers' perceptions of culture and social inclusion', *Journal of Poverty and Social Justice*, **18**(2), 169–83.

Meppem, T. and R. Gill (1998), 'Planning for sustainability as a learning concept', *Ecological Economics*, **26**, 121–37.

Morcol, G. (2001), 'What is complexity science? Postmodernist or postpositivist?', *Emergence: A Journal of Complexity Issues in Organizations and Management*, **3**(1), 104–19.

Morel, N., B. Palier and J. Palme (eds) (2012), *Towards a Social Investment Welfare State?*, Bristol: The Policy Press.

Morelli, C.J. and P.T. Seaman (2005), 'Universal versus targeted benefits: The distributional effects of free school meals', *Environment and Planning C: Government and Policy*, **23**(4), 583–98.

Moss, S. (2002), 'Agent based modelling for integrated assessment', *Integrated Assessment*, **3**(1), 63–77.

Nicolaides, A. and L. Yorks (2008), 'An epistemology of learning through', *E:Co*, **10**(1), 50–61.

Parsons, W. (2001), 'Modernising policy-making for the twenty first century: The professional model', *Public Policy & Administration*, **16**(3), 93–110.

Parsons, W. (2002) 'From muddling through to muddling up: Evidence-based policy making and the modernisation of British government', *Public Policy & Administration*, **17**(3), 43–60.

Perkins, D., P. Smyth and L. Nelms (2004), 'Beyond neo-liberalism: The social investment state?', *Social Policy Working Paper No. 3*, Melbourne/Fitzroy, Australia, Centre for Public Policy/ Brotherhood of St Laurence.

Prigogine, I. and I. Stengers (1984), *Order out of Chaos*, London: Bantam Books.

Quinn, R.E. (1988), *Beyond Rational Management*, San Francisco, CA: Jossey-Bass.

Raver, C.C. (2012), 'Low-income children's self-regulation in the classroom: Scientific inquiry for social change', *American Psychologist*, **67**(8), 681–9.

Robinson, J. (1982), 'Energy backcasting: A proposed method of policy analysis', *Energy Policy*, **10**(4), 337–44.

Robinson, J. (2003), 'Future subjunctive: Backcasting as social learning', *Futures*, **35**(8), 839–56.

Sanderson, I. (2009), 'Intelligent policy making for a complex world: Evidence and learning', *Political Studies*, **57**(4), 699–719.

Schön, D.A. (1979), 'Generative metaphor: A perspective on problem-setting in social policy', in A. Ortony (ed.), *Metaphor and Thought*, Cambridge: Cambridge University Press.

Schön, D.A. (1983), *The Reflective Practitioner*, New York: Basic Books.

Seel, R. (1999), 'Complexity and organisation development: An introduction', accessed 24 September 2013 at http://www.new-paradigm.co.uk/complex-od.htm.

Shine, K.T. (2006), 'Regenerating unpopular social housing estates: Can complexity theory help to achieve best possible solutions?', *Housing, Theory and Society*, **23**(2), 65–91.

Shine, K.T. and B. Bartley (2011), 'Whose evidence base? The dynamic effects of ownership, receptivity and values on collaborative evidence-informed policy making', *Evidence & Policy: A Journal of Research, Debate and Practice*, **7**(4), 511–30.

Sitterle, V.B. and W. Kessler (2012), 'Knowledge exchange and integrative research approach', *Information, Knowledge, Systems Management*, **11**(1), 169–93.

Smith, A.C. and F. Graetz (2006), 'Complexity theory and organizing form dualities', *Management Decision*, **44**(7), 851–70.

Smyth, E. and J. Banks (2012), 'High stakes testing and student perspectives on teaching and learning in the Republic of Ireland', *Educational Assessment, Evaluation and Accountability*, **24**(4), 283–306.

Smyth, E. and S. McCoy (2009), *Investing in Education: Combating Educational Disadvantage*, Dublin: ESRI.

Stacey, R. (1995), 'The science of complexity: An alternative perspective for strategy', *Strategic Management Journal*, **16**(6), 477–95.

Stacey, R. (2000), 'The emergence of knowledge in organisations', *Emergence*, **2**(4), 23–9.

Stacey, R. (2001), *Complex Responsive Processes in Organizations: Learning and Knowledge Creation*, London: Routledge.

Stacey, R. (2003), 'Learning as an activity of interdependent people', *The Learning Organisation*, **10**(6), 325–34.

Stewart, J. and R. Ayres (2001), 'Systems theory and policy practice: an exploration', *Policy Sciences*, **34**(1), 79–94.

Talbot, C. (2003), 'How the public sector got its contradictions – the tale of the paradoxical primate: Integrating the idea of paradox in human social, political and organisational systems with evolutionary psychology', *Human Nature Review*, **3**, 183–95.

Talbot, C. (2005), *The Paradoxical Primate*, Exeter: Academic Imprint.

Talbot, C. (2008a), 'Irish public service reforms in international perspective: A competing public values approach', paper presented at the Institute of Public Administration conference 'A public service for the future? The OECD challenge', Dublin, 29 May.

Talbot, C. (2008b), 'Measuring public value: a competing values approach', paper prepared for the *Work Foundation*, London.

Tyfield, D. (2012), 'A cultural political economy of research and innovation in an age of crisis', *Minerva*, **50**(2), 149–67.

Wagenaar, H. (2007), 'Governance, complexity, and democratic participation: How citizens and public officials harness the complexities of neighborhood decline', *The American Review of Public Administration*, **37**(1), 17–50.

Wangel, J. (2011), 'Exploring social structures and agency in backcasting studies for sustainable development', *Technological Forecasting and Social Change*, **78**(5), 872–82.

Whitehurst, G.J. (2013), 'Can we be hard-headed about preschool? A look at universal and targeted pre-K', Washington, DC, *Blog post of The Brookings Institution*, 23 January, accessed 24 September 2013 at http://brookings.edu/blogs/brown-center-chalkboard/posts/2013/01/23-prek-whitehurst.

Whitt, R.S. (2008), 'Adaptive policymaking: Evolving and applying emergent solutions for US communications policy', *Fed. Communications Law Journal*, **61**, 1 June, 483.

Wilsford, D. (2010), 'The logic of policy change: Structure and agency in political life', *Journal of Health Politics, Policy and Law*, **35**(4), 663–80.

Zimmerman, B.J. (2008), 'Investigating self-regulation and motivation: Historical background, methodological developments and future prospects', *American Educational Research Journal*, **45**(1), 166–83.

Zins, J.E., M.R. Bloodworth, R.P. Weissberg and H.J. Walberg (2004), 'The scientific base linking social and emotional learning to school success', in J.E. Zins, R.P. Weissberg, M.C. Wang and H.J. Walberg (eds), *Building Academic Success on Social and Emotional Learning: What Does the Research Say?*, New York: Teachers Press, Columbia University, pp. 1–22.

12. The role of models in bridging expert and lay knowledge in policy-making activities

Sylvie Occelli and Ferdinando Semboloni

INTRODUCTION

As with other societal organizations, governmental systems, currently exposed to the same social, economic, geopolitical and institutional pressures, will need to adapt and make major changes to their systems. Since the last decade, a number of processes have been at work in concocting a radically different policy background that has also been popularized as e-government and e-governance transformations (Gil-Garcia, 2012; Scholl, 2003). Transformations associated with responsibility enhancement are of utmost relevance, as they in fact urge public administration to improve its overall production processes by means of ICT, while also designing, managing and accounting for policy actions in innovative ways (Peach, 2004; Johnston, 2010; OECD, 2011; van Veenstra et al., 2010).

Increasingly, incomplete knowledge and uncertainty are recognized as a main source of the problems faced by policy-making. Not only does there exist an intrinsic difficulty in knowing the whole and detailed aspects of a social system, but also the complexity of social patterns makes behaviour unpredictable (Conklin, 2006). To cope with this problem, various solutions have been proposed such as, for instance, risk analysis, portfolio differentiation, and so on.

Actually, no matter how policy- and governmental-change processes are viewed, they turn out increasingly to demand knowledge (Occelli, 2010). This requirement should not be unexpected, however, since in the current debate knowledge and learning are widely acknowledged as being foundational to the very features of tomorrow's society. In addition, in the implementation of complex policy, the ability to learn new ideas appears as a necessary and distinct resource (Moss, 1999; Tait and Richardson, 2011).

So far, science has been the unwarranted champion in knowledge processes and its role in supporting a knowledgeable society of tomorrow is expected to further enhance its importance. Scientific thinking, however, has not been immune to the effects of changes in society. The transformations occurring in the field are by no means less deep than those observed in society (Healy, 1999). They herald, in fact, something different from the kind of paradigm shift advocated by Kuhn (1970) whenever novel theories are consolidated. Uncertainty, urgency, ethical values and the relevance of the issues at stake for society often make (policy) situations unsolvable by means of neat scientific expertise. In many cases, an extended peer community is called for in which scientific experts and stakeholders (including decision-makers and citizens) engage in dialogue and leverage their own knowledge in order to cope with the policy problems at hand (Maier-Rabler and Huber, 2011; OECD, 2011).

Whether the goals of this community should be locally based and set by specific population needs, such as those related to the delivery of education or health services, or motivated by wider concerns of society, such as those caused by climate change, is an open question. A related and even more thorny question is how expert and lay knowledge would (or should) be shared and how their mutual understanding could be leveraged in supporting policy processes in the extended peer community (Lupia, 2012).

The main argument of this chapter is that, in current increasingly ICT-mediated human organizations, models can be effective vehicles for supporting policy processes in the community. To illustrate this, in the next section the functional and cognitive mediator roles conventionally played by models in policy-making are recalled (Morrison and Morgan, 1999; Occelli, 2001; 2002; 2006; Semboloni, 2007a; 2007b). The functional aspects mainly concern the syntactic, semantic and design components of a modelling activity; the cognitive deal with the links between the process of abstraction involved in model-building and the general, socio-economic, cultural and institutional context in which the modelling activity takes place. Building upon the experience gained in the model-based policy studies carried out in Piedmont and Tuscany over the last decades, these roles are reappraised and the main pitfalls in model usage reviewed in the next section. In 'New requirements in policy-modelling' we suggest that, in order to overcome some of the current difficulties of policy-making, a reappraisal of the roles models play that also supports the extended peer community is necessary. A suggestion is made that these roles can be viewed as boundary objects, which enable relationships between the scientific, policy and societal worlds to be bridged. Finally, in the conclusion, we stress the cognitive mediation role of models in supporting policy processes, and emphasize the capability of models as connectors between expert and lay knowledge, making sense of social processes and helping communities to improve their organizations.

MODELLING IN POLICY: PROBING THE CONNECTIONS

Models and Policy Decision-making

Having insight into phenomena and obtaining general explanations are the main features of conventional modelling in the natural sciences. Models make it possible to derive a simplified account of phenomena, and enable an observer to better apprehend the ways they are produced. The view stems from the two classical approaches underpinning model development in science, namely, the realistic approach aimed to give substance, credibility and truthfulness to the observations of the real world and the axiomatic approach allowing a consistent analytic framework to be set up. The cognitive perspective discussed later in this chapter can be related to a projective approach, according to which, modelling is used both to uncover interpretive keys for problem definitions (Pidd, 1996; Umpleby, 2011) and engage theory into actions (Morin and Le Moigne, 1999).

In policy discourse, models typically have been regarded as a special kind of knowledge, aimed at reproducing in the social sciences the same characteristics of a 'laboratory' as seen in the natural sciences. In the social sciences, alike, models apply an organized set of concepts (a theory) which posit certain agents' behavioural rules and make it possible to investigate how they work (perform) for the social system as a whole. In particular, changing some parameters of the model structure allows analysts to gain insight into the system of global behaviour. When applied to a study area, furthermore, models make it possible to expose and/or investigate the most sensitive aspects of system behaviours and probe the system for likely future states.

When used to address urban change in particular, conventional model applications were carried out by conveniently entering scenarios which, say, posited overall growth rates for population and transportation costs or looked at the impact of certain policy actions (such as the road-building or constraints to land utilization) and then investigating likely outcomes on the study system. However, more often than not, this approach proved unsatisfactory in assisting the policy process.

What this system failed to address was the complex web of interactions among goals and interests underlying policy decision-making as a whole. In particular, it proved unable to deal with three main aspects (Head, 2008): (a) the inherently political and value-based nature of policy debate and decision-making; (b) the fact that information is perceived and used in different ways by actors looking through different 'lenses'; and (c) the complexity of the arrangements of networks, partnerships and collaborative governance when regarded against the current organization and functioning of the public sector.

Indeed, negotiation in decision-making is increasingly recognized as a part of the process of social learning (Pahl-Wostl and Hare, 2004). In this context, the conventional role of the model as a predictive tool is questioned and novel roles aimed at favouring a more inclusive process of system learning have gained momentum. Besides, model tools and applications can generate opportunities for social dialogue and discussion, thus producing an environment of shared knowledge among the various actors (stakeholders) involved in the decision-making process.

In fact, decisions involving public issues can no longer be regarded as linear courses of action undertaken by perfectly informed actors. Rather, policy decisions emerge as negotiated outcomes from within the (growing) social arena, often following a conflict situation. In other words, policies result from a process of social negotiation that entails two processes: learning by all the actors involved and decision-making.

By using dialogue and negotiation, communities and stakeholders improve their understanding of the different points of view in order to find a shared solution. Social learning thus arises in relationship to policies around societal issues through positive experiences of participation, so that learning and decision-making are strictly tied, in the sense of belonging to the same process and continuously interacting.

In this context, decision-making can be viewed as a sequence of negotiations among many actors who analyse situations, define problems, search for alternative solutions and finally implement those solutions. In this participatory process, the crucial role is that of the stakeholders, who represent the interests of individuals or a group of individuals involved in a particular issue. It is from the negotiations among these stakeholders that the decision emerges as an attractor of the process (Semboloni et al., 2004; Semboloni, 2007a; 2007b).

Functional and Cognitive Mediation Roles of Models in the Policy Domain: a Framework

From the argument above it is apparent that an evolution of the role of models in policy-making has taken place, although no unique statement of what a model is and the functions that it performs can be distinguished.

The role of models in bridging expert and lay knowledge 193

The first general statement according to Dupuy (2000: 29) maintains that:

a model is an idealization, usually formalized in mathematical terms, that synthesizes a system of relations among 'elements whose identity and even nature, are up to a certain point a matter of indifference, and which can, as a result, be changed, replaced by other elements, analogous or not, without [the model's] being altered'.

A second, more formal statement is offered by Josilyn and Turchin (1993) who point out that, for a system concerned with survival, the most immediate kind of a model is a metasystem which implements a form-preserving (homomorphic) relation between the states of the modelled system (the world) and those stemming from a modelling system (via a representational function).

On a conceptual ground, the idea that models correspond with processes of abstraction and knowledge representation is acknowledged in the Artificial Intelligence field. Arising from this idea is what we consider as the third statement about models: to grasp the basics of what a knowledge representation is, some researchers in the field (for example Davis et al., 1993) contend that this can be understood by considering the roles that knowledge representation entails whenever it is applied to a certain task (Table 12.1).

How each role is instantiated in a representation and the rationale for that reveals what the representation would command about how to view the world. Being aware of these roles turns out to be useful, primarily because they would inform a conscious choice of the properties of the knowledge representations required for a certain task. Eventually, the identification of the properties of a representation would stem from the different roles that a general knowledge representation would play when applied to a situation.

Table 12.1   The roles of knowledge representation

Roles	Contents	Implications and questions raised
A surrogate	It is a stand-in for the things that exist in the world	Intended identity and fidelity
A set of ontological commitments	A view in order to focus on the things in the world in which we are interested	Definition of the sets of concepts offered as a way of thinking about the world
A fragmentary theory of intelligent reasoning	Identification of the fundamental concepts of intelligent reasoning	All representations are imperfect, and any imperfection can be a source of error
A medium for efficient computation	Representations should be computable	Representations offer a set of ideas about how to organize information in ways that facilitate recommended inferences
A medium of human expression	The means by which we express things about the world and communicate them for our use	How well does the representation function? How precise and adequate?

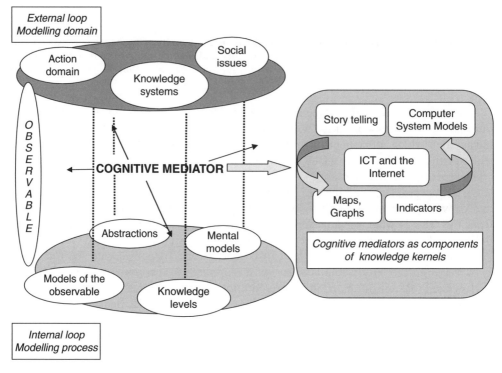

Figure 12.1   Modelling as a cognitive mediator

These arguments ultimately convey what is the fourth statement: that modelling works as a mediator between the human ability to conceptualize and the external world, while also being an autonomous artefact distinct from both of them (Morrison and Morgan, 1999).

Furthermore, when attention is paid to their implementation in practical cases, models can be viewed as cognitive mediator artefacts between so-called internal and external loops (Occelli, 2001; 2002). Both loops (see Figure 12.1) are centred on the observable – the phenomena observed in the world – which are subject to investigation. The lower loop, called the internal, refers to the conventional steps underlying a process of abstraction. This has its roots in the positivistic assumptions held in the mainstream of the social sciences. It proceeds from observation to the formulation of concepts and more formal models of it (the so-called encoding process), and terminates by referring back to the latter observed reality (through the so-called decoding process) (Rosen, 1985).

The upper (external) loop refers to the context to which the process of abstraction refers. It reminds us that modelling is not a stand-alone activity, but that it must engage with the context in which it develops (Occelli, 2001). A modelling domain, therefore, is defined by several societal and institutional factors characterizing the model-application area, such as the relevant policy issues raised in the local area, the available information (knowledge system), and the policy actions the modelling activity is expected to support.

It should be noted that the scheme accommodates and encompasses the statements about modelling introduced earlier. In particular, the first and second statements focus on the abstraction process dealt with by the internal loop of the modelling activity. As it specifically addresses the roles of cognitive mediation, the third statement emphasizes that many types of connections can be established between the external and internal loops of modelling. Finally, the fourth statement about the autonomy of models provides ground to the fact that, as a result of the increasing pervasiveness of ICT, different types of (model) artefacts can be developed, including maps, indicators, computer models and, more recently, storytelling. All of these artefacts are means for representing (instantiating) the knowledge hypotheses that observers embrace when investigating, interpreting and exchanging ideas about phenomena (Occelli, 2010).

On a conceptual ground, therefore, models can be regarded as cognitive mediator tools for testing, exploring, creating and communicating knowledge about the world in which agents live (Occelli, 2001; 2002; Occelli and Rabino, 2006).

When developing a model for practical application (see examples of cognitive mediators mentioned in Figure 12.1), three main components are leveraged. Their definition depends on the types of relationships between the external and internal loops of the modelling activity that a certain application will make use of. These components are:

1. A syntactic component, which entails a method of analysis that is a coherent set of steps of inquiry associated with an abstraction process (as typically commanded by the internal loop of a modelling process) and whose unfolding gives an explanation for how observed phenomena take place.
2. A representational (semantic) component, related to the interpretation of the phenomena as 'explained' by the syntactic component. The kind of conceptualization (the individuals' mental models and the system theoretical accounts) stemming from using the model, and its meaningfulness according to the adopted value system, are crucial aspects of this component.
3. A design component, associated with the model's purpose given the system conditions in which the model application occurs (that is, the aims of the model development, data availability, ICT familiarity, computer skills, expected results, and so on). This component plays a crucial role both in instantiating the model application and steering the accompanying learning process.

SOME INSIGHTS FROM ITALIAN REGIONAL CASE STUDIES

The potential of the cognitive mediation role played by models has also been observed in two research experiences performed at IRES and Florence University. At IRES, where the study is policy oriented, three main approaches were identified into the use of models carried out in the last two decades (see Table 12.2) (Occelli and Landini, 2002).

1. Intelligent packages: address specific developmental questions (population and economic growth at regional and sub-regional levels) which have been studied extensively in mainstream regional literature. They exploit consolidated analytic techniques, make use of large time-series regional data and have been incorporated into customized packages. Simulation, restricted mainly to the internal loop, plays a major role as an analytic dimension, supporting algorithmic procedures, and making the operational implementation of the model possible.

196 Handbook on complexity and public policy

Table 12.2 Stylized model approaches used in the regional studies at IRES

Descriptive keys (see Figure 12.1)	Intelligent package	Decision support system	Cognitive models
Social issues (policy-relevant phenomena)	Costs and benefits of sector growth in local areas	Efficiency and effectiveness of urban dynamics in the region	Human decision-making in a spatial environment and formation of collectives
Action domain	Sub-regional distributions of socioeconomic variables	Reasoning about socioeconomic scenarios in a spatial framework	Awareness of action spaces in a spatial environment
Observable	Region as a mix of sector activities and population types	Region as a system of interacting activities	Region as a self-organized system inhabited by social, autonomous, cognitive agents
Abstractions (theoretical bases)	Spatial equilibrium and utility maximization approaches	Integration of input–output and spatial-interaction approaches in a multi-level framework	Activity space of agents in a time–space framework, complexity approach
Mental models	Not relevant	Implicitly included	Included
Knowledge levels – see Moss (1999)	Low–medium granularity and high formalization	Medium granularity and medium formalization	High granularity, low formalization
Model of the observable	Regional system model, urban-sector partial equilibrium models	Stock-flow models, operational urban-system models	Multi-agent models of spatial movements and behaviour in an (artificial) environment

2. Decision-support systems: primarily are aimed at assisting analysts for policy-making. Typically they are hybrid systems integrating three components: a core system describing the urban structure; an information component for data retrieval and output evaluation; and a graphic interface for output visualization. In these system models, the role of simulation is not limited to the internal functioning of a component but, because of the 'what if' perspective, a set of links with the model's external domain are established, although mainly associated with the technological backcloth characterizing the architecture of the information system.

3. Cognitive models: the investigation of human decision-making and action in an uncertain and changing environment is the main purpose of these models. Human cognitive abilities are (tentatively) addressed, and simulation is an intrinsic dimension underpinning this type of model.

Notwithstanding the experience of models at IRES, it has been noticeable that model effectiveness per se has often been in doubt or scarcely perceived by the policy institutional setting (Occelli, 2007). Greater appreciation has been shown for the indicator-modelling more directly associated with monitoring human activity, such as that carried out as part of observations maintained at IRES on behalf of the regional government departments into, for example, population, employment and other socio-economic phenomena in regional sub-areas.

The potential of the cognitive approach is even more apparent in the model applications carried out at Florence University. The CityDev model (Semboloni et al., 2004; Semboloni, 2006; 2007a) is an urban, multi-agent simulation that includes many aspects of social organization, such as monetary fluxes of the urban system, which has been conceived as a comprehensive urban simulation that allows for the participation of human users. In essence, the proposed approach considers reality to be founded on the basic dichotomy of subject–object, whereby the subject or agent as the basic engine produces, transforms and consumes the objects (goods), which are conceived as passive entities. Nevertheless, these objects, especially real estate, change the environment of the subjects, thus influencing their decisions. Therefore, CityDev is based on agents, goods and markets, which allow agents to trade the goods and allow the user to generate scenarios.

Agent-based simulations are recognized as a means to improve the analysis of social processes. This is because they make it possible to account for the subjective viewpoints of the actors involved in each process. On the one hand, modelling may become part of the process of social learning and, on the other, once built, the multi-agent simulation, in connection with the role-playing approach, can be used for participatory simulation. In a participatory simulation, stakeholders take control of the software agents in order to improve the rules of the simulation and to generate scenarios in which the subjective view of actors is included.

It is for these reasons that the CityDev model has been conceived as an interactive system available on the web, in order to allow a participatory simulation (Semboloni, 2007b). In the participatory simulation of the CityDev model, a human user plays the role of an exogenous factor in the model dynamics. He/she may generate new agents and control these agents by using his/her own strategies, instead of those provided by the model for the software agents. In fact, each agent being an instance of the general group, the generation of an agent is simply an instantiation. If managed by the model, agents are processed as usual; otherwise, they are not processed and the computer simulation waits for the decision of a human user in order to get the value of the concerned variables; for instance, on which plot a land-developer wishes to build. However, once a human user has managed these agents, they are processed just like the other agents in the following phases of the simulation. The result is that agents managed by a human user and those managed by the computer compete at the same level, and the comparison of their performances, such as the amount of profit earned, is allowed.

The results show that the multi-agent method places the human actor at the centre of urban theory. In addition, the participation of human users proposes a new type of support for decision-making, in which participatory simulation is embedded in a process of social learning and negotiation, and contributes to the establishment of a final decision.

198 *Handbook on complexity and public policy*

NEW REQUIREMENTS IN POLICY-MODELLING

Opportunities and Lingering Barriers

Progress in ICT has contributed significantly to the increasing popularity of the cognitive mediating role of modelling in connection with the use of decision support systems and, since the 1990s, with GIS tools, impact analysis and scenario building (for example, see Batty, 1994; 2004; 2009).

ICT-driven methodological advancements have been considerable, and made available a wide set of 'intelligent packages' and 'decision support systems' (see Table 12.1), which probably have no previous precedent. It is not only the case that in the past decade the very notion of policy-modelling has enjoyed increasing popularity but that this has also been promoted by a number of European projects concerned with the opportunity to innovate current policy-practices and improve service delivery.

Indeed, the new model capabilities to assist policy activities have resulted, mainly, from the development of the syntactic component, and have not been accompanied by similar advancements in the other two components involved in the cognitive mediating role of models.

With regard to the semantic component, for example, the fact that, in most policy-relevant phenomena reflexivity is a constitutive property of human organizations, has rarely been made explicit in cognitive-oriented model applications, although it has been acknowledged on a conceptual level in dealing with individual and societal behaviours (see Dautenhahn, 2002; Poli, 2010).

As is the case with many other living systems, human organizations are complex adaptive systems. They share with the former the overarching property that their constitutive agents exist and behave in such a way as to contribute to the system to which they belong, without necessarily acting upon or having any control impinging on them. Additional properties, widely documented in more recent complexity literature (for example Miller and Page, 2007), are that: their pattern of connectivity is crucial to their survival; they co-evolve with their surrounding environment and thus increase their variety in order to improve their co-evolution possibilities; and finally, their evolutionary history is irreversible and their future unpredictable.

Furthermore, human organizations have some specific features which distinguish them from other living organizations. First, they not only build and update models to react to the surrounding environment, but also construct models for thinking about the world and acting within it. Second, they can deliberately modify themselves and their environment. Third, to maintain social, economic and environmental sustainability, human organizations purposefully inscribe into organizations and the environment their own thinking models, which acquired knowledge they pass on to future generations (Tsoukas, 2005).

To some extent, these very features also explain some of the problems encountered in designing the inquiry approach whenever modelling activities have to be applied in policy-making. Building on the experiences gained in the Italian regional case studies, the following difficulties can be mentioned.

First, the perception of the innovative impact that models can have in policy-making is still limited. A major hindrance can be attributed to cultural inertia, which hinders any

The role of models in bridging expert and lay knowledge 199

willingness to undertake innovative policy actions recommended by model applications. In addition, often, when implementing policy initiatives, the urgency of the issues makes the use of models impractical, perhaps because time limitations mean building an appropriate tool is impossible, or because the evidence provided by the available analytic tools does not answer the type of questions to be tackled.

Second, a language gap exists between the scientific and policy environments. Typically, discussion of models takes place in the academic environment where the methodological aspects of model applications are the main interest for the scientific community and not the practical consequences (for example, see the discussion of the technological fixers in Lupia, 2012).

Finally, we can mention the fact that the context in which the modelling activity takes place (the modelling domain in Figure 12.1) may not be sufficiently mature to realize its potential. Limiting factors include the difficulties many organizations experience in keeping up with the rapid pace of technological improvements, such as the deficiency in the regulative frameworks for accommodating ICT-based actions and the lack of a cultural setting capable of appreciating the quest for innovative policy actions.

Last, but by no means least, more often than not the goals of policy-makers and those of analysts are different or not aligned, that is, the former aim to build consensus while the latter are in search for solutions to societal problems.

The Cognitive Mediation Roles of Models as Boundary Objects

To overcome some of the difficulties mentioned above, increasing awareness about the cognitive mediator roles that models can play in enhancing policy capabilities will be a major undertaking.

One possibility is to leverage the (relative) autonomy of the model artefacts and to view their working as boundary objects between the scientific, policy and societal worlds. Boundary objects are 'abstract or physical artefacts which reside in the interfaces between organizations or social communities and have the capacity to bridge perceptual and practical differences among diverse communities in order to reach common understanding and effective cooperation' (Gal et al., 2004: 194.). Boundary objects, in fact, help overcome barriers between practices, facilitate mutual learning in heterogeneous environments, and/or spur novel learning endeavours for coping with new practices (Forgues et al., 2009).

Whereas this understanding does deserve deeper insight, in the following, some preliminary comments are offered which may help future research.

First, it helps to pay greater attention to the point of view of the policy activity (and more generally, to the features of the policy domain of Figure 12.1). How the cognitive mediator model deploys itself within the decoding process is a question largely overlooked at present and is likely to have increasing relevance in the near future. As clearly illustrated by the IRES studies, in most policy applications, the functional and cognitive mediator role of models have been addressed mainly from the point of view of the internal loop of the modelling activity. Indeed, modelling is usually undertaken by an analyst (an expert) whose primary goal is to observe the world and understand a certain phenomenon (identifying its features and processes). There is a need to extend this view and better account for the decoding process in the modelling activity. Typical questions raised in this respect are: (a) how the decoding process of modelling is affected (improved) by the changes/advancements occurring in the internal loop of the modelling activity (that is, the diffusion of agent-based modelling, the development of collaborative methodologies for decision-making and the implementation of social simulation platforms); and (b) how model applications indeed are of assistance to policy actions when they are implemented in real situations.

Second, being at the interface among different worlds, the communication function accredited to models is an important facet of a boundary object (see Thorne, 2003). A friendly communication of the modelling results is crucial when communities are participating in the policy decision-making process. It is true that developments in visualization techniques, computer interfaces and friendly computation tools are likely to favour the growth of a model-oriented attitude among a more general and diversified public. Nonetheless, the impact of policy-oriented models on people's learning capability is still largely unexplored. What is apparent, however, is that model applications entail engagement by a variety of stakeholders, including businesses, households and local communities.

Usually, the model's designer/builder has capabilities distinct from those who will use it; lay people are likely to have expectations of the model's usage and capability that differ from those of the analysts who built it. In addition, expectation and experience may vary across users according to individuals' characteristics and their engagement in the developmental path of policy-making. The issue of users' expectations and experience is an important one to address if modelling is expected to assist policy-makers in communicating their principles and goals to people, while also connecting them with clear and resonant images that will motivate communities to support the policy endeavour. Attention should be paid, therefore, to the social meaningfulness of modelling exercises and their ability to provide sense to these images.

Lastly, improvements to the components of the cognitive mediation role of models, that is, in their syntactic, semantic and design components, will help consolidate their working as boundary objects.

As for the syntactic component, one possibility is to better align among people forms of abductive processes, as these typically account for the formation of inferences that go from data simply describing something to a hypothesis that best explains the data. Abduction, in fact, is a kind of theory-forming or interpretive inference that has been recognized as a common source of knowledge both in science and ordinary life (Josephson and Josephson, 1996). Although the potentials of modelling for leveraging abductive reasoning can easily be grasped on a conceptual ground, how these can be brought into being in the model-decoding phase is an open research question. A possibility, for example, is to develop the arguments discussed in Shank and Cunningham (1996), which builds upon Peirce's general theory of signs, and suggests a categorization derived from considering the type of propositions associated with inferences, the relations signs have to the object (namely the use of icon, index and symbol) and the relation signs have to their ground (the use of tone, token and type). The interest of this categorization is that it accounts for different learning stages according to available knowledge (in relation to the observer) and information (in relation to the observed system), which can be levered by using the model as a cognitive artefact. Conceptually, therefore, the abduction type allows for some of the links connecting the model's external and internal loops to be made explicit.

For the semantic component, as emphasized above, there is a need to have more effective interpretations of the model's working and outcomes in order to improve intelligibility. In this respect, for example, representing model output as a narrative has advantages over more traditional modes of communication such as statistics and graphs. Narratives are much richer in content and context, supporting types of information that would not be included in standard output such as qualitative knowledge (Guhathakurta, 2002). ICT-enhanced narratives can provide a rich medium for describing the causal relationships among observed or modelled processes while also better connecting expert and lay knowledge (Foth et al., 2007; Reitsma, 2010; Snowden, 2010).

But it is the component concerned with the design of a model application that deserves more attention, as this is responsible for the relationship between the external and internal loops of the modelling activity. It concerns the definition and management of the two crucial phases of modelling activity: (a) the encoding, and, specifically, providing legitimization to the issues to be modelled; and (b) decoding, ensuring that social bodies with responsibility for policy-making are more appreciative of the modelling activity as a main source of innovation in the whole policy-organizational process (better alignment between governmental departments, raising levels of education in ICT among civil servants, developing model-based service portfolios, and so on). As pointed out in recent research by Fuerth and Faber (2012), developing the knowledge component of the project demands that greater attention is paid to the overall policy-production process:

- That the knowledge requirements necessary for the developmental path of the policy over time are in place: that is, accounting for the specific issues raised in problem recognition; policy guidance; and impact assessments for changes as actions are progressively implemented.
- That the functions certain policy actions can have are recognized depending on timescale (the strategic/long-term, tactical/medium-term and operational/short-term).
- That the entire knowledge system of the project is embedded within all the different departments of the organization that has policy-making responsibility.

CONCLUDING REMARKS

In an epoch where ICT-mediated human organizations are the norm (see Whitworth, 2009) the cognitive mediation role of models for supporting policy processes in an extended peer community is decisive. As recently re-emphasized in the complexity debate (see Geyer and Rihani, 2010), ICT–human mediation is a strategy that makes it possible to continuously move back and forth between recognition of the needs of the community and the solutions which can be put in place to meet those needs (Johnson, 2009).

But not only that; by creating the opportunity to bridge expert and lay knowledge, models provide an epistemology for making sense of social processes among members of communities. This gives modelling a novel, and to some extent unanticipated role, which can leverage the organizational capability of a community.

As already pinpointed, in order to successfully instantiate organizational capabilities, co-evolutionary changes should take place; on the one hand in the encoding–decoding

202 *Handbook on complexity and public policy*

processes underlying model activities and, on the other, in the ways in which policy-making activities are implemented in governmental systems (Occelli, 2012). This requires both understanding and engagement by all stakeholders, thus making the external and internal modelling loops more effective.

REFERENCES

Batty, M. (1994), 'A chronicle of scientific planning: The Anglo-American modelling experience', *Journal of the American Planning Association*, **60**(1), 7–16.

Batty, M. (2004), 'Dissecting the streams of planning history: Technology versus policy through models', *Environment and Planning B: Planning and Design*, **31**(3), 326–30.

Batty, M. (2009), 'Commentary. Spatial thinking and scientific urban planning', *Environment and Planning B: Planning and Design*, **36**, 763–8.

Conklin, J. (ed.) (2006), 'Wicked problems and social complexity', in *Dialogue Mapping: Building Shared Understanding of Wicked Problems*, Chichester: John Wiley, pp. 3–40; see also the CogNexus website http://cognexus.org or http://cognexusgroup.com/about-us2/resources/#WPSC, accessed 1 November 2013.

Dautenhahn, K. (ed.) (2002), *Human Cognition and Social Agent Technology*, Amsterdam: John Benjamin.

Davis, A., H. Shrobe and P. Szolovits (1993), 'What is a knowledge representation?', *Association for the Advancement of Artificial Intelligence*, **14**(1), 17–33.

Dupuy, J.-P. (2000), *The Mechanization of the Mind*, Princeton, NJ: Princeton University Press.

Forgues, D., L. Koskela and A. Lejeune (2009), 'Information technology as boundary object for transformational learning', *Journal of Information Technology in Construction*, **14**, 48–58.

Foth, M., G. Hearn and H. Klaebe (2007), 'Embedding digital narratives and new media in urban planning', Proceedings of Digital Resources for the Humanities and Arts Conference, Dartington, UK, accessed 4 November 2013 at http://eprints.qut.edu.au/8813/1/8813.pdf.

Fuerth, L.S. and E.M.H. Faber (2012), 'Anticipatory governance practical upgrades: Equipping the executive branch to cope with increasing speed and complexity of major challenges', Washington, DC: Elliot School of International Affairs, George Washington University.

Gal, U., Y. Yoo and R.J. Boland (2004), 'The dynamics of boundary objects, social infrastructures and social identities', Cleveland, OH: Case Western Reserve University, *Sprouts Working Papers on Information Systems*, **4**(11), 194–206.

Geyer, R. and S. Rihani (2010), *Complexity and Public Policy: A New Approach to 21st-Century Politics, Policy and Society*, Routledge: London.

Gil-Garcia, J.R. (2012), *Enacting Electronic Government Success: An Integrative Study of Government-wide Websites, Organizational Capabilities, and Institutions*, New York: Springer.

Guhathakurta, S. (2002), 'Urban modeling as storytelling: Using simulation models as a narrative', *Environment and Planning B: Planning and Design*, **29**(6), 895–911.

Head, B.W. (2008), 'Three lenses of evidence-based policy', *The Australian Journal of Public Administration*, **67**(1), 1–11.

Healy, S. (1999), 'Extended peer communities and the ascendance of post-normal politics', *Futures*, **31**(7), 655–69.

Johnson, J. (2009), 'Embracing design in complexity', in A. Katerina, J. Johnson and T. Zamenopoulos (eds), *Embracing Complexity in Design*, Abingdon: Routledge, pp. 193–204.

Johnston, P. (2010), 'Transforming government's policy-making process: Why encouraging more and easier citizen input into policy-making is not enough', *JeDEM*, **2**(2), 162–9.

Josephson, J.R. and S.G. Josephson (eds) (1996), *Abductive Inference: Computation, Philosophy, Technology*, Cambridge, MA: Cambridge University Press.

Josilyn, C. and V. Turchin (1993), 'Model', in F. Heylighen, C. Joslyn and V. Turchin (eds), 'Principia Cybernetica Web', Brussels: Principia Cybernetica, accessed 2 November 2013 at http://pespmc1.vub.ac.be/MODEL.html.

Kuhn, T. (1970), *The Structure of Scientific Revolutions*, Chicago, IL: University of Chicago Press.

Lupia, A. (2012), 'Can evolving communication technologies increase civic competence?', in P. Parycek, N. Edelmann and M. Sachs (eds), *CeDEM12: Proceedings of the International Conference for E-Democracy and Open Government*, Krems: Edition Donau-Universität, pp. 17–24.

Maier-Rabler, U. and S. Huber (2011), '"Open": The changing relation between citizens, public administration, and political authority', *JeDEM*, **3**(2), 182–91.

The role of models in bridging expert and lay knowledge   203

Miller, J.H. and S.E. Page (2007), *Complex Adaptive Systems: An Introduction to Computational Models of Social Life*, Princeton, NJ: Princeton University Press.

Morin, E. and J.-L. Le Moigne (1999), *L'Intelligence de la Complexité*, Paris: L'Harmattan.

Morrison, M. and M.S. Morgan (eds) (1999), 'Models as mediating instruments', in *Models as Mediators: Perspectives on Natural and Social Science*, Cambridge: Cambridge University Press, pp. 10–37.

Moss, S. (1999), 'Relevance, realism and rigour: A third way for social and economic research', *CPM Report* No. 99-56.

Occelli, S. (2001), 'La cognition dans la modélisation: Une analyse préliminaire', in N.V. Paugam-Moisy, H.V. Nyckees and J. Caron-Pargue (eds), *La Cognition Entre Individu et Société: Proceedings of the ARCo'2001 Conference*, Paris: Hermes, pp. 83–94.

Occelli, S. (2002), 'Facing urban complexity towards cognitive modelling. Part 1: modelling as a cognitive mediator', paper presented at the XII European Colloquium on Theoretical and Quantitative Geography, St-Valery-en-Caux, 7–11 September, 2001, accessed 17 December 2013 at http://cybergeo.revues.org/4179?file=1.

Occelli, S. (2006), 'Technological convergence vs. knowledge integration', paper presented at Les Journées Annuelles Transdisciplinaires de Réflexion au Moulin d'Andé, Colloques ASFCET, 13–14 May 2006, accessed 21 December 2013 at http://www.afscet.asso.fr/soAnde06.pdf.

Occelli, S. (2007), 'Assessing an urban model application to the Piedmont region: Cui prodest?', paper presented at the STOREP meeting, 3–4 June, Pollenzo.

Occelli, S. (2010), 'Reconciling tempus and hora: Policy knowledge in an information wired environment', paper presented at IC3K 2010, 2nd International Joint Conference on Knowledge Discovery, Knowledge Engineering and Knowledge Management, Valencia, 23–28 October, accessed 1 November 2013 at http://www.scitepress.org/DigitalLibrary.

Occelli, S. (2012), 'ICT for policy innovation: Empowering the policy production process', in M. Campagna, A. De Montis, F. Isola, S. Lai, C. Pira and C. Zoppi (eds), *Planning Support Tools: Policy Analysis, Implementation and Evaluation, Proceedings of the Seventh International Conference on Informatics and Urban and Regional Planning INPUT 2012*, Milan: FrancoAngeli, e-book.

Occelli, S. and S. Landini (2002), 'Le attività di modellizzazione all'Ires: Una rassegna e prime considerazioni', *WP 160*, Turin: IRES.

Occelli, S. and G.A. Rabino (2006), 'Cognitive modeling of urban complexity', in J. Portugali (ed.), *Complex Artificial Environments*, Berlin: Springer, pp. 219–34.

OECD (2011), *Government at a Glance 2011*, Paris: OECD Publishing.

Pahl-Wostl, C. and M. Hare (2004), 'Processes of social learning in integrated resources management', *Journal of Community & Applied Social Psychology*, **14**(3), 193–206.

Peach, I. (2004), 'Managing complexity: The lessons of horizontal policy-making in the Provinces', lecture at the Saskatchewan Institute of Public Policy, Regina: University of Regina, Saskatchewan, 8 June.

Pidd, M. (1996), *Tools for Thinking: Modelling in Management Science*, New York: Wiley.

Poli, R. (2010), 'The complexity of self-reference: A critical evaluation of Luhmann's theory of social systems', Dipartimento di Sociologia e Ricerca Sociale, *Quaderno 50*, Trento.

Reitsma, F. (2010), 'Geoscience explanations: Identifying what is needed for generating scientific narratives from data models', *Environmental Modelling & Software*, **25**(1), 93–9.

Rosen, R. (1985), *Anticipatory Systems*, New York: Pergamon.

Scholl, H.J. (2003), 'E-Government: A special case of business process change', *Proceedings of the 36th Hawaiian International Conference on System Sciences* (HICSS36), Waikoloa, HI, pp. 1–16.

Semboloni, F. (2006), 'The CityDev project: An interactive multi-agents urban model on the web', in J. Portugali (ed.), *Complex Artificial Environments*, Berlin: Springer, pp. 155–64.

Semboloni, F. (2007a), 'The multi-agent simulation of the economic and spatial dynamics of a poli-nucleated urban area', in S. Albeverio, D. Andrey, P. Giordano and A. Vancheri (eds), *The Dynamics of Complex Urban Systems, An Interdisciplinary Approach*, Berlin: Springer, pp. 409–27.

Semboloni, F. (2007b), 'The management of urban complexity through a multi-agent participatory simulation', *DISP – The Planning Review*, **43**(170), 57–70.

Semboloni, F., J. Assfalg, S. Armeni, R. Gianassi and F. Marsoni (2004), 'CityDev, an interactive multi-agents model on the web', *Computers, Environments and Urban Systems*, **4**, 45–64.

Shank, G. and D.J. Cunningham (1996), 'Modeling the six modes of Peircean abduction for educational purposes', paper presented at Seventh Midwest AI and Cog-Sci Conference, 26–28 April 1996, accessed 4 November 2013 at http://www.cs.indiana.edu/event/maics96/Proceedings/shank.html.

Snowden, D. (2010), 'Narrative research', accessed 4 November 2013 at http://narrate.typepad.com/100816-narrative-research_snowden-final.pdf.

Tait, A. and K.A. Richardson (eds) (2011), *Moving Forward with Complexity*, Proceedings of the 1st International Workshop on Complex Systems Thinking and Real World Applications, Litchfield Park, AZ: Emergent Publications.

Thorne, S.L. (2003), 'Artifacts and cultures-of-use in intercultural communication', *Language Learning & Technology*, **7**(2), 38–67.

Tsoukas, H. (2005), *Complex Knowledge*, New York: Oxford University Press.

Umpleby, S. (2011), 'Second-order economics as an example of second-order cybernetics', *Cybernetics & Human Knowing*, **18**(3–4), 173–6.

van Veenstra, A.F., M. Janssen and T. Yao-Hua (2010), 'Towards an understanding of e-government induced change: Drawing on organization and structuration theories', in M.A. Wimmer, J.-L. Chappelet, M. Janssen and H.J. Scholl (eds), *EGOV 10 Proceedings of the 9th IFIP WG 8.5 International Conference on Electronic Government*, Berlin: Springer, pp. 1–12.

Whitworth, B. (2009), 'The social requirements of technical systems', in B. Whitworth and A. de Moor (eds), *Handbook of Research on Socio-Technical Design and Social Networking Systems*, Hershey, PA: IGI, pp. 3–22.

13. Modelling complexity for policy: opportunities and challenges
Bruce Edmonds and Carlos Gershenson

INTRODUCTION

For policy and decision-making, models can be an essential component, as models allow the description of a situation, the exploration of future scenarios, the valuation of different outcomes and the establishment of possible explanations for what is observed. The principle problem with this is the sheer complexity of what is being modelled. A response to this is to use more expressive modelling approaches, drawn from the 'sciences of complexity' – to use more complex models to try and get a hold on the complexity we face. However, this approach has potential pitfalls as well as opportunities, and it is these that this chapter will attempt to make clear. Thus, we hope to show that more complex modelling approaches can be useful, but also to help people avoid 'fooling themselves' in the process.

The chapter is basically in three parts: a general discussion about models and their characteristics that will inform the subsequent decision and help the reader understand their potential and difficulties, then a brief review of some of the available techniques, and ending with a review of some models used in a policy context. It thus starts with an examination of the different kinds of model that exist, so that these kinds might be clearly distinguished and not confused. In particular it looks at what it means for a model to be formal. A section follows on the kinds of uses to which such models can be put. Then we look at some of the consequences of the fact that what we are modelling is complex and the kinds of compromises this forces us into, followed by some examples of models applied to policy issues. We conclude by summarizing some of the key dangers and opportunities associated with using complex modelling for policy analysis.

ABOUT MODELS AND MODELLING

The Formality of Models

A *model* is an abstraction of a phenomenon. A *useful model* has to be simpler than the phenomenon but to capture the relevant aspects of it. However, what needs to be represented in a model and what can be safely left out is often a matter of great subtlety. Since what is relevant changes with context, some models will be useful in some circumstances and useless in others. Also, a model that is useful for one purpose may well be useless for another. Many of the problems associated with the use of models to aid the formulation and steering of policy derive from an assumption that a model will have value for a different purpose or in a different context from the one the model was established and

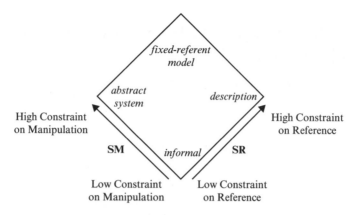

Figure 13.1 Two dimensions of formality

validated for. In other words, the value of a model is generally seen to be in its representation (for example simulation code) and not in its social embedding.

Generally speaking, all of our knowledge or descriptions can be thought of as models: they are abstractions of the phenomena with which we deal. These abstractions are required in order to understand and communicate about the complex phenomena that we have to deal with. However, models, in this most general sense, are not necessarily either precise or formal. Indeed, most of the models we use in everyday life are informal and couched in a language that is open to a considerable degree of interpretation. The formality (or otherwise) of a model has implications for its use and interpretation. Two dimensions of formality can be distinguished:

1. the extent to which the referents of the representation are constrained, e.g. by definition ('specificity of reference' or SR);
2. the extent to which the ways in which instantiations of the representation can be manipulated are constrained, for example by rules of logical deduction ('specificity of manipulation' or SM).

These two dimensions are illustrated in Figure 13.1.

For example, an analogy expressed in natural language has a low SR since what its parts refer to are reconstructed by each hearer in each situation. For example, the phrase 'a tidal wave of crime' implies that concerted and highly coordinated action is needed in order to prevent people being engulfed, but the level of this danger and what (if anything) is necessary to do, must be determined by each listener. In contrast to this, a detailed description is where what it refers to is severely limited by its content, for example 'Recorded burglaries in London rose by 15 per cent compared to the previous year'.

A system of abstract logic, mathematics or computer code has high SM since the ways these can be manipulated are determined by precise rules: what one person infers from them can be exactly replicated by another. This is in contrast to a piece of natural language which can be used to draw inferences in many different ways, only limited by the manipulators' imagination and linguistic ability. However, just because a representation has high SM does not mean that the meaning of its parts is well determined. Many

Modelling complexity for policy: opportunities and challenges 207

simulations, for example, do not represent anything we observe directly, but are rather explorations of ideas. We, as intelligent interpreters, may mentally fill in what it might refer to in any particular context, but these 'mappings' to reality are not well defined. Such models are more in the nature of an analogy, albeit one in formal form – they are not testable in a scientific manner since it is not clear as to precisely what they represent. Thus simulations, especially agent-based simulations, can give a false impression of their applicability because they are readily interpretable (but informally). This does not mean they are useless for all purposes. For example, Schelling's abstract simulation of racial segregation did not have any direct referents in terms of anything measurable[1], but it *was* an effective counter-example that can show that an assumption that segregation must be caused by strong racial prejudice was unsound. Thus such 'analogical models' (those with low SR) can give useful insights; they can inform thought, but they can not give reliable forecasts or explanations as to what is observed.

Formal models are a key aspect of science, since scientific models aim at describing and understanding phenomena. Their formality is important because that means that both their inference and meaning is (a) checkable by others and (b) stable. This makes it possible for a *community* of researchers and others to work collectively with the same models, confident that they are not each interpreting them in different ways. As the above discussion should have made clear, they can be formal in (at least) two different ways. For example data is a formal model of some aspect of what we observe, in the sense that it abstracts but in a well defined way – its meaning is precise. Data is not formal in terms of SM, however, and one could make very different inferences from the same set of data. Usually 'formal modelling' means that the inference from a model is well specified, in other words it is a representation with high SM. Thus scientific modelling is often associated with mathematics or computer simulation. However, in order to connect the formal inference to data it has to be formal in the SR sense as well; there needs to be a precise mapping between its parts and processes to what is observed, which is usually[2] done in terms of a map to some data.

The Use of Models

There are many purposes for models, including: as a game, an aesthetic construction, or an illustration of some idea.[3] Most scientific models claim to be *predictive*, that is they should allow us to obtain information about the future of the phenomenon before it occurs. For example, one can calculate and predict a ballistic trajectory aiming at a target using a mechanical model. However, on closer examination, many are more concerned with two other goals: *explanation* or *exploration*, with that of prediction being left as a theoretical possibility only.[4] There are many scientific models that are not predictive, or which only predict abstract properties. For example, the Gutenberg–Richter law describes the distribution of earthquake intensities, but this does not tell us when the next earthquake might be or how intense it might be. Darwin's theory of evolution does not tell us what will evolve next, or even the reasons why what has evolved did so, but it does predict the relationship between genetic distance and the length of time since species diverged. Unfortunately many reports about models are not clear as to their purpose in this regard, indeed many seem to deliberately conflate different purposes. Whilst models may have more than one goal, one should be wary of a model that was developed and

208 *Handbook on complexity and public policy*

tested for one purpose but is now being used for another. A clear case of this is where a model is designed to establish a theoretical counter-example (such as in the Schelling case discussed below) but then is later claimed to be for prediction (albeit in a modified form).

There is another, very basic, distinction in terms of the way models are used in practice. That is between models that (a) represent something observed and ones which (b) are a component of an adaptive process of decision-making.

In the former case, there is a well-defined mapping between the model and observational data/measurements, and the model is judged as to the extent of its error in its predictions of its target phenomena. Here the model is, to different degrees and ways, either correct or not. In this case an examination of the model can tell us something about the structure of what is modelled, for example by exploring 'what-if' questions using the model. For sake of clarity we call this a 'representational' model.

In the latter case, the model is part of a decision-making process to select strategies for action. It takes (processed) inputs from the world, for example indicators of success, and the model is changed depending on how well it is doing (for example by depreciating the parts of the model that resulted in a poor indicator). Outputs from the model are used in the determination of interventions. Here the model is continually being adapted according to events; it somehow encodes past successes or failures for different courses of action given different observations of the world. Here a useful model may not represent any aspect of the world at all, but just be a useful intermediary in the process of decision-making. However, if the process of adaption is effective it may come to encode knowledge as to what works and what does not. We call this an 'adaptive' model.

It may well be that an adaptive use of a model is more effective in a particular setting, particular if a considerable period of adaption has occurred, in effect training the model (given the decision-making structure it is embedded in) using a considerable amount of feedback from its policy environment. If the model is sufficiently flexible (that is, it has many adjustable internal parameters) that it could indicate the correct action from the available inputs (derived from observations of the environment) then, with enough training, the model will eventually do so. However, this kind of model adaption means that it is probably finely tuned to the particular situation and will not be useful by others in similar situations. Nor is it likely to be much use in exploring what would happen in cases not yet observed; so if the situation changes in some fundamental way, the model may well give totally the wrong answers. Furthermore, it might not be apparent from an inspection of the adapted model, why it works.

Representational models are usually hard to develop, taking considerable time and effort, often by a team of experts somewhat separate from those making policy decisions. Such models usually rely on some theory of the system being modelled, whose assumptions may be explicit. This kind of model, if it validates well, might have some validity outside its original test situation, and moreover, its assumptions and structure might give clues as to when and how it might reliably be used. If the situation changes in a way that is explicitly encoded into the model, one might be able to change its settings to suit the new situation.

In practice, models are often used with a mixture of adaptive and representational models, with adaptive models encapsulating some theory and being somewhat representational, and representational models undergoing some process of model adaption over time. In this case it is wise to know which aspects of one's model are representational and which have been 'tuned' to the particular situation or set of data.

Models are limited, and using them carelessly can have counterproductive consequences. One of the main limitations is due to the complexity of their subject matter, which is what we discuss next.

Complexity and its Implications for Modelling

That society is complex may seem an obvious statement. Still, the ways in which it is complex have implications for the use of models for the planning and execution of policy. One problem is the lack of agreement as to what 'being complex' means. There are dozens of different definitions of complexity (Edmonds, 2000). Frequently the word is used as a kind of negative. When available techniques (or accepted techniques) fail, we call what we were trying to analyse 'complex'– in this case it is a 'dustbin concept', a category to use when others fail. Here, in order to obtain a common understanding of the term, we can use its etymology. Complexity comes from the Latin *plexus*, which means 'interwoven'. Thus something complex is difficult to separate out into separate components or processes. This is because of *relevant interactions* (Gershenson and Heylighen, 2005; Gershenson, 2013a). Interactions are relevant when the future of an element of a system is partially determined by the interactions; in other words, if one eliminates these interactions then the future will be significantly different.

Traditionally, models have been reductionist, in the sense that they study phenomena in isolation. By definition, interactions are excluded. Either an element is modelled in isolation, or a whole system is modelled, averaging the properties of its elements. In other words, traditionally phenomena are modelled at a single scale. This approach is suitable for simple systems, but it is not sufficient for complex ones, where the properties of the system are a consequence of the interactions of the elements. This requires models to be multi-scale (Bar-Yam, 2004), and interactions must be modelled to relate different scales.

In terms of policy models, interactions need to be included in the model if the future projections are not to be distorted. Simple models that do not include interactions will be unreliable. Such interactions carry important implications for modelling complex systems. These all make the ideal of the assessment of the impact of policy interventions using a model difficult.

First, it implies that elements cannot be studied in isolation. Different social processes can interact to produce effects different from those caused by each process singly. The outcome from a population that is both disaffected and has access to an effective medium for dissemination of views (for example Twitter) might well be very different from that of a disaffected population with only local gossip or a satisfied population with something like Twitter. The impact of this is that separately analysing the impacts of different factors upon the outcome might well be misleading: one has to consider the outcomes from the whole system. This leads to the problem that one might well not know how much one needs to include in a model adequate for one's purposes, and that an approach that starts with the simplest possible model and then experimentally adds processes one at a time, might never get you to an adequate model (Edmonds and Moss, 2005).

Secondly, it implies that interactions generate *novel information*, which is not present in

210   *Handbook on complexity and public policy*

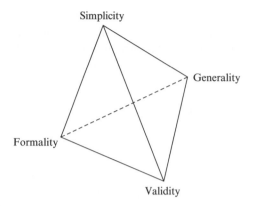

Figure 13.2   Some of the tensions implicit in modelling complex systems

initial or boundary conditions. This new information inherently limits the predictability of a complex system. In other words, the results are at least partially 'caused' by processes within the system, and not by external factors that can be controlled for. At best, this may mean that one has to make do with a broad distribution of outcomes as a forecast, or the prediction of 'weaker', second-order properties of the outcome (for example the volatility of the focus outcomes, or what will *not* happen). At worst, it may mean that there is no well-defined distribution of outcomes at all, with any measures upon the outcomes from a model being due to artefacts (for example model size).

The result of such difficulties means that any policy model is inevitably a compromise between different desirable modelling goals. Figure 13.2 illustrates some of these tensions in a simple way.

These illustrated desiderata all refer to the model that is being used. *Simplicity* is how simple the model is, the extent to which the model itself can be completely understood. Analytically solvable mathematical models, most statistical models and abstract simulation models are at the relatively simple end of the spectrum. Clearly a simple model has many advantages in terms of using the model, checking it for bugs and mistakes (Galán et al., 2009) and communicating it. However, when modelling complex systems, such as those policy makers face, such simplicity may not be worth it if gaining it means a loss of other desirable properties. *Generality* is the extent of the model scope: in how many different kinds of situation could the model be usefully applied. Clearly *some* level of generality is desirable, otherwise one could only apply the model in a single situation. Authors are often rather lax about making the scope of their models clear – often implying a greater level of generality that can be substantiated. *Formality* is what was called specificity of reference (SR) above. Models where the meanings of the model parts are well-defined have such formality; those which do not, and are more in the way of a model of ideas about some target system rather than the system directly, have less of this. Finally *validity* means the extent to which the model outcomes match what is observed to occur – it is what is established in the process of model validation. This might be as close a match as a point forecast, or as loose as projecting qualitative aspects of possible outcomes.

What policy makers want, above all, is validity, with generality (so they do not have to keep going back to the modellers) and simplicity (so there is an accessible narrative to build support for any associated policy) coming after this. Formality for them is not a virtue but more of a problem; they may be convinced it is necessary (so as to provide the backing of 'science'), but it means that the model is inevitably somewhat opaque to them and not entirely under their control. Modellers, usually, have very different priorities. Formality is very important to them so that they can replicate their results and so that the model can be unambiguously passed to other researchers for examination, critique and further development (Edmonds, 2000). Simplicity and generality are nice if you can get them, but one cannot assume that these are achievable (Edmonds, 2012). Validity *should* be an overwhelming priority for modellers; otherwise they are not doing any sort of empirical science. However, they often put this off into the future, preferring the attractions of the apparent generality offered by analogical models (Edmonds, 2001; 2012). Relatively simple models that explore ideas rather than relate to any observed data that give the illusion of generality are, unfortunately, common.

Another ramification of the complexity of what is being modelled is in the goal of modelling: what sort of purpose the model is suitable for. One of the consequences of this is that prediction of policy matters is hard, rare, and only obtained as a result of the most specific and pragmatic kind of modelling developed over relatively long periods of time.[5] It is more likely that a model is appropriate for establishing and understanding candidate explanations of what is happening, which will inform policy making in a less exact manner than prediction, being part of the mix of factors that a policy maker will take into account when deciding action. It is common for policy people to want a prediction of the impact of possible interventions 'however rough', rather than settle for some level of understanding of what is happening; however, this can be illusory. If one really wanted a prediction 'however rough' one would settle for a random prediction[6] dressed up as a complicated 'black box' model. If we are wiser, we should accept the complexity of what we are dealing with and reject models that give us ill-founded predictions.

One feature of complex systems is that they can result in completely unexpected outcomes, where due to the relevant interactions in the system, a new *kind* of process has developed, resulting in qualitatively different results. It is for this reason that complex models of these systems do not give probabilities (since these may be meaningless, or worse be downright misleading) but rather trace some (but not all) of the possible outcomes. This is useful, as one can then be as prepared as possible for such outcomes, which otherwise would not have been thought of.

The effective use of models for policy formulation will thus involve a clear focus as to its purpose and its manner of use combined with some compromise between the factors discussed above. However, the extent and impact of such compromises should be openly and honestly made, as a proper balance is necessary for reliable uses of the model. It is probable that a combination of related models, each making different compromises, might be a productive way forward. However, this requires extra work and care.

We now look at a number of different approaches, commenting upon the compromises and properties of each.

212   *Handbook on complexity and public policy*

SOME TOOLS AND APPROACHES

System Dynamics

System dynamics is an approach to modelling that represents a system in terms of a set of interconnected feedback loops (Forrester, 1971). It models these in terms of a series of flows between stocks plus additional connections between variables and flows. Crucially, it allows the representation of delays in such feedback and that the outcomes of some variables can control/affect the rate of other flows. These flows and relationships can then be simulated on a computer and (more recently) visualized. Its advantages are that a complex set of feedback relationships can be explored and hence better understood. However, in practice, the variables it deals with are themselves abstract entities, often representing abstract and aggregate quantities. This approach is not well suited to the modelling of systems where internal heterogeneity is significant in terms of determining the outcomes.

Network Theory

Networks naturally describe complex systems, representing elements as nodes and their interactions explicitly as links. Only in the last decade, there has been an explosion in the scientific exploration of networks and their application to a broad range of domains. Network theory has its roots in graph theory as proposed by Euler in the eighteenth century. However, it is only recently that its use has become widespread, in part because of the large computing power and big data sets available.

Networks are useful for representing the *structure* of systems, indicating how elements interact. However, they can also represent the *function* of systems, with nodes representing states and directed links representing transitions. Relating the structure and function of systems is one of the most common questions for understanding systems, that is, how do changes in the structure affect the function of a system? Network theory can be used to study both structure and function using the same formalism. Also, adaptive and temporal networks (Gross and Sayama, 2009; Holme and Saramäki, 2012) have been used to study the change in time of network structure.

From the study of different natural and artificial networks, it has been found that most of them do not have a trivial topology, that is, there is a relevant organization in their structure. Still, several modelling approaches assume homogeneous topologies, as in cellular automata (see below), or even a so-called 'well mixed' population, that is, there is no structure considered (only the macro state). It has been shown that structure (micro scale) plays a crucial role in the dynamics of such systems. For example, the same system may change its dynamics drastically depending on whether or not the local structure is considered (Shnerb et al., 2000).

Network models can be useful to study several aspects related to policy and decision-making. For example, random agent networks (RANs) were proposed to model organizations such as bureaucracies (Gershenson, 2008), showing how a few modifications to the structure of an organization can improve its performance considerably. In general, 'computing networks' (Gershenson, 2010) can be used to study and relate adaptability at different scales. Since policy and decisions are usually made over changing and uncertain scenarios, adaptability is a desired property of models. However, the more networks change and the more complex the interactions represented over the links become, the less classic network theory is applicable, and the closer to an individual-based model one has.

Information Theory

Claude Shannon (1948) proposed information theory in the context of telecommunications. He was interested in how a message could be transmitted reliably over unreliable media. He proposed a measure of information (equivalent to the Boltzmann–Gibbs entropy in thermodynamics) where information is minimal for regular strings, as new symbols do not carry new information. Shannon's information is maximal for random strings, as new symbols carry all the new information, that is, they are not predictable. Several other measures have been derived from Shannon's information, such as mutual information, predictive information, excess entropy, and information transfer, among others (Prokopenko et al., 2009).

Information theory has been used repeatedly to measure complexity. However, there are two different views. One view implies a similarity of information to complexity, where maximum randomness (Shannon information) would have maximum complexity. A more popular view poses that complexity is maximal when a balance between regularity (order) and randomness (chaos) is reached (Langton, 1990; Kauffman, 1993).

Recently, measures of complexity, emergence, self-organization, homeostasis, and autopoiesis were proposed based on information theory (Fernández et al., 2014). These measures are fast to compute and simple enough to be used by people without a strong mathematical background, but can give insights into the dynamics of systems. It has been argued (Edmonds, 1999) that there is not one such measure that can always be used, but rather one has to choose a measure that gives meaningful results for the kind of system that one is considering. Thus this approach assumes that one has understood the target system sufficiently to select the appropriate measure.

For decision-making, it is vital to identify which type of dynamics are followed by systems and their components, as different decisions should be made depending on regular, complex or chaotic dynamics. Used correctly, these measures can provide this information precisely and thus aid in knowing how to respond to change in the systems.

Cellular Automata

Cellular automata (CA) can be seen as a particular type of network. Each cell (node) has a state that depends on the states of its neighbours (links) and its own previous state. Different CA models can have different number of states and consider different numbers of neighbours. Cells can also be arranged in one dimension (array), two dimensions (lattice), three, or more dimensions.

Perhaps the most popular CA is Conway's 'Game of Life' (Berlekamp et al., 1982). Each cell can have one of two states: '0' (dead) or '1' (alive). Rules consider how many of the eight closest neighbours are alive. For a live cell to continue living, it must have two or three living neighbours (in any configuration). More than three or less than two neighbours implies that the cell will die in the next time step. New cells are born in empty cells when they have exactly three neighbours. With these simple rules, several

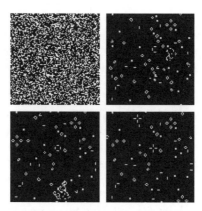

Notes: Evolution of the Game of Life from a random initial condition (a), where white cells are 'alive' and black cells are 'dead'. After 410 steps (b), certain stable structures have been formed, but there are still some active zones. After 861 steps (c), some structures have been destroyed and some new ones have been created. Activity continues in the lower part of the lattice. After 1416 steps (d), the dynamics is periodic, with stable and oscillatory structures. Images created with NetLogo (Wilensky, 1999).

Source: Figure initially published in Gershenson (2013b).

Figure 13.3 An example of evolution in the Game of Life from a random initial condition at four time stages

complex structures emerge: stable structures of different shapes, oscillators of different periods, moving structures (gliders, spaceships), eaters, glider guns, and so on. The structures emerging with the simple rules of the Game of Life can be used even to build a Universal Turing Machine.[7] An example of the dynamics of the Game of Life is shown in Figure 13.3.

Cellular automata have been used in several urban and land-use models (e.g. Portugali, 2000; Batty, 2005). However, 'pure' CA models tend to be too formal and abstract, so they have been found to be more useful in hybrid models, in many cases combining CA with agent-based modelling.

Individual- and Agent-based Modelling

Individual-based modelling is given when social actors or entities are represented by separate 'objects' within a computational simulation. Each object can have different properties, so this technique can represent heterogeneous collections of individuals. The interactions between the actors are represented by messages between the objects of the simulation. Thus, the mapping between what is observed and the model can be very much more straightforward with such simulations: each object modelling its corresponding actor.

When the objects in the simulation have internal processes representing their learning or decision-making processes so that these processes could be usefully interpreted as cognition, we call the computational objects 'agents' since they can act somewhat independently – they have a simple form of agency. When the agents are of this form,

Modelling complexity for policy: opportunities and challenges 215

one has the technique of Agent-Based Modelling (ABM). This technique is very flexible and puts few constraints upon the modeller, so the simulations that result are difficult to characterize in general but are of various kinds. An accessible introduction to the approach is (Gilbert and Troitzsch, 2005) and a more comprehensive guide (Edmonds and Meyer, 2013).

In particular, simulations differ greatly as to their level of detail, ranging from highly abstract and relatively simple simulations to very specific and complicated ones. The key difference here is whether the driver for model development is simplicity or relevance (Edmonds and Moss, 2005). Agent-based modelling has now been applied to a large number of policy-relevant subjects, including (to take an arbitrary sample of recent applications): energy infrastructure siting (Abdollahian et al., 2013), password behaviours within an organization (Renaud and Mackenzie, 2013), mobile banking adoption (Wei et al., 2013), China's housing market (Zhang et al., 2013) and return migration (Biondo et al., 2013).

The problem with ABM lies not in its expressiveness but in the complexity of its models (which means that it may be hard to understand the models themselves) and establishing the relationship between the models and what they represent (Moss and Edmonds, 2005).

EXAMPLES OF MODELS AND THEIR USE

Club of Rome's 'Limits to Growth'

In the early 1970s, on behalf of an international group under the name 'The Club of Rome', a simulation study was published (Meadows et al., 1972) which attempted to convince humankind that there were some serious issues facing it, in terms of a coming population, resource and pollution catastrophe. To do this they developed a system dynamics model of the world. Thus, this is a fairly simple kind of model that does not explicitly represent the parts of a system or its interactions, but rather the feedback cycles between key global factors. It was important to the authors to go beyond simple statistical projections of the available data, since that missed out the crucial delays in the usually self-correcting feedback processes. The results of the simulations were a set of computed curves showing, for example, pollution, population, and so on. The results indicated that there was a critical point in time coming and that a lot of suffering would result, even if humankind managed to survive it.

The book had a considerable impact, firmly establishing the idea that it was possible that humankind could not continue to grow indefinitely. The book presented the results of the simulations as predictions: a series of what-if scenarios. Whilst they did add caveats and explore various possible versions of their model, the overall intent of the book was unmistakable: if we did not change our lifestyles, disaster would result.

The authors clearly hoped that by using a simulation they would be able to make the potential feedback loops real to people. Thus this was a use of simulation to illustrate an understanding that the authors had. It was thus a model of ideas rather than directly of any such data. It did not, and could not, make predictions about what will happen in the future, but rather illustrate some possibilities. However, the model was not pre-

216   *Handbook on complexity and public policy*

sented as such, but as something more scientific in some sense. It was the presentation as 'scientific' that made this book such a challenge but also what laid it open to criticism (for example Cole et al., 1973). An examination of the model showed that some of its parameters were very sensitive and thus had to be 'tuned' to get the published results (Vermeulen and de Jongh, 1976). In other words, whilst the models had an illustrative and exploratory purpose, they were presented and criticized as if they were a predictive model.

The book made a considerable impact upon the general consciousness of the problem, and did act to get people questioning previously-held assumptions (that we could keep on growing economically and physically). However, it was also largely discredited in the eyes of other modellers due to its perceived lack of 'rigour'. This was somewhat unfair as the alternatives were no better in terms of validity or generality. However, a lack of humility in terms of its results and the relative simplicity of their model did lay it open to such attacks. The predictions of the book have not yet come to pass, but it is not clear that a similar future critical point and attendant suffering has been avoided.

Schelling's Model of Racial Segregation

In addition to a host of simpler, analytically expressed models (similar in kind to the Club of Rome's systems dynamics models), Schelling developed what we might recognize as a simple agent-based model (Schelling, 1971; 1978). It did represent individuals and their neighbourhoods explicitly, albeit abstractly in a 2D grid with black and white 'counters' representing the people. The simulation was very simple: counters were distributed randomly to start with, then each counter that had less than a given percentage (c) of like neighbours moved to a new empty spot. The simulation showed that segregation emerged even with relatively low levels of racial bias (values of c down to 30 per cent). This did not relate to any particular data but was rather a counter-example to the idea that the observed segregation must be due to strong racial prejudice. In other words this was intended to be an exploration of ideas to inform policy rather than a direct representation of what was happening. It produced an understanding of possible segregation processes, and so influenced local policies in Chicago away from focusing on prejudice as a cause of the extreme segregation they had.

Employment in an Arctic Community

Berman et al. (2004) consider eight employment scenarios defined by different policies for tourism and government spending, as well as different climate futures, for an ABM case study of sustainability in the Arctic community of Old Crow, in Canada. Scenarios were developed with the input of local residents, tourism being a policy option largely influenced by the autonomous community of Old Crow (stemming from their land rights), and attracting great local interest. Here the policy options were addressed as a certain type of scenario, embedding the behaviour of actors within a few possible future contexts. The simulation here ensured the consistency of the scenarios, and helped to integrate the various inputs into a coherent whole.

The merit of this model is that it can improve the reckoning of human and social

Modelling complexity for policy: opportunities and challenges   217

factors and information into the issues at stake, allowing the exploration of some real possible outcomes. The drawback is the multiplication of uncertainties, not least of which is that we do not convincingly know how social actors might adapt to new circumstances (even if the policy options are relatively concrete).

A Detailed Model of HIV Spread and Social Structure

Alam et al. (2007) investigate the outcomes indicated by a complex and detailed model of a particular village in the Limpopo valley of South Africa. This model in particular looks at many aspects of the situation, including: social network, family structure, sexual network, HIV spread, death, birth, savings clubs, government grants and local employment prospects. It concludes with hypotheses about this particular case, showing that complex destructive synergies between the spread of HIV and the breakdown of social structure were possible, and could be exacerbated by the influx of workers from outside due to the granting of mining concessions. This does not mean that these outcomes will actually occur, but this does provide a focus for future field research and may provide thought for policy makers. Unfortunately in this case, the conclusions of this study were not what the local authority wanted to hear, and so the findings were ignored. This was a model with a high degree of validation, but a very specific and complex model taking three years to develop.

Evaluating Pandemic Preventive Measures

Bajardi et al. (2011) combined in a model networks and agents to model the epidemic spread of the H1N1 influenza in 2009. Nodes represent regions, which are linked by the commercial flights between their major airports. At each node, agents can be susceptible, exposed, infected or can recover (Anderson and May, 1992). Adjusting different parameters, the global spread of the disease could be reproduced. Travel restrictions were imposed as a preventive measure, reducing air travel by 40 per cent. Simulations showed that travel restrictions are ineffective in preventing the spread of the disease. Comparing with scenarios with no travel restrictions or with even more stringent travel restrictions, the authors found that the disease reached a peak almost on the same day.

This is a theory-based model intended for predictive purposes. Its validity depends upon the approximations and assumptions in the model, including the characteristics of the social network used in the model.

CONCLUSION: SOME PROSPECTS AND DANGERS FOR COMPLEX POLICY MODELLING

All models have limits – not only limits in the accuracy of their predictions but in the expression of the situations under which such projections are based. The nature of complex situations means that attempting to use models to aid policy formulation is susceptible to some particular dangers and pitfalls, including:

218 *Handbook on complexity and public policy*

- confusing a model that has exploratory or explanatory purposes for one that is predictive;
- preferring a 'black box' model that seems to give definite predictions despite neither understanding it nor knowing that it is reliable for this purpose;
- trying to use a model that has been adapted within a highly specific situation out of its original context;
- attempting very general or simple models of policy issues probably means sacrificing direct validity for an indirect, analogical relationship only.

Complex models of the kind described here (and elsewhere in this book) have the potential to express a broader range of kinds of situation than previous approaches. They are thus not so limited by the kind of 'brave' assumptions that bedevil models where analytic results are deemed necessary. They are also ideal for the exploration and 'laying bare' complex dynamics. For this reason, they are prospective as an important tool in the exploration and consideration of policy options. In particular, more descriptive models can be directly related to what they are modelling, allowing a greater range of data and input to be utilized in their specification and validation (both high SM and SR in terms of the above discussion).

Given the fact that models of complex systems will offer a limited predictability, it is advisable to complement this lack of predictability with adaptability (Gershenson, 2007). This will enable decision makers to take the best choice for the specific circumstances that are faced, as adaptability implies a distinction of current circumstances that purely predictive models do not consider. For example, it might be far more useful to use the understanding of possibilities revealed by a complex model to design visualizations of the available data, than to provide forecasts of dubious reliability, thus enabling policy makers to 'steer' policy effectively.

ACKNOWLEDGEMENTS

Bruce Edmonds acknowledges the support of the EPSRC under grant number EP/H02171X/1. Carlos Gershenson was partially supported by SNI membership 47907 of CONACyT, Mexico.

NOTES

1. Subsequent elaborations of this model have tried to make the relationship to what is observed more direct, but the original model, however visually suggestive, was not related to any data.
2. It is possible to directly 'wire' something like a computational process to reality via sensors and actuators, as happens in programmed trading; in this case it is not always clear the extent to which the model is a representation of anything observed, but more an embedded participant in it.
3. Epstein (2008) lists 16 different reasons; Edmonds et al. (2013) considers reasons that are specifically focused upon understanding human society.
4. It is common for papers describing them to list prediction as 'future work' when the model is more fully developed.
5. For an account of actual forecasting and its reality, see Silver (2013).
6. Or other null model, such as 'what happened last time' or 'no change'.

Modelling complexity for policy: opportunities and challenges 219

7. To explore the Game of Life and other interesting CA, the reader is advised to download Golly at http://golly.sourceforge.net.

REFERENCES

Abdollahian, M., Z. Yang and H. Nelson (2013), 'Techno-social energy infrastructure siting: Sustainable energy modeling programming (SEMPro)', *Journal of Artificial Societies and Social Simulation*, **16**(3), 6.
Alam, S.J., R. Meyer, G. Ziervogel and S. Moss (2007), 'The impact of HIV/AIDS in the context of socioeconomic stressors: An evidence-driven approach', *Journal of Artificial Societies and Social Simulation*, **10**(4), 7.
Anderson, R.M. and R.M. May (1992), *Infectious Diseases of Humans: Dynamics and Control*, Oxford: Oxford University Press.
Bajardi, P., C. Poletto, J.J. Ramasco, M. Tizzoni, V. Colizza and A. Vespignani (2011), 'Human mobility networks, travel restrictions, and the global spread of 2009 H1N1 pandemic', *PLoS ONE*, **6**(1), e16591.
Bar-Yam, Y. (2004), 'Multiscale variety in complex systems', *Complexity*, **9**(4), 37–45.
Batty, M. (2005), *Cities and Complexity*, Cambridge, MA: MIT Press.
Berlekamp, E.R., J.H. Conway and R.K. Guy (1982), *Winning Ways for Your Mathematical Plays*, volume 2: *Games in Particular*, London: Academic Press.
Berman, M., C. Nicolson, G. Kofinas, J. Tetlichi and S. Martin (2004), 'Adaptation and sustainability in a small Arctic community: Results of an agent-based simulation model', *Arctic*, **57**(4), 401–14.
Biondo, A.E., A. Pluchino and A. Rapisarda (2013), 'Return migration after brain drain: A simulation approach', *Journal of Artificial Societies and Social Simulation*, **16**(2), 11.
Cole, H.S.D., C. Freeman, M. Jahoda and K.L. Pavitt (eds) (1973), *Models of Doom: A Critique of the Limits to Growth*, New York: Universe Books.
Edmonds, B. (1999), *Syntactic Measures of Complexity*, PhD thesis, MMU: Centre for Policy Modelling.
Edmonds, B. (2000), 'Complexity and scientific modelling', *Foundations of Science*, **5**, 379–90.
Edmonds, B. (2001), 'The use of models: Making MABS actually work', in S. Moss and P. Davidsson (eds), *Multi Agent Based Simulation* (Lecture Notes in Artificial Intelligence, 1979), Berlin: Springer, pp. 15–32.
Edmonds, B. (2012), 'Complexity and context-dependency', *Foundations of Science*, **18**(4), 745–55.
Edmonds, B. and R. Meyer (eds) (2013), *Simulating Social Complexity: A Handbook*, Berlin: Springer.
Edmonds, B. and S. Moss (2005), 'From KISS to KIDS: An 'anti-simplistic' modelling approach', in P. Davidsson et al. (eds), *Multi Agent Based Simulation 2004* (Lecture Notes in Artificial Intelligence, 3415), Berlin: Springer, pp. 130–44.
Edmonds, B., P. Lucas, J. Rouchier and R. Taylor (2013), 'Human societies: Understanding observed social phenomena', in B. Edmonds and R. Meyer (eds), *Simulating Social Complexity: A Handbook*, Berlin: Springer, pp. 709–49.
Epstein, J.M. (2008), 'Why model?', *Journal of Artificial Societies and Social Simulation*, **11**(4), 12.
Fernández, N., C. Maldonado and C. Gershenson (2014), 'Information measures of complexity, emergence, self-organization, homeostasis, and autopoiesis', in M. Prokopenko (ed.), *Guided Self-Organization: Inception*, Berlin: Springer, in press.
Forrester, J. (1971), 'Counterintuitive behavior of social systems', *Technology Review*, **73**(3), 52–68.
Galán, J.M. et al. (2009), 'Errors and artefacts in agent-based modelling', *Journal of Artificial Societies and Social Simulation*, **12**(1).
Gershenson, C. (2007), *Design and Control of Self-organizing Systems*, Mexico: CopIt Arxives, available at http://tinyurl.com/DCSOS2007.
Gershenson, C. (2008), 'Towards self-organizing bureaucracies', *International Journal of Public Information Systems*, **4**(1), 1–24.
Gershenson, C. (2010), 'Computing networks: A general framework to contrast neural and swarm cognitions', *Journal of Behavioral Robotics*, **1**(2), 147–53.
Gershenson, C. (2013a), 'The implications of interactions for science and philosophy', *Foundations of Science*, Early View.
Gershenson, C. (2013b), 'Facing complexity: Prediction vs. adaptation', in A. Massip and A. Bastardas (eds), *Complexity Perspectives on Language, Communication and Society*, Berlin, Heidelberg: Springer, pp. 3–14.
Gershenson, C. and F. Heylighen (2005), 'How can we think the complex?', in K. Richardson (ed.), *Managing Organizational Complexity: Philosophy, Theory and Application*, Greenwich, CT: Information Age Publishing, pp. 47–61.
Gilbert, N. and K. Troitzsch (2005), *Simulation for the Social Scientist* (2nd edn), Maidenhead: Open University Press.

220 *Handbook on complexity and public policy*

Gross, T. and H. Sayama (eds) (2009), *Adaptive Networks: Theory, Models and Applications*, Berlin, Heidelberg: Springer.

Holme, P. and J. Saramäki (2012), 'Temporal networks', *Physics Reports*, **519**(3), 97–125.

Kauffman, S.A. (1993), *The Origins of Order*, Oxford: Oxford University Press.

Langton, C. (1990), 'Computation at the edge of chaos: Phase transitions and emergent computation', *Physica D*, **42**(1–3), 12–37.

Meadows, D.H., D. Meadows, J. Randers and W.W. Behrens III (1972), *The Limits to Growth: A Report for the Club of Rome's Project on the Predicament of Mankind*, New York: Universe Books.

Moss, S. and B. Edmonds (2005), 'Sociology and simulation: Statistical and qualitative cross-validation', *American Journal of Sociology*, **110**(4), 1095–131.

Portugali, J. (2000), *Self-organization and the City*, Berlin, Heidelberg: Springer-Verlag.

Prokopenko, M., F. Boschetti and A.J. Ryan (2009), 'An information-theoretic primer on complexity, self organisation and emergence', *Complexity*, **15**(1), 11–28.

Renaud, K. and L. Mackenzie (2013), 'SimPass: Quantifying the impact of password behaviours and policy directives on an organisation's systems', *Journal of Artificial Societies and Social Simulation*, **16**(3).

Schelling, T.C. (1971), 'Dynamic models of segregation', *Journal of Mathematical Sociology*, **1**(2), 143–86.

Schelling, T.C. (1978), *Micromotives and Macrobehavior*, New York and London: Norton.

Shannon, C.E. (1948), 'A mathematical theory of communication', *Bell System Technical Journal*, **27**, 379–423 and 623–56.

Shnerb, N.M., Y. Louzoun, E. Bettelheim and S. Solomon (2000), 'The importance of being discrete: Life always wins on the surface', *Proceedings of the National Academy of Sciences*, **97**(19), 10322–4.

Silver, N. (2013), *The Signal and the Noise: Why Most Predictions Fail – but Some Don't*, New York: Penguin.

Vermeulen, P.J. and D.C.J. de Jongh (1976), 'Parameter sensitivity of the "Limits to Growth" world model', *Applied Mathematical Modelling*, **1**(1), 29–32.

Wei, X., B. Hu and K.M. Carley (2013), 'Combination of empirical study with qualitative simulation for optimization problem in mobile banking adoption', *Journal of Artificial Societies and Social Simulation*, **16**(3), 10.

Wilensky, U. (1999), NetLogo, Center for Connected Learning and Computer-Based Modeling, Northwestern University, Evanston, IL, available at http://ccl.northwestern.edu/netlogo/.

Zhang, H., Y. Wang, Y. Lin, Y. Zhang and M.J. Seiler (2013), 'Simulation analysis of the blocking effect of transaction costs in China's housing market', *Journal of Artificial Societies and Social Simulation*, **16**(3), 8.

14. Using agent-based modelling to inform policy for complex domains
Mirsad Hadzikadic, Joseph Whitmeyer and Ted Carmichael

INTRODUCTION

Complex Systems and Policy

Complex systems have been a topic of study in the natural sciences for decades. Practitioners in physics, chemistry, biology, mathematics, meteorology and engineering (Braha et al., 2006; Flake, 1998; Gell-Mann, 1994; Gleick, 1988; Nicolis and Prigogine, 1989; Strogatz, 2000; von Bertalanffy, 1969; Callebaut and Rasskin-Gutman, 2005) have used the concept of complex systems to explain phenomena as diverse as phase transitions in physical matter, immune system functions, and weather patterns. These systems have been modeled using, for the most part, the concept of dynamical systems and nonlinear equations.

Recently, social scientists have started experimenting with complex systems tools developed in physics, mathematics and engineering in order to better understand the nature of issues that our society is facing today. Unlike natural systems, social systems involve the active participation of system elements (they possess 'will').

Subsequently, tremendous progress has been made in applying the methodology of complex adaptive systems (CAS) to economics, sociology, transportation, warfare, decision making and other disciplines (Axelrod and Cohen, 1999; Bonabeau et al., 1999; Buchanan, 2002; 2007; Capra, 1982; 2002; Dooley, 1997; Durlauf and Young, 2001; Epstein, 2006; Gell-Mann, 1994; 1995; Gribbin, 2004; Innes and Booher, 1999; Johnson, 2007; Hazy et al., 2007; Holland, 1992a; 1992b; 1995; 1998; Kauffman, 1993; 1995; 2000; 2008; Khlebopros et al., 2007; Kohler and Gumerman, 2000; Kollman et al., 2003; Krugman, 1996; Langton, 1995; Suleiman et al., 2000; Waldrop, 1993).

We note some specific examples. Marten used self-organization and emergent properties of CAS to understand and explain a human–ecosystem interaction (Marten, 2001). Sawyer defined societies as complex systems and simulated social emergence with artificial, agent-based societies (Sawyer, 2005). Tesfatsion used agent-based systems to simulate and model economies (Tesfatsion, 2003). Pascale, Milleman and Gioja asserted that complex adaptive systems are the law of nature and suggested that they should become the new law of business (Pascale et al., 2000). Buchanan used CAS principles to explain why catastrophes happen (Buchanan, 2002). Farnsworth used CAS to explain the water-cycle in systems of romantic culture (Farnsworth, 2001). Harrison applied CAS to provide a better understanding of complexity in world politics (Harrison, 2006). Innes explained some aspects of consensus-building by using a system built on CAS principles (Innes and Booher, 1999). Levin defined ecosystems and the biosphere as complex

222 *Handbook on complexity and public policy*

adaptive systems (Levin, 2003). Miller modeled social life using CAS (Miller and Page, 2007). Resnick explored parallel worlds of turtles, termites and traffic jams using the CAS paradigm (Resnick, 1994). Robb wrote a fascinating book outlining the 'new war' based on the basic tenets of complex adaptive systems (Robb, 2007), while Sageman used similar concepts to understand terror networks (Sageman, 2004). Sornette explored the relationship between stock markets and complex systems (Sornette, 2003).

Recently, there have been interesting developments in applications of CAS to policy. For example, a new *Journal of Policy and Complex Systems* has been announced. Maroulis et al. applied complexity theory to educational policy research (Maroulis et al., 2010). OECD issued in 2009 a report on applications of Complexity Science on Public Policy (OECD, 2009). Peake elaborated on policy making as design in complex systems (Peake, 2010). However, we are still at the very beginning stage of the effective use of CAS in policy simulation, modeling and evaluation. This is primarily because policy makers are not used to building models of phenomena under consideration. Agent-based methodology and complex adaptive systems have the potential to change that. So, what are CAS and ABM?

Complex Adaptive Systems

Complex adaptive systems are systems that consist of interconnected parts. These parts are usually referred to as agents, hence the term agent-based modeling (ABM), often used for the most widely adopted method for simulating and modeling complex systems. The power of complex adaptive systems comes from both the simplicity of agents and the complexity of their connections. Agents exhibiting simple behaviors, enriched through their interactions with other agents in the system, enable system designers to demonstrate that a few parameters can frequently lead to tremendously complicated behavior by the system as a whole. Such systems are often described as systems in which 'the whole is greater than the sum of its parts'.

However, agents' behavior is not static – it can change over time. Some of the agents' rules provide agents with the ability to modify the rules that they are 'born' with. Agents can also improve their movements. These improvements can come about either through a (full or partial) imitation of more successful agents in the society or through random mutations of their behavior. As a result, these agents adapt over time – hence the term complex adaptive system.

Agents are either of the same kind (homogeneous) or diverse (heterogeneous). They are often simple in that they have a few rules defining their behavior. This is what distinguishes them from intelligent agents, which have a pre-defined, complex set of behaviors. As agents move about in the environment they interact with each other according to their function rules. These transactions between agents represent the connectivity among the agents. If we represent the whole system as a network, therefore, then, we would say that the agents correspond to nodes, while the transactions correspond to edges or links of the network. It is these interactions (transactions, edges) between agents that represent the true power of CAS. Even though the agents are fairly simple, the complex patterns of the agents' interactions produce immensely intricate and diverse networks that often produce surprising results.

Complex adaptive systems have some interesting characteristics. First, they can be used for simulating non-linear phenomena that are harder to capture using a mathematical framework. Non-linear phenomena imply that at a certain point in the life of a system a small change in the input variable causes a dramatic (non-linear) change in the output variable. A tipping point is an example of a non-linear change. Many interesting phenomena today can be described as non-linear, including the growth of the Internet, development of the brain, escalation of military conflicts, uneven development of countries, performance of the stock market, rate of adoption of the health-care reform, or the rate of spreading of a particular virus.

Second, they exhibit self-organization, which is a property of systems whose internal organization increases in complexity over time without the explicit guidance of an exogenous force. Biological organisms are founded on this principle, which can be seen in the formation of cells, tissues and organs. This principle is also frequently behind the formation and growth of markets, companies, universities, societies and countries.

Third, self-organizing systems usually, but not always, exhibit emergent properties. Emergent properties are complex patterns that have not been anticipated at the design stage of the system. They are properties of the system, even though their presence surprises the designer. For example, consciousness is an emergent property of the brain. Until recently, the resilience of markets was thought of as a property emergent from selfish market agents acting in their own interests. Now, after the most recent meltdown of financial markets, this has been called into question, such that some government officials believe that some form of a market regulation might be needed to ensure the long-term viability of these markets.

Designing an Agent-based Model

Designing a CAS is not easy. It is more of an art than a science. This difficulty in building a CAS stems from the fact that the world is extremely complex and we cannot possibly replicate it in silicon to any degree of completeness, at least not at the current level of computational and communication technologies. Consequently, we have to make choices for all elements of a CAS system, including:

- *Agents*: One of the most important decisions we need to make concerns the type and the number of agents that we need to introduce in the system. Is there a need for more than one kind of agent? In other words, is this a homogeneous or a heterogeneous system? Then we need to decide how many agents of each kind we need to ensure that we capture both the granularity and the intricacy of the situation in order to devise an accurate enough model of the world that we are trying to simulate. The general rule is that we should introduce the *minimum* required number and type of agents while still capturing the essence of the situation, thus not compromising the predictive power of the simulation.
- *Attributes*: For each agent type we need to determine the number and the type of agent attributes. These attributes collectively define agents' characteristics, ranging from attitudes and preferences to capabilities and resources. It is clearly not simple to determine which attributes are important enough to be represented in the system. A considerable amount of research deals with the issue of attribute relevance, especially in the machine learning research, but this issue has not

224 *Handbook on complexity and public policy*

been resolved. Eichelberger and Hadzikadic have reported promising results using the CAS framework itself to estimate attribute relevance (Eichelberger and Hadzikadic, 2006). We hope that future research along these lines will provide satisfactory results.

- *Rules*: Agents act based on their pre-specified behavior. This behavior is defined by the set of rules that each agent follows. The rules fall into several categories, including: moving, performing specific functions, and learning. Most of the rules are of 'IF Conditions – THEN Actions' variety. However, other types of formalism are possible, including: genetic operators, neural networks and mathematical functions.
- *Environment*: Every system functions within an environment, examples of which include a geographic space (e.g., a country, a manufacturing plant layout, the Internet, a human tissue), a political landscape (e.g., contending ideologies, warring factions, intrastate conflicts, interstate disputes), an economic unit (e.g., a country, a company, an oil field, an industry sector), a social environment (e.g., relationships, group dynamics, viral advertising), or an abstract space (e.g., ideas, publications, laws).

Environments can be open or closed. Open environments (open systems) allow for the infusion of new resources into the 'world', thus allowing the system to expand and grow. Closed environments (closed systems) do not allow for new resources to be added into the world, so the system basically deals with the redistribution of existing resources only. Agents can communicate not only among themselves, but they can also communicate with the environment itself. This interaction with the environment normally takes the form of the consumption of new resources, or the form of modifying the status of some aspect of the environment (for example overusing a patch of land, thus making it less productive in the next cycle).

- *Adaptation*: Agents are 'brought into' the environment with pre-determined characteristics and behavior. This predetermination happens either randomly or with the purpose of mimicking some starting conditions of the situation being simulated. Regardless of how the agents obtain their 'self', they can improve over time. They do that in many different ways. One way to improve is to learn from one's more successful neighbours. As most agents continually move around, they change the 'demographic structure' of their neighbourhood. Another way to improve performance is to learn from one's own experience. A third way to improve is to learn from a group of agents that are selected based on some criterion (for example, best performers, most similar agents).

These three ways define from whom the agents learn. We still need to address the issue of what it is that the agents are learning, and how they are doing it. The agents can learn to improve both their characteristics and their behavior. For example, they can learn the optimal distribution of their retirement savings across a wide range of investment mechanisms, including money markets, mutual funds, stocks and options. If they learn just the percentages of asset allocations, then we say that they have improved one of their attributes, namely the distribution of investments attribute. However, if they learn the

Agent-based modelling for complex domains 225

rule that lets them decide over time how to change their asset allocation, then we say that they learned a rule for (more) optimal asset allocation.

- *Fitness function*: There are two ways to build a CAS. One way is to build it as we described it so far and then push the Start button to see what happens with it. The value of such a system is to understand where it will end up after a certain period of time (that is, 'define the beginning and calculate the end'). The other way is exactly the opposite: define the 'end game' and program the system to do some 'number crunching' to compute what the process should be for getting the system to the specific end state that is of a particular interest to us. This is interesting in situations where we would like to devise a set of policies for accomplishing a specific goal for the society.

This is accomplished by defining the so-called fitness function. Most States define tax revenues as the fitness function that they want to maximize. So, an example of a meaningful question would be: what can we do, as a State, to reach \$200 bn in sales by 2020? How is the fitness function used by the system? It is done through the learning/ adaptation mechanism of agents. The agents try to maximize their own profitability/tax collection, expecting that the total fitness function is simply the aggregate of all individual contributions. The other option for utilizing the fitness function is for the agents to evaluate their own actions in light of their contribution to the overall fitness function. This option is difficult to implement, as it is non-trivial to measure an agent's contribution to the system fitness function since this function is a result of many agent interactions in every cycle. The first approach is obviously much easier to implement and it is the favourite approach for most CAS implementations.

- *Locality principle*: This is an important property of CAS systems. It indicates that agents only see other agents who happen to be in their neighborhood. The user defines the 'radius' of the neighborhood. Obviously, if the radius is set to its maximum value, then agents can see the whole environment. The benefit of the locality principle is that it encourages the proliferation of diverse, competing solutions to the problem. Since agents improve their performance (solution to the problem) by learning from other agents in the neighborhood (plus some random modification), they will never learn from distant agents, unless these agents travel by chance through the neighborhood of the target agents. This, in turn, ensures that individual agents constantly learn, innovate and develop localized solutions to the problem. The set of all best local solutions represents a rich set of possible actions to be taken by the end user at the conclusion of every step/cycle.
- *Self-organization*: People, ants and bees, among other living beings, self-organize when such action improves efficiency (and thus complexity) of the system. This is often expressed in terms of the division of work, and it happens without a perceived authority responsible for giving orders. Sometimes this self-organization needs to be institutionalized as part of the permanent structure, which usually leads to a hierarchy of levels of self-organization. Companies and governments provide examples of such institutionalization of self-organization.

226   *Handbook on complexity and public policy*

The objective then is to have agents (attributes, rules) represented in the computer in such a way that it enables the self-organization of agents. For example, the agent attributes might include a 'field' representing the group the agent belongs to. If we allow groups to form links themselves, then we have a way of representing any organization of arbitrary complexity. This can be enhanced by allowing one agent to serve as the 'representative' of its 'colleagues', should the need arise to exchange services at the group level. Alternatively, we can design agents to find it beneficial to perform only one of the several tasks they are endowed with, and purchase the remaining functionality with the resources they already accumulated.

- *Emergence*: Once we define the attributes and the behaviors of agents and start running them in parallel, it is hard to predict the overall behavior of the resulting system. The complexity of the phenomena exhibited by the system as a whole often surprises us. If we did not know the make-up/design of the system, then we would never be able to deduce it from the behavior of the system. That is why the phrase 'the whole (system) is greater than the sum of its parts (agents)' is so apt in describing the essence of a non-linear system. The secret is in the interactions among agents, not in their own 'architecture'. Sometimes these interactions lead to avalanches and sometimes they lead to no movement at all. Hence the term non-linear systems. These surprising behaviors at the system level are often described as emergent properties.

Can we predict or anticipate them? No. If we know that they will happen then we built them (the surprises) into the agent architecture and/or the agent rules (that is, the outcome is predetermined). This would indicate that the system is not a complex adaptive system at all. It is truly hard to design CAS environments. It is often a process of trial and error. Consequently, it is extremely important to build the system in a fundamentally sound way – to address the main phenomena under consideration with as few rules and attributes as possible. Then we can tweak the available parameters to simulate the situation at hand.

Policy in the Context of Complex Systems

Policies represent the intent of policy makers to change the value of certain political, social, economic, military or other indicators that reflect the current state of the society. In democratic societies the intent is always positive, for the good of the society. However, because of the complexity of the society and the interconnections between social, economic, political and cultural variables, it is not clear that intended outcomes of the policies will be achieved. In fact, it is often the case that many policies accomplish negative results in the long term.

CAS and agent-based technologies offer a promising set of technologies and tools for evaluating policies at the design stage, long before they get implemented in a society. However, as has already been indicated earlier in this section, it is not a straightforward task to create an ABM simulation or a model of the phenomenon intended to be regulated with the proposed policy. The proceeding section on the challenges of designing an ABM model of the target policy can be a useful framework for analyzing the design and implementation of each policy element.

Agent-based modelling for complex domains 227

- *Agents*: This decision requires both domain expertise and technical know-how. The domain expertise is needed because it is the experts from many disciplines who will decide on the segments of population, opinion influencers and policy enforcers that need to be introduced as agents in the model. The technical know-how is needed to help the experts disambiguate between representing each agent type as a separate entity as opposed to an attribute of a more general agent type. For example, instead of introducing 'healthcare provider, payer and consumer' agent types, the system designer could find it more efficient to introduce an attribute 'agent role' with values 'provider, payer and consumer' for a general agent type called 'participant.'

The next decision has to do with the number of agents of each type that need to be introduced in the model, primarily to guarantee that a sufficient number of interactions among the model agents will be carried out, thus guaranteeing that the space of possible agent actions and reactions will be explored. This increases the likelihood that the policy will be properly evaluated.

- *Attributes*: For each model agent type we need to determine the number and the type of agent attributes. These attributes outline the space of possible agent opinions, preferences, demographics and assets. It is not unusual for the policy maker to insist that every little bit of information on each constituent is important. This is especially true in medicine and defence. However, research in Machine Learning (Hadzikadic et al., 1996) has demonstrated that introducing attributes that are not relevant for the question being asked actually negatively influences the quality of the overall predictive accuracy and utility of the model.

Another choice to be made by policy makers and designers is the distinction between procedural rules of behavior, say IF (agent X cannot afford policy Y) THEN (agent X avoids participating in the healthcare system), and attribute-based rules of behavior, say IF (agent X cannot afford policy Y) THEN (agent X does Z) where Z is an attribute of the agent X defined at the beginning of the simulation either randomly or according to certain starting criteria. This is important because it is much easier to have agents change and adapt their behavior under attribute-based rules than under procedural ones.

- *Rules*: Rules define how agents behave. Therefore it is imperative to have policy makers understand not only the most likely reactions the population may have to certain policies being enacted, but also all possible reactions of the population so that they (policy makers) do not get blindsided with sudden shifts in population attitudes in response to particularly effective marketing campaigns or previously unknown alliances among the strategic partners. Often these rules of possible behavior are allocated to the population of agents either randomly or according to certain demographic characteristics uncovered through opinion research and analysis. In the first case the ABM analysis will uncover the set of rules that need to be enforced/promoted in order to accomplish the target goal/fitness function, which is the successful adoption of the policy. In the case of a model that starts with the understanding of the current opinions and behaviors of the population

228 *Handbook on complexity and public policy*

under consideration, the final outcome of the simulation is the likelihood that the desired policy outcome will indeed be reached.

- *Environment*: Each policy has a target environment, be it a geographic space (e.g. a state), an industry (e.g. financial services, healthcare), a social group (e.g. those with income below a certain level), or an abstract space (e.g. ideas, publications, laws). Environments can be open or closed. As was stated earlier in this section, open environments allow for the infusion of new resources into the environment, while closed environments do not. It is easier to model a closed environment; however, such models tend to ignore exogenous events and the influx of resources from the outside that may change the outcome of the enacted policy.

Special interest groups, the prevailing culture, economic conditions, or political or social uncertainty can create conditions that are best implemented as an 'environment' affecting all agents. Agents can interact with the environment, just as they can interact among themselves. By doing so they change the environment, just as the environment can influence the agents themselves.

- *Adaptation*: Agents can improve their probability of 'success' regardless of whether they were 'brought into' the environment with random attributes and rules of behavior or with the purpose of mimicking some starting conditions of the situation being simulated. People often learn from their neighbors, friends and colleagues. They also learn from their own past. Not surprisingly, they also 'learn' from commercials crafted to influence target populations. In any case, agents adjust their response to various policies based on all the influences around them. Policy designers and modellers must build into the simulation system a realistic method of agent adaptation, both at the level of characteristics (attributes) and behavior (rules).
- *Fitness function*: If the policy designers know exactly what needs to be done but they do not know what effect the policy will have on the target population, then a forward-simulation from the reality-mimicking beginning to the end of the policy duration period is what is needed. In such an implementation, agents do not need to adjust their behavior to see if they can improve their actions to contribute to the system. However, if the policy designers are at the stage of exploring concrete steps that need to be proposed in the policy, then the CAS/ABM system must have a system-level fitness function (the function evaluating the overall 'quality' of the system) that guides the agents in their understanding of how they need to change in order to increase the overall fitness of the whole system. Examples of such a fitness function are the gross domestic product (GDP), the number of people with health insurance, or the percentage of eligible workers who are employed. It is the resulting behavior/characteristics of agents that will provide a clue to the policy designers as to the kind of actions that might need to be proposed in the policy in order to achieve the target state as expressed by the system fitness function.
- *Locality principle*: We all know the saying 'all politics is local'. This is exactly what the locality principle indicates: what works in one environment may or may not be the best solution in another. This creates a diversity of solutions that can be compared and contrasted on the market of ideas. Ultimately, the system explores

a large space of possible solutions, thus allowing a high quality solution to emerge. This solution is then likely to be replicated frequently. On the grand scale, the world is simply a laboratory of ideas in which each sub-laboratory (country) is finding its own way to solve economic, social, educational, political and other issues. The situation repeats itself on a smaller scale if the policy designers choose to focus on a particular country.

- *Self-organization*: People self-organize because it allows them to maximize their resources. This is how information exchanges work as well. Everybody has a circle of trusted friends whose advice and presence they value. People one is familiar with make one comfortable. This self-organization then has an effect on how people perceive the utility of various policies. This, in turn, determines how effective the implementation of these policies will be. Although this principle sounds simple in nature, in reality it is very hard to accomplish it in a model without explicitly coding it in the program (which is of course not acceptable). Therefore modellers must be both careful and clever in their effort to replicate self-organization in a model.

- *Emergence*: While self-organization happens at the agent-level, emergence is a property of the system itself. Examples include resilient economy, sustainable eco-system, or efficient healthcare system. Most policies are concerned with enabling positive features of the system to emerge. This is exactly why CAS/ABM simulations are invaluable: they allow policy designers to evaluate concrete steps that would eventually find themselves in a policy that has been vetted and evaluated at the simulation level.

These principles have been applied in the 'Actionable Capability for Social and Economics Systems' (ACSES), a Defense Advanced Research Project Agency (DARPA)-funded seedling of the battle for the hearts and minds of the population in the specific region of Afghanistan (Hadzikadic et al., 2013). The following section provides an illustrative example from the ACSES effort that brings to life some of the ideas presented here.

EXAMPLE: ACSES

Project Description

The ACSES project was a pilot project funded by DARPA to demonstrate the feasibility of constructing a CAS simulation model that could incorporate real-time or near-real-time data and estimate the probable futures of different decisions. The specific scenario modeled was citizen allegiance during the Taliban insurgency in Afghanistan. The model incorporated four kinds of agents: citizens, Taliban, government fighters and coalition fighters. The simulation also implicitly involved leaders of various ethnic groups in that the user specified the allegiance-related orders and some characteristics of the leaders, which then affected the citizens' allegiance. The leaders were not represented visibly in the simulation display. Each time step each citizen agent chose its allegiance: to be pro-Taliban, pro-government, or neutral. The outcome of principal interest was the geographical distribution of citizen allegiance.

230 *Handbook on complexity and public policy*

The environment was the country of Afghanistan, which was realized in the simulation as a map divided into a grid of 865 squares, called 'patches'. Two features of Afghanistan potentially relevant for citizen agent behavior were the distribution of ethnic groups and the distribution of opium cultivation. These were included as aspects of the environment, therefore, taken from publicly available data.

For the rules defining how the citizen agents cast their allegiance, a utility function model was used in order to allow flexibility and adaptability. Specifically, the utility function model allowed for a variety of theories of social behavior to be implemented for the citizen agents, with the user able to choose the theory. To facilitate our analyses, we specified 12 discrete theories, but the user could also combine these along a continuum of possibilities. In addition, we built into the model Markov models instantiating four alternative theories of how, over time, citizens may change in characteristics important for the social theories. The rules defining how the fighters behave were given simply as a Markov decision model, because their aggregate behavior was not the main focus of the simulation model. Their possible behaviors were to move, to fight or to flee. The probabilities in the decision model were affected by the relative numbers of citizens supporting sides locally as well as the different local concentrations of fighters.

The attributes of the citizen agents were their allegiance, location and ethnic group membership. The variables in the utility functions that governed their behavior derived from these attributes. Thus, citizens' location determined their economic situation, the level of violence they experienced, and what their neighbors were doing. Their ethnic group membership determined their underlying ideology and the leader to whom they might listen.

In operation, the model worked as follows. Prior to running the model, the user chose the social theory to be operative, whether the run was to use the violence data (and, therefore, be a calibration or testing run), and any deviations from the default settings for other parameters, including the leader attributes of resources and ideology and citizen ideologies in various regions. After starting the simulation, the user was able to make certain changes as the simulation ran. The important changes included the addition of any of the three kinds of forces to any specific region, thus modeling a troop surge; changes in leader resources; and alterations of leader ideology. The real-world counterpart especially of the latter might be a change of leader, perhaps due to assassination, imprisonment, and so on. While running, the simulation display allowed the user to switch between views of the distribution of agents over the map or distribution of allegiance and also showed ever-lengthening graphs of the time history of the numbers of the various kinds of agents.

Expected Outcomes (Questions Answered, Policies Evaluated)

Fundamentally, the ACSES project had a policy orientation, which is clear from noting the three phases of such a project: (1) construction of the model; (2) calibration and validation of the model using data; and (3) use of the model in evaluating policy. The latter includes examining the effect of user-specified interventions, but it also involves exploration of the parameter space and the search for possible surprising effects and relationships, phenomena which CAS models are particularly well-suited to uncover. Because the ACSES model was only a pilot project, we accomplished the three phases

Agent-based modelling for complex domains   231

incompletely. Nevertheless, the ACSES project went far enough to show that the enterprise is feasible and, in fact, to suggest how it may be done effectively.

The specific policies that could be evaluated using the ACSES model concerned (1) replacement of local and ethnic leaders, and (2) resource allocation in the form of either fighters committed or material resources allocated to the aforementioned leaders. Obviously, the choices available, even in the specific context of citizen allegiance during the Taliban insurgency in Afghanistan, were by no means exhaustive. Nevertheless, they were enough to show that this could be a valuable aid to policy-making.

We should note that the third phase, use of the model for evaluating policy, makes sense only after the model has been constructed, calibrated and validated, which includes selecting the social theory that works best in the context. Nevertheless, results concerning the social theory may not be completely decisive, or their applicability may be in question, in which case policies can be evaluated contingently. For example, the effects of replacing the local leader – operationalized in the simulation model by changing the leader's ideology, say, from pro-Taliban to neutral or pro-government – depended on the social theory implemented. That is, replacement of the local leader could have a large effect or virtually no effect at all. If calibration of the model is contingent, therefore – one theory appears to work best for one ethnic group and another theory for a different ethnic group, say – then the policy evaluations would be similarly contingent.

The section on Surprise below suggests the value, in use of a model such as the ACSES model, of going beyond addressing specific policy questions into policy exploration. For example, an analyst might want to estimate the possible effects of putting a given surge of fighters into a particular region under certain conditions – say, Taliban control and so forth – and could use the ACSES model to get some projections. A prior policy exploration consisting of systematic investigation of effects of varying surge sizes, as given in the Threshold Response analysis, however, also would provide invaluable guidance to the analyst. Namely, it suggests that the analyst should search for an efficient minimum size of the surge, that is, a size much below which the surge would be ineffective but above which increases would have little effect.

Cobb–Douglas Function and Implementation

A variety of utility functions have been devised; we used the Cobb–Douglas utility function, which has been employed to instantiate a variety of social theories through both formal and simulation modeling (Coleman, 1990). The Cobb–Douglas utility function specifies utility as the product of preferences (that is, values, interests, motivations), each raised to a fractional power or 'weight'. The simulation model uses the utility function by calculating, at each time step for each citizen agent, the expected utility of each allegiance choice. The agent chooses the allegiance that has the highest expected utility. A given social theory may be implemented by including relevant values in the utility function and assigning them appropriate weights. The Cobb–Douglas model has several desirable characteristics. Losses affect agents more than gains (that is, the second derivative of U with respect to any given value is negative), a well-established characteristic of human decision-making (Tversky and Kahneman, 1981). The model is easy to expand or contract, depending on the values that are relevant according to the social theories used and the situation in question. The model is also simple and transparent.

232 *Handbook on complexity and public policy*

Equation (14.1) gives the specific version of the Cobb–Douglas utility function implemented for citizen agents in the simulation model:

$$U = (1-L)^{w_L} (1-c)^{w_c} (1-I)^{w_I} (1-E)^{w_E} (1-V)^{w_V} (1-F)^{w_F} (1-R)^{w_R} \tag{14.1}$$

L (loyalty to leader), C (coercion), I (ideology), E (economic welfare), V (security against violence), F (influence of close associates), and R (repression and social influence for defying repression) are the deviations from an ideal state for each of seven different preferences or motivations. The weights, w_X for motivation X, give the relative importance of the different motivations to the agent and the relative effect they have on U. For modeling citizen choice of allegiance for the ACSES project, we chose to implement three alternative theories of 'followership' – that is, how citizen choice is affected by their leader – which could be combined with one or both of two social influence theories. This gave the user the option of $3 \times 2 \times 2 = 12$ discrete theories. Each theory was implemented by fixing a particular configuration of the weights of the motivations in the utility function. Thus, blending the theories by choosing intermediate weight values was also an option for the user.

The three followership theories we labeled *Legitimacy*, *Coercion* and *Representative*. Under *Legitimacy* theory, the citizen wants to follow the orders of the leader because the leader is legitimate or has inspiring or charismatic qualities (French and Raven, 1959; Weber, 1968[1922]). Under *Coercion* theory, the citizen wants to follow the orders of the leader because the citizen believes the leader will reward following orders or punish failure to follow them (French and Raven, 1959; Levi,1988; Machiavelli, 1985[1513]). Under *Representative* theory, the citizen follows a leader only if that leader advocates what the citizen wants. Thus, the leader does not affect the likely behavior of the citizen, although the leader may be able to affect how successful that behavior is (Marx, 1990 [1867]; Wickham-Crowley, 1991). The two social influence theories come from the literature on social movements. Under one, which we labeled simply *Social Influence*, agents prefer to perform the same behavior as those close to them (McPhail, 1994). Under *Repression* theory, the more local forces there are that oppose a behavior the less likely agents are to perform the behavior but this effect is lessened if other agents are successfully defying the repression (Marwell and Oliver, 1993; Opp and Roehl, 1990).

Let us consider, for example, the contribution of ideology to a given citizen's choice of allegiance. Suppose his ideology is 0.3 on a scale of 0 to 1, where 0 indicates fully pro-Taliban and 1 fully pro-government. In other words, he is somewhat sympathetic to the insurgency. I, his deviation from his ideal ideological state, for supporting the government would be $I = |1 - 0.3| = 0.7$; for supporting the Taliban, $I = |0 - 0.3| = 0.3$; and for being neutral, $I = |0.5 - 0.3| = 0.2$. The contributions of the ideology component to the utility function, equation (14.1), and, thus, to the choice of allegiance would be $(1 - I)$, that is, 0.3, 0.7 and 0.8, respectively, raised to the weight w_I. Clearly, the effect of the citizen's ideology would push him most strongly to be neutral, but the effects of the other motivations would have to be factored in, of course, with their weights.

Table 14.1 shows the weights for the different motivations in the utility function, that is the values of w_i in equation (14.1), that instantiate the three followership theories, as well as two sample combinations with a social influence theory.

Agent-based modelling for complex domains 233

Table 14.1   Motivation weights instantiating different social theories

Theory	Value						
	L	C	S	E	I	SI	R
Legitimacy	0.6	0.1	0.0	0.2	0.1	0	0
Coercion	0.1	0.5	0.2	0.1	0.1	0	0
Representative	0.05	0.05	0.3	0.3	0.3	0	0
Legitimacy + Social Influence	0.36	0.06	0	0.12	0.06	0.4	0
Coercion + Repression	0.06	0.3	0.12	0.06	0.06	0	0.4

Surprise

Clustering

To demonstrate use of the ACSES model as well as its ability to generate CAS phenomena, we discuss two effects that fall under the label of surprise. That is, these effects are not prescribed by the rules of the system, nor can they be readily deduced from the rules. The first surprising effect is the clustering of allied fighter agents. The essence of the process seems to be that fighter agents are more likely to survive in large groups of like agents. The consequence is the emergence of large clusters of like fighter agents in the environment. We emphasize that no behavioral rule tells same-side fighter agents to group together or even to seek out like agents. Nevertheless, clustering happens.

Below we show images from three simulation runs. The black dots represent Taliban fighters and the white dots government fighters (CF and AGF in the figure's legend). The other shades represent the citizens according to their choice of allegiance: dark grey for pro-Taliban, middle grey for neutral, and light grey for pro-government. The first image (Figure 14.1) is from the initial set-up for the run. The citizens are placed according to demographic information on Afghanistan, while the government fighters and Taliban fighters (1500 agents each) are allocated across Afghanistan randomly. Notice that the placement or grouping of the white or black agents (the government and Taliban fighters) follows no discernible pattern.

In the first simulation run, replacement of the Taliban and government fighters did not match their attrition rates. After 50 time steps in the simulation (Figure 14.2), some patterns begin to emerge. By 200 time steps, the numbers of fighter agents have stabilized, and the clustering patterns are more pronounced (Figure 14.3).

Note that by time step 200, strongholds, that is, areas of clustering, have emerged for both factions, particularly for the Taliban. The Taliban agents dominate mainly in the Eastern-Central area, but they also control the region that extends north-west from there, up to the northern border of Afghanistan.

For the other two independent simulation runs we show (Figures 14.4 and 14.5), the infusion of new agents was manipulated so that the number of each faction at the end of the simulations would approximately equal the initial settings, that is, 1500 for each faction. In these runs we can see that the Taliban seems to always dominate the eastern-central portion of the map, but sometimes the Taliban cluster extends north-west from there, and sometimes it extends south-west. In Figure 14.4, Taliban presence seems to consist of two distinct groups rather than a single large one. Nevertheless, all runs

234 *Handbook on complexity and public policy*

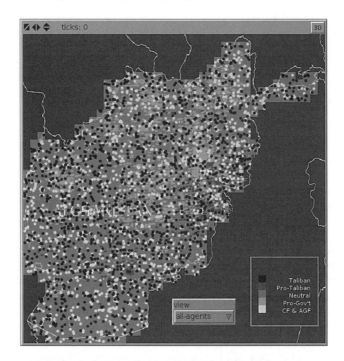

Figure 14.1 Initial agent configuration for all simulation runs

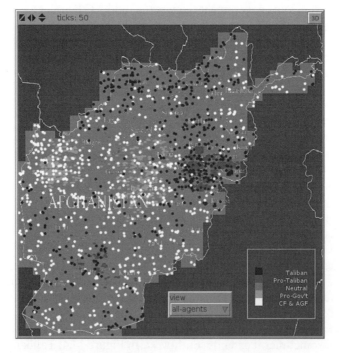

Figure 14.2 Results after 50 time steps

Agent-based modelling for complex domains 235

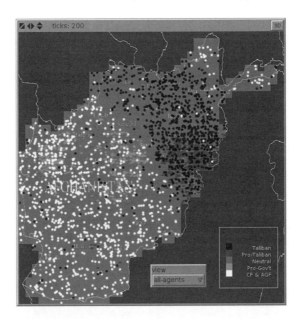

Figure 14.3   Results after 200 time steps

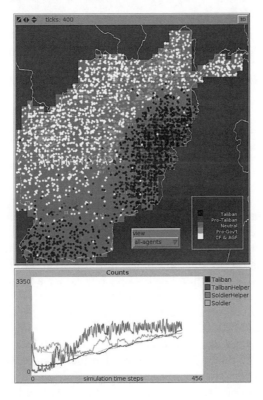

Figure 14.4   Results from Run 2

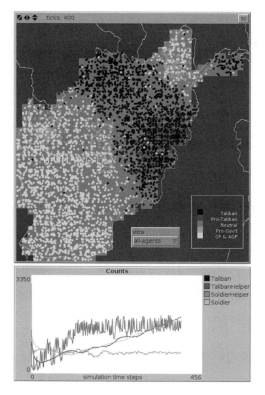

Figure 14.5 Results from Run 3

demonstrate the clustering phenomenon for both Taliban and government fighters, a phenomenon that was neither explicitly programmed into the simulation nor an obvious consequence of the agent models.

Threshold response
We also ran an experiment to explore the effects different surge levels of coalition forces in one particular region (Ghazni) might have on the number of citizens with pro-government allegiance. We ran this experiment with one set of settings from the experimental design changing only the size of the surge. Specifically, we set the social theory to one where the citizens choose their allegiance not in response to the leadership's direction (that is, neither Coercion nor Legitimacy) but rather are motivated primarily by preferences for a better economic situation and for less violence (that is, Representative). Citizens also prefer to conform in their allegiance to the citizens around them (that is, Representative + Social Influence).

We systematically increased surge levels by from 40 to110 by steps of 2 or 3, running each level either 30 or 60 times. There was considerable variance within each set of runs, but some patterns did emerge from our analysis.

One threshold effect concerned the success of surges. We defined a surge as successful if the number of pro-government citizens at the end of the run was greater than the

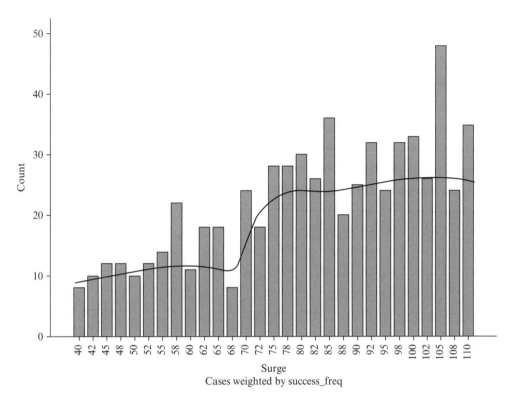

Figure 14.6 Successful surges produced in the surge experiment

number of government fighters. We found that when the surge level reaches approximately 70 government fighters, the number of successes increases dramatically. With a few exceptions, when the surge is less than 70, the number of successes tends to be around 10 to 15 out of every 60 runs; when the surge level is greater than 70, then the success rate tends to be 25 to 30 out of every 60 runs. Some sort of threshold is apparently reached at the 70-fighter surge level that allows the system to greatly increase the success rate. A plot of the number of runs that met the criterion for a successful surge (Figure 14.6) shows clear evidence of this threshold effect.

Figure 14.7 shows another threshold pattern, more complex and interesting and more challenging to explain. This graph shows a set of histograms of counts of the number of pro-government citizens ('soldier helpers') at the end of each run. There is one histogram for the surge values of 40, 50, 60, 70, 80, 90, 100 and 110.

At a surge level of 40, most of the simulation runs result in somewhere between 0 and 40 pro-government citizens in the Ghazni region, with very few runs resulting in more than 40 pro-government citizens. The same is generally true for surge levels of 50 and 60. At surge levels of 70 and 80, far fewer runs end up with the number of pro-government citizens in the 0–40 range, and more runs fall in the 41–80 range. We can also see an increase in runs that result in 120–300 pro-government citizens. Finally, on the back left wall of the 3D histogram, we see that the highest surge levels usually end up with either

238   *Handbook on complexity and public policy*

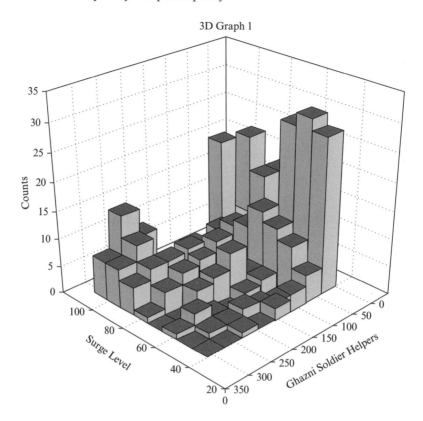

Figure 14.7   Number of pro-government citizens as a function of surge level

a very low number of pro-government citizens or a very high number of pro-government citizens; rarely do these surge levels result in a number of pro-government citizens in the middle ranges.

In general, we can see that the trend pushes toward a greater and greater degree of separation between success and failure. At the higher surge levels, this threshold is more pronounced, resulting in either a low or a high number of pro-government citizens but rarely a quantity in between.

SUMMARY AND CONCLUSIONS

This chapter presented an overview of complex adaptive systems and Agent-Based Modelling concepts and techniques. It then reviewed the characteristics of CAS and ABM that naturally lead to simulation and modeling of policy design, evaluation and implementation. The last section presented a concrete example that embodied principles of CAS and ABM in policy design and evaluation in a DARPA-funded ACSES project.

The ACSES example demonstrates the 'art more than science' point made earlier in the chapter. From a design standpoint, for example, the model was to incorporate a model of the agent that was reasonable from the standpoint of social science but both adjustable and adaptable. The introduction presented some of the general considerations in designing the agent; the ACSES model shows just how much it is art – such as designer decisions about the form of the model, the content and its extent, and to which agents it should apply.

From the policy or outcome standpoint, the user could explore, within the possibilities the design allowed, to see what seemed to 'work best'. From a policy standpoint, the model certainly does not predict. It does suggest that certain outcomes may be achievable under some conditions, if some general policies are followed. The ACSES model also points to the possibility under some circumstances of outcomes that the user may not have intended and perhaps may not have anticipated. As discussed in the introduction, there are some general outcomes of this nature that arise in complex systems and, therefore, in CAS models, including self-organization and emergence. Emergence did occur as an outcome in the CAS model, but where it was seen was not anticipated.

It may be unsettling that CAS models are not more predictive than they are. The outcome of an effective CAS model is more an identification of 'possible futures', as one of our colleagues likes to say, possibly weighted by likelihoods, than it is predictions. Yet especially regarding policy development and analysis, this actually may be truer to the real world than any predictive model would be. Setting effective policy itself can be said to be more art than science, in that it cannot rely on deterministic prediction, but must be flexible in response to different possible outcomes and even must expect the occasional unexpected. A satisfactory CAS model will be helpful with exactly these difficulties in policy analysis.

REFERENCES

Axelrod, R.M. and M.D. Cohen (1999), *Harnessing Complexity: Organizational Implications of a Scientific Frontier*, New York: Free Press.
Bonabeau, E., M. Dorigo and G. Theraulaz (1999), *Swarm Intelligence: From Natural to Artificial Systems*, New York: Oxford University Press.
Braha, D., A.A. Minai and Y. Bar-Yam (2006), *Complex Engineered Systems: Science Meets Technology*, Berlin, New York: Springer.
Buchanan, M. (2002), *Ubiquity: Why Catastrophes Happen*, New York: Three Rivers Press.
Buchanan, M. (2007), *The Social Atom: Why the Rich get Richer, Cheaters get Caught, and your Neighbor Usually Looks Like You*, Bloomsbury, New York: Holtzbrinck Publishers.
Callebaut, W. and D. Rasskin-Gutman (2005), *Modularity: Understanding the Development and Evolution of Natural Complex Systems*, Cambridge, MA: MIT Press.
Capra, F. (1982), *The Turning Point: Science, Society, and the Rising Culture*, New York: Simon and Schuster.
Capra, F. (2002), *The Hidden Connections: Integrating the Biological, Cognitive, and Social Dimensions of Life into a Science of Sustainability*, New York: Doubleday.
Coleman, J.S. (1990), *Foundations of Social Theory*, Cambridge, MA: Harvard University Press.
Dooley, K.J. (1997), 'A complex adaptive systems model of organization change', *Nonlinear Dynamics, Psychology, and Life Sciences*, 1(1), 69–97.
Durlauf, S.N. and H.P. Young (2001), *Social Dynamics*, Brookings Institution Press, Cambridge, MA: MIT Press.
Eichelberger, C.N. and M. Hadzikadic (2006), 'Estimating attribute relevance using a complex adaptive system', Foundations of Intelligent Systems, *Lecture Notes in Computer Science*, **4203**, 671–80.
Epstein, J.M. (2006), *Generative Social Science: Studies in Agent-based Computational Modeling*, Princeton, NJ: Princeton University Press.
Farnsworth, R. (2001), *Mediating Order and Chaos: The Water-cycle in the Complex Adaptive Systems of Romantic Culture*, Amsterdam, New York: Rodopi.

240 *Handbook on complexity and public policy*

Flake, G.W. (1998), *The Computational Beauty of Nature: Computer Explorations of Fractals, Chaos, Complex Systems, and Adaptation*, Cambridge, MA: MIT Press.

French, J.R.P. and B.H. Raven (1959), 'The bases of social power', in D. Cartwright (ed.), *Studies in Social Power*, Ann Arbor, MI: The University of Michigan.

Gell-Mann, M. (1994), *The Quark and the Jaguar: Adventures in the Simple and the Complex*, New York: W.H. Freeman.

Gell-Mann, M. (1995), 'Complex adaptive systems', in H.J. Morowitz and J.L. Singer (eds), *The Mind, The Brain, and Complex Adaptive Systems, Santa Fe Institute Series*, Boulder, CO: Westview Press, pp. 11–24.

Gleick, J. (1988), *Chaos: Making a New Science*, New York: Penguin.

Gribbin, J.R. (2004), *Deep Simplicity: Bringing Order to Chaos and Complexity*, New York: Random House.

Hadzikadic, M., S. O'Brien and M. Khouja (eds) (2013), *Managing Complexity: Practical Considerations in the Development and Application of ABMs to Contemporary Policy Challenges*, Berlin: Springer.

Hadzikadic, M., B. Bohren, A. Hakenewerth, J. Norton, B. Mehta and C. Andrews (1996), 'Concept formation vs. logistic regression: Predicting death in trauma patients', *Artificial Intelligence in Medicine*, **8**, 493–504.

Harrison, N.E. (2006), *Complexity in World Politics: Concepts and Methods of a New Paradigm*, Albany: State University of New York Press.

Hazy, J.K., J.A. Goldstein and B.B. Lichtenstein (eds) (2007), *Complex Systems Leadership Theory: New Perspectives from Complexity Science on Social and Organizational Effectiveness (Exploring Organizational Complexity)*, Mansfield, MA: ISCE Publishing.

Holland, J.H. (1992a), *Adaptation in Natural and Artificial Systems: An Introductory Analysis with Applications to Biology, Control, and Artificial Intelligence*, Cambridge, MA: MIT Press.

Holland, J.H. (1992b), 'Complex adaptive systems', *Daedalus*, **121**(1), 17–30.

Holland, J.H. (1995), *Hidden Order: How Adaptation Builds Complexity*, Reading, MA: Addison-Wesley.

Holland, J.H. (1998), *Emergence: From Chaos to Order*, Reading, MA: Addison-Wesley.

Innes, J.E. and D.E. Booher (1999), 'Consensus building and complex adaptive systems', *Journal of the American Planning Association*, **65**(4), 412–23.

Johnson, N.F. (2007), *Two's Company, Three is Complexity: A Simple Guide to the Science of all Sciences*, Oxford: Oneworld.

Kauffman, S.A. (1993), *The Origins of Order: Self Organization and Selection in Evolution*, New York: Oxford University Press.

Kauffman, S.A. (1995), *At Home in the Universe: The Search for Laws of Self-organization and Complexity*, New York: Oxford University Press.

Kauffman, S.A. (2000), *Investigations*, Oxford, New York: Oxford University Press.

Kauffman, S.A. (2008), *Reinventing the Sacred: A New View of Science, Reason and Religion*, New York: Basic Books.

Khlebopros, R.G., V. Okhonin and A.I. Fet (2007), *Catastrophes in Nature and Society: Mathematical Modeling of Complex Systems*, Hackensack, NJ: World Scientific Publishing Company.

Kohler, T.A. and G.J. Gumerman (2000), *Dynamics in Human and Primate Societies: Agent-based Modeling of Social and Spatial Processes*, New York: Oxford University.

Kollman, K., J.H. Miller and S.E. Page (2003), *Computational Models in Political Economy*, Cambridge, MA: MIT Press.

Krugman, P.R. (1996), *The Self-organizing Economy*, Cambridge, MA: Blackwell Publishers.

Langton, C.G. (1995), *Artificial Life: An Overview*, Cambridge, MA: MIT Press.

Levi, M. (1988), *Of Rule and Revenue*, Berkeley, CA: University of California Press.

Levin, S.A. (2003), 'Complex adaptive systems: Exploring the known, the unknown and the unknowable', *Bulletin-American Mathematical Society*, **40**(1), 3–20.

Machiavelli, N. (1985[1513]), *The Prince*, Chicago, IL: University of Chicago Press.

Maroulis, S., R. Guimerà, H. Petry, M.J. Stringer, L.M. Gomez, L.A.N. Amaral and U. Wilensky (2010), 'Complex systems view of educational policy research', *Science*, October, **330**(6000), 38–9.

Marten, G.G. (2001), *Human Ecology: Basic Concepts for Sustainable Development*, London, Sterling: Earthscan Publications.

Marwell, G. and P. Oliver (1993), *The Critical Mass in Collective Action. A Micro-Social Theory*, New York: Cambridge University Press.

Marx, K. (1990[1867]), *Capital*, Harmondsworth: Penguin.

McPhail, C. (1994), 'The dark side of purpose: Individual and collective violence in riots', *Sociological Quarterly*, **35**, 16–32.

Miller, J.H. and S.E. Page (2007), *Complex Adaptive Systems: An Introduction to Computational Models of Social Life*, Princeton, NJ: Princeton University Press.

Nicolis, G. and I. Prigogine (1989), *Exploring Complexity: An Introduction*, New York: Freeman.

OECD Global Science Forum (2009), 'Applications of complexity science for public policy: New tools for

Agent-based modelling for complex domains 241

finding unanticipated consequences and unrealized opportunities', accessed September 2009 at http://www.oecd.org/sti/sci-tech/43891980.pdf.

Opp, K.-D. and W. Roehl (1990), 'Repression, micromobilization, and political protest', *Social Forces*, **69**, 521–48.

Pascale, R.T., M. Millemann and L. Gioja (2000), *Surfing the Edge of Chaos: The Laws of Nature and the New Laws of Business*, New York: Crown Business.

Peake, S. (2010), 'Policymaking as design in complex systems: The international climate change regime', *Emergence: Complexity and Organization*, **12**(2), 15–22.

Resnick, M. (1994), *Turtles, Termites, and Traffic Jams: Explorations in Massively Parallel Microworlds*, Cambridge, MA: MIT Press.

Robb, J. (2007), *Brave New War: The Next Stage of Terrorism and the End of Globalization*, Hoboken, NJ: John Wiley & Sons.

Sageman, M. (2004), *Understanding Terror Networks*, Philadelphia, PA: University of Pennsylvania Press.

Sawyer, R.K. (2005), *Social Emergence: Societies as Complex Systems*, New York: Cambridge University Press.

Sornette, D. (2003), *Why Stock Markets Crash: Critical Events in Complex Financial Systems*, Princeton, NJ: Princeton University Press.

Strogatz, S.H. (2000), *Nonlinear Dynamics and Chaos: With Applications to Physics, Biology, Chemistry, and Engineering*, Cambridge, MA: Westview Press.

Suleiman, R., K.G. Troitzsch and G.N. Gilbert (2000), *Tools and Techniques for Social Science Simulation*, Heidelberg, New York: Physica-Verlag.

Tesfatsion, L. (2003), 'Agent-based computational economics: Modeling economies as complex adaptive systems', *Information Sciences*, **149**(4), 262–8.

Tversky, A. and D. Kahneman (1981), 'The framing of decisions and the psychology of choice', *Science*, 211, 453–8.

von Bertalanffy, L. (1969), *General System Theory: Foundations, Development, Applications*, New York: G. Braziller.

Waldrop, M.M. (1993), *Complexity: The Emerging Science at the Edge of Order and Chaos*, New York: Simon & Schuster.

Weber, M. (1968[1922]), *Economy and Society*, New York: Bedminster Press.

Wickham-Crowley, T.P. (1991), *Guerrillas and Revolution in Latin America: A Comparative Study Insurgents and Regimes Since 1956*, Princeton, NJ: Princeton University Press.

PART III

APPLYING COMPLEXITY TO LOCAL, NATIONAL AND INTERNATIONAL POLICY

15. Local government service design skills through the appreciation of complexity
Catherine Hobbs

INTRODUCTION

Following seven years of research and 29 years working in local government in road safety, transport and environmental planning, what compelling thoughts have brought me at this point to question the potential links between local government management and complexity? For me, the learning approach to be gleaned from an appreciation of complexity represents a hidden gem which can help us to disconnect those things we persistently pretend are connected, and connect those things where connection is often (sometimes conveniently) denied. Adopting this learning approach leads us straight to a rich store of knowledge about complexity and systems thinking which is ready and waiting to be applied to the ongoing capacity-building demand of local government service design and developed further in the twenty-first century. Many resourceful people are needed to attend to this gem, to polish the darkened stone, and my doctoral research about local government capacity-building, applied systems thinking and leadership will play but a tiny part in what I believe to be a larger process of evolution in the theory and practice of public governance.

This chapter challenges why there is an apparent disconnection between the process of public service reform and an appreciation of complexity, particularly when the timing of the connection is so apt. First, the plea for a more fundamental reform in public service is summarized, suggesting that complexity represents the first stepping stone in rising to this challenge. Secondly, some indication of the substantial body of literature is presented; this includes significant debates about public service futures, relevant academic research themes, and the degree to which complexity and systems thinking have been linked with local government. There is much more simply waiting to be drawn together and, in the meantime, a learning leadership would be conducive to the change beginning to happen in local government. This takes us to a consideration of the barriers which may be preventing progress and what future insights may lie somewhat tantalizingly ahead, with a conclusion that many researchers, policy actors and practitioners could help in polishing this hidden gem over the coming years.

THE PLEA FOR FUNDAMENTAL REFORM: RISING TO THE CHALLENGE OF PUBLIC VALUE VIA COMPLEXITY?

For a number of decades, local government has been charged with handling a series of cost-cutting reforms whilst also having to show evidence of improvement. As local government is responsible for essential services such as education, social services, transport,

246 *Handbook on complexity and public policy*

housing, public health, emergency planning and waste management, the stakes are high at individual, community and environmental level. This dominant theme of public service reform has led to literature ranging from formal enquiries and professional debates, through academia to much 'think tank' commentary. An underlying theme of these reform exercises often relates to the need for systems thinking and for the breaking down of 'silo' working. A significant international project (Bourgon, 2010) examined how the role of government for the twenty-first century should be re-defined in a post-industrial era away from being mechanistic and efficiency-focused, as government still fails to address the complex problems we face. The characteristics required to achieve such a change would be bottom up (rather than top down), outward looking (rather than inward looking), addressing root causes (rather than symptoms), skilled at relationships (rather than processes), change accepting (rather than resistant), networked (rather than siloed), personal (rather than impersonal) and enabling and co-producing (rather than doing). It concluded that this calls for a paradigm shift which will not be achieved at the level of specific problems and issues. This suggests that the practice of local government needs to move beyond being project and service focused, and that the skills of appreciating wider contexts, and the capacity to move thinking between levels of detail and abstraction are required.

'Getting more for less' has been an aim for local government for some time, and the demands of localism have placed local government as being primarily accountable to its own communities, rather than focusing on compliance with central government prescription. The case for systemic change has been highlighted by Bovaird (2011), with local government identified as a key link between personal behaviour and global effects. Bovaird stresses the importance of purpose coming first, rather than structure and the influence of political control. In reaction to what has been perceived as the piecemeal development of local government policy, a Commission on the Future of Local Government was led by Leeds City Council, incorporating the public, private and voluntary sectors (Commission on the Future of Local Government, 2012). This resulted in a set of propositions, commitments and calls to action, aiming to focus on the concept of civic enterprise and 'enterprising councils'. It was felt that opportunities had been missed for local government to be a force for change in an age of austerity. Solutions had been largely structural ones, rather than focusing on ways of working, culture and values. There had been a variety of operating models including pure commissioning hubs, a re-invention towards a social enterprise role, or simply a reduction in services offered. The fragmented and pilot-study-saturated state of affairs led to a plea for the development of a public service strategy (Bichard, 2013). Bichard points out that 'the reforms which have so far been offered . . . do little to tackle the fundamental flaws which currently exist. As such, they offer a depressing future in which the present inadequate model continues much as it is but in reduced circumstances' (Bichard, 2013: 3). Bichard asks a series of questions, concluding, 'the fact that such fundamental questions remain, at best, only partially answered indicates how far we are from creating a modern sustainable approach to the delivery of services for the public good' (Bichard, 2013: 4).

In focusing on effectiveness to the public, rather than considering efficiency in isolation, the concept of public value in relation to UK public service reform has been around at least since Kelly and Muers (2002). This Cabinet Office paper defined public value as 'the value created by government through services, laws, regulation and other actions' (ibid.: 4). Government has a stewardship role towards future generations that makes government distinctly different from private profit-oriented companies. In this respect, judging government by cost efficiency is a weak barometer of success. The paper called for a more holistic approach to accountability rather than a focus on process-driven audit regimes and narrow measures of efficiency. Following the work of Moore (1995), Benington and Moore (2010) defined a strategic triangle of public value to include the authorizing environment, operational capacity and public value outcomes. Public managers need to make sense of the new context of working as a way of guiding strategic thinking, with a suggestion that 'the new paradigms may include thinking about government and public services less as machines or structures and more as "complex adaptive systems" (for example the language of cultures and organisms rather than of levers and cogs)' (Benington and Moore, 2010: 14).

The current juxtaposition of budget cuts, getting more for less, and aspirations to deliver a public value approach amidst rising demographic demand can be dismissed as an impossibility, in which case a retrenchment is the only answer, or it can be taken as an opportunity to think and work differently, starting from the practical position of wherever people are. This is precisely where a complexity approach can be the stepping stone to help face the demands made, and there is much published material to help with this sense-making process.

A SUBSTANTIAL BODY OF LITERATURE TO HELP WITH SENSE-MAKING

This section covers three aspects of the literature. First, a further flavour of public service futures exercises and debates is presented. Secondly, some idea is provided of the extent to which complexity and systems have been linked with public service both beyond and within the UK. Finally, what of the field of complexity itself, and which ideas or insights could be embodied in local government policy, research and practice?

First, the changes outlined have led to many futures exercises and debates. A 'local vision' initiative (Office of the Deputy Prime Minister et al., 2006) incorporated a comprehensive Delphi survey of expert policy makers and a series of futures events, considering local governance challenges for the timescale towards 2015. The study concluded that the functions of local governance point towards a new concept of local government compared with current understandings, with a move beyond silo working at local level. Principles for 2020 Public Services were established by an inquiry of the Royal Society of Arts (Commission on 2020 Public Services, 2010). This pointed to the need for what was described as 'three mutually reinforcing, systemic shifts' as follows:

A shift in culture: from social security, to social productivity.
A shift in power: from centre to citizens.
A shift in finance: reconnecting financing with the purposes of public services. (Ibid.: 3)

It was suggested that debates should go beyond being narrow and service focused, to begin with citizens. Furthermore, with the changing relationship between the roles of central and local government, the Political and Constitutional Reform Committee

248 *Handbook on complexity and public policy*

(2013) pointed to a cultural shift in favour of local autonomy, making the exploration of different ways of working at local level very timely indeed.

Further debate about the difficult fiscal situation (RSA, 2012), suggested that a new starting point is needed to define a public service which is rooted in social citizenship, is productive and collaborative, integrated and relational. This means that incentives should be created for collaboration and integration. It suggested that ideas are needed to help deal with the challenge of 'doing more for less' which is values and needs driven, rather than compliance-based and target driven.

Futures thinking within the professions has also been in evidence. For example, the Chartered Institute of Public Finance and Accountancy has, with the Public Management and Policy Association and Accenture, produced papers about redefining local government and public service reform (Oyarce, 2011; Oyarce et al., 2012). This work included interviews with chief executives across England, with common themes including the need to take risks, the value of localism, and the importance of the local context. Despite the variety of approaches to current challenges, 'Notably, there is agreement that the introduction of systemic changes and the subsequent adoption of this altered paradigm require a shift in the role of leadership, the role of the organization and its relationship with partners' (Oyarce and Kirkman, 2011: 26). There is some suggestion that local authorities are being held back by the need to demonstrate a linear relationship between investment and outcomes, whereas the third sector is freer to take a multi-disciplinary approach to good effect. In order to make the necessary strategic cuts in a radical (rather than operational) sense, systems approaches are required, taking time to plan, and may require front-end investment (Talbot and Talbot, 2011). Overall, the significant upheaval to local government will require that long-held norms must be challenged (Boardman, in Oyarce et al., 2012).

There has therefore been a focus on innovation as a way of encouraging transformation within local government. Nesta describes itself as 'the UK's innovation foundation', and one of its projects was to develop a Public Services Lab with the Local Government Association, incorporating the 'creative councils' initiative. This programme aims to support those councils who are endeavouring to achieve long-term radical change at the local level. A series of reports (for example Gillinson et al., 2010; Christiansen and Bunt, 2012; Mulgan and Leadbeater, 2013) provide commentary on new thinking about public governance. The key challenge remains, however, of how to take examples of innovation to scale through to the mainstream – an acknowledged weakness of the public sector (Leadbeater in Benington, 2011). Christiansen and Bunt (2012) suggest that there is an uncomfortable interaction between the new perspective and existing decision-making structures. This highlights the importance of instilling a general understanding of exploring creative processes from the mainstream itself to encourage a gradual shift in the culture of decision-making.

Secondly, to what extent has complexity been linked with public service reform? It has notably been highlighted in areas beyond the UK such as the OECD countries together (OECD Global Science Forum, 2009), New Zealand (Eppel et al., 2011), Singapore (Ho, 2012), Finland (Haveri, 2006), the USA (Boal and Schultz, 2007; American Society for Public Administration and Governance of Complex Systems, 2013), the Netherlands (Teisman and Klijn, 2008) and Canada (Homer-Dixon, 2010). The OECD workshop heralded complexity science as an exciting, interdisciplinary field, including consideration of new ways of thinking for public policy, highlighting the importance of influence and likelihood, in addition to causation and control. Limits to rational reform processes will give way to the importance of political leadership due to complex and changing circumstances (Haveri, 2006). In view of the dynamic nature of phenomena, it is better to create circumstances for improvement rather than expect direct improvements (Teisman and Klijn, 2008; Homer-Dixon, 2010). Complexity thinking can provide useful insights to gain a more holistic understanding of public management (Eppel et al., 2011), and the challenge for resilient government is to move beyond efficiency, developing capacity to deal with an uncertain future (Ho, 2012). The first international seminar on Complexity and Policy Analysis took place in 2008, and the International Research Society for Public Management considered complexity at its 2013 conference, indicating the timely relevance of wide debates regarding this approach.

Within the UK, systems and complexity were linked with public management some time ago (Blackman, 2001; Medd, 2001; Mulgan, 2001). Mulgan identified that systems thinking was 'largely foreign to the everyday practice of government' (Mulgan, 2001: 23), with its focus instead on causal relationships, inputs and outputs. He also identified an excess of jargon in systems thinking as a problem, and that there are barriers to the use of complexity and systems thinking in public policy, such as time, the need for accountability, lack of institutional capacity to think, and uncertainty. He added a warning that the moral agenda should not be overlooked in systems theory, to avoid a tendency towards technocracy. At the same time, new patterns of governance emerging in the public and voluntary sector required that management science reject universal theory as impossible; an emergent style connecting complexity thinking and management science could help match the new demands of governance (White, 2001). A landmark paper was developed by Chapman in 2002 and subsequently updated (Chapman, 2004) identifying that, because the current system had failed, governments must learn to think differently, pointing clearly towards the techniques of systems and complexity thinking. This demands a shift from mechanistic thinking towards systemic learning, including self-reflection. In linking complexity theory with public services management, 'a new mindset is needed' (Haynes, 2003: 151), involving creative managers who can manage tensions and contradictions. Incorporating values into problem-structuring methods in public policy making represents a challenge (White, 2003), as does the prospect of encouraging policy makers to learn about these fields as being relevant to their work, as the use of systems approaches has been hampered by lack of demand from decision makers (Mabey, 2004).

The appropriateness of applying complexity theory has also been questioned, concluding that it should be developed further using insights from psychology and social theory (Houchin and MacLean, 2005). A range of contributors have subsequently considered the linkages between complexity science and society, including the health sector, social theory, politics and policy (Bogg and Geyer, 2007). There is a wider indication that others are thinking more in depth about complexity approaches in the context of public service (Bovaird, 2008; Butler and Allen, 2008; Haynes, 2008; Geyer and Rihani, 2010; Cairney, 2010; Meek, 2010; Rhodes et al., 2011; Geyer, 2012; Landini and Occelli, 2012). These papers develop a variety of ideas to contribute to applicability in the public sector, though this most commonly relates to general arguments, national policy or the health sector, with little direct reference to local government. Geyer and Rihani (2010)

250 *Handbook on complexity and public policy*

identify complexity and public policy as a new approach to twenty-first-century politics, policy and society, concluding that the enlightenment from this approach will result in public policy which is more humane. They question why the field of complexity has been slow to move towards the social sciences and public policy, identifying the problem of disciplinary silos even within the social sciences, and accepting that complexity, with its resident uncertainty, is a difficult concept to sell. They maintain that complexity approaches may be resisted by political elites, though local policy actors intuitively understand complexity because they are caught between the order demanded of policy elites (for example clear outcomes and accountability), and the messy reality of the everyday.

A new way of thinking is therefore required which accepts government as being part of a complex adaptive system, with the limits of reductionist thinking accepted and the importance of a systems thinking approach recognized (Benington, 2011). This approach was notably adopted by the Munro Review of Child Protection (Munro, 2011), which also recommended adopting a systems methodology for case reviews. Hallsworth (2011) has called for 'System Stewardship' as the future of policy making, referring to 'strategic systems leadership'. He presents policy making as an ongoing interaction between purpose, design and realization. In this way, judging the level at which a policy problem should be tackled is itself a crucial role for policy makers. Overall, this potential both home and abroad suggests that momentum is gathering to the credibility of linking complexity science with public policy in the twenty-first century and, as the spotlight falls on local autonomy and a stewardship role for local government, the time is good to apply such thinking at local government level.

Thirdly, what of the field of complexity itself; which ideas could be embodied in practice? The transdisciplinary field of complexity is a developing one, but with disparate roots stemming from hard rational approaches (for example Santa Fe Institute, 2013) and softer more interpretive ones (Shaw, 1997; Boal and Schultz, 2007). It has a range of ideas which are relevant to practice, such as the concept of 'fuzz and chaos' (Morcol, 1996) or emergence (Seel, 2000; 2006; Stacey, 2005; Huaxia, 2007; Letiche et al., 2011), applied to organization studies (Anderson, 1999; Warfield, 1999; Mitleton-Kelly, 2003; McMillan, 2004; Tsoukas, 2005; Maguire et al., 2006; Johannessen and Kuhn, 2012), leadership (Snowden and Boone, 2007; Stacey, 2012) and strategic management (Stacey, 2007; Rosenhead, 1998; Lissack, 2002; Stacey, 2012; Levy, 1994). It has been suggested that complexity theory lacks coherence for application to practice (Johnson and Burton, 1994; Ortegón-Monroy, 2003). Yet the divisions of the complexity research community between the hard functionalist sociological paradigm, and the soft interpretive/emancipatory paradigms[1] are much the same within systems science, operational research and management science. Notably, moves have been made to explore the fields of systems and complexity as overlapping areas of research, given that both communities should be working towards making a positive difference (Richardson et al., 2006; Midgley and Richardson, 2007; Richardson et al., 2007). These paradigmatic differences should therefore not be seen as a block to applicability, indicating that the application of complexity to local government is itself a significant area for future research. In particular, any such change emerging within local government would crucially demand a particular style of leadership, and this is considered next.

A LEARNING LEADERSHIP?

Significantly, there has been a push to improve leadership skills across public service settings to address whole system challenges (Benington and Hartley, 2009). In doing so, collaboration inevitably remains a significant theme (Sullivan, 2010). Furthermore, combining leadership with collaboration has led to the development of the concept of integrative leadership (Crosby and Bryson, 2005; Crosby, 2010; Crosby and Bryson, 2010; Bono et al., 2010; Fernandez et al., 2010; Morse, 2010; Page, 2010; Silvia and McGuire, 2010). Integrative leadership is defined as 'bringing diverse groups and organizations together in semi-permanent ways – and typically across sector boundaries – to remedy complex public problems and achieve the common good' (Crosby and Bryson, 2010: 211). Fernandez et al. (2010) considered integrated leadership in public sector settings in the US, suggesting that integrated leadership is different from previous theories in that it has synthesized leadership theories into a workable framework. This synthesis results in five leadership roles essential for the success of leaders in the public sector:

- task-oriented;
- relations-oriented;
- change-oriented;
- diversity-oriented;
- integrity-oriented.

In accepting that the local government challenges go beyond task-oriented management through traditional hierarchical structures, there is a need to take on the other roles of integrative leadership, which are linked with an approach which demands ongoing learning. Those with a sense of curiosity for new ways of working, or finding out from others how to tackle problems and issues, could be the true leaders to influence change. In this respect, Communities of Practice networks (Wenger, 1999) have been to the fore, enabling motivated people to seek help from each other about new ways of thinking or working. In this way, learning from other practitioners is as likely, if not more likely, to be a real influence to changes in practice than the academic literature. There is a danger, however, that activities such as Communities of Practice fail to link with the body of scholarly knowledge (and vice versa). Matching academic endeavour and practitioner curiosity as an ongoing learning process could have a powerful potential to help find the different ways of working which are demanded. Yet there are barriers to progress in the field of practice, and some of these are considered next.

OVERCOMING THE BARRIERS: CHALLENGING ASSUMPTIONS

First and foremost, our own culture creates barriers. Bateson, whose work spanned different scientific disciplines suggested that, rather than seeking detailed 'certainty', we should be focusing on the connectedness between things, 'the pattern which connects' (Bateson, 1979: 7). We are inevitably dealing (partially) with and trying to make sense of multiple versions of the world. This puts emphasis on the need for a wider perspective

252 *Handbook on complexity and public policy*

than individual disciplines, and the generation of a complex and creative level of understanding. For example, one of the principles underlying the work of Elias is that the combination of human action is often unplanned and unintended (Elias, 2000). Elias also argues that humans can only be understood in their interdependencies with each other, through social networks. These multiple versions of the world and unintended consequences make everything 'messy'. A more holistic, though untidy approach may be more honest in our pursuit of understanding complex social issues. Furthermore, it has been suggested that culture could be taken as a root metaphor in organizational analysis (Smircich, 1983). This puts the focus of attention on interactional dynamics, and the higher mental functions of human behaviour in terms of language and the creation of meaning. What becomes important is the expressive, non-rational qualities of the organization, and what is subjective is of great importance. Yet it could be argued that the current local government culture suppresses these values. Smircich posited that, if culture is taken to be a root metaphor, this leaves room for ambiguity. It is this approach which helps us to move away from taken-for-granted assumptions, or at least to healthily question these assumptions and values through which organizational analysis is carried out.

In particular, the performance management culture has been in evidence in local government at least since the 1990s. It has focused on target-setting, measurement and the monitoring of achievement through set 'milestones'. It is easier to find adverse criticism of governance by targets rather than the opposite. This has not been unique to the UK context, being a key element of government reforms across the OECD member states (Greiling, 2006). Greiling's review of whether performance measurement increases the efficiency of public service found a sceptical picture. For example, transaction costs when contracting out public services are resource-consuming, there is a risk of window-dressing, 'gaming' the results, and compliance with external accountability rather than improved services for the client. Yet 'organizational slack' may be a good thing as this allows for experiments and learning, which could then go on to improve services in a more genuine way. It is suggested that performance measurement should be linked with purposes of strategic planning rather than processes of accountability. A frequently cited paper (Bevan and Hood, 2006) described the degree to which the 'gaming' of targets had been taken to the extreme, outlining unintended consequences of the performance-by-measurement culture. The failure of the target-driven reform regime has also been highlighted by Seddon (2008) who claims, ironically, that this costly regime has itself created a system which is systemically incapable of doing the right thing. Sanderson (1998) questioned the extent to which performance measurement provides an acceptable basis for a full evaluation of local government renewal, suggesting that an approach taking multiple viewpoints, rather than a purely technical approach, would more truly enhance capacity to address the complexity of society's problems, and also help develop a more meaningful interaction with the public. More recently, a 30-year review considered what has been learned through 'governance by numbers' (Jackson, 2011). Jackson highlighted the naïve goal model, the multi-dimensional concept of public sector performance, the importance of allocative efficiency, and of what is left out (equity, environmental impact, ethics); the need for project/programme-level assessments, the distributional aspects of public spending, that little is known about the interface between academia and practice, that consultants' and trainers' literature is prescriptive and that, with limits to 'the linear path' there is a lack of a general theory of public sector performance. Despite a significant cull of the UK national performance indicators, the habit of devising measurable targets prevails.

As the demand for innovation goes hand in hand with learning, organizational learning is also relevant. A systematic literature review was carried out as a long-term evaluation of a Beacon Scheme to improve performance in local government (Rashman et al., 2008; 2009). The model created through this research emphasized a two-way process of knowledge sharing and the importance of interaction in the knowledge transfer process. This requires inter-organizational learning as well as organizational learning, and an understanding of both the immediate context and the wider policy and practice context. Within the wider context, cultural features will affect the capacity to learn. Networks, however, with a high degree of consensus may block the learning process due to lack of challenge. In common with previous findings, power is identified as an influential factor favouring or inhibiting learning. The fast pace of change in the public sector also renders learning difficult. Overall, this suggests that a narrow organizational context for learning, with a mixture of power structures, consensus and a fast pace of change, could all harbour hidden assumptions which are actually blocking the learning process. The development of practitioner forums for interaction across organizational boundaries was recommended. Further support for purpose-driven work groups was proposed by Moynihan and Landuyt (2009), with such groups preferably to include dissenting views.

A significant barrier to overcome is a limited understanding of what it means to be an expert, which is normally associated with 'knowing' techniques, or 'knowing' the answers. What is required is recognition of the value of remaining uncertain, leading to a style of collaborative way-finding, which is also of genuine value to the tackling of society's complex challenges.

A shift in thinking is required beyond the assumed levers of change to overcome these cultural barriers. Whereas consensus, power and the fast pace of change are sometimes acting as inhibitors to learning, nevertheless there is a genuine appetite for learning which could be situated by linking academic learning with everyday practice. Insights from complexity can provide techniques with the potential to deliver public value and evolve towards the common good, but only if local government management can escape the imposed 'fire-fighting' nature of the year-on-year budget cuts, and be drawn somehow towards a complexity approach.

THE GEM OF COMPLEXITY: FUTURE INSIGHTS AND CONCLUSION

Acknowledgement of complexity seems unpalatable when the focus is upon measurable proof of success, so the term 'complexity' is used colloquially to denote intractable problems over which we are helpless to exert an influence, rather than as a healthy approach to problem-solving. Yet the 'second order' complexity of Tsoukas and Hatch (2001) suggests that it relates to how we organize our thinking about complexity, as well as complexity being a feature of the system under study. Geyer and Rihani believe that complexity theory could reconcile opposing views, acting as a bridge between the rationalist orderly paradigm and its antithesis of post-modernism (Geyer and Rihani,

254 *Handbook on complexity and public policy*

2010). So, the persisting mechanistic and orderly model of local government would need to give credibility to uncertainty, unpredictability and multiple perspectives, as local government is constantly grappling with problems which are complex, such as poverty or environmental sustainability. This suggests that complexity should be re-branded as an active resource to improve thinking, rather than merely being a passive label for an intractable problem.

Despite complexity's disparate roots and often difficult terminology, there is every indication that there are creative ways of thinking about complexity, moving towards the social sciences and public policy, in which case there is much scope to apply it within local government. Rather than avoiding complexity, we can put complexity to work and get it to work for us. This was well recognized by Churchman (1977), who referred to the 'bright side of complexity' providing a role for planners in considering the interaction of problems, how we must consider future generations, and the importance and uniqueness of the individual, the self. This means a capability of spanning global with individual thought, and resisting the more usual embedded hierarchical and corporate categorizations. Local politics may also be freed up from apparently intractable problems by working in a spirit of 'principled localism' (Blair and Evans, 2004), perhaps taking sustainability (the integration of economic, social and environmental issues) as the key principle of local leadership, providing directional powers rather than prescriptive 'solutions'. Overall, this could create a healthier scene at the local level, returning to the idea of a sustainable approach for the public good (Bichard, 2013). This requires leaving the illusory territory of Plato's cave and emerging in a spirit of collaborative way-finding.

More specifically, techniques through insights from complexity are outlined in this book and elsewhere, considering how to create the circumstances for improvement, which support such an approach; fitness landscapes (Geyer and Rihani, 2010), Cynefin framework (Snowden and Boone, 2007), complexity cascade (Geyer and Rihani, 2010; Geyer, 2012), emergence, (Seel, 2000; 2006; Stacey, 2005), changing conversations (Shaw, 2002), the Stacey diagram (Geyer and Rihani, 2010). Furthermore, considering linkages with the field of systems thinking, emancipatory and community-based methods of improvement, such as Critical Systems Heuristics (Ulrich, 1983; Reynolds and Holwell, 2010), Boundary Critique or Community Operational Research (Midgley, 2000; Midgley and Ochoa-Arias, 2004; Midgley and Richardson, 2007) are of relevance, whilst well established Soft Systems Methodologies (Checkland and Scholes, 1990; Checkland and Poulter, 2006) can help with participatory approaches. Lean systems thinking and associated methods also help to focus on efficiency and effectiveness for the customer (Seddon, 2008).

There is an impetus to explore complexity's applicability to public policy beyond the UK, and this could also be pursued further within UK local government, whether within policy, research or practice. The initial challenge is to see organizations not only as hierarchical structures, but also as networks of human relationships, thereby creating a rich source of possibility of combined expertise. In addition to providing potential to release positive energy at the local level, an appreciation of complexity also adds resilience to decision-making. A valuable aspect of complexity approaches is that they guard against us becoming 'prisoners of the proximate' (McMichael, 1999) by expanding our thinking. In the face of budget cuts, there is a natural tendency to become more inward-looking in this fight for survival, thus overlooking the external environment. Yet Ashby's law of requisite variety states that 'only variety can destroy variety' (Ashby, 1956). It is therefore desirable to amplify variety, whilst attenuating the variety of the wider system or environment. In local government, this could be interpreted as the need to amplify variety and to attenuate demand for local government services. Yet funding is being reduced, therefore potentially reducing variety, whilst demand for services is increasing, due in particular to the ageing population. Whilst consideration needs to be given to the 'de-marketing' of public services to attenuate demand, the creative amplification of variety could be approached through complexity and systems skills in the design of local government services.

In order to escape the vicious circle of repeating the same mistakes, it is important to take time to explore and ask a range of questions. Instead of asking the popular questions: 'what tools are available to fix this problem, how can we afford it, and how can we measure the results?', more appropriate questions are: 'how can we understand the nature of the problem more fully, where are its boundaries, who do we involve, how do we talk about it, can we afford not to act, and how do we avoid unintended consequences?' In this way, the meaning of 'resource' goes beyond the accountancy sheet, galvanizing human capability with a sense of purpose. Furthermore, as well as challenges of the here and now, which will inevitably always be our starting point, we must imagine the critical voice of future generations. This moves thinking beyond 'what works now?' to 'what should we be doing under current and future circumstances?' In taking up these ideas, difficulties in terminology are likely to be resolved to reduce the gap between academia and practice, creating the desired stronger link between scholarly rigour and relevance (Pettigrew, 1997; 2011). There is the possibility of a sound future for the application of complexity to local government, if a feasible connection can first be made with practitioners.

In conclusion, this chapter has highlighted the relevance and significance of a complexity approach to local government based on many years as a local government practitioner, a Master's dissertation and the first year of doctoral research. Chapman (2004) concluded that it is the failure of existing thinking which encourages adoption of systems thinking. If public policy cannot be allowed to fail, then a way has to be found to negotiate a change when the timing is more conducive. Ten years on from Chapman, the evidence of the time 'becoming right' is compelling, for the following reasons:

- there is continued emphasis on public value, the need expressed for a public service strategy, a growing significance for the role of local government and a local sense of 'place';
- a range of authoritative inquiries has shown an appetite to find different ways of working;
- there is evidence at home and abroad that complexity and systems thinking could potentially step up to this capacity-building role;
- a style of local government leadership could combine leadership with collaborative skills.

The cultural barriers remain, which are wedded in particular to the performance management regime of measurement. Faced with the immediate challenges of cost-cutting, the current popularity of peer group learning, professional specialisms, and a succession

256 *Handbook on complexity and public policy*

of innovative pilot studies, the essential puzzle remains: how best to connect supply (the broad range of relevant scholarly knowledge) with demand (a new sustainable model for local government which itself achieves sustainability) so that this approach of valuing complexity can flourish in the mainstream through an acceptance of multi-disciplinary working methods of service design which are tailored to the locality. My research will explore the possibilities of connecting the range of complexity and systems thinking skills with local government practice, but the 'gem' of complexity needs polishing by many, whether researchers, policy actors or practitioners. In the future decades and certainly beyond my working life, there is much to be done by many, for the stakes are high. There is considerable scope for further research, both within the UK and abroad, to help move on to a phase in local government policy which will harness the richness of human energy and values, in the true spirit of sustainability.

ACKNOWLEDGEMENTS

The basis of this chapter has been sourced from study for a Master's dissertation at the Institute of Local Government Studies (INLOGOV), University of Birmingham in 2008 (MSc Local Governance), and the first year of doctoral research at Hull University Business School (HUBS) between 2012 and 2013. With thanks to Steven Griggs (INLOGOV), Alberto Franco, Amanda Gregory, Giles Hindle and Gerald Midgley (HUBS) for their sound academic guidance. Thanks also to Robert Geyer for his assured support over a number of years in encouraging me to pursue these compelling interests.

NOTE

1. See Jackson (2003). This provides a presentation of paradigmatic groupings for systems techniques although, as with other fields of endeavour, there is some degree of flexibility in how techniques can be used from different paradigmatic stances. See also Kuhn's original treatise on paradigms Kuhn (1962). Kuhn was concerned with the process of revolution rather than categorization, so such ideas are indeed constantly evolving.

REFERENCES

American Society for Public Administration and Governance of Complex Systems (2013), 'Challenges of Making Public Administration and Complexity Theory (COMPACT) Work', conference at La Verne, CA, USA, 5–8 June.
Anderson, P. (1999), 'Complexity theory and organization science', *Organization Science*, **10**(3), 216–32.
Ashby, W.R. (1956), *An Introduction to Cybernetics*, New York: Wiley.
Bateson, G. (1979), *Mind and Nature: A Necessary Unity*, Cresskill, NJ: Hampton Press.
Benington, J. (2011), *New Horizons for Local Governance*, Warwick: Warwick Business School and LARCI.
Benington, J. and J. Hartley (2009), *Whole Systems Go! Leadership Across The Whole Public Service System*, Ascot: National School of Government.
Benington, J. and M.H. Moore (2010), *Public Value: Theory and Practice*, Maidenhead: Palgrave Macmillan.
Bevan, G. and C. Hood (2006), 'What's measured is what matters: Targets and gaming in the English public health care system', *Public Administration*, **84**(3), 517–38.
Bichard, Lord (2013), 'Editorial: the need for a public service strategy', *Public Money & Management*, **33**(1), 3–4.

Blackman, T. (2001), 'Complexity theory and the new public management', *Social Issues*, **1**(2), Special Issue: Complexity Science and Social Policy.

Blair, F. and B. Evans (2004), *Seeing the Bigger Picture: Delivering Local Sustainable Development*, York: Joseph Rowntree Foundation.

Boal, K.B. and P.L. Schultz (2007), 'Storytelling, time, and evolution: The role of strategic leadership in complex adaptive systems', *The Leadership Quarterly*, **18**(4), 411–28.

Bogg, J. and R. Geyer (eds) (2007), *Complexity, Science and Society*, Oxford, New York: Radcliffe Publishing.

Bono, J.E., W. Shen and M. Snyder (2010), 'Fostering integrative community leadership', *The Leadership Quarterly*, **21**(2), 324–35.

Bourgon, J.E. (ed.) (2010), *A Public Service Renewal Agenda for the 21st Century: The New Synthesis Project*, London: Public Governance International.

Bovaird, T. (2008), 'Emergent strategic management and planning mechanisms in complex adaptive systems', *Public Management Review*, **10**(3), 319–40.

Bovaird, T. (2011), 'The future of local authorities: Revolutionary, revolving in the grave or just going round in circles?', in C.M. Oyarce (ed.), *Redefining Local Government*, London: Public Policy and Management Association, CIPFA.

Butler, M. and P. Allen (2008), 'Understanding policy implementation processes as self-organizing systems', *Public Management Review*, **10**(3), 421–40.

Cairney, P. (2010), 'Complexity theory in public policy', paper presented at the Political Studies Association Conference, Edinburgh University, 29 March–1 April.

Chapman, J. (2004), *System Failure: Why Governments Must Learn to Think Differently*, 2nd edn., London: Demos.

Checkland, P. and J. Poulter (2006), *Learning for Action: A Short Definitive Account of Soft Systems Methodology and its use for Practitioners, Teachers, and Students*, Chichester: Wiley.

Checkland, P. and P. Scholes (1990), *Soft Systems Methodology in Action*, Chichester: Wiley.

Christiansen, J. and L. Bunt (2012), *Innovation in Policy: Allowing for Creativity, Social Complexity and Uncertainty in Public Governance*, London: NESTA.

Churchman, C.W. (1977), 'A philosophy for complexity', in H.A. Linstone and W.H. Simmonds (eds), *Managing Complexity*, Reading, MA: Addison-Wesley.

Commission on 2020 Public Services (2010), *From Social Security to Social Productivity: a Vision for 2020 Public Services*, London: 2020 Public Services Trust.

Commission on the Future of Local Government (2012), *Final Report*, CFLG.

Crosby, B.C. (2010), 'Leading in the shared-power world of 2020', *Public Administration Review*, **70**(1), S69–S77.

Crosby, B.C. and J.M. Bryson (2005), 'A leadership framework for cross-sector collaboration', *Public Management Review*, **7**(2), 177–201.

Crosby, B.C. and J.M. Bryson (2010), 'Integrative leadership and the creation and maintenance of cross-sector collaboration', *The Leadership Quarterly*, **21**(2), 211–30.

Elias, N. (2000), *The Civilising Process*, London: Blackwell.

Eppel, E., A. Matheson and M. Walton (2011), 'Applying complexity theory to New Zealand public policy: Principles for practice', *Policy Quarterly*, **7**(1), 48–55.

Fernandez, S., Y.J. Cho and J.L. Perry (2010), 'Exploring the link between integrated leadership and public sector performance', *The Leadership Quarterly*, **21**(2), 308–23.

Geyer, R. (2012), 'Can complexity move UK policy beyond "evidence-based policy making" and the "audit culture"? Applying a "complexity cascade" to education and health policy', *Political Studies*, **60**(1), 20–43.

Geyer, R. and S. Rihani (2010), *Complexity and Public Policy: A New Approach to 21st Century Politics, Policy and Society*, London: Routledge.

Gillinson, S., M. Horne and P. Baeck (2010), *Radical Efficiency: Different, Better, Low Cost Public Services*, London: NESTA.

Greiling, D. (2006), 'Performance measurement: A remedy for increasing the efficiency of public services?', *International Journal of Productivity and Performance Management*, **55**(6), 448–65.

Hallsworth, M. (2011), 'System Stewardship: The Future of Policy Making?', Working Paper, Institute for Government.

Haveri, A. (2006), 'Complexity in local government change', *Public Management Review*, **8**(1), 31–46.

Haynes, P. (2003), *Managing Complexity in the Public Services*, Maidenhead: Open University Press.

Haynes, P. (2008), 'Complexity theory and evaluation in public management', *Public Management Review*, **10**(3), 401–19.

Ho, P. (2012), 'Coping with complexity', in *Government Designed for New Times*, McKinsey & Co.

Homer-Dixon, T. (2010), 'Complexity science and public policy', *John L. Manion Lecture*, Ottawa, National Arts Centre.

258   *Handbook on complexity and public policy*

Houchin, K. and D. MacLean (2005), 'Complexity theory and strategic change: An empirically informed critique', *British Journal of Management*, **16**(2), 149–66.
Huaxia, Z. (2007), 'Exploring dynamics of emergence', *Systems Research and Behavioral Science*, **24**(4), 431–43.
Jackson, M.C. (2003), *Systems Thinking: Creative Holism for Managers*, Chichester: Wiley.
Jackson, P.M. (2011), 'Governance by numbers: What have we learned over the past 30 years?', *Public Money & Management*, **31**(1), 13–26.
Johannessen, S.O. and L. Kuhn (eds) (2012), *Complexity in Organization Studies*, London: Sage.
Johnson, J.L. and B.K. Burton (1994), 'Chaos and complexity theory for management: Caveat emptor', *Journal of Management Inquiry*, **3**(4), 320–28.
Kelly, G. and S. Muers (2002), *Creating Public Value: An Analytical Framework for Public Service Reform*, London: Strategy Unit, Cabinet Office.
Kuhn, T. (1962), *The Structure of Scientific Revolutions*, Chicago and London: The University of Chicago Press.
Landini, S. and S. Occelli (2012), 'Innovative public policy – the role of complexity science', *E:CO*, **14**(4), vii–xiii.
Letiche, H., M. Lissack and R. Schultz (2011), *Coherence in the Midst of Complexity: Advances in Social Complexity Theory*, New York: Palgrave Macmillan.
Levy, D. (1994), 'Chaos theory and strategy: Theory, application, and managerial implications', *Strategic Management Journal*, **15**(2), 167–78.
Lissack, M. (2002), *The Interaction of Complexity and Management*, Westport, CT: Greenwood Publishing Group.
Mabey, N. (2004), *System Thinking and System Dynamics in Public Policy Making: Some Experiences of the Prime Minister's Strategy Unit*, London: Cabinet Office.
Maguire, S., B. McKelvey, L. Mirabeau and N. Oztas (2006), 'Complexity science and organization studies', in S. Clegg, C. Hardy and T. Lawrence (eds), *Handbook of Organization Studies*, (2nd edn), Thousand Oaks, CA: Sage, pp. 165–214.
McMichael, A.J. (1999), 'Prisoners of the proximate: Loosening the constraints on epidemiology in an age of change', *American Journal of Epidemiology*, **149**(10), 887–97.
McMillan, E. (2004), *Complexity, Organizations and Change*, London: Routledge.
Medd, W. (2001), 'Critical emergence: Complexity science and social policy', *Journal of Social Issues*, **1**(2).
Meek, J.W. (2010), 'Complexity theory for public administration and policy', *Emergence: Complexity and Organization*, **12**(1), 1–4.
Midgley, G. (2000), *Systemic Intervention: Philosophy, Methodology, and Practice*, Berlin: Springer.
Midgley, G. and A.E. Ochoa-Arias (2004), *Community Operational Research: OR and Systems Thinking for Community Development*, Berlin: Springer.
Midgley, G. and K.A. Richardson (2007), 'Systems thinking for community involvement in policy analysis', *Emergence: Complexity & Organization*, **9**(1/2), 167–83.
Mitleton-Kelly, E. (2003), *Complex Systems and Evolutionary Perspectives on Organisations: The Application of Complexity Theory to Organisations*, Bingley: Emerald Group.
Moore, M.H. (1995), *Creating Public Value: Strategic Management in Government*, Cambridge, MA: Harvard University Press.
Morcol, G. (1996), 'Fuzz and chaos: Implications for public administration theory and research – book reviews', *Journal of Public Administration Research and Theory*, **6**(2), 315–25.
Morse, R.S. (2010), 'Integrative public leadership: Catalyzing collaboration to create public value', *The Leadership Quarterly*, **21**(2), 231–45.
Moynihan, D.P. and N. Landuyt (2009), 'How do public organizations learn? Bridging cultural and structural perspectives', *Public Administration Review*, **69**(6), 1097–105.
Mulgan, G. (2001), 'Systems thinking and the practice of local government', *Systemist*, **23**, 23–9.
Mulgan, G. and C. Leadbeater (2013), *Systems Innovation*, Discussion Paper, NESTA.
Munro, E. (2011), *The Munro Review of Child Protection: Final Report – A Child-Centred System*, London: Department of Education.
OECD Global Science Forum (2009), *Applications of Complexity Science for Public Policy: New Tools for Finding Unanticipated Consequences and Unrealized Opportunities*, Erice, Sicily: OECD.
Office of the Deputy Prime Minister, The Tavistock Institute, SOLON Consultants & Local Government Information Unit (2006), *All Our Futures: The Challenges for Local Governance in 2015*, London: ODPM.
Ortegón-Monroy, M.C. (2003), 'Chaos and complexity theory in management: An exploration from a critical systems thinking perspective', *Systems Research and Behavioral Science*, **20**(5), 387–400.
Oyarce, C.M. (ed.) (2011), *Redefining Local Government*, London: Public Policy and Management Association, CIPFA.
Oyarce, C.M. and E. Kirkman (2011), 'Leadership and the courage to change in local government: Interviews with local authority chief executives', in C.M. Oyarce (ed.), *Redefining Local Government*, London: Public Policy and Management Association, CIPFA.

Oyarce, C.M., Accenture, F. Boardman, Public Management and Policy Association (2012), *Public Service Reform in the UK: Revolutionary or Evolutionary?*, London: CIPFA.

Page, S. (2010), 'Integrative leadership for collaborative governance: Civic engagement in Seattle', *The Leadership Quarterly*, **21**(2), 246–63.

Pettigrew, A.M. (1997), 'The double hurdles for management research', in T. Clarke (ed.), *Advancement in Organizational Behaviour: Essays in Honour of D.S. Pugh*, London: Dartmouth Press, pp. 277–96.

Pettigrew, A.M. (2011), 'Scholarship with impact', *British Journal of Management*, **22**(3), 347–54.

Political and Constitutional Reform Committee (2013), *Political and Constitutional Reform: Third Report. Prospects for Codifying the Relationship between Central and Local Government*, London: UK Parliament.

Rashman, L., E. Withers and J. Hartley (2008), *Long-term Evaluation of the Beacon Scheme. Organizational Learning, Knowledge and Capacity: A Systematic Literature Review for Policy-makers, Managers and Academics*, DCLG, London: Warwick Business School.

Rashman, L., E. Withers and J. Hartley (2009), 'Organizational learning and knowledge in public service organizations: A systematic review of the literature', *International Journal of Management Reviews*, **11**(4), 463–94.

Reynolds, M. and S. Holwell (2010), *Systems Approaches to Managing Change: A Practical Guide*, Berlin: Springer.

Rhodes, M.L., J. Murphy, J. Muir and J.A. Murray (2011), *Public Management and Complexity Theory: Richer Decision-Making in Public Services*, London: Routledge.

Richardson, K.A., W.J. Gregory and G. Midgley (eds) (2006), *Systems Thinking and Complexity Science: Insights for Action*, proceedings of the 11th ANZSYS Managing the Complex V Conference, Christchurch, New Zealand, 5–7 December, 2005, Litchfield Park, AZ: ISCE Publishing.

Richardson, K.A., W.J. Gregory and G. Midgley (2007), 'Editorial introduction to the special double issue on Complexity Thinking and Systems Theory', *Emergence: Complexity & Organization*, **9**(1/2), vi–viii.

Rosenhead, J. (1998), 'Complexity theory and management practice', Operational Research working papers, LSEOR 98.25, Department of Operational Research, London School of Economics and Political Science, London.

RSA (2012), 'Fiscal fallout', *RSA debate*, London, available at http://www.thersa.org/events/audio-and-past-events/2012/Fiscal-Fallout-the-challenge-ahead-for-public-spending-and-public-services.

Sanderson, I. (1998), 'Beyond performance measurement? Assessing "value" in local government', *Local Government Studies*, **24**(4), 1–25.

Santa Fe Institute (2013), 'About the Santa Fe Institute', accessed 21 August 2013 at http://www.santafe.edu/about/>.

Seddon, J. (2008), *Systems Thinking in the Public Sector: The Failure of the Reform Regime . . . and a Manifesto for a Better Way*, Axminster: Triarchy Press Limited.

Seel, R. (2000), 'Culture and complexity: New insights on organisational change', *Culture and Complexity: Organisations and People*, **7**(2), 2–9.

Seel, R. (2006), *Emergence in Organisations*, accessed 29 April 2013 at http://doingbetterthings.pbworks.com/f/RICHARD+SEEL+Emergence+in+Organisations.pdf.

Shaw, P. (1997), 'Intervening in the shadow systems of organizations: Consulting from a complexity perspective', *Journal of Organizational Change Management*, **10**(3), 235–50.

Shaw, P. (2002), *Changing Conversations in Organizations: A Complexity Approach to Change*, London: Routledge.

Silvia, C. and M. McGuire (2010), 'Leading public sector networks: An empirical examination of integrative leadership behaviors', *The Leadership Quarterly*, **21**(2), 264–77.

Smircich, L. (1983), 'Concepts of culture and organizational analysis', *Administrative Science Quarterly*, **28**(3), 339–58.

Snowden, D.J. and M.E. Boone (2007), 'A leader's framework for decision making', *Harvard Business Review*, **85**(11), 68–76.

Stacey, R. (2005), *Experiencing Emergence in Organizations Local Interaction and the Emergence of Global Pattern*, London: Routledge.

Stacey, R. (2012), *Tools and Techniques of Leadership and Management*, London: Routledge.

Stacey, R.D. (2007), *Strategic Management and Organisational Dynamics: The Challenge of Complexity to Ways of Thinking About Organisations*, New York: Prentice Hall.

Sullivan, H. (2010), 'Collaboration matters', inaugural lecture, College of Social Sciences, University of Birmingham.

Talbot, C.R. and C.L. Talbot (2011), 'Local government strategies in an age of austerity', in C.M. Oyarce (ed.), *Redefining Local Government*, London: Public Policy and Management Association, CIPFA.

Teisman, G. and E. Klijn (2008), 'Complexity theory and public management', *Public Management Review*, **10**(3), 287–97.

Tsoukas, H. (2005), *Complex Knowledge*, Oxford: Oxford University Press.

260 *Handbook on complexity and public policy*

Tsoukas, H. and M.J. Hatch (2001), 'Complex thinking, complex practice: The case for a narrative approach to organizational complexity', *Human Relations*, **54**(8), 979–1013.

Ulrich, W. (1983), *Critical Heuristics of Social Planning: A New Approach to Practical Philosophy*, Chichester: Wiley.

Warfield, J.N. (1999), 'Twenty laws of complexity: Science applicable in organizations', *Systems Research and Behavioral Science*, **16**(1), 3–40.

Wenger, E. (1999), *Communities of Practice: Learning, Meaning, and Identity*, Cambridge: Cambridge University Press.

White, L. (2001), '"Effective governance" through complexity thinking and management science', *Systems Research and Behavioral Science*, **18**(3), 241–57.

White, L. (2003), 'Voices and values: Thinking about OR and systems in public policy making', in *Proceedings of the Critical Management Conference*, July, Lancaster University.

16. Managing complex adaptive systems to improve public outcomes in Birmingham, UK
Tony Bovaird and Richard Kenny

INTRODUCTION

This chapter outlines an innovative and challenging approach to modelling how public policy can impact upon the main outcomes which public agencies seek to achieve in a metropolitan area. It draws upon the longstanding literature on cause-and-effect mapping to identify the pathways to outcomes in a key service area and to explore how the relative cost-effectiveness of alternative pathways to outcomes can be calibrated. It then discusses how some aspects of the public sector's interventions cannot be modelled convincingly in this way, because they have the characteristics of complex adaptive systems. The consequences of this for public policy are then explored.

The chapter outlines how strategy maps were developed for a wide range of economic programmes in Birmingham City Council in order to help city decision makers to:

- better understand the City's current resource allocation patterns and the range of potentially interesting future options;
- highlight areas in which resource allocation is poorly evidenced;
- structure debate about the likely impacts of major budget reductions on outcomes.

The specific case study considered here is the 'Succeed Economically' outcome area (see Figure 16.1) in the city of Birmingham in the UK, which has the second largest city council in Europe. The strategy map in Figure 16.1 embodies a set of economic outcomes which complements the 'health and wellbeing' and 'cleaner, greener, safer' outcomes maps, which together make up the full set of quality of life outcomes agreed for the Birmingham city area by the City Council and its partners. The public sector interventions that are currently the main focus of the mapping work carry with them specific value for money requirements that must be fulfilled. It was anticipated therefore that evidence about impacts and outputs on outcomes would be known.

While the figure may be difficult to read, it should be clear from Figure 16.1 that the approach has modelled the key relationships and interdependencies between interventions, costs, outputs and outcomes across the city using the best available evidence. It is a whole system approach that sets out to develop full cause-and-effect chains. The model is strongly outcome-focused. By working backwards from outcomes it highlights the pathways by which those outcomes are currently being achieved and which alternative pathways, on current evidence, might be more cost-effective.

A key focus is on finding pathways which enable better outcomes by reducing recurring social and administrative costs through earlier intervention and in ways that avoid future costs through prevention. This means shifting systematically towards investing in

Succeed Economically: Level 3 Conceptual Model

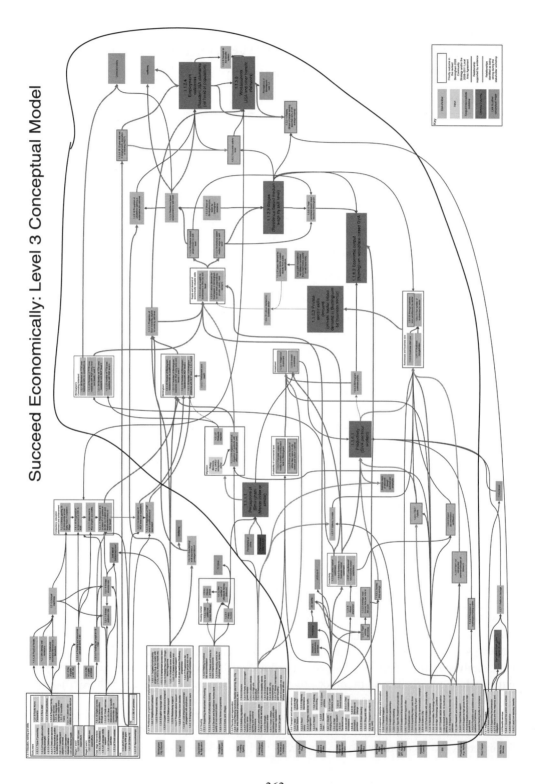

Succeed Economically: Worklessness programmes and employment support

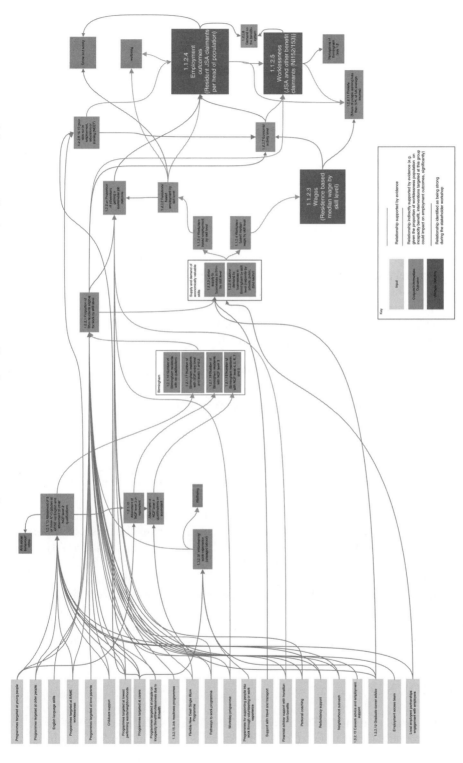

Note: Copies of this figure are available from the authors upon request.

Figure 16.1 The 'Succeed Economically' pathway to outcomes

264 *Handbook on complexity and public policy*

productive activities, like building social capital or creating jobs. It requires an explicit approach to balancing supply and demand within, for example, labour and housing markets.

On the back of a whole city system mapping of public sector interventions designed to impact on the Birmingham economy, amounting to an annual spend of £7.5 billion (EKOSGEN, 2010), the project undertook a series of 'deep dives' to quantify the key relationships within specific cause-and-effect chains. Decision-makers decided that, on balance, the weaker side of the Birmingham economy at the time was the demand side. So the main arena for detailed analysis has been on programmes to stimulate the demand side of the economy – concentrating on business support innovation and enterprise, city marketing and the visitor economy, attracting inward investment and digital connectivity.

The project has identified financial inputs of £312 million of annual investment within these fields, both capital and revenue. Over 200 research/evaluation and business case reports were reviewed, in order to pull out evidence on the input–output–outcome relationships in these programmes. As the model was developed, we tested it by seeking views from a wide range of stakeholders and a panel of academic 'experts' on the strength of the underlying relationships in the pathways to higher-level outcomes.

However, in the course of conducting this analysis, it has become clear that cause-and-effect pathways cannot be constructed for all aspects of the outcomes being sought by public sector agencies in the city. Some outcomes are the result of complex adaptive systems (CAS), in which the interdependencies between the contributing variables are so rich that the results of their interaction cannot be predicted in any detail in a model. However, there are still some policy levers which can influence the overall shape of the potential outcomes which can be achieved in these areas, although the actual outcomes cannot be predicted, nor *a fortiori* the time trajectory through which they will occur. Moreover, it may be possible to model some outcomes only to a limited extent, while recognizing that they are partly the result of the operation of CAS – here we have to apply several methodologies in order to understand fully how outcomes emerge in the city.

The modelling approach reported here is being developed to support city leaders in the development of their priorities in reconfiguring city-wide programmes and budgets for the medium to longer term. It is intended as a tool for improving the quality of debate around the outcomes from public intervention, rather than giving a definitive 'answer'. This chapter argues that the modelling approach can help in this debate – but it is also important for policy makers to be aware of the limitations of their knowledge and their ability to use cause-and-effect models to predict. Where they are dealing with complex adaptive systems, policy makers are in danger of making claims for their favoured interventions which have no substance. The chapter considers what it is legitimate for policy makers to do in these circumstances, together with the implications for strategy and organizational learning which arise from the rather more modest claims that policy can make when dealing with complex adaptive systems, where attribution issues are particularly problematic.

BACKGROUND

Birmingham is a city of 1.1 million people in the West Midlands region of the UK. Originally a manufacturing oriented city at the heart of the Industrial Revolution, it has suffered from the fast decline of manufacturing since the 1960s and is now a modern services-based city with a small but significant advanced manufacturing sector, but suffering levels of unemployment which are relatively high compared to the more prosperous South-East of Britain. The city council budget is £3.1 billion annually and its decisions aim to fit with and complement the other £4 billion of public spend annually in the city. The current fiscal crisis in the UK means that the City Council has had to achieve a 28 per cent reduction in its revenue budget over four years from 2010.

The first phase of Modelling Birmingham (2010–11) was a partnership between the City Council, the University of Birmingham and KPMG. This phase succeeded in setting out the overall model architecture. The second phase, with the University of Warwick and University of Birmingham, drilled down into the cause-and-effect relationships to produce strategy maps for key parts of the model. The third phase, with the original partners, calibrated a sample of these in-depth strategy maps with quantitative data, developing algorithms to identify interdependencies between parts of the model, and testing the confidence that city policymakers can have in these relationships when forecasting outcomes from changing resource allocations in the city.

At the highest level of policy in the city, it has been recognized that current decision making, especially on resource allocation, is partially 'flying blind', without the evidence needed to work out priorities between major spending programmes. Of course, this is not the case only in Birmingham – it appears to apply right across local and central government and all public agencies. To some extent this reflects the political nature of decision-making in central and local government but the lack of evidence to support the development of public policy and programmes is a major concern in determining value for money. The weak evidence base for UK decision-making is confirmed by the academic literature on 'what works' in the public sector (Davies et al., 2000, Cartwright and Hardie, 2012), in spite of the Blair government's push towards evidence-based policy targets and a central delivery unit and, more recently, the coalition government's initiative to set up six research-based 'what works' centres across public policy systems. The Birmingham model was intended to help to rectify this – in particular, it was intended:

- to better inform the understanding of City Council Directorates (and their partners) about the effects of their interventions;
- to highlight the areas in which this understanding is contested or poorly evidenced, so that further work can be done to produce a clearer understanding; and to structure the debate about the likely impacts of different kinds of intervention on outcomes;
- to determine possible alternative interventions and pathways that would provide better return on outcomes, particularly reducing costs in light of the financial pressures;
- to provide greater scope for understanding and factoring in third and private sector inputs to see how they could improve outcomes, replacing or working with existing or redesigned public sector interventions and mobilizing user and community co-production of outcomes.

It was recognized from the start that inputs from all parts of the public, private and third sectors are relevant to modelling outcomes achieved in the city, although modelling

266 *Handbook on complexity and public policy*

has so far focused on public sector interventions. As the model develops, partners are already taking ownership of parts of its evolution, for example Marketing Birmingham (in relation to attracting leisure visitors, business tourism and inward investment). This ownership and engagement has enabled the development of 'bottom-up' strategy maps that provide a greater degree of granularity and a deeper sense of the practical realities that 'front line' public sector workers deal with regularly. These strategy maps have enabled new and emerging inputs, including third and private sector inputs, to be incorporated into the approach. This has enabled decision-makers and practitioners to have a clearer picture of how the whole system is changing and to speculate on the likely implications of reconfiguring the inputs, both in terms of their quantity and the mix between sectors, providing scope for greater synergies, which will yield improvements in both outcomes and efficiency.

CONCEPTUAL FRAMEWORK

The main approach taken to modelling outcomes in Birmingham has essentially been cause-and-effect mapping, drawing on the evolving literature from hierarchies of objectives (Ansoff, 1965; Bovaird and Mallinson, 1988; Bovaird, 2012), the 'logical framework' ('logframe' or 'logic model') in international development studies (Rosenberg and Posner, 1979), the analytic hierarchy (Saaty, 1980), 'programme theory' (Bickman, 1990) 'strategy maps' (Kaplan and Norton, 2001; 2004), 'policy maps' based on cognitive mapping (Eden and Ackermann, 2003), 'theory of change' (Sullivan and Stewart, 2006), 'outcome- (or results-)based accountability' (Friedman, 2007), 'value creation maps' (Marr et al, 2004) and 'systems maps' (Mulgan, 2009).

Following this approach, strategy maps have been developed for a wide range of programmes in Birmingham City Council, in order to the help decision-makers in the City Council and its partner agencies to develop an understanding of the City's collective investment patterns. These have particularly brought alive the issue of 'dependencies'. In traditional business cases dependencies are in practice 'latent' and soon forgotten once the case is approved. Mapping existing interventions and future strategy enables different intervention owners and agents to see how they contribute to overall outcomes. Done well, this can also show what depends on specific interventions and what needs to work both before, after and in parallel, for those interventions to succeed. For example, in spite of longstanding demands for a 'skilling up' of the Birmingham labour force, analysis quickly shows there has been a dysfunctional skills equilibrium in the past, where businesses in reality demand low skills, which the education and training sectors have met. It is not enough to disrupt this sub-optimal and static scenario by raising aspirations of employers and growing the size of the knowledge economy – this has simply highlighted that demand by firms such as JLR for high engineering skills cannot currently be met in the region. It therefore makes little sense trying to stimulate new demand attracting advanced engineering companies and suppliers into the region until there is a clear labour skills supply strategy and operational capability – through apprenticeships, for instance – to be able to meet future demand. Modelling the labour market and identifying the prerequisite interventions and associated pathways highlights this stark reality and draws attention to the barriers and bottlenecks.

Managing complex adaptive systems to improve public outcomes 267

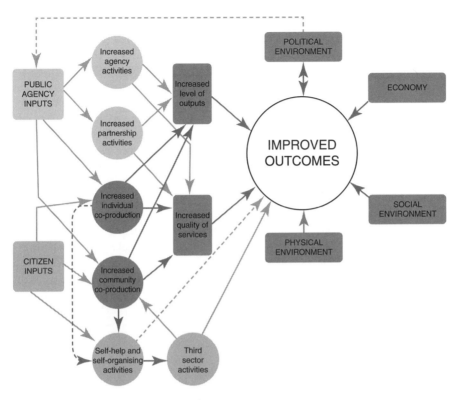

Note: Adapted from Bovaird and Tholstrup (2010).

Figure 16.2 The conceptual framework for the Birmingham model

In summary, the approach has modelled the key relationships and interdependencies between interventions, outputs, outcomes and inputs across the city using the best available evidence. It is a whole system approach that develops full cause-and-effect chains. The model is strongly outcome-focused. By working backwards from outcomes it is intended to provide greater understanding of how those outcomes can be achieved, what alternative pathways might be more cost-effective and what is the best available evidence to demonstrate impact of current council and other public sector activities. The overall conceptual framework, as set out in the original brief for partners, is set out in Figure 16.2.

Within the model in Figure 16.2, it was originally intended to show relationships between several outcomes:

- the outcomes (economic, social and environmental) included within the UK National Indicator set (about 170 nationally available performance indicators);
- the outcomes reported by local strategic partners (over and above the National Indicators) to government departments;
- other outcome indicators collected by local strategic partners for local purposes.

268　*Handbook on complexity and public policy*

In the event, some of these data sources became problematic during the course of the study; for example, collection and reporting of data for the UK National Indicator set became non-mandatory after 2010, so that significant parts of this data set were no longer available.

Because the study coincided with the onset of financial austerity in UK local government, there has in practice been a strong focus on cost avoidance (a financial consequence of improved outcomes achieved), for example through earlier intervention that reduces future demand by preventing and overcoming problems. There was also a drive within the modelling process to explore the potential for shifting systematically towards investing in productive economic activities, such as building social capital, which would help to reduce recurring social and administrative costs, such as those caused by ill-health or crime.

Although the need was recognized to distinguish between policy interventions which have a specifically area-based dimension – for example housing, local environmental improvement and so on – and those which are intended to help people wherever they may live in the city, most of the modelling has focused on city-wide analysis.

MODELLING PROCESS

The project has so far had four phases:

Phase 1: Conceptual Design of the Model

The main stages involved in the design phase were:

- developing 'strawman' strategy maps (for example Figure 16.1) for each of the high-level outcomes specified by Birmingham City Council and its partners: succeed economically, be healthy, and live in a clean, green and safe, city covering the whole city system;
- stakeholder workshops and one-to-one discussions with key stakeholders;
- gathering research evidence from literature and Academic Challenge workshops;
- input and support from Whitehall, particularly HM Treasury.

Phase 2: Developing the Strategy Maps

The main focus for detailed analysis has been on programmes to stimulate the demand side of the economy, concentrating on innovation, enterprise, city marketing, digital connectivity and transportation improvement, covering in total financial inputs of £312 million in annual investment, both capital and revenue. Views were sought from a wide range of stakeholders on the strength of the underlying relationships in the pathways to higher-level outcomes.

An example of one of the early strategy maps which was drawn up (for Marketing and Attracting Investment) is shown at Figure 16.3. These maps have now been taken on by the 'intervention holders' and early feedback has emphasized that this mapping

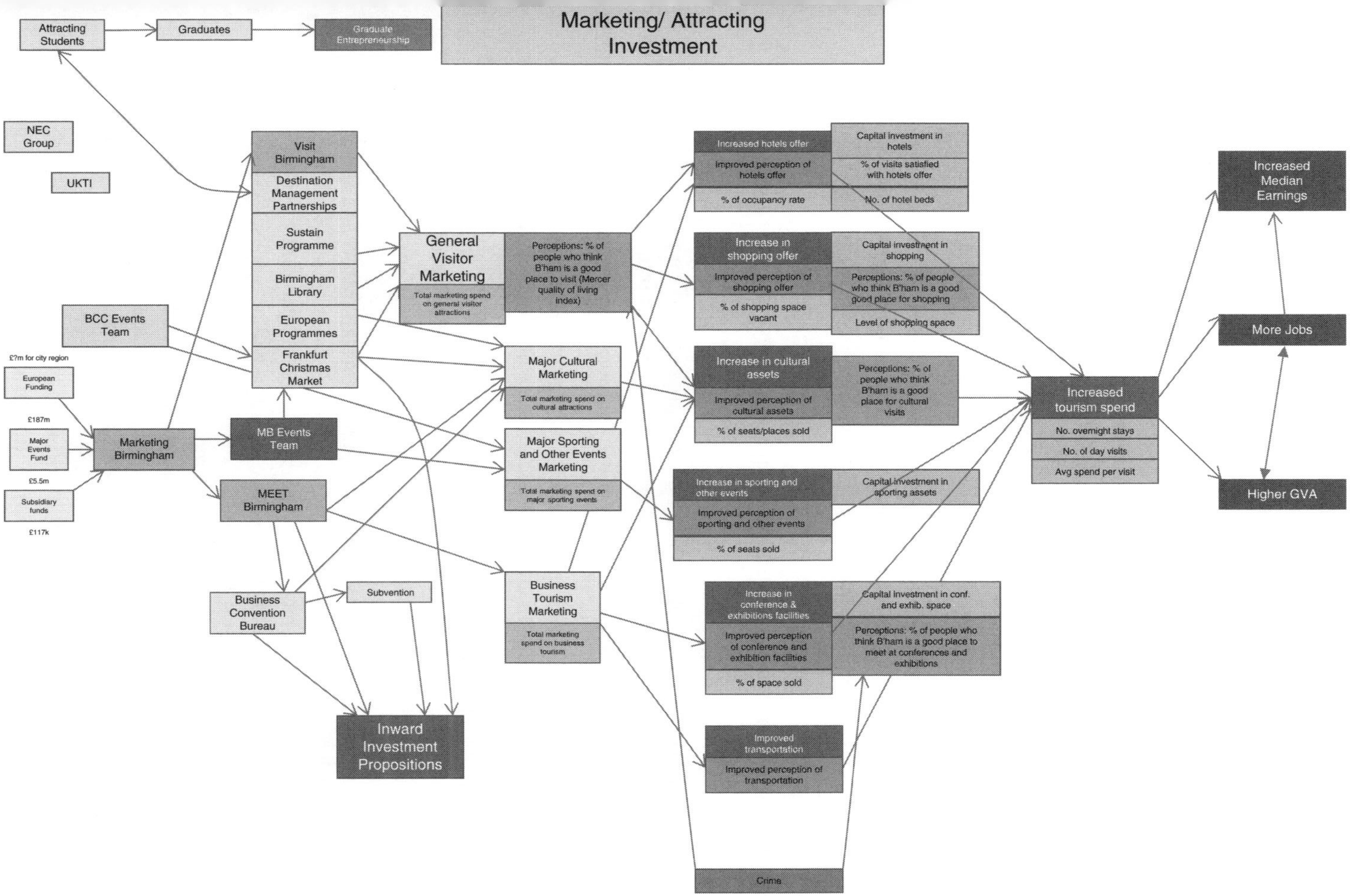

Source: Bovaird and Kenny (2013).

Figure 16.3 A strategy map for marketing and attracting investment

270   *Handbook on complexity and public policy*

activity has helped to clarify key strategic choices facing each of the programmes being examined. However, the full benefits of this approach can clearly only be reaped if and when the key relationships in these strategy maps can be quantified and if the most cost-effective pathways to outcomes can be identified.

Phase 3: Quantifying Links in the Strategy Maps

Small groups of involved staff have explored existing quantitative evidence in relation to the links in the strategy maps, making a central estimate of the value of the link (calling on the experience of the intervention holders to interpret the evidence and fill in gaps) and then entering the coefficients in an Excel spreadsheet. This has so far been done for all the links in Figure 16.4 (which is a later version of the Marketing and Attracting Investment programme).

The implication of the analysis was that the initial £100K of extra public sector spending was likely to have most impact (in terms of jobs created) if channelled through the strategy of direct marketing for high-volume job creation businesses. This was not the strategy which was receiving the highest volume of spending in Birmingham at the time this analysis was undertaken.

Phase 4: Rolling Out and Expanding the Modelling Approach

The 'Marketing and Attracting Investment' mapping and quantification has not been completed in full. Whilst there remains an outstanding need to complete quantification of the model, the analysis was developed far enough to demonstrate the need to redirect strategy.

This stage also included the integration of another strategy map: the visitor economy. This work explored ways of linking this set of 'demand management' interventions to another part of the overall Succeed Economically map, where the supply of major visitor attractions (conference centres, hotels and restaurants and so on) was being planned and supported – but this integration of cause-and-effect maps was not finalized.

Now attention needs to switch to other maps within the 'Succeed Economically' outcome area, and then eventually to the other aspects of the City Council's programme, including 'Be Healthy' and 'Cleaner, Greener, Safer City'.

In this rollout, the different types of input–outcome relationship in the model may need to be modelled in rather different ways eventually. For example, the economic impact of service activities will be analysed partly using local economic multiplier methodology, while environmental impact of service activities will be analysed partly through its ecological footprint. However, the value of an outcomes-based approach is that the multiplicity of these 'outcome currencies' emerges as they do in the 'trade-offs' in real life, and the strategy mapping approach brings greater visualization of these 'trade-offs' and synergies than in other approaches. These approaches may well benefit from being developed jointly with other like-minded urban local authorities, since many of the models and much of the evidence is likely to apply across such authorities.

However, the project has also thrown up the need, in some aspects of public policy, for the use of complexity sciences – to which we turn after considering the strengths and limitations of this approach.

It's also about Pathways!

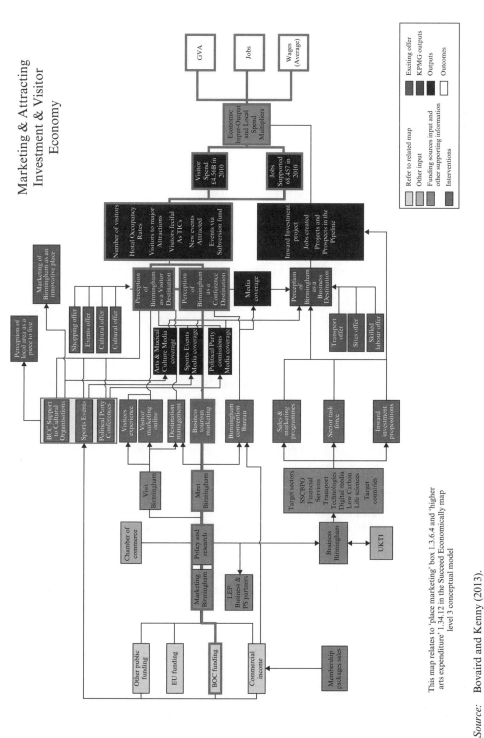

Source: Bovaird and Kenny (2013).

Figure 16.4 Pathways to outcomes in marketing and attracting investment

STRENGTHS AND LIMITATIONS TO THE MODELLING APPROACH

The conceptual model which has been developed in this work to date (as set out in Figure 16.2) differs significantly from the Value for Money models which have traditionally been at the centre of UK performance management systems. In particular, it features:

- user and community co-production as well as public agency partnership;
- outcomes across the full range of dimensions of public, social and private value;
- the service environment, economic markets and the social and political environment are included as important contextual factors affecting outcomes in the model.

However, the model also has a number of limitations which will need to be tackled over time:

- It needs *marginal* coefficients in each link of a pathway, not *averages*, but these have been hard to determine.
- Some pathways refer to city-wide relationships, although they are likely to vary significantly between *neighbourhoods and stakeholder groups* (particularly priority groups, as determined in political programmes).
- Many interventions are particularly aimed at priority groups within the different populations affected by a pathway.
- Confidence in the strength of relationships for *whole pathways* will be much lower than for individual links (or for the short pathways typically modelled in medical 'metrics') – this will require a search for 'key intervention points' and identification of any 'thresholds' embedded within a pathway.
- Every link in the model could (and may need to be) unpacked into a similar more detailed cause-and-effect map.
- There will need to be careful analysis of the sensitivity of all relationships to changes in the underlying baseline.
- The modelling approach has to be kept proportionate to the likely costs and the potential gains in knowledge and therefore potential benefits of a more elaborate model.

Some outcome pathways are too complex to be modelled by these essentially linear pathways to outcomes – mainly because the constituent variables are so interrelated. This requires the use of complexity sciences rather than cause-and-effect mapping.

MODELLING APPROACHES AND KNOWLEDGE DOMAINS

In the project we undertook a 'sense-testing' of the first round of mapping and calibration, by bringing together a range of interested stakeholders and discussing the quantified estimates with them. It became clear that the modelling process has thrown up some major questions about the evidence base available to city policy makers in relation to

Managing complex adaptive systems to improve public outcomes 273

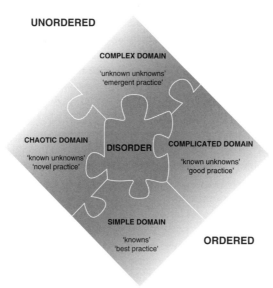

Source: Bovaird (2013); adapted from Snowden and Boone (2007).

Figure 16.5 Domains of knowledge available to decision makers

economic strategy. A large evidence base has been analysed – over 200 separate documents have been reviewed in detail but very little useable evidence found. Most of these documents are either research reports, often interim or final evaluations of specific projects or programmes, or business cases which provide appraisals in advance on the imputed input–output–outcome relationships which are forecast to obtain in these programmes. This macro-assessment has suggested that most 'evaluations' have not been based on 'cause-and-effect' models. Rather, they have sought mainly to identify 'input–outcome' ratios, without probing the logic of the intervening links. The rigour of this approach is thrown into question when it transpires that there is actually little evidence in relation to these intervening links – this highlights that often we are dealing in practice with a 'black box' approach to evaluation, where the internal links between inputs and outputs or outcomes are assumed rather than calibrated and tested. Moreover, it has become clear that most 'business cases' drawn up in the public sector have a strong 'wish factor' attached (the 'optimism bias') and relatively thin evidence.

In addition, it became clear that there were certain aspects of the economic strategy which simply did not fit this 'cause-and-effect' approach and which seemed more akin to complex adaptive systems – this was most evident in relation to the innovation support programme.

We found that a useful way of categorizing the evidence base for modelling was the Cynefin framework (Snowden and Boone, 2007; Bovaird, 2013), which envisages five domains of knowledge in which theorizing can take place (see Figure 16.5). The first four domains are as follows.

Simple domain, in which the relationship between cause and effect is widely believed

274 *Handbook on complexity and public policy*

to be obvious, or at least is derived from solid data that allows for predictable outcomes, based on linear and replicable relationships. In this domain, we are dealing with 'knowns' and we can expect to apply *best* practice. Some parts of our strategy maps in the economy programme seemed to belong to this knowledge domain, for example some of the relationships in Marketing and Attracting Investment (Figures 16.3 and 16.4), where advertising (for example in specific parts of the trade press) could be expected to produce a predictable level of enquiries.

Complicated domain, in which there is a general understanding of the problem to be tackled but the specific relationship between cause and effect in relation to any given decision, although 'knowable' in principle, requires tailored analysis and/or the application of expert knowledge. Many of the other elements of the economy programme seemed likely to fit within this knowledge domain, for example effects of building conference centres and other business visitor infrastructure on business tourism numbers; with sufficient data and appropriate techniques, patterns are likely to be identified which can then be used to predict likely outcomes. In this domain, we can expect to apply *good* practice.

Complex domain, in which the relationship between cause and effect can only be perceived in retrospect, but not in advance. The analysis involved largely relies on informal qualitative modelling ('soft modelling') or structured problem-solving methods. Here, the best we can do is to sense *emergent* practice. This analysis can inform decision-makers but only at quite a broad level of decision and with no prediction of the details. This much more exploratory approach to modelling can only point to potential ways forward, which will need to be tested, often based on emergent practices which are already evident but whose outcomes can not yet be predicted with any confidence. Often the qualitative analysis will override the recommendations suggested by quantitative analysis; decisions will be based on expert insight, rather than on quantitative evidence, because the latter is unreliable, incomplete or difficult to interpret. Identifying, making sense of and responding to the emergent possibilities (Weick and Quinn, 1999) poses problems of an entirely different order of magnitude to the analysis required in the previous two knowledge domains.

Of course, an optimist is entitled to point out that the phrase 'unpredictable' here may only mean 'unpredictable for the moment'. For example, for the moment we are acutely conscious that we have no idea how to predict the possible effect of solar flares on public infrastructure (the subject of a major scare in the USA during 2012), much less on what this will do to the ability of terrorists to harness solar flares for their purposes. However, none of these links in the logic chain seems inherently 'unknowable', just 'unknown for the present'. (By the same token, a pessimist is entitled to point out that the relationships currently 'behaving' in predictable ways may be influenced by intervening variables as yet undiscerned, changes in which could wholly disrupt these measured relationships in the future.)

However, it is also possible that we are dealing with a system in which the overall envelope of outcomes is pre-determined by the characteristics of the system but the interconnectedness of the agents is such that specific outcomes cannot ever be predicted. While this interdependency is a good thing in terms of public policy, as it means that many policy initiatives have effects on several outcomes at the same time, it presents a problem for modelling, as it means that it is hard to distinguish clearly what is leading to what.

There are many different ways in which a system might behave in such a fashion. They include systems in which events are:

- predictable in terms of their occurrence conforming to a probability distribution but not predictable in terms of WHEN they will occur (for example 'black swans', whose frequency often follows a power function);
- predictable within an envelope of potential outcomes at system level (so-called 'strange attractors' in complex adaptive systems) (for example weather patterns, pandemics and so on).

Systems which exhibit these characteristics are classified as *complex adaptive systems* (CAS). Because of the non-linear relationships which guide behaviours of agents in a CAS, all trajectories of the system may eventually lead to radical outcome changes (Bovaird, 2008). However, these non-linear relationships simultaneously mean that the pathway to these transformation points is unpredictable, albeit conforming to the 'strange attractors' which provide outer limits to system behaviour. Knowledge of what these 'strange attractors' look like is the only form of predictability potentially available for CASs. Consequently, planners and strategists can only hope, at best, either to join the game themselves, as 'within-system' players, or to take part in the setting of very outline 'meta-rules' which will determine the shape of these 'attractors'.

There is still considerable disagreement in the literature on the prevalence of CASs. There is a body of literature that refers to cities as a whole as complex adaptive systems (Cooke, 2006; Batty et al., 2012). As Batty et al. (2012) write: 'Cities are complex systems par excellence, more than the sum of their parts and developed through a multitude of individual and collective decisions from the top down to the bottom up'. Other authors (for example Schneider and Somers, 2006) see whole organizations as complex adaptive systems. Many others are still to be convinced that CASs are common in the realm of public policy.

Chaotic domain, in which we can discern no relationship between cause and effect at systems level, so that we cannot predict the events in which we are most interested. Here, the best we can do is to explore *novel* practice. Snowden and Boone (2007) suggest that people who find themselves in this domain will naturally revert to their own comfort zone in making a decision; that might, of course, be in any one of the other domains. In this domain, we typically find suggestions from policy analysts that, rather than any 'cause and effect' logic, we are dealing with person-dependent (or 'personality'-dependent) phenomena, which have to be understood as one-off events. None of the elements of the economy programme appeared to fit with this domain during the period of modelling, although it is likely that this domain would have been appropriate during the onset of the economic recession from 2008 to 2010.

Finally, in the fifth domain ('Disorder'), the context is wholly unknown. We can expect that politicians will tend to bluster rather than owning up to such a situation, making implausible statements of their confidence in propositions which are essentially indefensible. Moreover, it is not a situation that top managers find comfortable either, since their expertise is self-evidently of little value – they, like everyone else, cannot give a lead on what is going on. Fortunately, we did not find that any elements of the economy programme appeared to fit in this domain, although again it may well have been a relevant domain during the onset of the economic recession in 2008–10, before our study started.

276 *Handbook on complexity and public policy*

INNOVATION AS A COMPLEX ADAPTIVE SYSTEM

The treatment of 'blue skies' innovation in the Modelling Birmingham process was initially along similar lines to the other elements of the economy programme, drawing up a cause-and-effect map for the variables which were considered relevant by the different stakeholders involved (Figure 16.6). As such, it complemented a similar map for the wider programme of Business Development and Innovation which focused on much more 'close-to-market' innovation.

The discussion with stakeholders on this innovation model resulted in agreement that the strategy around innovation had to take into account the interrelationship between all the variables in this strategy map but that the set of these relationships was likely to be even denser than portrayed in this map. As set out in Figure 16.6, the discussion distinguished four separate strategies for promoting 'research- and invention-driven' innovation in the city:

- enablers/infrastructure;
- marketing of Birmingham as an innovative place;
- business networking;
- demand stimulation.

There are a number of current initiatives which can be grouped under each of these headings (and many initiatives contribute, at least partly, to more than one of these strategies). Each of these strategies tend currently to be pursued independently. However, they are more likely to be cost-effective if combined in appropriate mixes and possibly varying sequences.

The discussion agreed that there is currently insufficient evidence to calibrate the whole of this strategy map – or indeed of many of the links in the model. However, it highlighted two distinct potential options for moving the model forward, based on this strategy map:

- partial quantification over a period of time;
- conceptualization of innovation as a complex adaptive system.

Partial Quantification over a Period of Time

In this approach, some of the pathways through the model, from inputs to initiatives to outputs and outcomes, would be quantified over time, seeking out specific data which would throw light on the strength of links in those pathways. It was generally agreed that some pathways might well be appropriate for such analysis, though no pathway had yet got sufficient data for such a calibration to be undertaken. However, there was substantial scepticism as to whether all the pathways within the strategy map could eventually be calibrated in this way. In consequence, if some pathways within each strategy remained unquantified, it would continue to be difficult to assess which of the four strategies was the most cost-effective. Although some initiatives were likely to emerge as superior to others, which would at least allow some prioritization, this might well simply lead to sub-optimization, based on a narrow picture rather than a whole-systems view.

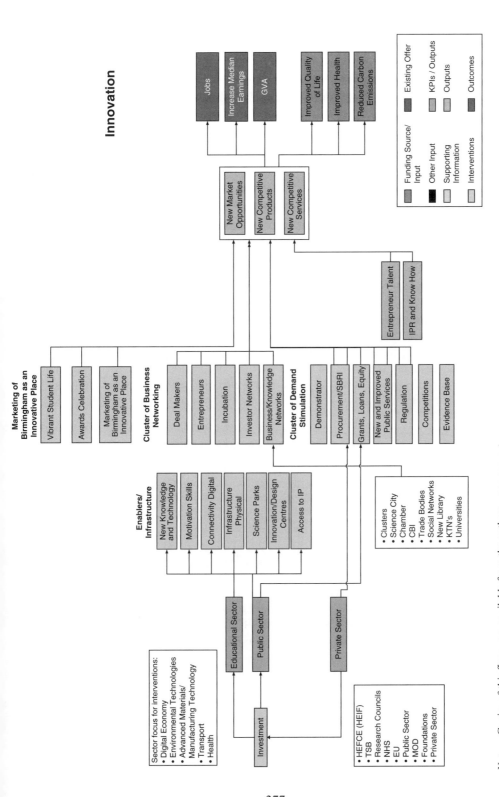

Figure 16.6 Initial cause-and-effect map for innovation

Note: Copies of this figure are available from the authors upon request.

278   *Handbook on complexity and public policy*

Conceptualization of Innovation as a Complex Adaptive System

This approach accepts that there is insufficient information on which to rank the cost-effectiveness of most pathways to outcomes. It starts from the understanding that, in practice, the outcomes are the result of a large number of closely connected variables, only some of which are within the control (or even influence) of local public sector agencies. Consequently, the best way forward is to engage in a meta-strategy of experimentation, using different combinations of the four basic strategies. These experiments might be conducted in a variety of ways in order to see what effects these changes have on outputs and outcomes, including:

- in different geographies (for example varying the mix of strategies between priority areas or between different districts in the city);
- with different priority sectors (for example digital firms, media sector, advanced manufacturing);
- over time (for example speeding up some strategies or slowing down others).

In each case, the time period needed to assess the success or failure of an experiment might be substantial, as the effects of innovation often take a considerable time to show up. This would not, in itself, be a reason for rejecting the experimental approach. If this approach is taken, then a TQM (Total Quality Management) of experimentation is needed, in which the experiments are properly designed and executed, so that lessons can be reliably learnt from them.

MOVING TO AN INNOVATION STRATEGY BASED ON COMPLEXITY

The stakeholders leading the modelling of innovation in the city agreed that neither of these quite different approaches has been adopted to date, which has hindered the learning of lessons from the many different innovation initiatives which have been undertaken in the city over the last few decades. It will be important to make a clear decision to take one approach or the other (or to mix them in some clear and explicit 'third approach').

In particular, it would be valuable to identify clusters of pathways that show such interdependencies so that they can be modelled together, rather than as independent pathways to outcomes. Where these act together as mutually reinforcing mechanisms, they may provide potential 'engines of growth' for outcomes in the city, for relatively little input. While the search for such synergy has been a longstanding activity in public policy, the recognition that this synergy might arise essentially from the dynamics of the system rather than non-linearities such as economies of scale is more recent.

NEXT STEPS IN MODELLING PLACES

The case study of Modelling Birmingham demonstrates a range of possibilities for a programme of modelling local places. The combination of mapping and modelling the 'pathways to outcomes' provides better visibility (through mapping) and increased understanding (through modelling) of the impact (both outputs and outcomes) of interventions in the whole city system. The model is intended to guide Birmingham City Council and its partners in their development of city-wide programmes and budgets in future years. It is designed to provide insight for the design and redesign of those programmes, within and across organizations and sectors, thereby informing the prioritizing of city-wide strategy and budgets.

It is recognized, of course, that in early iterations the results of the model, even for those issues belonging to the simple and complicated knowledge domains, will have too wide a range, within any sensible confidence intervals, to be used reliably for policy decisions. Nevertheless, the results may play a useful role in structuring the debate around the likely impacts of policy decisions. The model is therefore intended as a tool for improving the quality of discussion around the outcomes from public intervention in Birmingham, rather than giving a definitive 'answer'.

This highlights the need for 'proportionality': the search for ever more accurate predictive analytics can become highly cost-ineffective in areas of major uncertainty, even where we are not dealing with complex adaptive systems. We must accept that, in the current context, it is not possible to develop a comprehensive and integrated model of the Birmingham city system, even in the simple and complicated knowledge domains. A number of barriers stand in the way: the attribution problem – disentangling what causes what; the knowledge problem – the availability and organization of existing evidence; and political 'gaming', where the modelling process opens up and challenges the vested interests of multiple stakeholders.

However, this chapter demonstrates that these problems do not make 'modelling places' wholly intractable; there is indeed some scope, albeit limited, for an effective change management tool to support 'disruptive innovation'. This involves a hybrid multi-methods approach, combining a 'plausibility test', utilizing the knowledge domains from within the Cynefin Framework, based on (partially) quantitative pathway analysis, and a 'sensemaking test', which through multi-stakeholder-engagement probes for the boundaries of the complex systems in the city and seeks to enhance those connectivities which lead to positive disruptive innovation in the city; we discuss this in the final section.

The attribution and knowledge problems can be addressed through tools which make best use of the evidence base for cause-and-effect chains, while also identifying issues which require the tools for exploring the dynamics and non-linearities in complex adaptive systems. These tools include 'as is' mapping and modelling of existing interventions, positioning of interventions within the knowledge domains within the Cynefin framework, and mapping and, where possible, testing of emerging and potential interventions. In some contexts – but certainly not all – there are grounds to hope that 'big data' and improved data measurement and analysis tools will extend the possibilities for building useable causal models of the city. At the very least, the Modelling Birmingham project should be able to identify areas of city services where there are significant non-linearities, which can be mined to bring major improvements to outcomes for relatively small increases in inputs, and where relatively large budget cuts might result in relatively small-scale impacts on quality of life outcomes for citizens. This would represent a huge advance on what is currently available to policy makers. Indeed, it may be that the search

280 *Handbook on complexity and public policy*

for positive non-linearities is the major contribution which research can currently make to public management; understanding the intricacies of managing them, both in the complicated and complex knowledge domain, may require research over a much longer timescale.

It has always been recognized that modelling a city's outcomes is likely to require a multi-method approach, especially where non-linearities are involved. As well as developing the 'pathways to outcomes' approach outlined earlier, Birmingham City Council has worked with Cambridge University, University College London and Geofutures to trial the use of dynamic modelling for specifying pathways to outcomes. The area tackled in detail was the worklessness programme, which was modelled by a Cambridge University graduate student. The potential of dynamic modelling is still hard to gauge, given that the results from this approach are less intuitively easy to interpret by decision-makers than results which come from a more linear spreadsheet approach.

The organizational behavioural issue of 'political gaming' (which can sometimes also be party-political) is an inevitable part of the contestation of interests and power. Here, a multi-stakeholder approach can often help, in which plausible pathways to outcomes are co-produced, even as they are contested. Modelling Birmingham has demonstrated a process that directly captures the perceptions of different stakeholders on how 'disruptive innovation' can be achieved and subjects these differing perceptions to logical testing. However, problems of poor data will always mean that stakeholders will be able to find some ways of rationalizing strongly held viewpoints, however poorly supported by the evidence, while bounded rationality will always mean that even well-evidenced pathways may be ignored in the decision-making process.

However, for all its positive benefits in terms of informing debate, the model may have an even greater role in warning politicians and decision-makers about what they do NOT know in relation to pathways to outcomes. It should therefore help to identify those parts of the city that operate as complex adaptive systems and where it would be inappropriate for decision-makers to claim that specific interventions are likely to have predictable impacts on desired outcomes. Of course, for the moment, we cannot be sure whether the difficulties we have experienced in identifying and calibrating pathways in many of the strategy maps are due to the existence of this kind of 'complexity' – or simply lack of experience and expertise in undertaking predictive analytics.

Consequently, it is still unclear if it will ever be possible to develop an 'urban simulator', as originally desired by the leading decision-makers in Birmingham City Council, which will give city decision-makers a 'console' into which they can feed policy changes, mapping the outcomes changes which result, with confidence intervals. It may be that only much narrower insights will be feasible from modelling processes. Nevertheless, given the potential already identified in the project, Birmingham City Council is now engaging in wider collaboration with universities and the private sector to take this work forward. Indeed, the idea of the city simulator remains a major ambition of 'smart' cities around the world. The Modelling Birmingham approach won the IBM Smarter Cities Challenge in 2012 and other cities such as Portland, Oregon, have recently pursued a similar approach (Robinson, 2012). The next steps are now being led through the Warwick Institute for Science of Cities as part of the New York Center for Urban Science and Progress global network with a view to creating a dedicated Birmingham 'satellite', in the form of a Birmingham Institute for Urban Science.

CONCLUSIONS: LIVING WITH COMPLEXITY

For all the potential of Modelling Birmingham for dealing with issues in the simple and complicated knowledge domains, it is clear that the longer-term use of the approach is likely to involve much more attention to dealing with complex adaptive systems. We have illustrated this in detail with reference to the innovation programme. However, it can also be seen directly from Figure 16.3, where there are a series of 'infrastructure' boxes which are key inputs to the outcomes of activities in the Marketing and Attracting Investment programme – these cover key attractions in the city such as the hotel offer, the leisure offers, the conference and convention centre offer, the transportation system, and so on. For the moment, the modelling shown in Figures 16.3 and 16.4 has taken these as 'fixed assets', based on the current offer: the *ceteris paribus* approach. However, in the longer term these become part of the 'agenda'; changes in these offers become key policy targets, because they are likely to have a major effect on the success of business visitor marketing and investment decisions in the city. This introduces a level of recursion into the model which is currently not fully represented; the development of a larger and better offer in each of these infrastructure categories is an outcome from other strategy maps (for example in city planning, transportation, and so on), which in turn are influenced by the level of prosperity of city, which is highly influenced by this Marketing and Attracting Investment model. This, of course, opens up the exciting possibility of mutually reinforcing interventions – although these will not always be positive in direction.

Where we are dealing with highly interrelated variables in densely interwoven models, there is a strong likelihood of complex adaptive systems arising. Modelling of the partial relationships between variables along one pathway is then not only inadequate but potentially highly misleading. It is therefore important to insist that there are likely to be areas of city policy, such as economic innovation, where an 'urban simulator' will NEVER be available, because they are situated in the complex knowledge domain. This will apply in all systems where the behaviours of agents and the strategies they adopt owe at least as much to emergent complex interactions within the policy system as to the cognitive processes occurring in any agency (Bovaird, 2008). Complexity theory suggests that emergent strategies arising from interactive behaviour in such policy systems will be much more robust and reliable than mechanistic approaches to strategic planning in public organizations. Here, the key is to focus not on planning but on sensemaking, in order to understand the current environment better, providing the right conditions for positive change (including changing the most potentially constraining elements in the current environment), stimulating innovation and providing incentives for those behaviours which appear most correlated to the outcomes desired.

In such a context, it is also likely that some actors in the system, particularly self-confident local agencies with an ability to mobilize other actors behind their actions, will try to shape these emergent strategies into a 'meta-planning' approach, albeit heavily circumscribed in scope. This 'meta-planning' approach is different from traditional strategic planning, in that it does not involve the development of a single, preferred set of strategic actions but rather entails tracking how emerging situations offer the possibility of changing the 'opportunity map' facing the organization, and developing the capability of the organization to influence the overall system. It is this prospect which Modelling

282   *Handbook on complexity and public policy*

Birmingham now offers to its proponents – but there are many obstacles to overcome before such a promise can be realized. For now, perhaps the most important lesson of complexity theory for Modelling Birmingham is that we should beware of placing too much confidence in deterministic models of economic, social and political interventions and against over-elaborate analysis of single agency interventions in policy making, strategic management and public governance within policy systems whose interactions are, at best, only partially understood.

Finally, major tensions can arise from recognizing that some elements of a local strategy operate in the complexity domain, and hence can only be subject to very limited influence by public policy. The model developed in Birmingham is expected by politicians and top policy makers to improve performance and outcomes; indeed, for them the model would otherwise be a waste of money. Accepting a 'meta-planning' role only, and forswearing detailed intervention on issues that are politically important, will require a degree of self-control which may be beyond many top decision-makers. This may lead them to ignore or downplay the complexity factor in the local economy. Of course, any intervention they select in these circumstances is likely to be largely unsuccessful – but that may only become evident much later and, often, can be blamed on other factors. Consequently, we must expect that local decision-making on issues located in the complex policy domain is likely to be characterized by over-optimistic attempts at intervention. Only a move to more honesty and humility about the potential of public intervention would provide an antidote to such mistakes – and there are few signs that these virtues are about to become more widespread in UK local or national government policy.

ACKNOWLEDGEMENTS

The authors would like to acknowledge the contributions made to the development of Figures 16.1, 16.4 and 16.5 by several staff of KPMG and to Figure 16.6 by Phil Extance, Pro-Vice Chancellor of Aston University. We are grateful to Birmingham City Council for permission to reproduce Figures 16.1, 16.4 and 16.6. We have also benefited from helpful comments from participants at the Public Management Research Conference, Madison USA in June 2013 and, subsequently, by the editors of this volume.

REFERENCES

Ansoff, I.H. (1965), *Corporate Strategy*, New York: McGraw-Hill.
Batty, M., K.W. Axhausen, F. Giannotti, A. Pozdnoukhov, A. Bazzani, M. Wachowicz, G. Ouzounis and Y. Portugali (2012), 'Smart cities of the future', *European Physical Journal: Special Topics*, **214**, 481–518.
Bickman, L. (ed.) (1990), *Advances in Program Theory: New Directions for Program Evaluation*, San Francisco, CA: Jossey-Bass.
Bovaird, T. (2008), 'Emergent strategic management and planning mechanisms in complex adaptive systems: the case of the UK Best Value initiative', *Public Management Review*, **10**(3), 319–40.
Bovaird, T. (2012), 'Attributing outcomes to social interventions: gold standard or fool's gold in public policy and management?', *Social Policy and Administration*, **48**(1), 1–23.
Bovaird, T. (2013), 'Context in public policy: implications of complexity theory', in C. Pollitt (ed.), *Context In Public Policy And Management: The Missing Link?*, Cheltenham, UK and Northampton, MA, USA: Edward Elgar Publishing.

Managing complex adaptive systems to improve public outcomes 283

Bovaird, T. and R. Kenny (2013), 'Modelling Birmingham: using strategy maps to compare the cost-effectiveness of alternative outcome pathways', paper presented at the 11th PMRA Conference, Madison, Wisconsin, USA, 20–22 June, available at http://www.union.wisc.edu/pmra2013/Paper%20Submissions/Renamed/Modelling%20Birmingham%20Using%20Strategy%20Maps%20to%20Compare%20the%20Cost-Effectiveness%20of%20Alternative%20Outcome%20Pathways.pdf.

Bovaird, T. and I. Mallinson (1988), 'Setting objectives and increasing achievement in social care', *British Journal of Social Work*, **18**, 309–24.

Bovaird, T. and J. Tholstrup (2010), 'Collaborative governance between public sector, service users and their communities', in E. Bohne (ed.), *Repositioning Europe and America for Economic Growth: The Role of Governments and Private Actors in Key Policy Areas*, Berlin: LIT Verlag.

Cartwright, N. and J. Hardie (2012), *Evidence-Based Policy: A Practical Guide to Doing it Better*, Oxford: Oxford University Press.

Cooke, P. (2006), *Complex Adaptive Innovation Systems: Relatedness and Transversality in the Evolving Region*, Abingdon: Routledge.

Davies, H., S. Nutley and P. Smith (2000), *What Works? Evidence-based Policy and Practice in Public Services*, Bristol: Policy Press.

Eden, C. and F. Ackermann (2003), 'Cognitive mapping expert views for policy analysis in the public sector', *European Journal of Operational Research*, **152**(3), 615–30.

EKOSGEN (2010), *Birmingham Public Expenditure and Investment Report*, Birmingham: Be Birmingham.

Friedman, M. (2007), *Trying Hard is Not Good Enough: How to Produce Measurable Improvements for Customers and Communities*, IFS and Victoria, BC: Trafford Publishing.

Kaplan, R. and D. Norton (2001), *The Strategy-Focused Organisation*, Boston, MA: Harvard Business School Press.

Kaplan, R. and D. Norton (2004), *Strategy Maps: Converting Intangible Assets into Tangible Outcomes*, Boston, MA: Harvard Business School Press.

Marr, B., G. Schiuma and A. Neely (2004), 'The dynamics of value creation: mapping your intellectual performance drivers', *Journal of Intellectual Capital*, **5**(2), 312–25.

Mulgan, G. (2009), *The Art of Public Strategy: Mobilising Power and Knowledge for the Common Good*, Oxford: Oxford University Press.

Robinson, R. (2012), 'Pens, paper and conversations. And the other technologies that will make cities Smarter', *The Urban Technologist*, available at http://theurbantechnologist.com/2012/12/06/pens-paper-and-conversations-and-the-other-technologies-that-will-make-cities-smarter/.

Rosenberg, L.J. and L.D. Posner (1979), *The Logical Framework: A Manager's Guide to a Scientific Approach to Design and Evaluation*, Washington, DC: Practical Concepts Inc.

Saaty, T.L. (1980), *The Analytic Hierarchy Process*, New York: McGraw Hill.

Schneider, M. and M. Somers (2006), 'Organizations as complex adaptive systems: implications of complexity theory for leadership research', *Leadership Quarterly*, **17**, 351–65.

Snowden, D.J. and M. Boone (2007), 'A leader's framework for decision making', *Harvard Business Review*, November, 69–76.

Sullivan, H. and M. Stewart (2006), 'Who owns the theory of change?', *Evaluation*, **12**(2), 179–99.

Weick, K.E. and R.E. Quinn (1999), 'Organizational change and development', *Annual Review of Psychology*, **50**, 361–86.

17. Brazil and violent crime: complexity as a way of approaching 'intractable' problems
Kai Enno Lehmann

INTRODUCTION

Brazil has the unenviable reputation of being one of the most violent countries in the world. In 2011, no fewer than 43 913 homicides were recorded in the country, a rate of 22.33 homicides per 100 000 inhabitants.[1]

Nowhere has this problem been more acute than in the country's two biggest cities, São Paulo and Rio de Janeiro, whose metropolitan areas account for roughly 15 percent of the entire Brazilian population. Since these two cities also represent the major economic hub of the country (in the case of São Paulo) and the country's major tourist destination (in the case of Rio de Janeiro), what happens in these cities generally attracts more attention than what happens elsewhere.

Yet these two cases are also interesting because of the ways in which the authorities have tried to deal with the problem of violent crime and the seemingly positive results this has achieved. In the case of São Paulo, murder rates have more than halved over the last 15 years, whilst in Rio de Janeiro the installation of so-called *Unidades Policiais Pacificadoras* (UPPs, Pacifying Police Units) in some of the city's most violent shanty towns (or 'favelas') since 2009 has also led to (or coincided with) a significant reduction in homicide cases.[2]

The aim of this chapter is to analyze and compare these two policies from a complexity perspective. It will be argued that one of the reasons for their relative success has been the fact that violent crime has been seen as a self-organizing process. As such, instead of 'eradicating' the perpetrators of these crimes in the hope that this will reduce crime or focusing on *one* particular variable in the hope of moving the problem to an 'acceptable' state, the policy aim has been to change the *patterns* of self-organization which helped sustain the conditions in which violent crime flourished, allowing for the *emergence* of new processes of development, and responding to the local boundary conditions of each particular area. In short, they have encouraged 'adaptive action', as termed by Eoyang and Holladay (2013). However, a recent upsurge of violence, especially in São Paulo, as well as cases of people going missing after being picked up by police in some of the supposedly pacified favelas in Rio de Janeiro raises the question of whether a sustainable process has been established.

After an analysis of each case and a comparison between them, the chapter will ask what can be learned from these experiences and whether the experiences can serve as a blueprint for further policy development.

THE CONTEXT: VIOLENT CRIME AS A LINEAR PROBLEM

In 1975, the murder rate in Rio de Janeiro was less than 15 homicides per 100 000 inhabitants. By 1995, it had reached 64.9 (Carneiro, 2010a). In São Paulo the situation was even worse, with the official homicide rate reaching a staggering 69.1 per 100 000 inhabitants by 1999. In so doing, both cities were amongst the most violent on the planet and even managed to considerably outstrip the national murder rate, which rose from 11.7 homicides per 100 000 inhabitants in 1980 to 28.9 per 100 000 inhabitants in 2003.[3]

The causes of this increase have been the subject of debate for many years.[4] As Carneiro (2010a) has pointed out, they include political incompetence linked to 'traditional' problems such as corruption, and the 'reach' of the state. Within this context, the performance of the Military Police has received particular attention in both Rio de Janeiro and São Paulo. In the case of Rio de Janeiro, Carneiro (2010b) identifies four principal problems which have seriously undermined the ability of the police force to fight violent crime effectively. First is internal police corruption. Second, there is a general problem of indiscipline within police ranks 'with orders simply not being followed' (Carneiro (2010b: 64). Third, poor training leads to poor operational conduct. Lastly, there simply are not enough police officers to deal with the multitude of issues within an often very difficult physical terrain. In São Paulo, Cardia (2000) identified similar problems, though, as Denyer-Willis and Tierney (2012) have shown, the exact expression of them in São Paulo was a little different from in Rio de Janeiro, of which more will be said below.

Linked to these particular institutional problems are deep structural ones of both the Brazilian economy and society at large, which are very evident in virtually every urban center in Brazil. Chief amongst them is the vastly unequal distribution of wealth, as demonstrated by the example of Rio de Janeiro's biggest shanty town, Rocinha. Rocinha is separated from a wealthy neighboring community by one road which marks 'a 9-fold difference in employment, a whopping 17-fold difference in income and a 13-year difference in life expectancy' (Goldstein and Zeidan, 2009: 288). Such problems are the result of 'formal and informal mechanisms which preserve the existing power structures in all its forms', in particular the police, the judiciary, the political structures, as well as the educational system (Guimarães, 2008: 16).

At the beginning of the 1980s, these structural factors had combined with some specific circumstances to produce a steep increase in violence: 'The combination of a weak state, economic crisis inherited from the military which led to harsh economic restructuring, and the expansion of the drugs trade, led to an increase in violent crime' (Leu, 2008: 3).On top of that, both Rio de Janeiro and São Paulo were facing rapid urbanization. For example, at the start of the twentieth century, São Paulo was a small city of approximately 100 000 inhabitants. By the 1990s, it had become one of the biggest conurbations in the world, with approximately 20 million inhabitants. This expansion was as unplanned as it was rapid and was accelerated in both cases by the mass migration of people from the north to the south of the country in search of economic opportunity. The mass migration underlined the structural problem outlined above, a problem which has been reflected to a significant degree in the two cities under consideration in this chapter.[5] The unplanned nature of this expansion led to the 'favelalization' of city space, manifested in both the growth of existing shanty towns and the emergence of new

286 *Handbook on complexity and public policy*

ones. In both cities, therefore, the structural feature of economic inequality led to a particular type of city expansion which reinforced existing divisions – both physically and economically – and which has been seen as a key factor in violence in some studies, such as Richardson and Kirsten (2005). In fact, as one specialist in urban crime remarked to the author, the way Brazilian cities constantly evolve lets people know who belongs to which 'class', reinforcing social divisions. According to several studies, it is this *disparity* in income and opportunity which explains, at least in part, the explosion in violent crime, more so than income levels (and therefore poverty) per se, not just in Brazil but for the whole of Latin America.[6]

In Rio de Janeiro's case, the above factors interacted with a massive expansion of the drugs trade which was often organized by armed groups. Taking advantage of the state's weaknesses, the armed groups installed themselves in favelas and established alternative power structures for their particular areas, underpinned by their own 'laws' and 'code of conduct' (breaks of which were often summarily punished through torture or death), the provision of some basic social services, and other functions traditionally regulated by the state, such as television or the internet, as Carneiro (2010b) has shown.

Over the years, there has been an intense debate in Rio de Janeiro about how to confront this problem. Whilst it is beyond the scope of this chapter to go into this debate in detail, it is possible to distinguish between two broad policy approaches which can be termed 'containment' and 'confrontation' and which were applied sequentially, depending on the political constellations of the state.

As a policy, 'containment' essentially limited the number of incursions into areas dominated by drug gangs with the argument that a more confrontational approach within geographically very difficult and heavily populated terrain would lead to the loss of innocent life. Mostly applied during the governorship of Lionel Brizola (1983–86 and 1991–94), the policy was based on the premise that crime was, above all, a social problem which had to be treated within the context of the ongoing process of democratization and human rights, meaning that the preservation of peace for the largest possible part of the population of the city would take precedent over the apprehension of criminals.[7]

The alternative approach of 'confrontation' was mostly associated with governor Alencar (1995–99). According to his argument, containment had led to large parts of the city being virtually controlled by heavily-armed drug gangs as they were essentially 'police-free' zones into which the state did not venture. To counter this trend, Alencar created incentives (pay rises, promotion and so on) for the police to kill or apprehend as many drug dealers as possible, as well as for seizing drugs and weapons.[8] Whilst the homicide rate initially dropped and the quantity of weapons and drugs seized by the police soared, the number of innocent people killed rose sharply whilst the structural causes of the problem remained unaddressed. The policy had no appreciable effect on the levels of drug consumption and therefore drug trading. As a result, dead drug dealers were simply replaced by new ones.

In São Paulo, a similar process can be traced. In conjunction with the growth of the illicit drugs market, there was also a rise in organized crime, best represented by the formation of the *Primeiro Comando da Capital* (PCC) in 1993, on which a bit more will be said below. Just like any other urban area of Brazil, the vastly unequal distribution of wealth represented (and represents) its own challenges.

In policy terms, the response of the authorities in São Paulo to the increase in violent

Brazil and violent crime: complexity for 'intractable' problems   287

crime, particularly during the 1990s, was much more tilted towards confrontation. This was due to a number of factors. First, there has always been – and there continues to be – strong public pressure for such a policy approach. One former commander of the Military Police argued that 'for the public, a good bandit is a dead bandit'.[9] This pressure for 'action' was amplified by the culture of the Military Police which, during the period of military dictatorship, was specifically trained *not* to interact with the public, as de Souza (2009) observed. As such, according to Monjardet (1996), the force is trained to be primarily *reactive* and has little capacity to respond to changing circumstances.

That this should be so is perhaps hardly surprising bearing in mind the principal responsibilities assigned to it. First and foremost, the Military Police – be it in Rio de Janeiro, São Paulo or elsewhere – is charged with intervening in disorder at 'street level'. In other words, it is the Military Police which will confront any type of violent disorder, though it is not responsible for investigating such disorder or bringing the perpetrators to trial (de Souza, 2009: 41).

According to the same author, the combination of the above factors had a significant impact on the way in which policing is done. Policing by officers 'on the beat' is offensive in its posture, 'especially in those regions considered more dangerous or vulnerable' (de Souza, 2009: 43). As shown by Denyer-Willis and Tierney (2012: 4), in the case of São Paulo these regions are on the periphery of the city and are marked by poverty, reinforcing once again the social divisions already mentioned above.

The policy approach in São Paulo, then, was very similar to that of Rio de Janeiro in terms of 'confrontation'. Yet, despite the similarities in policy, it is important to note some crucial differences between the two cases.

First, there is the geographic component: in Rio de Janeiro, whilst drug gangs were operating in a number of slums, it is noteworthy that these slums were geographically dispersed, with some being located in the middle of some of Rio de Janeiro's wealthiest neighborhoods. In São Paulo, by contrast, as Denyer-Willis and Tierney (2012) have observed, control of drug gangs over particular areas has been concentrated in the periphery of the vast city. At the same time, whereas the key criminal factions of Rio de Janeiro are, in most cases, the specific result of the drugs trade, the PCC in São Paulo emerged out of the local prison system, one of its initial demands being the improvement of prison conditions for inmates.[10] Within this context, 'just like everything else, even crime is more organized in São Paulo than in Rio de Janeiro', according to one policy advisor and specialist on violent crime in São Paulo.[11] According to the same advisor, the PCC has well-established and coherent structures, which extend until today into the prison system, and do not depend on one particular leader or other: 'They are resilient and have some support'. This assertion is backed up by research conducted by Denyer-Willis and Tierney (2012).

Therefore, despite the undoubted similarities which exist between Rio de Janeiro and São Paulo in terms of the problems confronted, there were and are crucial differences which had a significant impact over the years on the evolution of policy in the two cities. These differences will be assessed below. During the 1990s, however, what both had in common was their utter failure in bringing violent crime under control. It will be argued that a key explanation for this failure had to do with the definition of the nature of the problem confronted by authorities, and what eventually led to a reversal of the trend in both cases over the last few years.

288 *Handbook on complexity and public policy*

APPLYING COMPLEXITY: ADAPTIVE ACTION AS A WAY OF CHANGING PATTERNS OF SELF-ORGANIZATION

The key problem with the policy approaches outlined above was the definition of the problem as 'complicated'. As Chapman (2002) has noted, in a complicated system 'it is possible to work out solutions and implement them' (Chapman, 2002). For instance, in the case of Rio de Janeiro, one policy ('containment') saw the solution as being the physical containment and concentration of drug dealers in particular areas and that this would lead to the rest of the city living in relative peace. The key determinant variables, then, were the geographical location of the problem as well as the group of people being responsible for the problem (drug dealers). In the second policy, the solution was the physical elimination of drug dealers. At the same time, the policy also had a specific focus on particular areas (the favelas), essentially leaving the rest of the city untouched.

In São Paulo, similar thinking can be detected. In basic terms, as shown above, the idea of confrontation was to kill as many drug dealers as possible. At the same time, just as in Rio de Janeiro, there was a spatial and geographical component to the problem, with concerted efforts being made to leave the problem confined to the periphery of the city, as Denyer-Willis and Tierney (2012) have shown.

Yet, as Lehmann (2012) has shown in relation to Rio de Janeiro, the problem of violence is not complicated, but complex. Peirce (2008: 86) has drawn similar conclusions in her much broader study about urban violence in Brazil, arguing that 'policy solutions must be more complex, moving beyond the assumptions of the 'rational criminal' model'. In other words, the problem of violent crime is characterized by:

- a number of elements or phenomena;
- emergence and sensitivity to initial conditions – its development is at best partially predictable;
- parts of the system are reducible whilst others are not;
- the elements of the system form coherent patterns over time;
- the system is open to its environment and therefore capable of adaptation and survival.[12]

As a result, 'the relationship between cause and effect is uncertain and there may not be agreement on the fundamental objectives [of any given policy]' (Chapman, 2002: Foreword). Since such systems self-organize, 'policies and interventions have unpredictable consequences [whilst complex systems] also have remarkable resilience in the face of efforts to change them' (Chapman, 2002: Foreword). Therefore, policy outcomes are unpredictable.

Yet the policies outlined above were simply not able to account for this unpredictability. In both cases there was an attempt to *simplify* the policy landscape by focusing almost exclusively on a couple of key variables. Little thought was given to the fact that the complex adaptive systems within which policies were being applied *interacted* with other such systems. For instance, as Denyer-Willis and Tierney (2012) have shown, just looking at some of the areas of the periphery of São Paulo studied by them, one can detect dynamic interaction between the PCC, local agents of the state, such as the police, and the local population. There was also interaction between that population and the state, as well as interactions between those areas of the city on the periphery and those closer to the center, as well as the richer and poorer parts of the population. All interviewees also pointed to the tensions between the Civil and Military Police. These tensions, however, are crucial and long-standing. As one former commander of the Military Police in São Paulo put it, 'the tensions between the Military and the Civil Police go back a long time. The PM (Military Police) in São Paulo is over 100 years old and has a very different culture to the Civil Police, but these cultures are deeply ingrained'.[13] As a result, there is, amongst other things, no tradition of exchanging information and, according to the former commander, mutual suspicion. One former state minister agreed, claiming that there was still more corruption in the Civil Police: 'They solve a crime and then negotiate with the criminal to let him go'.[14]

Even with this one example, it becomes clear that the landscape within which violent crime would have to be confronted is very different from the simplified one implied by the policies outlined above. Rather than dealing with a problem which depends on a couple of variables, one is confronted with a problem made up of multiple variables that interact across various levels of analysis. One is dealing not just with a self-organizing complex adaptive system but a highly interdependent one which, in the words of one former minister, was highly disorderly: 'In São Paulo, disorder was everywhere, whether it was fly-pitching in the main shopping street or the growth of favelas across the city. Violent crime has to be seen in this context'.[15]

How did both cities manage to reverse the trend of rising violent crime? As will be shown now, essentially both began – at different times and within different circumstances – policies that encouraged the emergence of different patterns of self-organization and began to define the problem of violent crime as one embedded within a broader context and as an *ongoing* process, rather than one which would eventually arrive at an end-point. In short, both began to think in terms of 'adaptive action', as Eoyang and Holladay (2013) have termed this process. In practice, this meant that they started to ask different questions to inform policy, which led to different answers.

ADAPTIVE ACTION IN SÃO PAULO AND RIO DE JANEIRO: CHANGING PATTERNS

The first big change occurred in relation to *what* the problem was that was being encountered. One former minister put it thus: 'We were confronting disorder. [The main shopping street] used to be full of fly-pitchers and unlicensed traders and we removed them. We [also] removed [slums].' According to him, the message being sent out was that *disorder* would not be tolerated in the hope that this would have an impact on violent crime, based on the assumption that the loss of a sense of impunity would lead to a reduction in such crimes.[16]

Recognition of the need to redefine the problem being confronted also extended to the Military Police force, at least according to one of its former commanders. There was a need to 'put more value on life': 'I kept insisting that our job was to protect the population, not to kill [criminals]'.[17]

This change in and of itself though would not have been enough. There was also an urgent need to create understanding and trust between the population and the police in

290   *Handbook on complexity and public policy*

particular. According to the former commander, this included the need to communicate to and with the population as to what the police were doing and why they were doing it.

With this emphasis on creating trust, the aim was to shift perceptions of and expectations about the police from fear to one that the police had a positive role to play in *creating order*. According to one former minister at state level, achieving such a goal inevitably meant treating the problems confronted in a broader fashion, linking the fight against crime to the provision of social and basic services, such as education. In other words, at least according to this minister, the approach taken was holistic and again attempted to shift the way that the state was seen by many to a perception that the state existed to *facilitate* the life of its citizens, a feeling which was often very limited or absent. Critical to this was the interaction that was facilitated between the state and civil society, which 'today is common: between the police, the city council, the church etc.'.[18] In summary, then, the problem of violent crime was seen as *part* of a much broader context which looked at issues concerning the role of the state.

So, what does such a redefinition mean for policy-makers and implementers? According to the interviewees for this research, the key conclusion drawn from this redefinition was the need to act holistically and across institutional and bureaucratic boundaries across all possible levels of analysis. For instance, for the former minister, as hinted at in the paragraph above, there was a need to coordinate policies across areas of public safety, education, social services and education, to name but a few. For the former police commander, this need for coordination meant the beginning of a process of re-approximation between the Military and the Civil Police.[19]

Coordination was also identified as key in terms of the message being sent out across time and space. As one policy advisor pointed out, crucial to the recent decline in crime was the consistency of message which became, at least apparently, self-sustainable for a time. With a consistent message, backed up by – as will be shown below – specific investments and clear strategic actions, 'the message that was being [received] by the population was that things were getting better. The *feeling* of insecurity diminished'.[20]

This consistency of message translated into clear policies. At a strategic level there was, according to all interviewees, a coordinated effort to invest in equipment, communication and the criminal justice system. For instance, data-sharing systems were created which allowed the police force to become much more agile and proactive in *preventing* crime from occurring. For the former commander of the PM, this investment, together with that in new equipment and the expansion of the prison system, allowed the force to respond much better to the population's demands. It also allowed for tailored responses to particular local circumstances. For instance, as the commander pointed out, areas like *Jardim Angela*, considered to be one of the most dangerous urban areas in the world during the 1990s,[21] required a different type of policing than some areas in the center of the city. Investment in new technology, according to this argument, allowed the police to finally respond to these differing local boundary conditions.[22]

Finally, there was also recognition that there was an urgent need for better communication, not just between the different agents of the state but also between the state and the society it governed. As the former police commander argued, it was critical that the police explained to the population its role and its initiatives and that the media be used to doing so: 'We had to change the narrative about the police and I was very keen to go public about what we're doing'. Only this way, according to him, was it possible

Brazil and violent crime: complexity for 'intractable' problems 291

to address the problem of trust identified as one of the key impediments to fight crime effectively. As part of this process, police training was changed significantly towards community policing and more intensive study: 'We now train the police up to university-level education'.[23]

Communicating also meant interacting with other police forces across the world to learn and exchange experiences: 'We have been to New York, Toronto, Tokyo ... lots of places to see how things are done', as both the former minister and the former police commander pointed out, arguing that there was recognition of the need to do things differently and a willingness to learn from those who had also managed to reverse decades-old patterns of crime, such as in the case of New York in particular.

In consequence, several key initiatives were started in order to *transmit* the new message being developed. One of them was to 'de-militarize' society. As one policy advisor pointed out, 'there were various campaigns encouraging people to hand in their guns. The objective was to take guns out of circulation irrespective of *who* had them'.[24] Between 2004 and 2005, no fewer than 110000 guns were taken out of circulation in São Paulo, according to the Institute *Sou da Paz* (2006).

As already pointed out above, a second key plank of the policy strategy was to tackle disorder in *all* spheres of society, be it through the removal of fly-pitchers, the urbanization or removal of shanty towns, or taking guns off the street. For the policy advisor, the campaigns to take guns off the street was the most important amongst these initiatives because it sent out a message regarding the priorities that the authorities had: 'Illegal guns came off the street, [either] handed in, or discovered by the police or they just stayed indoors, either way they were off the streets, this reinforced the message about security'.[25]

For the former minister, the coherent approach to disorder was critical: 'We wanted to change the sense of impunity [irrespective of the crime committed] in the hope that this [would have an impact] on violent crime'. As shown, this was backed up by heavy investments in the police, the prison system and other initiatives, identified as critical by all those who participated in this research, arguing that, regardless of the efficiency of how that money was spent, in combination with a much more high-profile public presence of the police in the media, it reinforced the *message* that fighting crime was a priority for the government, something which had broad public support.[26]

The results certainly seemed to vindicate the policies pursued. From the 2000s onwards, homicide rates in São Paulo declined significantly whilst, according to the former minister and the policy advisor, the city today does not possess 'no-go' areas for the state.[27]

In Rio de Janeiro, the process of change took longer. Hindered by years of political back and forth, and the subsequent alternation of anti-crime policies already outlined above, it was not until 2008 that the government of new governor Sergio Cabral began to reformulate policy, leading to its flagship policy, the establishment of so-called *Unidades Policiais Pacificadoras*, or Pacifying Police Units (UPPs) in some of Rio de Janeiro's most violent shanty towns. The aim of these units was first to establish a permanent presence of the state in areas where it had been absent, in some cases for decades. Secondly, the aim was, in the words of one of those who developed the policy, 'to establish the rights of everybody to come and go when he would like to where he would like to', that is, to establish freedom of movement for the population so far under the arbitrary control of drug gangs.[28]

292 *Handbook on complexity and public policy*

To do so, it was felt that the *base* of these gangs had to be eroded. As a result, UPPs were established in those shanty towns that were strategically important for the entry of both drugs and arms into the city. As one strategic commander of the UPPs put it: 'The point was to take their territory away. Without territory and without arms, they simply do not have the influence they used to have. Outside their communities, a drug-dealer does not have the power he has inside'.[29]

This conclusion led to another critical decision, which was *not* to focus on the apprehension of drug dealers. The establishment of UPPs was announced prior to the arrival of police forces, essentially giving drug dealers the chance to disperse, hide and flee. The government has justified this approach by arguing that 'the aim is to establish peace in these communities and that means avoiding confrontation'.[30] This, however, means that drug-trafficking often continues in the affected areas. However, according to one commander, the aim of the policy was to 'de-militarize' the area, not to stop drug trafficking: 'There is a market here, so there will always be a supply. The key point is to preserve life'.[31]

These, then, represent significant changes in terms of defining *what* the problem is and what that means. Shifting the focus from drug dealers to arms is one critical change of pattern in that it says that the key problem encountered is not just drugs but violence and that, just as in São Paulo, the preservation of life should become the priority over the elimination of drug dealers.

According to one person closely involved in the development of the policy, a second key issue was the image of the police, which perpetuated the pattern of violence and state ineffectiveness in parts of the city: 'For many people in the favelas we now occupy, their only experience of the state was that of violence [between police and drug dealers]. Therefore, we need to change that image by guaranteeing their rights, enabling the provision of services and generally [creating order]'.[32]

This, though, would mean a significant change in the approach taken by the police force and in the organizational culture that sustained this approach, which would require enormous investment in training and manpower. UPPs are staffed by newly-recruited police officers, subject to a different training program with heavy focus placed on community policing. This was key since, for many citizens, the permanent presence of the state was a new thing: 'Where we are, there was no effective presence of the state for 30 years so clearly there is mistrust. [We need to] establish a relationship of trust with the population'.[33]

In order to win this trust, the commander of one of the UPPs argued that his officers needed to show 'practical results'. By that he meant not dead drug dealers, but 'taking weapons out of the area' and, crucially, allowing for the installation of basic services, such as gas, electricity, water and the like. To establish priorities, it was critical to engage with the local population. Said the commander: '[The population] know what they need much better [than we do]. We need to respond to their wishes otherwise they will not [trust us]. There needs to be constant feedback and we have consciously engaged with [representatives of the community] to show them that we are here for them.'[34]

This approach has had a positive impact on the image of the Military Police, according to both anecdotal evidence and public opinion surveys. 'We are seen in a different light here. People used to [associate the police] with violence, now they perceive us to be here to help and to stop violence', as one unit commander put it.[35] Studies done by the

Brazilian Institute of Social Research (IBPS) underscore this impression, with 86 percent of people saying UPPs had made their area 'much better' or 'better', whilst 79 percent said that the presence of the police had eliminated the presence of armed gangs in their area. This led to 80 percent of respondents saying that the image of the Military Police had become 'much better' or 'better' (IBPS, 2010). In terms of hard statistical numbers, since the inauguration of the first such unit in 2009, the number of homicides declined from 2155 to 1209 per year at the end of 2012, according to the *Instituo de Segurança Pública de Rio de Janeiro*.[36]

On the face of it, therefore, there has been a clear downward movement in terms of homicides in both cities after the adoption of a new type of policing policy. In both cases, the focus was far more on changing patterns than on 'killing bad guys'. Changing patterns included changing the *perception* both of the police force about what it does and how it works, and changing the *perception* of individual police officers in relation to what their job is and how they see their job. These patterns have changed through simple strategic leadership that aimed at setting out and maintaining clear and consistent policy goals which were disseminated throughout the system and facilitated through clear and strategic investments in training, equipment and tactics, often as a result of learning from the best practices across the world. At the same time, at the local, operational level, commanders were given the autonomy and freedom to act according to the local boundary conditions which they encountered.

However, recent events, particularly in São Paulo, have raised questions over the sustainability of these policies and, in the context of complexity, the maintenance of the coherent patterns that have underpinned the process of self-organization.

ERODING PATTERNS OF PROGRESS? RECENT PROBLEMS IN SÃO PAULO AND RIO

Critics of the policies in Rio and São Paulo can be divided into several distinct, yet overlapping groups: those that have questioned the degree of progress that has been made; those that question whether the fall recorded in crime over recent years can be linked to specific policies; those that argue that any progress made is being eroded by current policy and political failures; and those that have questioned whether the policies were ever designed to change the underlying causes of violent crime in the first place. Each will be briefly looked at in turn.

Within the first group, several analysts have questioned whether the progress made in terms of recorded crime is really all that significant. Within this group, one can find those that point to the difficulties in accurately recording crime, bearing in mind the fact that the state often has shown itself to be incompetent or incapacitated to act in particular areas, especially those controlled by parallel power structures. In Rio de Janeiro, despite the tremendous efforts that have been made to pacify some slums, many are still under the control of drug gangs or, increasingly, militias, as Cano and Duarte (2013) have pointed out. In other words, there may not have been a profound sea-change in the way the city functions, quite apart from the fact that, with many others still not under state control, recorded violent crime does not necessarily reflect actual levels of these crimes. This argument can be underscored by recent events of people going missing in areas already pacified and the police being blamed for such disappearances by part of the population, as well as state authority being openly challenged in others, as has happened recently in Rocinha and the Complexo de Alemão (a very large collection of previously very violent favelas in the north of Rio de Janeiro) respectively. These events serve to emphasize the continued problem of mistrust between population and state.[37]

In response, those charged with developing and implementing the policy have pointed out that, whilst there are still significant problems to be overcome, 'we now have a waiting-list for UPPs',[38] and – for all the problems that persist – the thought of the state having a permanent presence in areas that used to be absolute no-go areas for its agents only some years ago – such as Rocinha or the Complexo de Alemão – represents indeed a sea-change.

The second group of analysts has questioned whether the fall in recorded violent crime can be linked specifically to the policies outlined above or are the result of an overlapping set of circumstances to which the policies may or may not have contributed. Certainly, in their review of the literature on the fall in violent crime in São Paulo, Carneiro et al. (2010) identify such a broad range of possible factors that it would be difficult to isolate particular government policies as being *the* decisive variable. To be fair to those responsible for developing these policies, they have, on the whole, not claimed that theirs is the solution to all problems. For instance, the secretary for public security in Rio de Janeiro, Beltrame, has repeatedly stated that UPPs will only work in conjunction with a broader approach to public order, which involves the community as well as all levels of government, and will only make a significant difference within the particular areas in conjunction with other initiatives.[39] This includes the launching of a package of social measures called 'UPP Social'.[40]

In São Paulo, as already pointed out above, the state government claims to have done significant work in terms of combating disorder, social exclusion and other problems associated with crime.

The third group has focused on the erosion of the patterns which have, according to them, underpinned the fall in violent crime over recent years. According to one policy advisor in São Paulo, the recent upsurge in violence can be explained by the 'disastrous administration of public security' in the city, in particular the failure to stem the influence of imprisoned gang leaders on their members on the outside.[41] The former commander of the PM agrees, albeit in somewhat more diplomatic terms: 'There has been a change towards confrontation again, giving more value to the elite forces that do the killing [of criminals], at the expense of ordinary policemen'. This, according to him, has led, amongst other things, to a reaction from elements of the organized crime network, leading to a spiral of action and reaction, including the re-emergence of police death squads, something which the public like but 'which undermines the basis of the state of rights'.[42]

By way of contrast, one former minister interviewed argued that the recent upsurge of violence, particularly in São Paulo, was to be expected: 'There was always going to be a reaction and I think we are seeing that now.' At the same time, 'it is fairly simple to get the homicide rate from 30 to 10 [per 100 000 inhabitants, as has happened in São Paulo], but quite difficult to get it from 10 to 8. Setbacks are normal.' This same minister also heavily criticized the media for the way that they report crime, especially in areas of the middle and upper social classes: 'If you just had the TV on and did not pay particular

Brazil and violent crime: complexity for 'intractable' problems   295

attention you would think that violent crime was an epidemic which affected all parts of the city the whole time. It is not like that but it feeds a sense of insecurity'.[43]

Yet a further group of critics points out that such setbacks will continue to occur because the underlying causes feeding crime have not been tackled and, in fact, have been reinforced by elements of the current policies. For instance, one community leader in one of Rio's most violent shanty towns pointed out during a seminar attended by the author that the majority of pacified slums in Rio lie in a circle close to the major tourist destinations of the city and close to some of the city's wealthiest neighborhoods. As such, the pacification of these *entrenches* the divisions between those areas of the city considered safe and those that are not, pointing out that even before pacification the richest areas of the city were also the safest, despite the existence of violent favelas in their midst.[44] Equally, in São Paulo, the policies have done very little to reverse the divisions of wealth between the richer parts close to the city center and those poorer at the periphery of the city. In fact, according to Denyer-Willis and Tierney (2012), in some instances the provision of security in those peripheral areas has essentially been contracted out by the state to the very criminal gangs the state is meant to confront. In Rio de Janeiro, equally, one unit commander admitted that there was no way of knowing how many people had chosen or managed to enter the formal economy as a result of pacification at the same time as living costs – such as rents – in pacified areas have shot up significantly.[45] Following this argument, it is therefore doubtful whether any drop in violent crime – welcome though it is – is sustainable, since it does not change underlying patterns that underpin crime in the country at the higher levels of analysis.

CONCLUSIONS

The last section should have given a clear idea that the problem of violent crime in Brazil generally – and its urban centers in particular – is a long way from being resolved, and that the patterns which have sustained this issue for so long have not yet been definitively changed. Brazil remains an extremely violent country.

However, it is also undeniable that considerable progress has been made and that significant parts of Rio and São Paulo – as the focus of this chapter – are today much safer than they were 10 years ago. As several analysts have shown, this is due to a number of factors. Yet I would argue that recognition of the problem encountered as complex and multi-faceted – as recognized in both cases analyzed here – represents significant progress. Equally, the fact that in both cases there were efforts to change the *context* within which self-organization occurs represented a massive step forward from the previous belief that one can simply – and literally – eliminate the problem. Thirdly, recognition that initial progress is much easier than sustainable progress later on – as displayed by the former minister interviewed – bodes well in terms of setting expectations about the policy at realistic levels.

The key challenge that emerged out of this research, then, is to make progress sustainable regardless of election outcomes. Both a minister in the Rio de Janeiro state government and the former commander of the Military Police in São Paulo argued that this is difficult in a culture which is geared towards short-termism and in which the media has a huge role in setting the public agenda. In such circumstances, practical results are the

296 *Handbook on complexity and public policy*

key to convincing the population of the advantages of continuing the current course. As shown, particularly in São Paulo, there was a preoccupation of those interviewed about the chances of making progress sustainable in view of the new direction currently being taken by parts of the government and the public clamor for 'action'.

Dealing with highly complex adaptive systems such as these, such setbacks should, I would argue, be expected and require action across all levels of society and government, bearing in mind the deep structural features of Brazilian society which also help sustain the pattern of violence. Changing those will be a process which will last decades, not months or years. As such, many challenges still lie ahead.

Finally, I think one can sketch out some broader lessons for complex public policy issues from the above case studies. First, objectives and expectations for addressing any complex problem have to be realistic. No given policy – however well thought out and implemented – will, by itself, definitively 'resolve' any given issue. The aim should be to influence and change the patterns which have led to the situation being addressed.

In order to be able to influence and change established patterns, it is critical that policy-makers ask *questions* rather than provide ready-made answers. Policy-makers need to be clear *what* patterns they are confronting, what these patterns *mean* in terms of setting objectives and generating options for actions and what they can then realistically *do* to change these patterns, something that was clearly done in the cases of Rio de Janeiro and São Paulo in relation to violent crime. Such a process of questioning what one is seeing and doing does not end but, rather, is continuous.

Thirdly, a key lesson is that public policies are made for someone. The participation and feedback of these users is crucial to the success of any policy. Therefore, communication needs to be a critical part of policy development, implementation and adjustment.

Finally, feedback mechanisms are critical in order to achieve coherence across any system. The objectives of a policy need to be *scaled* across the system in the sense that, at any given level of the system, actors (or agents) are working in the pursuit of commonly defined objectives. Violent crime generates very destructive patterns across time and space and only concerted action across all levels of the system will enable the possibility of changing towards more generative patterns. The same rule applies to any other area of public policy, be it health, education or anything else. The role of policy-makers is to enable the emergence of such patterns and to sustain them across time and space once established.

NOTES

1. See http://tuliokahn.blogspot.com.br/p/world-homicides-homicidios-no-mundo.html for the full statistics.
2. For a very detailed breakdown of the figures see, again, http://tuliokahn.blogspot.com.br/p/world-homicides-homicidios-no-mundo.html.
3. Numbers from http://mapadaviolencia.org.br/mapa2012.php.
4. This section is adapted from Lehmann (2012).
5. For details, see Santos and Silveira (2008).
6. See, for instance, Felbab-Brown (2011).
7. See Sento-Sé (2002).
8. See Carneiro (2010b).

9. Interview with former commander of the Military Police of São Paulo.
10. For the history and evolution of the PCC, see Nunes Dias (2013).
11. Interview with policy advisor.
12. Adapted from Lehmann (2012).
13. Interview with former commander of the Military Police of São Paulo.
14. Interview with former minister in state government of São Paulo.
15. Interview with former minister of the state government of São Paulo.
16. Interview with former minister in state government, June 2013.
17. Interview with former commander of the Military Police, June 2013.
18. Interview with former minister in state government.
19. Interview with former commander of the Military Police.
20. Interview with policy advisor.
21. See World Bank (2013).
22. Interview with former commander of the Military Police of São Paulo.
23. Ibid.
24. Interview with policy advisor.
25. Ibid.
26. Interview with former minister of São Paulo state government.
27. Interview with policy advisor.
28. Interview with minister in Rio de Janeiro state government.
29. Interview with UPP commander.
30. Interview with minister in Rio de Janeiro's state government.
31. Interview with UPP commander.
32. Interview with minister in Rio de Janeiro state government.
33. Interview with UPP commander.
34. Ibid.
35. Ibid.
36. See http://urutau.proderj.rj.gov.br/isp_imagens/Uploads/201212capital.pdf, accessed on 2 August 2013.
37. See http://g1.globo.com/rio-de-janeiro/noticia/2013/08/caso-amarildo-pode-por-em-risco-pacificacao-na-rocinha-diz-pm.html, accessed on 14 August 2013.
38. Interview with member of the Rio de Janeiro state government.
39. See, for instance, http://cbn.globoradio.globo.com/rio-de-janeiro/2013/08/02/BELTRAME-NEGA-DESGASTE-NA-POLITICA-DAS-UPPS-MAS-RECONHECE-PROBLEMAS-LOCAIS.htm, accessed on 14 August 2013.
40. See http://www.uppsocial.org/ for more information.
41. Interview with policy advisor in São Paulo.
42. Interview with former commander of the Military Police in São Paulo.
43. Interview with former minister of state government in São Paulo.
44. Community leader from a favela in Rio de Janeiro during a seminar discussing the pacification strategy in the city.
45. Interview with UPP commander in Rio de Janeiro.

REFERENCES

Cano, I. and T. Duarte (2013), *No Sapatinho: A Evolução das Milícias no Rio de Janeiro*, Rio de Janeiro: Fundação Heinrich Böll.
Cardia, N. (2000), 'Urban violence in São Paulo', *Comparative Urban Studies Occasional Paper Series*, **33**, Washington, DC: Woodrow Wilson Centre.
Carneiro, L.P. (2010a), 'Mercados ilícitos, crime e segurança pública: temas emergentes na política brasileira', *CLP Papers*, **5**.
Carneiro, L.P. (2010b), 'Mudanças de guarda: as agendas da segurança pública no Rio de Janeiro', *Revista Brasileira de Segurança Pública*, **4**(7), 48–71.
Carneiro, L.P., B.P. Manso and T.N. Fonseca (2010), 'A queda do crime em São Paulo: revisão da Bibliografia', *Núcleo de Pesquisa de Políticas Públicas*, São Paulo: Universidade de São Paulo.
Chapman, J. (2002), *System Failure*, London: Demos.
Denyer-Willis, G. and J. Tierney (2012), 'Urban resilience in situations of chronic violence: case study of São Paulo, Brazil', *MIT Centre for International Studies*, Urban Resilience in Chronic Violence Project, Cambridge, MA.

298 *Handbook on complexity and public policy*

Eoyang, G. and R.J. Holladay (2013), *Adaptive Action: Leveraging Uncertainty in your Organization*, Stanford, CA: Stanford Business Books.

Felbab-Brown, V. (2011), *Bringing the State to the Slum: Confronting Organized Crime and Urban Violence in Latin America: Lessons for Law-enforcement and Policy-makers*, Washington, DC: Brookings Institution.

Goldstein, J. and R.M. Zeidan (2009), 'Social networks and urban poverty reduction: a critical assessment of programs in Brazil and the United States with recommendations for the future', in J. Goldstein, J.K. Hazy and J. Silberstang (eds), *Complexity Science and Social Entrepreneurship: Adding Social Value through Systems Thinking*, Litchfield Park, AZ: ISCE Publishing.

Guimarães, S.P. (2008), *Desafios Brasileiros na Era dos Gigantes*, Rio de Janeiro: Contraponto.

IBPS (2010), 'Pesquisa Sobre a Percepção Acerca das Unidades de Polícia Pacificadora', Rio de Janeiro: *Instituto Brasileiro de Pesquisa Social*, PR 004-10-UPP-25.01.

Instituto Sou da Paz (2006), 'Brazil: changing a history of violence', Communication, Rio de Janeiro: Sou da Paz, accessed 1 August 2013 at http://www.soudapaz.org/Portals/0/Downloads/An%C3%A1lise%20do%20Resultado%20do%20Referendo%20Popular_ingl%C3%AAs.pdf.

Lehmann, K. (2012), 'Dealing with violence, drug trafficking and lawless spaces: lessons from the policy approach in Rio de Janeiro', *Emergence*, **14**, 51–66.

Leu, L. (2008), 'Drug traffickers and the contestation of city space in Rio de Janeiro', *E-Compós*, **11**(1), 1–16.

Monjardet, D. (1996), *Ce que fait la police. Sociologie de la Force Publique*, Paris: La Découverte.

Nunes Dias, Camila C. (2013), *PCC: Hegemonia nas Prisões e Monopólio da Violência*, Rio de Janeiro: Saraiva.

Peirce, J. (2008), 'Divided cities: crime and inequality in urban Brazil', *Paterson Review*, **9**, 85–98.

Richardson, L. and A. Kirsten (2005), 'Armed violence and poverty in Brazil: a case study of Rio de Janeiro and assessment of Viva Rio for the Armed Violence and Poverty Initiative', Report by the Centre for International Cooperation and Security, Bradford: University of Bradford.

Santos, M. and M.L. Silveira (2008), *O Brasil: Território e Sociedade no início do século XXI*, Rio de Janeiro: Record.

Sento-Sé, J.T. (2002), 'O discurso brizolista e a cultura política carioca', *Varia Historia*, **28**, 85–104.

de Souza, L.A.F. (ed.) (2009), *Políticas de Segurança Pública no Estado de São Paulo*, São Paulo: Cultura Acadêmica.

World Bank (2013), *Making Brazilians Safer: Analyzing the Dynamics of Violent Crime*, Washington, DC: World Bank Publications.

18. Educating for equality: the complex policy of domestic migrants' children in China

Qian Liu

INTRODUCTION

For the last 30 years, China has been going through a massive economic and social transformation. Large-scale economic development has driven up the wealth and living standards of society and generated substantial internal migration. Most of this migration has been from the rural and agricultural area to the rapidly developing cities in China. Managing this migration has been a huge task and has led to increasing social strains – particularly in the growing urban areas where millions of migrants with limited social and economic rights put increasing pressure and demands on local and regional authorities. One of these main areas of pressure is education, in particular for the children of internal migrants.

This chapter will explore some of the complexities that confront Chinese policy makers who are trying to respond to the significant inequalities in educational opportunities for the children of internal migrants. To set the scene, the education system in China will be briefly introduced, focusing on recent policy developments, the three-level structure of Chinese education and the importance of the GAOKAO examination process. Next, the policies about migrant children education in Beijing will be described, focusing on the achievements and pitfalls of the policy implementation at the local level of Area C, a neighborhood in Beijing. During these discussions, various problems will be highlighted including: the various public, media and stakeholder debates that have taken place over the issue and the confusion and disagreements that have been caused by the vague definition of three key conceptions of the policies: who are the 'migrant children', what is the meaning of the 'main responsibilities' for the policies and what are the options for the 'higher-level education entrance examination'? In addition, various Chinese social and cultural factors will be discussed that have helped to shape the debate. These include: resource constraints; the practice of traditional bureaucratic politics; educational desire in the Chinese tradition; and growing awareness of human rights in modern China. All of these social and cultural factors are adding to the complexity of educational policy in China and the children of internal migrants in China are caught within this complex mix.

BUT FIRST, WHY USE COMPLEXITY?

Originating in some of the most significant developments in the natural sciences of the last few decades, the concept and theory of complexity has become increasingly influential in scholarly debates and policy decision making (Bousquet and Geyer,

300　*Handbook on complexity and public policy*

2011; OECD, 2009). Over the last 5–10 years, complexity thinking has been having a growing impact on Chinese academic debates (Chen, 2006; Chen, 2008; Foo, 2005). With its emphasis on viewing the world and policy-making in a more holistic, comprehensive and naturalistic perspective, complexity has become a significant challenger to the more orderly and traditional forms of policy making. For example, as demonstrated by the work of Cairney (2012), Geyer (2012) and Sanderson (2000) and others in the UK, complexity has begun to demonstrate the limits of the dominant 'evidence-based policy making' and 'audit culture' (Powers, 2007) with its emphasis on viewing policy making from a traditional orderly world view of causality, reductionism, predictability and determinism. On the other hand, complexity emphasizes that there are no such golden rules of order in human societies. Instead there is partial causality, reductionism and holism, predictability and uncertainty. In addition to this, human systems demonstrate clear elements of unpredictable emergent properties and narrative and interpretive uncertainty (Geyer, 2012). What this implies is that one must take a much more nuanced and holistic approach to understanding and evaluating policy making and the often uncertain and contradictory outcomes that these policies create.

SETTING THE SCENE IN CHINA

With the economic success and process of urbanization in China many people from rural communities have moved to towns in the provinces for economic reasons. According to the 2010 Chinese Census, anyone who does not hold household registration for at least six months is known as an internal 'migrant'. Many internal migrants, who typically work as construction workers, street venders, or garbage collectors, earn more money in the city than they can in their home town. However, they are often still relatively poor in relation to existing urban residents and, previously, internal migrants could only access limited social welfare systems in the cities. This internal migrant status had a significant impact on children. It was reported that in 2010 there were about 11.67 million internal migrant children receiving compulsory education across China, increasing 17.1 percent from the previous year. Among them, more than 58 percent were concentrated in the large cities in Eastern China (MOE, 2013). It is because of the movement of domestic migrants and their children throughout modern China that the education of migrant children has become a striking topic across the country. In many cases, children of internal migrants were not given access to local schools and were therefore forced into private or sub-standard education facilities. To address the problem of this growing inequality in education a series of official policies were adopted. Among these were the two milestone policy documents of 2001 and 2006.

THE 2001 POLICY DOCUMENT

In 2001, the State Council launched an official document titled 'Decision on Basic Education Reform and Development' (National State Council, 2001), of which the 12th item stated:

the government of the cities with high immigrations should take the main responsibility for implementing nine years of compulsory education for migrant children; and public schools should take the main responsibility to admit these children. According to the law, therefore, a multi-strategic effort should be applied to safeguard the right of compulsory education for the migrant children.

Later, this policy was abbreviated to the 'Two Main Responsibilities Policy'.

This policy had strong regional implications and implied that the responsibilities and costs of educating children of internal migrants should be carried by the regional/city governments of the urban areas with high migrations rather than the regions/cities undergoing net migration out of their areas. Moreover, children of internal migrants should be educated primarily in public schools rather than private ones. Regarded as a milestone, the policy put forward important steps towards resolving the issue of compulsory education for internal migrant children in China.

THE POLICY DOCUMENT IN 2006

In 2006, an updated version of the law on compulsory education was approved by the committee of the Tenth National People's Congress (Standing Committee of the Tenth National Congress, 2006). Chapter 2, item 16 stated: 'For children who receive compulsory education where their parents or guardians are working, away from the places of their household registration, the government where the children are living is obliged to provide equal opportunities for compulsory education.' So, the 'Two Main Responsibilities Policy' was then integrated into law and given much more importance and authority.

THE PRESIDENT'S REMARK IN NOVEMBER 2012

President Hu Jintao presented a report, regarded as one with high authority, to the Eighteenth National Congress of the Communist Party of China in 2012. It stated: 'We should support special education, increase subsidies to students from poor families, and ensure that children of rural migrant workers in cities have equal access to education. All this is designed to help all children gain required knowledge and skills.'

The president's remarks concerning the education of migrant children thus expanded the requirements of the initial policies of 2001 and 2006 from equal opportunity in compulsory education to one seeking to ensure educational equality in general. At the same time, the report was consistent with national concern about the issue. The report was then transferred from the Ministry of Education to all provincial governments in 2012, urging them to launch their own policies on migrant students applying for the higher-level education entrance examination (General Office of State Council, 2012).

302 *Handbook on complexity and public policy*

CHINA'S EDUCATION SYSTEM, GAOKAO AND FINANCIAL ISSUES OF MIGRANT CHILDREN EDUCATION

To better understand the context of these policies, it is crucial to have a basic understanding of China's education system. This system, with its lengthy history and multiple cultural aspects, shapes the specific situation of migrant children education in modern China.

THREE-LEVEL EDUCATION SYSTEM: COMPULSORY EDUCATION, HIGH SCHOOL EDUCATION AND HIGHER EDUCATION

China has a three-level education system for the general population: compulsory education, high school education and higher education. The 1986 National Law on Education stipulates that all Chinese citizens are obliged to receive nine years' free compulsory education. In most areas of China, primary school caters for students in grades 1–6 from age 6. Students then graduate to junior middle schools for another three years. When students finish compulsory education at around age 15, they receive a junior middle school diploma.

After compulsory education, some students will leave the education system and join the labor market.[1] But most students will continue to high school education. In the high school education period, there are two options: general high schools and occupational training schools. In principle, both of them can offer the graduates qualifications that enable them to apply for the higher education entrance examination. In Chinese this is called the 'GAOKAO'. The main difference between these two options is that the general high schools are oriented towards academic learning focused on the GAOKAO, while the occupational training schools focus on labor skills training, such as in vehicle repairing, hotel service, cookery and so on. In practice, this means the curriculum of occupational training schools does not fit the GAOKAO very well, even though the graduates are entitled to take the GAOKAO. Moreover, for high school education, the students' family will be responsible for the tuition fee, even though the government will support it indirectly.

After completing their high school education, some students will go out to work without taking the GAOKAO or after they have failed it. Others who pass the GAOKAO will go on to higher education. There are two types of higher education: 4-year university courses for bachelor's degrees and 2–3-year higher technical college courses for diploma degrees.

Because the curricula of general high schools and occupational training schools are so different, most occupational training school graduates have more chance of entering higher technical colleges than universities. At the same time, the graduates of higher technical colleges are entitled to apply for bachelor's degrees after another two more years of study. Besides that, adult schools and online learning can also offer various higher education certificate, diploma and bachelor degrees as long as the candidates pass the relative examinations (see Figure 18.1).

Educating for equality: domestic migrants in China 303

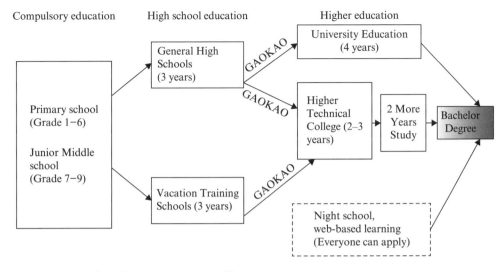

Figure 18.1 The education system in China

FINANCING THE EDUCATION OF MIGRANT CHILDREN

As discussed above, the Chinese government is responsible for financing nine years of compulsory education for all citizens, with province-, city- and county-level governments in charge of financing and resourcing education for children of registered families in their areas. Currently, migrants living outside their home-provinces cannot register for household residency in the cities where they are working, thus registration must remain in their home town. Theoretically, therefore, responsibility and financial resources for the education of migrant children should remain with the home town.

Besides this inconsistency in the allocation of educational resources and children's geographical status, other complicated issues around finance come into play. Similar to many other countries, China has both a public and a private school system. The latter are financed mainly through tuition fees from the students' families rather than government input. Private schools in China are one of two main types. One, the top-end, high-fee-paying schools are perceived as offering higher standards and broader curricula, including languages, science and sports. The second type of school is known as Minbanxiao (meaning: run by the local people). These are schools designed for the children of low-income families and are mainly located in the cities with large migrant populations. The migrant families can pay around 500 Yuan (about US$85) per year to send their children to these Minbanxiao. All private schools in China, both high-end and low-end, must hold a license to operate. However, in many migrant communities in the cities, there are a large number of illegal schools that have emerged to meet the educational needs of migrant children before the creation of the 'Two Main Responsibilities' policies. Most of these illegal schools are without government approval and are of generally low quality, but still contain many migrant children.

304 *Handbook on complexity and public policy*

GAOKAO: A KEY EXAMINATION FOR MOST CHINESE STUDENTS

After completing compulsory and high school education, migrant children, if they want to move into higher education at university level, have to face the GAOKAO. That is another striking topic for migrant families. Most Chinese families nowadays consider the GAOKAO as the most important examination for their children. Many scholars argue it is a cultural pattern embodied in the Chinese world view to this day. China has a history of imperial examinations dating back 1500 years to the Sui dynasty (Tian, 2009: 7).[2] Even during China's feudal period, people living in poverty were disadvantaged by this exam system. However, it did open doors for men of lower social status who, if successful at imperial examinations, were able to enter the official system. So for many parents, especially from disadvantaged groups, nowadays taking the GAOKAO is one of the opportunities for upward social mobility.

However, for the migrant children in China today, two main issues impact on their potential success in taking the GAOKAO. First, the exam can only be taken in the province where the student's families hold household registration. Second, as there is no standard GAOKAO assessment system across China, each province runs its own provincial-level credits and exam system. Consequently, university admittance procedures rank students' academic performance according to their home province. The key problem is that migrant students who enroll to sit the GAOKAO experience a disconnection between the exam system of their home province with that of the province to which the family migrated. This puts migrant students at a disadvantage and frequently leads to their failure in the exam.

FROM POLICY TO IMPLEMENTATION

As complexity theory implies, the aims of most policies should be to set core rules that provide stable general rights and opportunities among various stakeholders, while the exact boundaries of the rules remain flexible and are open to further discussion and feedback. Thus, the policy system is continually evolving in response to its changing environment, without designating a fixed end point. The key is not to become focused on a final endpoint/order and march blindly towards it, but to encourage the actors involved to adapt and adjust to the continual evolutionary changes.

In the case of education policies for the children of internal migrants in China, there are various stakeholders involved, including different levels of government, owners of Minbanxiao and illegal schools, as well as non-resident and resident families. The process of developing the 'Two Main Responsibilities' policy reflects the struggle and the imbalance among all these stakeholders. While it is generally appreciated that the current policy protects the rights of migrant children, nonetheless individuals, agencies and the media must find a pragmatic approach towards moving the policy forward and keeping it balanced in relation to the actual needs of the children. To demonstrate the complexity and difficulty of maintaining this balance, the chapter will now discuss the application of this policy to the Area C neighborhood in Beijing.

MIGRANT CHILDREN'S EDUCATION IN BEIJING

A brief overview of the education of migrant children in Beijing is necessary to provide context for the more detailed discussion of Area C.

As the capital city of China, Beijing is one of the most populated in China. The government of Beijing has tried to control the population size, but its policies did not work (Zhang and Hou, 2009). By the end of 2011, Beijing's population reached 20.18 million, about three times the population of New York. Among inhabitants of Beijing in 2011, around one-third were migrants from other provinces across the country. They are considered to be the main driving force behind Beijing's increasing population (Zhou, 2009). Consistent with the increase in the migrant population in Beijing, the rise in the number of migrant children receiving compulsory education became an increasingly important issue. It is reported that by the end of August 2013 there were 437000 migrant children in Beijing, 70 percent of them being enrolled in public schools. For the rest of the migrant children, more than 50000 migrant children studied in legal Minbanxiao, and more than 40000 studied in illegal schools (Xian, 2011).

PUTTING POLICY INTO PRACTICE IN BEIJING

The Beijing government launched a series of official documents and regulations in response to the State Council's decision on the 'Two Main Responsibilities' policy.

The first, a joint effort launched by multiple bureaux at city level in 2004, responded to the State Council's requirements in a direct way. Four main points are emphasized in this document.

1. Children of rural migrant workers with eligibility documents should not be expected to pay extra fees for temporary schooling in Beijing's public schools. If there are no spaces in public schools in the school district, admitting the students into legal Minbanxiao is one of the solutions. In that way, the migrant families pay only a limited amount of tuition fees, and the local government must input resources into them.
2. Governments of the administration area levels should take responsibility for the financial issues of the public schools in their area, rather than relying on city-level government.
3. Governments at administration area levels are also responsible for the quality control of Minbanxiao. Governments of administration areas have the authority to shut down illegal schools that cannot reach the basic standards of legal schools and make arrangements for the students accordingly.
4. Working mechanisms at administration area level are to be set up to solve the issues and assign responsibility to the multi-departments, including the public security department, the finance and education bureaux, and so on.

The importance of this document is in the way that it adapted requirements at a national level to those at city level, and clearly addressed the responsibilities of governments at administration area levels.

306   *Handbook on complexity and public policy*

In 2008, the Beijing Government integrated the compulsory education of migrant children into the strategic planning of Beijing city. In 2012, to respond to the Chinese President's remarks on equal education for migrant children, the policies at Beijing city level stipulated that migrant children who have been officially enrolled in the junior middle schools of Beijing for three consecutive years and whose parents have been issued with a legal residential record, employment record and social security record for three consecutive years in Beijing, are entitled to attend the occupational training schools' entrance exam. On this point, it should be noted that the policy stated that students with or without Beijing household registration will be treated equally (Beijing Government, 2013). This is a major change from the previous policy. It also opened the door for the migrant children who graduate from occupational training school to be able to apply for higher technical colleges in Beijing.

In summary, these three developments can be seen as the Beijing government's adoption of the policies and requirements of the national government. The hierarchical tendency is evident, with the national government asserting authority over provincial and city government and government at this level holding authority over lower levels of government. The Beijing government explicitly appointed the 'Two Main Responsibilities Policy' to the administration-level governments and also tried to make provision for post-compulsory education for migrant children in Beijing. This is what provided the environment for relatively successful implementation of the policy in Area C.

Beyond the series of documents and policies, other various actions further tangled the issues. For example, in August 2011, the government of Beijing determined to shut down 24 illegal schools across the city and placed 14000 children in public schools or legal Minbanxiao. Area C played an active role at this juncture.

IMPLEMENTING THE 'TWO MAIN RESPONSIBILITIES' POLICY IN AREA C: THE ACHIEVEMENT AND STRUGGLES

Background to Area C

Being the most populated of Beijing's city areas, Area C has 3.8 million people living in an area covering 470.8 square kilometers. There are more than 2 million migrants, exceeding the number of local citizens of the area (Wu, 2010). By the end of 2010, there were 87510 students without household registration receiving compulsory education in Area C's schools, amounting to about one in four migrant students in Beijing city. Among them, 78.3 percent were enrolled in public schools or legal Minbanxiao (the Education working committee of Area C, 2010). By March 2012, the percentage of migrant children enrolled in public school or legal Minbanxiao in Area C reached 90 percent (the Education Committee, 2012). It can be regarded as the achievement of the 'Two Main Responsibilities' policies at local level. The government of Area C also launched their plan to shut down all illegal schools within three years.

Policy in Practice and the Challenges Faced in Area C

In this context, this section looks at some of the echoes from the media, representatives of illegal schools, the voices of migrant families, and administration area-level governments will also be presented to show the unexpected difficulties in implementing the policies.

Echoes from the mass media

In general, the mainstream media reported the education policies and practice for migrant children in Beijing in a positive way. But some critical and negative opinions were spread. For example, in August 2012, while the education committee of Area C was working on closing illegal schools as planned, the popular media, such as Sohu.net, Sina. net and the newspaper Xinjing, posted a news article titled '6 elites jointly signed a letter to Ministry of Education to call for keeping Tongxin School for the migrant children'.[3]

Once the letter and the related news article were posted on the website they quickly spread on the Internet. This led to a confused public debate where many people began to worry that the government would deprive migrant children of the right to an education at the school of their choice. The government of Area C then came under attack from both the general public and the Ministry of Education. With the news media and social media playing an increasingly salient role in the shaping of public opinion, the policy became much more difficult to implement.

Concerns about illegal schools

Tongxin School is still open today; it suffers from poor infrastructure and unsafe conditions. Regardless, about 600 migrant children attend school there. However, in order to carry out the 'Two Main Responsibilities' policy, the government of Area C had intended to shut down Tongxin School and move students to Minban schools, about a ten-minute drive away, and which, being legal, would have better facilities and offer a safer environment.

Tongxin School is typical of illegal private schools in Beijing. Many private schools were formed in the 1990s. It is reported that the first private school for migrant children was set up in 1991. The numbers of newly opened schools of this kind increased from that time to about 15 per cent a year (Lv and Zhang, 2010). As the policy at the time blocked migrant children from attending public schools, the newly opened illegal private schools met the community's particular needs. For example, many were located locally, making it easier for parents and carers to pick up children from school. Sometimes the students, teachers and even the principal were from the same province and had similar socio-economic backgrounds, which may have meant that families faced less discrimination than they experienced at public schools (Lv and Zhang, 2010).

Thus, illegal schools generally had a longer and closer relationship with the migrant communities before the new policies were developed by the government. These schools accepted volunteer staff and raised funds among NGOs, charities and celebrities. In many ways, it would seem that they took advantage of media, celebrity and charitable contacts to survive, seeking support beyond the government and communicating with the media to put pressure on the government. On the other hand, one of the functions of private schools is to make money, thus the owners of the schools are motivated to protect their investment.

308 *Handbook on complexity and public policy*

This conflict raises the issue that the 'Two Main Responsibilities' policy cannot solve: historic disadvantage. However, migrant communities and illegal schools can and do obstruct the implementation of the policies in a number of ways.

Responses of migrant families

As evident beneficiaries, migrant families now enjoy better conditions toward ensuring their children receive compulsory education in Beijing. But there are still complaints from some migrant families. These complaints can be summarized according to three perspectives: disadvantages associated with applying to GAOKAO, the quality of compulsory education for migrant children and migrant children's right to free compulsory education.

For those migrant children who attend public schools, especially those with good academic performance, the main concern is application to GAOKAO in Beijing. According to the 2013 Beijing government policy, they can apply to higher technical colleges in Beijing but cannot attend the GAOKAO as Beijing citizens. The migrant families perceive this as further evidence of a continually unfair educational system.

For those who moved from illegal schools to legal Minbanxiao, some were concerned about the quality of education in these legal schools when they found that most of the teachers at the Minbanxiao schools had previously taught at illegal schools. At the same time, about 80 percent of parents of children attending Minbanxiao hope that their children will one day attend university. They regard it as the central ladder of upward social mobility. However, more than 50 percent of parents had no idea that their children could be given the opportunity to go to university after they had gained a diploma from occupational training schools in Beijing.

For migrant families whose children are still attending illegal schools with around 500 Yuan annual tuition fees, they hope to be able to take advantage of free compulsory education in public schools, but are concerned that there are no public schools or legal Minbanxiao nearby.

The government's dilemma

Under national and city policies, the government of Area C increased resources for the education of migrant children. It was reported in 2008 that government finance for this sector alone was 860 million Yuan; in 2009 it was 1170 million Yuan; and in 2011 it was 1470 million Yuan. More than 90 percent of students without Beijing household registration have been admitted to public schools or legal Minbanxiao. From this statistic, it is clear that fulfilling the requirements of the national and city governments, the government of Area C is taking the main responsibility for educating migrant children in the area very seriously.

Meanwhile, the Education Committee is engaged with the challenge of implementing the policies. These challenges include resource constraints, a lack of coordination with other stakeholders and sectors, insufficient communication with the media, and so on. With a limited budget, the infrastructure has been improved to prioritize students' safety, but funds were then lacking to employ teachers of high quality (Lv and Zhang, 2010). At the same time, as Minbanxiao are evaluated on a separate academic performance system to public schools, so the teachers at schools for migrant children were not motivated to improve the quality of their teaching. This performance, in turn, led to questions from migrant parents.

Educating for equality: domestic migrants in China   309

Moreover, a lack of coordination between different governmental administrations made it more difficult for the government of Area C to expand the number of public or Minbanxiao schools for the children of internal migrants; neither have they been able to coordinate public transportation sectors to plan more buses in the migrant communities. All of this has had a negative influence on policy implementation.

Though the main difficulties for taking the higher education exam have been resolved for migrant children, the migrant parents have remain focused on the GAOKAO and seem to have little knowledge of the opportunities that have become available due to the 2013 policy changes regarding the higher-level education entrance examination. These new options offer several pathways to work towards a bachelor's degree for migrant children in Beijing. Despite these new options, the local government of Area C has been caught between the central government's demands to quickly improve overall educational opportunities and equality for post-compulsory education and growing pressure from migrant families to bring these changes about in a flexible and pragmatic fashion.

'CREATIVE COMPLEXITY' AND EDUCATIONAL POLICY IN CHINA

How the 'Two Main Responsibilities' policy entered the public realm and was then implemented at national-, province-, city- and administration-area levels has been described above. Migrant families, governments at various bureaucratic levels, schools, media and the public form the different stakeholders. Complexity theory points out that when you are dealing with complex problems it is essential to involve as many stakeholders as possible and to involve them in the policy process as thoroughly as possible. It is only with this type of input that you can fully understand the problems and see how they change over time. As Geyer and Rihani point out, 'what is considered as a problem, and in what way it is a problem, depends heavily on who is making that distinction. . .. And crucially, the "problem" might change during the process of "solution"'(Geyer and Rihani, 2010: 68). In the case of educational equality policies in Beijing, China, the three key terms: migrant children; main responsibilities; and higher-level education entrance examination had multiple meanings among different stakeholder groups. It is these vague definitions that have led to the debates among various stakeholder groups which, in turn, have shaped the dynamic of the policies.

SOME OF THE PROBLEMS THAT EMERGED

Migrant Children

For the government of Area C, the first vague term, 'migrant children', was defined by them to mean children without Beijing household registration. But at national and city level, 'migrant children' meant the children of migrant workers and laborers. In other words, according to Area C's government, 'migrant children' included both the children from high-income as well as low-income families. While the high-income migrant group can easily send their children to public schools, this is not the case for

310   *Handbook on complexity and public policy*

the low-income migrant families. Therefore, the two groups had very different interests in response to the educational policy actions of the Area C government. Moreover, the data on migrant children presented by the Area C government did not give an accurate picture of the needs of its migrant children and could not ensure educational equality for the children of low-paid migrants. This is a key element of the requirements of national- , province- and city-level governments. 'Migrant children', therefore, are not a simple and cohesive population with clear and similar needs with regard to education policy.

Main Responsibilities

The second subtle concept is 'main responsibilities', which scholars argue is an implicit statement. For instance, what is the percentage of migrant children and their families that cities are required to accept? And what is the quality standard for compulsory education? As we have seen above, central government at national level passes the responsibility down to the city government level of those cities with high migrant populations. Then, governments at city level then pass the responsibility on to the local administration level (Gehong and Hong, 2012). This means that local governments do fulfill their obligatory responsibilities, such as moving a required percentage of migrant children into public schools. However, they do not take direct ownership of the responsibilities for implementing the policy. This lack of ownership allows problems to occur such as the willingness of local governments to tolerate low-quality teaching for migrant children (Shao, 2010). In turn, at national level it is difficult to collate strong evidence about to what extent these 'main responsibilities' have been accomplished (Zhao and Wang, 2012).

Higher-level Education Entrance Examination

The third contested term among stakeholders is that of the 'higher-level education entrance examination'. For high-income migrant families, the 'higher-level education entrance examination' equals the GAOKAO. A permit to apply to occupational training schools or higher technical schools will never satisfy them. While lower-income migrant families also have high aspirations for higher education, they lack detailed knowledge about the education system and have little sense of other options for the 'higher-level education entrance examination', including entrances to occupational training schools and higher technical colleges. National-level governments have only promoted educational equality in principle and governments at national level did not launch specific policies to coordinate across country; for example, a possible step would be to make the GAOKAO test comparable and transparent across the different provinces.

Clearly various stakeholders have different understandings of the key concepts of the policy, which shapes the uncertainty and complexity of its outcomes. Beyond these issues, another driving force in the construction of these concepts can be gleaned from several socio-economic and cultural tradition factors.

Educating for equality: domestic migrants in China 311

THE SOCIAL AND CULTURAL FACTORS EMBODIED IN THE AREA C CASE STUDY

The vague definition of the three key words in the policies can also be understood within the specific social and cultural context in China. First, financial constraints are an obvious reason for the governments' inability to arrange the necessary resources to set up quality schools within a short timeframe. Beyond that, the tradition of the strong centralized political system in China can be revealed under the previously discussed education policy changes. Governments of lower administration levels feel under great pressure to respond to the demands of the higher-level government. However, with limited resources, local governments needed to find alternative ways of implementing the particular requirements. To some extent, this explains why there are different definitions of 'migrant children' and 'main responsibility' among different levels of government. People are interpreting and manipulating the definition from their own standpoint. At the same time, the tradition of imperial examinations shapes Chinese people's strong educational desire to take the GAOKAO. All these factors mean that when local governments are focusing on compulsory education for migrant children, the pressure for greater access to the GAOKAO examination is already well-established and immediately forces local governments to take additional steps to promote educational equality. Finally, the special way of Chinese thinking also contributes to this complexity. Chinese are good at thinking of and seeing problems and changes as a process, which overlaps with some aspect of complexity theory (Chen, 2006). This cultural recognition of process and complexity shapes both how the Chinese government, through its various levels, carries out and implements its policies and how the Chinese societies respond to them. As complexity thinking would emphasize, it is a continually evolving multi-level process.

CONCLUSION

The core of complexity theory argues that societies are not fundamentally simple and orderly systems that can be directed by monolithic hierarchical structures, nor are societies wholly unpredictable, unknowable and ungovernable. Instead, the reality of everyday life is a combination of both order and disorder that results in bounded, but complex outcomes (Geyer and Rihani, 2010: 13–35). This chapter has tried to illustrate the complexity of the recent compulsory education policies for the ever-growing population of domestic migrant children in Area C in Beijing. The 'Two Main Responsibilities' policy from national-level government aimed to set a new order to deal with compulsory educational issues for the migrant children across China. From the Area C case, evidence suggests that the governments of cities with a high migrant population did accept these central demands and actively designated substantial financial and policy responses to tackle the issues. However, the different stakeholders involved had different interpretations of the key concepts, such as migrant children, main responsibility and the higher-level education entrance examination. These multiple understandings and blurred concepts led to a process of negotiation among the groups and various levels of government that is still ongoing. Though messy and uncertain, this openness and continual involvement of stakeholders is the best way to achieve the delicate policy balance

312 *Handbook on complexity and public policy*

between general central government policies and the needs of diverse local actors and stakeholders. From a complexity perspective, this keeps the policy in a zone of 'creative complexity', which avoids stifling order or destructive disorder. This zone does not guarantee success and is uncertain and difficult. However, as argued by Geyer and Rihani (2010: 56), in the long run it is a most successful strategy for policy action.

ACKNOWLEDGEMENT

This research was funded by Beijing Philosophy and Social Science Planning Project (code: 12SHB008): 'The study on the social integration of domestic migrants children in Beijing'.

NOTES

1. Although it is illegal to hire employees under the age of 16, hiring workers under this age happens in practice in the labor market in China.
2. There is a debate on the time of the emergence of imperial examinations in China among scholars. Here we draw on Tian's point to illustrate the long history of it.
3. In this paper it was reported that Cui Yongyuan, a popular talk-show celebrity in China, with another five scholars from the Chinese Academy of Social Sciences, Tsinghua University and Renmin University of China, wrote a joint letter to the Minister of Education to ask that Tongxin, an illegal private school for migrant children, be kept open. They stated three reasons for this. First, according to the Convention on the Rights of the Child, the benefit of migrant children in the community should be considered as the priority; secondly, Tongxin School was the reason that the community formed there; and third, that Tongxin School was established with the active participation of the migrant youth and should therefore be kept open for the aspirations of the migrant youth.

REFERENCES

Beijing Government (2012), 'The protocol for migrant children applying for higher education post their compulsory education in Beijing', Education Bureau, Reform and Development Bureau, Human Resource Bureau of Beijing Government, 30th December, accessed 1 January 2013 at http://report.qianlong.com/333 78/2012/12/30/2502@8413723.htm.
Bousquet, A. and R. Geyer (2011), 'Introduction: complexity and the international arena', *Cambridge Review of International Affairs*, **24**(1), 1–3.
Cairney, P. (2012), 'Complexity theory in political science and public policy', *Political Studies Review*, **10**(3), 346–58.
Chen, H. (2006), 'Chinese traditional science paradigm and complexity science', *Academic Forum*, **7**, 18–21.
Chen, Hongbing (2006), 'The consistence between Chinese traditional science paradigm and complexity theory', *Academic Forum*, **168**(7).
Chen, W. (2008), 'Complexity theory and education in China', Proceedings of the 2008 Complexity Science and Educational Research Conference, 3–5 February, University of Georgia.
Foo, C. (2005), 'Three kingdoms, sense making and complexity theory', *Emergence*, **7**(3–4), 85–94.
General Office of State Council (2012), 'The General Office of State Council forward the requirements from the Ministry of Education calling Local Governments arranging migrant children receiving education after the compulsory education and attending higher', 31 August, accessed 2 September 2012 at http://www.gov. cn/zwgk/2012-08/31/content_2214566.htm.
Geyer, R. (2012), 'Can complexity move UK policy beyond "evidence-based policy making" and the "audit culture"? Applying a "complexity cascade" to education and health policy', *Political Studies*, **60**(1), 20–43.
Geyer, R. and S. Rihani (2010), *Complexity and Public Policy: A New Approach to 21st Century Politics, Policy and Society*, London: Routledge.

Educating for equality: domestic migrants in China   313

Lv, S.Q. and S.Y. Zhang (2010), 'The education facing the difference between cities and rural area: an investigation about private schools for migrant children in Beijing,' *Strategy and Management*, **4**, 95–109.

Ministry of Education (2013), 'Brief background of education in China', 23 October, accessed 20 January 2014 at http://www.moe.gov.cn/publicfiles/business/htmlfiles/moe/s5990/201111/126550.html.

National State Council (2001), 'The state council on the decision of the elementary education reform and development', 29 May, accessed 2 September 2013 at http://www.moe.gov.cn/publicfiles/business/htmlfiles/moe/moe_719/200407/2477.html.

OECD Global Science Forum (2009), 'Applications of complexity science for public policy: new tools for finding unanticipated consequences and unrealised opportunities', September, available at http://www.oecd.org/science/sci-tech/43891980.pdf.

Powers, M. (2007), *Organized Uncertainty: Designing a World of Risk Management*, Oxford: Oxford University Press.

Sanderson, I. (2000), 'Evaluation in a complex policy world', *Evaluation*, **6**(4), 433–54.

Shao, Shulong (2010), 'Social stratification and the education for children of migrants: sociological analysis on the games of two main policies in education', *Education Development Research*, **4**, 6–12.

Stacey, R., D. Griffin and P. Shaw (2000), *Complexity and Management: Fad or Radical Challenge to Systems Thinking?*, London: Routledge.

The Education Committee of Area C (2012), 20 March, accessed 4 April 2013 at http://www.bjchyedu.cn/zwgk/mbjy/201210/t20121017_141513.html.

The Education Working Committee of Area C (2010), 'A report on migrants' children receiving compulsory education in Area C', accessed 4 April 2013 at http://www.bjchyedu.cn/zwgk/mbjy/201010/t20101022_101126.html.

The Standing Committee of the Tenth National Congress (2006), 'The compulsory-education law of the People's Republic of China', 29 June, accessed 5 August 2013 at http://www.moe.gov.cn/publicfiles/business/htmlfiles/moe/moe_619/200606/15687.html.

Tian, Jianrong (2009), *The Tradition and Movement of Imperial Examination in China*, Beijing: Education Science Press.

Wu, X.S. (2010), 'A report on city inspection and management in Area C', 8 June, available at http://chyrd.bjchy.gov.cn/rdhy/zrhyeshq/8a24f09529209e61012991ca94c40046.html.

Xian, Lianping (2011), 'The education committee responding that no children will drop from school', accessed 17 August 2011 at http://www.edu.cn/zong_he_news_465/20110817/t20110817_667118.shtml.

Zhang, Y.F. and J.W. Hou (2009), 'The evolution of Peking's floating population's certified administration and its empirical study,' *South China Population*, **24**(3), 35–43.

Zhao, Gehong and Hong Wang (2012), 'The policies analysis on the funds of migrant labor's children's compulsory education', *The New Orient*, **188**(1), 73–7.

Zhou, Jin (2009), 'The urban planning and population development in Beijing since the establishment of P.R. China', June, accessed 15 September 2013 at http://zhengwu.beijing.gov.cn/zwzt/jd90/xzlt/t1168488.htm.

19. The emergence of intermediary organizations: a network-based approach to the design of innovation policies
Annalisa Caloffi, Federica Rossi and Margherita Russo

POLICIES FOR INNOVATION NETWORKS AND THE EMERGENCE OF INTERMEDIARY ORGANIZATIONS

The importance of networking among heterogeneous organizations as a source of innovation is increasingly acknowledged within the scientific community. Some contributions (Nooteboom, 2000; Powell and Grodal, 2005) stress that the creative recombination of heterogeneous knowledge is an important source of innovation; others (Lane and Maxfield, 1997; Russo, 2000; Lane, 2009) focus on generative relationships characterized by heterogeneous competences, mutual and aligned directedness in contexts of joint action, as drivers of innovation processes; while yet others (Spence, 1984; Katz, 1986) suggest that networks foster innovation through the production and internalization of spillovers within the group of participants. In line with this growing consensus, policymakers increasingly implement interventions in support of networks among either small and large firms, or firms and universities, explicitly aimed at promoting innovation through joint R&D, knowledge transfer or technology diffusion. Nonetheless, our understanding of what network configurations most contribute to innovation, or indeed whether networks lead to innovation, and precisely how they do so, is still limited (Cunningham and Ramlogan, 2012).

Greater understanding of what factors support the formation of innovation networks and their successful performance would help policymakers improve the design of policy interventions. In their recent review of the literature on the effectiveness of innovation policies, Cunningham and Ramlogan (2012) propose several elements that contribute to the success of innovation networks: strong network management and leadership, coupled with transparent and efficient administrative processes; established connections and relationships, which can drive the formation of new networks; the ability of network participants to actively manage their relationships, which often depends upon prior experience and network management competencies. The networks' objectives are more easily achieved when the policy instruments facilitate network formation and development, for example by supporting various types of intermediary organizations that create ties across different organizations.

Several studies have acknowledged that the success of policies in support of innovation networks depends on the involvement of intermediary organizations (Bessant and Rush, 1995; Hargadon and Sutton, 1997; Cantner et al., 2011; Kauffeld-Monz and Fritsch, 2013): their role suggests that the production of knowledge spillovers is not necessarily a spontaneous process, nor is their absorption automatic. This is particularly true for policies aimed at micro and small firms: here, the presence of intermediaries may facilitate the

Intermediary organizations: a network-based approach to innovation 315

exchange of knowledge and competencies among agents (such as small firms, large firms and universities) that have different languages, organizational cultures, decision-making horizons, systems of incentives and objectives (Howells, 2006; Russo and Rossi, 2009).

Therefore, improving our understanding of which organizations are more likely to play intermediary roles in the context of innovation networks carries useful policy implications. Such knowledge can be used to identify the most appropriate organizations to involve in the policy interventions, and to set up more successful networks by promoting collaborations with the most appropriate intermediaries.

But identifying who network intermediaries are is not straightforward, as the identity of the organizations that play this role is likely to vary according to the network's characteristics. Indeed, intermediaries are usually best identified on the basis of their behaviour in the network, rather than a priori on the basis of their 'mission' or economic activity. In this chapter we aim to identify intermediaries according to their relational positioning within networks of relationships, as an emergent result of the involvement of organizations in multiple networks. We adopt therefore a complexity approach to understanding social organization, according to which micro-level interactions among individual agents give rise to emergent meso-level structures whose behaviour in turn influences the actions of individuals by providing constraints and opportunities for action (Dopfer et al., 2004; Lane, 2009).

In particular, we use Social Network Analysis (SNA) to map the micro-level interactions and detect the meso-level structures (such as network communities) whose presence affects the behaviour of individuals. Complexity approaches to the study of social organizations and SNA do not coincide, nor do they imply one another. However, all studies that use SNA must epistemologically recognize the role played by the structure of inter-individual interactions in constraining individual behaviour. When used in the analysis of complex social systems, SNA can provide useful tools to empirically identify higher-order structures emerging from the micro-level interactions among individual agents.

We claim, as described in greater detail in the following sections, that the meso-level network structure affords agents the opportunity to act as intermediaries, either by bridging structural holes in the network or by connecting different network communities. To identify these 'bridging organizations', or intermediaries, we experiment with two different measures developed in SNA: the brokerage index and a measure of inter-cohesion, respectively. We argue that the different measures identify different types of intermediary positions, which are linked to different ways to manage flows of knowledge within the network and hence to different roles in the innovation process.

The chapter is structured as follows. In the second section we review some of the literature on innovation intermediaries and we discuss how knowledge flows within networks, introducing two network measures that can be used to identify organizations that play intermediary roles. We also discuss how different measures may identify intermediaries that perform different functions in the network. In the third section we describe some of the main features of our dataset and our empirical strategy. In section 4, we present our empirical results, aimed at identifying different types of intermediaries and analysing their features. In section 5, we derive some conclusions and implications for policy.

316   *Handbook on complexity and public policy*

IDENTIFYING INTERMEDIARIES IN INNOVATION NETWORKS

The Features of Intermediaries

In recent years, numerous strands of research[1] have highlighted the important role played in innovation processes by organizations that facilitate connections between other organizations that are engaged in the invention, development and production of new products, processes and services. These connecting organizations have been identified with a variety of terms, such as 'intermediaries', 'knowledge (or technology, or innovation) brokers', 'bricoleurs', 'boundary organizations', 'superstructure organizations', 'innovation bridges', and others (Howells, 2006). Intermediaries enable the formation of appropriate partnerships, facilitate the realization of innovation projects and even support the appropriate dissemination and implementation of their results.

In the context of policies for innovation networks, greater awareness of which organizations can act as intermediaries, and of the roles of different intermediaries, could help policymakers design more targeted interventions, and potential beneficiaries to set up more successful networks. Within a comprehensive review of the literature on innovation intermediaries, Howells (2006) suggests that – while it is not yet possible to identify a specific body of literature on innovation intermediaries (exploring their functioning, organization, performance and theoretical rationales) – a consensus has emerged around a few general themes.

First, the functions of innovation intermediaries extend beyond the roles of 'information clearinghouses' and 'matchmakers' between potential collaborators. Intermediaries often engage in long-term collaborations with other organizations, which sometimes lead to further innovation processes, to new relationships, and to new services. The functions of intermediaries have therefore been found to be many and diverse, including:

- facilitating relationships between organizations, by identifying potential partners for innovation projects (Shohert and Prevezer, 1996) and helping to compensate firms that have a poor advice network and lack connections to socially distant organizations (McEvily and Zaheer, 1999);
- acting as 'superstructure' organizations that provide collective goods to their members and facilitate and coordinate information flows between them (Lynn et al., 1996; Russo and Whitford, 2009);
- supporting innovation processes by helping package the technology to be transferred between firms (Watkins and Horley, 1986), selecting suppliers to make components for the technology (Watkins and Horley, 1986), adapting technological solutions available on the market to the needs of individual users (Stankiewicz, 1995), and acting as knowledge repositories, able to provide solutions that are new combinations of existing ideas (Hargadon and Sutton, 1997).

Second, intermediaries do not always operate on a simple one-to-one basis, but they are increasingly involved in more complex relationships, such as many-to-one-to-one, one-to-one-to-many, many-to-one-to-many, or even many-to-many-to-many collaborations. For example, Provan and Human (1999) contrast two different examples of

Intermediary organizations: a network-based approach to innovation 317

innovation intermediaries, one of which engaged in one-to-one relationships with the members of its networks, while the other engaged in many-to-many interactions, being primarily involved in stimulating collective discussions and interactions among network members. Similarly, Russo and Whitford (2009) describe how both types of relational behaviours were adopted by the same intermediary organization, which provided both one-to-one services to its members as well as opportunities for simultaneous interactions between several members (these different activities are described by the authors respectively as 'switch' and 'space' functions). Although these functions were provided by the same organization, they involved different parts of the organization and responded to different revenue generation models.[2]

However, our understanding of which organizations are best suited to play these roles in innovation networks, and whether there are different types of intermediaries each with different roles and specificities, is still limited. When analysing organizations that play an intermediary role in innovation processes, one important question is how to identify them. The literature often conflates intermediaries with service providers, because providing services is one of the most important tasks of intermediaries (Shohert and Prevezer, 1996). However, this is not always the case. For example, it is well known from studies of industrial clusters that the role of 'knowledge gatekeepers' (Allen, 1977), absorbing external knowledge, translating it and transmitting it to other organizations within the cluster – this way performing an important intermediary function in the cluster's innovation processes – is very often played by large leading firms (Morrison, 2008). Moreover, the organizations that perform intermediary functions may be different in different economic sectors, in different areas, or even in different innovation networks. Therefore, it is not possible to simply identify certain types of organizations that should 'naturally' play the role of intermediaries based, for example, on their stated economic activity or mission. A more exploratory approach is needed, with the objective to detect what the organizations are that actually mediate relationships between other organizations in a specific context. In this chapter, we exploit information about relationships within an innovation network, in order to identify, using methodologies based on SNA, who are the actors that occupy positions in the network that allow them to mediate between other actors, or groups of actors: we assume that the position of an actor within a network of relationships is likely to influence the functions that it performs. This is not an unrealistic assumption given that numerous studies have demonstrated that an organization's network position affects its opportunities for shared learning, knowledge transfer and information exchange (Burt, 1992; Provan and Human, 1999; Nooteboom et al., 2005) and hence its success in developing innovations (Nooteboom, 2000; Tsai, 2001; Graf and Krüger, 2011).

SNA provides a powerful analytical tool to highlight features of innovation networks in general, and of intermediaries within such networks in particular, and it has flourished in the last few years (Fritsch and Kauffeld-Monz, 2010; Gilsing et al., 2008). This approach requires the analyst to possess precise and comprehensive data about the relationships among the participants in innovation processes so as to be able to construct a fairly reliable network of the relationships between them. However, the analysis focuses on the presence of relationships and not on the quality and nature of these relationships. This prevents us from developing rich analyses of how relationships evolve, carrying different functionalities and leading the participants to explore new directions. These

318   *Handbook on complexity and public policy*

aspects have been investigated through different methodologies such as ethnography and case study research (see for example Hargadon and Sutton, 1997; Morrison, 2008; Parolin, 2010).

Brokers and Intercohesive Agents

An organization's positioning within a network of relationships strongly affects its ability to manage and control communications within that network, and consequently the exchange of information and knowledge among network participants. How such inter-organizational knowledge flows are structured shapes the way in which production and innovation processes are distributed between organizations. Consequently, there is a link between an organization's positioning within the network and the extent and nature of its contribution to innovation processes.

In his seminal contribution, Burt (1992) suggested that a node that spans a structural hole in a network – that is, a node that creates a bridge between two otherwise non-connected parts of that network – enjoys the opportunity to broker the flow of information between people, and to control the projects that bring together people from opposite sides of the hole. People on either side of the structural hole have access to different flows of information; hence, the node that bridges the structural hole (an actor that can be termed a 'broker') creates a connection that allows the transmission of non-redundant knowledge between the two sides. According to Burt (1992), a broker can enjoy numerous benefits from its network position. First, the broker can learn early on about activities in different groups, therefore it can spread new ideas and behaviours; second, as it has many diverse contacts, it is more likely to be included in the discussion of new opportunities; third, the more diverse its range of contacts, the more it is attractive for other actors who want to become part of its network; finally, it has the opportunity to control the flow of communication between other actors and hence can exploit this power to its own advantage.[3]

According to McEvily and Zaheer (1999), actors that occupy a broker position – that is, actors that are at the centre of a network of non-overlapping ties – can exploit access to non-redundant knowledge which allows them to build better competitive capabilities. Compared with actors who are only connected to one side of the structural hole, brokers have better knowledge of their environment and can gain access to a wider spectrum of information and knowledge. The more cohesive a network is (that is, the greater the connectedness among nodes), the fewer the structural holes and the fewer the opportunities for brokerage. Combining the analysis on agents' position in a network and agents' basic features, Gould and Fernandez (1989) propose a finer distinction between different types of brokerage positions according to the nature of the actors that are connected through the broker.

The identification of intermediary positions in a network may also come as a result of analyses aimed at discovering the emergence of meaningful communities, that is network sub-groups populated by agents that are more intensively connected to each other than to the rest of the larger network. Stark and Vedres (2009) use the term 'intercohesive nodes' to identify intermediaries that are embedded in different communities at the same time. Differently from the ideal-typical broker, which provides bridges between actors that are not directly connected to each other, intercohesive nodes bridge communities of actors, some of which may be connected to each other. While brokers may mediate between different communities, but do not belong to any of them, intercohesive nodes are insiders to multiple social groups.[4]

Intermediary Positions and Innovative Contexts

Following the line of enquiry that we have developed so far, we explore whether agents that occupy broker or intercohesive positions in a network can play different intermediary roles in distributed innovation processes involving several organizations and can be linked to different forms of learning and innovation. Some studies suggest that different intermediaries may support different learning dynamics within the networks.

In their analysis of the impact of network embeddedness on firm novelty creation and absorption, Gilsing et al. (2008) discuss the relation between agents' position in a network, exploitation or exploration learning dynamics and agents' cognitive distance. Assuming that exploitation processes most often occur when environmental conditions are stable, while exploration processes strongly characterize agents' activity in turbulent contexts, the authors recall that the literature on innovation networks has often shown that networking among similar agents is beneficial for knowledge exploitation processes (Nooteboom, 1999; Nooteboom et al., 2005). In fact, partners that have similar technological knowledge, expertise and beliefs are able to understand each other quickly and easily learn from others. This easy and fast dissemination of information and knowledge is most appropriate for an effective implementation of exploitation processes (Gilsing, 2005). On the other hand, what matters most for the realization of exploration processes is a certain degree of cognitive distance between the agents: distant agents bring different pieces of knowledge within the network, which can be recombined in new and original ways (Nooteboom et al., 2005).

Taking a structural perspective, and drawing on Burt's (1992) concept of structural holes and brokerage positions, Gilsing et al. (2008) observe that brokers are more likely to be in a good position to engage in knowledge exploration processes. Assuming that agents engage in homophilous behaviour – this is, they tend to form ties with similar others (McPherson et al., 2001) – and, therefore, that similar agents are likely to be part of the same group, the authors argue that brokers, who are connected to different groups of partners, are in a position to recombine different knowledge and then engage in exploration processes.

Intercohesive agents, instead, belong to multiple cohesive subgroups at the same time. By the same homophily assumption, overlapping cohesive subgroups that share many connections are likely to be formed by similar agents. As a consequence, intercohesive agents will play a role in coordinating similar agents. In light of the contribution of Gilsing et al. (2008), this type of agent is more likely to be found in stable contexts, where it can best exert its role in knowledge exploitation processes.

If these arguments hold, we can hypothesize that different intermediaries support the implementation of different types of learning and innovation processes, which take place in different contexts. We expect to find that brokers more often engage in turbulent contexts, where learning processes are mainly explorative; while intercohesive positions are more often occupied by organizations that engage in learning processes that are mainly exploitative, as happens in the case of stable contexts.

320 *Handbook on complexity and public policy*

In the following sections we will perform an exploratory analysis to understand whether brokers and intercohesive agents actually have these characteristics. Our empirical analysis concerns the implementation of a regional policy intervention to support innovation in small firms. The involvement of intermediaries with knowledge diffusion and knowledge recombination capabilities can be particularly important when policies are aimed at small businesses, whose internal knowledge and skills, and ability to participate in innovation projects, are limited.

DATA AND METHODOLOGY

The Dataset

The empirical analysis focuses on a set of innovation networks set up thanks to funds competitively allocated by the Tuscany Region (Italy). In 2002, the regional government launched a set of policy initiatives designed to support joint innovation projects performed by networks of heterogeneous economic actors, with the ultimate objective to promote non-transitory forms of collaboration among the small and medium-sized firms (henceforth: SMEs), the universities and the research centres based in the region.[5]

These policy initiatives were funded through two main European Regional Development Funds (ERDF) funding schemes: the Single Programming Document (SPD) 2000–2006 and the Regional Programmes of Innovative Actions ('Innovazione Tecnologica in Toscana' 2001–2004: hereafter RPIA-ITT-2002, and 'Virtual Enterprises' 2006–2007: hereafter RPIA-VINCI-2006).[6] The programmes were implemented between 2002 and 2008.

Within these funding schemes, the regional government launched nine calls for innovation projects to be realized by networks of cooperating organizations. Some of these programmes allowed agents to participate in only one project per programme, while others allowed multiple participations. In what follows we will consider only the five programmes that admitted simultaneous multiple participations (the RPIA-ITT-2002 programme, and four waves of the SPD programme measure 1.71 implemented in 2004, 2005, 2007 and 2008) because it is precisely within this type of programme that we can see the emergence of intermediaries such as brokers and intercohesive nodes, mediating the relationships among agents participating in different projects.

Overall, 1362 different organizations were involved in these five programmes, submitting 225 project proposals. Out of these, 141 projects (62.7 per cent) were granted funding. In what follows we will consider organizations participating in funded and/or non-funded projects. Their main characteristics are listed in Table 19.1.

The characteristics of the networks, including the nature of the organizations involved, were often shaped by the tender requirements, which, especially in the early stages of the policy period, imposed numerous constraints on the composition of the admissible networks (Rossi et al., 2013). Some tenders explicitly mandated the involvement of certain types of knowledge-intensive business service providers (KIBS) that should have played the role of intermediaries: namely, innovation centres (usually public or public–private agencies) and private business services providers. The presence of some KIBS (be they public, private or mixed) was advocated in order to introduce some interfaces between

Table 19.1   Participating organizations by type

Type of organization	Participating organizations		Funds received (000€)	
	n.	%	n.	%
Enterprises	860	63.1	12300.0	35.0
Universities & research centres	116	8.5	7316.8	20.8
Private research companies	23	1.7	537.6	1.5
Innovation centres	37	2.7	6191.9	17.6
Private service providers	76	5.6	3784.5	10.8
Associations	97	7.1	2738.0	7.8
Chamber of Commerce	11	0.8	802.2	2.3
Local governments	92	6.8	691.7	2.0
Other public bodies	50	3.7	815.4	2.3
Total	1362	100.0	35178.0	100.0

Note: The table shows the number of organizations that have taken part, both in funded and non-funded project proposals, in the observed policy programmes. Given that these programmes allowed multiple participations, some of the 1362 participants have been involved in more than one project.

the manufacturing SMEs – particularly those operating in low-tech sectors – and private, academic and government research organizations.[7] Besides their interface role, the KIBS (together with a broader set of agents) were also supposed to act as catalysts for knowledge dissemination, and to reach out to SMEs.

The programmes mostly encouraged the implementation of process innovations, and targeted a mix of sectors and technologies. Thanks to the information we have collected in previous analyses and evaluations of these programmes, we have identified two main types of projects: those focusing on technological environments characterized by relative stability, and those focusing on technological environments characterized by a fast rate of change. The former include all those projects involving low and medium-low-tech sectors, or focusing on the diffusion of well-established technologies (53 projects, or 24 per cent), while the latter include projects involving high or medium-high-technology sectors (115 projects, or 51 per cent).[8] The remainder, 57 projects (25 per cent), could not be classified based on the available information.

Empirical Strategy

Our objective is to explore the extent to which some organizations have played intermediary roles in mediating the relationships between other organizations involved in the policy programmes, and to ascertain whether different types of intermediary positions in the programme networks are occupied by organizations with different characteristics and that operate in different technological environments, suggesting that different learning dynamics are taking place.

In order to identify broker and intercohesive positions and to examine their characteristics, we have analysed the set of participants and projects by means of SNA.

First, we have considered the set of relationships activated in each of the observed programmes and we have constructed five two-mode networks (one for each programme),

322   *Handbook on complexity and public policy*

where each organization is connected to the project(s) in which it participates. Second, the five two-mode networks have been transformed into as many one-mode undirected networks in which the participating organizations are connected to each other through co-membership in innovation projects. The organizations participating in more than one project create connections between the other organizations participating in these projects. Then, we have focused on two different types of intermediary positions, as identified by means of two different SNA measures: brokers and intercohesive nodes.

In SNA terms, a broker is a 'go-between' for pairs of other agents that are not connected directly to one another. If A, B and C are three agents and A and C are not linked without the intermediation of B, B is a broker. For the analysis of brokers, we refer to the normalized brokerage index that is implemented in the Ucinet software (Borgatti et al., 2002). Considering a node's immediate neighbourhood (all nodes to which the node is directly connected), the normalized brokerage index is the ratio between the pairs of nodes that are not connected to each other, and the overall number of pairs of nodes in that node's neighbourhood. An agent is considered as a broker when its normalized brokerage index is greater than zero.

Given the structure of our data, where agents are connected through their co-participation in the same innovative projects, we note that brokers are agents that participate in more than one project at the same time (in the same programme), and hence bridge different project partnerships. Hence the interesting policy question is whether this type of agent can be an important vehicle for the exchange and the absorption of knowledge among different innovation projects.

In order to detect intercohesive nodes, we have used the clique percolation algorithm developed by Palla et al. (2005), included in the CFinder software, which aims to find meaningful network subgroups. The algorithm identifies communities as groups of adjacent k-cliques (where a k-clique is a set of nodes, each of which is connected to at least other k nodes): two k-cliques are adjacent if they have $k - 1$ vertices in common. The idea underlying such communities is that, for a social group to be cohesive, it is not necessary for all members of the group to interact with all others (as in a k-clique) but there can be cohesion even if some actors interact with only $k - 1$ others. We have identified all the communities in the network that are formed as groups of adjacent k-cliques, with k varying depending on the policy programme considered.[9] An intercohesive agent is then identified as an agent that belongs to two or more communities. Even intercohesive agents perform a bridging role among different projects. In fact, the communities they connect are formed by groups of agents that are more intensely connected among each other than with the rest of the network (in our case, the network is the policy programme).

Table 19.2 shows the distribution of nodes according to whether they have positive normalized brokerage index (they are brokers) and whether they belong to more than one community (they are intercohesive), in at least one of the five programmes considered. In particular: 69 nodes are brokers in at least one of the five programmes, but never intercohesive (see the column 'Pure brokers (B)' in Table 19.2) and 197 nodes are intercohesive in at least one of the five programmes, (see the column 'Intercohesive agents (I)' in Table 19.2).[10]

After we have identified brokers and intercohesive agents, we have defined a number of variables illustrating their features.

Intermediary organizations: a network-based approach to innovation 323

Table 19.2   Number of brokers and intercohesive nodes by type of agent

Nature	Brokers (B)	Intercohesive agents (I)	Total intermediaries (B + I)	Total agents
Enterprises	28	61	89	860
Universities & research centres	3	30	33	116
Private research companies	2	5	7	23
Service centres	3	16	19	37
Service providers	8	13	21	76
Associations	11	27	38	97
Chamber of Commerce	0	9	9	11
Local governments	12	28	40	92
Other public bodies	2	8	10	50
Total	69	197	266	1362

Note: Brokers and intercohesive nodes are calculated on the basis of the individual programme. 'Brokers' include agents that are brokers in at least one of the observed programmes, but never intercohesive, while 'Intercohesive agents' belong to more than one community at the same time in at least one of the five programmes. The third column reports the sum of Brokers and Intercohesive agents, which is the total number of intermediaries. The last column reports the total number of agents involved in the five observed programmes.

In order to describe the type of technological environment in which the agent is embedded, we have used the information on the technological field of the projects in which the agent participates (see the previous section). Then we have defined the variable *turbo_pct* measuring the share of the agent's projects which focus on technological environments characterized by a fast rate of change, and the variable *stable_pct* measuring the share of the agent's projects which focus on relatively stable technological environments. This variable, as well as the others we have defined, are described in Table 19.3.

We have tried to specify a model that helps us to account for the possible sample selection bias due to the fact that intermediaries in general (either brokers or intercohesive nodes) could have a number of features that distinguish them from the whole population;[11] this needs to be accounted for when modelling an agent's likelihood to be a broker as opposed to an intercohesive node, so as to remove the influence of characteristics that are typical of intermediaries in general, rather than of specific types of intermediaries. Therefore, in order to identify the brokers' characteristics, we estimate a probit model with sample selection (Heckman two-stage probit) on the 1362 agents. In the first stage, we estimate the probability that an agent is an intermediary (*either* broker or intercohesive) or not *(neither* broker nor intercohesive), using 1362 observations. In the second stage we estimate the probability that an agent is a broker or an intercohesive agent, using 266 observations. The analysis is very exploratory, and focuses on the behaviour and on the characteristics of agents within the policy programmes.[12] In the main equation, we seek to determine what is the probability that an agent is a broker (rather than an intercohesive node) given a set of characteristics of the agent (nature) and of the projects (be they funded or not) in which it is involved (turbulent or stable technological

324   *Handbook on complexity and public policy*

Table 19.3   Basic descriptive statistics of agents' characteristics

Variable	Description	Total population N obs=1362		Intermediaries N obs=266	
		Mean	St.Dev	Mean	St.Dev
Intermediary	(S) Dependent variable in the selection equation. Dummy variable equal to 1 when the agent is an intermediary (broker or intercohesive) and 0 otherwise	0.195	0.397	1.000	0.000
Broker	(M) Dependent variable in the main equation. Dummy variable equal to 1 when the agent is a broker and 0 when it is an intercohesive node. The agent is a broker when the brokerage index, as calculated by the software Ucinet, is >0, and the agent is not an intercohesive node	0.051	0.219	0.259	0.439
N_projects	(S) Total number of projects participated in by the agent. The variable is calculated based on the 9 policy programmes issued in 2000–06 (see Caloffi et al., 2011), not only on the 5 programmes that admitted multiple participation	2.093	2.491	5.342	4.108
Turbo_pct	Share of projects participated in by the agent, which focused on technological environments characterized by a fast rate of change. The variable is calculated on the total number of projects, funded and not funded, in all programmes in which the agent participated	0.270	0.407	0.223	0.257
Stable_pct	Share of projects participated in by the agent, which focus on relatively stable technological environments. The variable is calculated on the total number of projects, funded and not funded, in all programmes, in which the agent participated	0.510	0.446	0.555	0.315
Share_fin	Share of projects participated in by the agent, which were funded	0.665	0.418	0.661	0.277
Avg_dur	Average duration of the project, expressed in days (non-funded projects had a duration of zero days)	342.977	226.836	441.400	151.142

Intermediary organizations: a network-based approach to innovation 325

Table 19.3 (continued)

Variable	Description	Total population N obs=1362		Intermediaries N obs=266	
		Mean	St.Dev	Mean	St.Dev
Enterprise	Dummy variable equal to 1 when the agent is an enterprise and 0 otherwise	0.631	0.483	0.335	0.473
University – research centre	Dummy variable equal to 1 when the agent is a university or research centre and 0 otherwise	0.085	0.279	0.124	0.330
Private research company	Dummy variable equal to 1 when the agent is a private research company and 0 otherwise	0.017	0.129	0.026	0.160
Innovation centre	Dummy variable equal to 1 when the agent is an innovation centre and 0 otherwise	0.027	0.163	0.071	0.258
Private service provider	Dummy variable equal to 1 when the agent is a private service provider and 0 otherwise	0.056	0.230	0.079	0.270
Association	Dummy variable equal to 1 when the agent is an association and 0 otherwise	0.071	0.257	0.143	0.351
Chamber of Commerce	Dummy variable equal to 1 when the agent is a chamber of commerce and 0 otherwise	0.008	0.090	0.034	0.181
Local government	Dummy variable equal to 1 when the agent is a local government and 0 otherwise	0.068	0.251	0.150	0.358
Other	Dummy variable equal to 1 when the agent is a public body or another type of agent not included in the previous classes, and 0 otherwise	0.037	0.188	0.038	0.191
Medium	Dummy variable equal to 1 when the agent is a medium-sized firm and 0 otherwise	0.012	0.108	0.008	0.087
Small	Dummy variable equal to 1 when the agent is a small-sized firm and 0 otherwise	0.051	0.221	0.038	0.191
Micro	Dummy variable equal to 1 when the agent is a micro-sized firm and 0 otherwise	0.181	0.385	0.109	0.312

Notes: (S) identifies the variables that are included in the selection equation only. (M) identifies the variables that are included in the main equation. All the other variables (except for *other*, which we have displayed only for clarity) are included both in the selection and in the main equation.

326 *Handbook on complexity and public policy*

environment, average project duration, share of projects that are funded). In the selection equation, we consider a number of variables that could influence whether the agent becomes an intermediary, namely its nature, the number and technological features of projects in which it participates, and the other features that we have included in the main equation (average project duration, share of projects that are funded).

EMPIRICAL RESULTS

Table 19.4 displays the main results of the Heckman probit model that we have estimated.

Our hypothesis that broker positions are more often occupied by organizations that engage in turbulent contexts and that intercohesive agents are more often found in stable environments, finds support in the data. In fact, as we can see from Table 19.4, the variable *turbo_pct* is positive and (weakly) significant for brokers. Intermediaries in general seem to be more likely than non-intermediaries to be engaged in turbulent contexts, but this depends on brokers in particular. In fact, the increase in probability attributed to a one-unit increase in the variable *turbo_pct* is larger for brokers (0.783) than for intermediaries in general (0.434). Given that the latter are the sum of brokers and intercohesive agents, this means that the contribution from the intercohesive agents is lower than the contribution from the brokers.

As for the agents' nature, we find that intermediaries in general are more likely to be (local) associations and local governments. These agents participated in the observed projects less frequently than others, but when they did, they mobilized their local communities.

Besides associations and local governments, no other agents have a significant probability to be intermediaries. Hence, service providers do not preferentially play an intermediary role between agents involved in different projects. Performing such a role may require the mobilization of technological and scientific knowledge that service providers do not always have, and which may reside instead in organizations like university or industry research centres, private research companies, innovation centres, and firms. It may also require skills such as the ability to interact and communicate with a variety of organizations possessing different cultures and languages (Russo and Rossi, 2009) and to mobilize different professional and social networks, which may be found in organizations like business associations, Chambers of Commerce, local governmental bodies, innovation centres, private service providers, and even firms. Hence, intermediary roles are played by a large variety of agents, varying from project to project.

Intermediaries are also involved in longer projects and less likely to receive public funds. Becoming an intermediary, as well as actually playing an intermediary role, can take some time, in order to get to know the various types of agents, create connections among them and maintaining such connections. Therefore, we can understand why the impact of the variable *avg_dur*, although very small, is positive. Intermediaries are agents participating in many project proposals (even simultaneously): this is why the incidence of failures (projects that are not selected for funding), may be higher than that of other types of actors who play no role as intermediaries.

Intermediary organizations: a network-based approach to innovation 327

Table 19.4 Regression results

Variable	(1) Broker	(2) Intermediary
N_projects		**1.032*****
		(0.068)
Turbo_pct	**0.783+**	**0.434***
	(0.477)	(0.262)
Stable_pct	0.149	**0.441***
	(0.410)	(0.242)
Share_fin	0.145	**−1.089*****
	(0.391)	(0.261)
Avg_dur	0.000	**0.002*****
	(0.001)	(0.000)
Enterprise	0.528	0.385
	(0.478)	(0.337)
University – research centre	−0.502	−0.268
	(0.547)	(0.406)
Private research company	0.696	0.694
	(0.701)	(0.557)
Innovation centre	0.762	0.034
	(0.535)	(0.392)
Private service provider	−0.002	−0.090
	(0.564)	(0.479)
Association	0.503	**0.630***
	(0.494)	(0.380)
Chamber of Commerce	−5.637	0.386
	(19.985)	(0.665)
Local government	0.486	**1.003*****
	(0.492)	(0.373)
Medium	−9.921	−0.216
	(0.000)	(0.181)
Small	−6.422	**−3.837*****
	(34.436)	(0.425)
Micro	**−0.488+**	−0.216
	(0.297)	(0.181)
Constant	**−1.614*****	**−3.837*****
	(0.584)	(0.425)
Athrho		**0.575*****
		(0.189)
Observations	266	1362

Notes:
Number of obs: 1362 in the selection equation and 266 in the main equation.
Standard errors in parentheses. Significance levels: *** $p < 0.01$, ** $p < 0.05$, * $p < 0.1$, +$p < 0.15$.
LR test of indep. eqns. (rho = 0): $chi^2(1) = 11.88$ Prob > $chi^2 = 0.0006$.

328 *Handbook on complexity and public policy*

CONCLUSIONS

Our analysis has tried to identify the peculiar features of intermediaries in the context of innovation policy programmes. Several studies (Bessant and Rush, 1995; Hargadon and Sutton, 1997; Cantner et al., 2011; Kauffeld-Monz and Fritsch, 2013) have acknowledged that intermediary organizations support the formation and successful management of innovation networks. However, their features and the role they play in practice are still under-investigated.

The exploratory analysis presented here takes a step in this direction. Focusing on a set of policy programmes that allowed organizations to participate in more than one project, thus creating bridges between projects, we have tried to identify *ex post* what are the main features of different types of intermediaries based on an analysis of their positions within networks of relationships. We have observed that brokers and intercohesive agents have different features. The former – linking agents that are not connected among each other – are more likely to be found in technologically turbulent environments, while the latter – bridging cohesive communities of network agents – operate in more stable contexts. Drawing on the analysis of Gilsing et al. (2008) we could presume that brokers play a more incisive role than intercohesive agents in knowledge exploration processes, while intercohesive agents are more likely than brokers to engage in knowledge exploitation processes.

Intermediaries in general are more likely to be local governments or local associations. However, besides this, it is not possible to clearly identify organizations that, by nature, are more likely to be either brokers or intercohesive agents: different innovation networks may require different organizations to mediate relationships between the other participants. This finding calls for further research into what types of knowledge and competencies are needed in order to (effectively) facilitate and manage different types of innovation networks.

NOTES

1. Howells (2006) identifies four main sources: '(a) literature on technology transfer and diffusion; (b) more general innovation research on the role and management of such activities and the firms supplying them; (c) the systems of innovation literature; and (d) research into service organizations and more specifically Knowledge Intensive Business Services (KIBS) firms' (Howells, 2006: 716).
2. In particular, 'switch' services were priced on a mark-up–on-cost basis, while 'space' services were included in the annual association fee paid by members. The latter type of services indirectly stimulated the interaction among the members of the organization and enhanced their demand for additional switch services.
3. A classic study on the exploitation of a brokering position to maintain political power is the work by Padgett and Ansell (1993) on Cosimo de Medici. The authors have shown how the main sponsor of the Italian and European Renaissance gained his power thanks to a strategy aimed at creating links between different élites (through business relationships or marriages), and then exploited this position in his favour, playing an intermediary role between different groups.
4. However, it is important to keep in mind that there is no antinomy between the two definitions, since they relate to different aspects. We will come back to this issue in the empirical section.
5. Similar initiatives eliciting the growth of self-organized cooperation networks in research and development have been promoted in other European regions (Eickelpasch and Fritsch, 2005).
6. The empirical research was carried out over an extended time span, starting from 2004, since the authors had participated in the monitoring and analysis of three specific regional programmes implemented

Intermediary organizations: a network-based approach to innovation   329

during this period, namely the RPIA-ITT (see Russo and Rossi, 2009), the RPIA-VINCI and the SPD line 1.7.1, 2005–2006 (see Bellandi and Caloffi, 2010).

7. In some cases the call for tender explicitly required the presence of a minimum number of service centres (a particular kind of KIBS), while in other cases the tender simply responded to the general objective to promote 'networks among enterprises, research centres and universities, innovation centres and other public and private organisations' for innovation and innovation-diffusion purposes.
8. Projects in biotech, geothermal energy, optoelectronics, nanotech, new materials and multiple technologies have been classified as belonging to turbulent environments, while projects in ICT applications to traditional sectors, mechanics and organic chemistry have been included in the group of projects in stable environments.
9. The value of k should be determined on the basis of the peculiar features of the network under observation. Since our networks are made up of projects in which everyone is connected to everyone, it is very likely that the algorithm identifies exactly these groups of agents (projects) as communities. To identify meaningful subgroups that are not a mere duplication of projects, for each programme we have chosen k equal to the size (number of participants) of the smallest project in that programme, minus 1. In this way, we are sure to find subgroups that do not coincide with the projects.
10. When an agent is both broker and intercohesive, we have classified it as intercohesive, because we are interested in observing the differences between a 'typical' broker – which mediates amongst organizations that are not linked – and other types of intermediary positions which are more similar to that represented by the intercohesive agents. All intercohesive agents also have a positive brokerage index; however, in all five programmes, their brokerage index is on average lower than that of the pure brokers.
11. For instance, intermediaries in general could be more likely than the non-intermediaries to operate in turbulent environments or to have a homophilous environment. Therefore, this feature should not be considered as typical of either brokers or intercohesive nodes.
12. Obviously, being an intermediary can be influenced by a number of events happening outside of the policy framework. Therefore, our analysis is partial. However, we take this as a first attempt to identify a number of features that can be typical of the different types of intermediaries.

REFERENCES

Allen, T.J. (1977), *Managing the Flow of Technology: Technology Transfer and the Dissemination of Technological Information within the R&D Organization*, Cambridge, MA: MIT Press.

Bellandi, M. and A. Caloffi (2010), 'An analysis of regional policies promoting networks for innovation', *European Planning Studies*, **18**(1), 67–82.

Bessant, J. and H. Rush (1995), 'Building bridges for innovation: the role of consultants in technology transfer', *Research Policy*, **24**(1), 97–114.

Borgatti, S.P., M.G. Everett and L.C. Freeman (2002), *Ucinet for Windows: Software for Social Network Analysis*, Harvard, MA: Analytic Technologies.

Burt, R.S. (1992), *Structural Holes: The Social Structure of Competition*, Cambridge, MA: Harvard University Press.

Caloffi, A., F. Rossi and M. Russo (2011), 'Promoting successful innovation networks: a methodological contribution to regional policy evaluation and design. The case of Tuscany's innovation policies 2000–2006', *Materiali di Discussione*, Dipartimento di Economia, Universita' di Modena e Reggio Emilia, no. 657, June.

Cantner, U., A. Meder and T. Wolf (2011), 'Success and failure of firms' innovation co-operations: the role of intermediaries and reciprocity', *Papers in Regional Science*, **90**(2), 313–29.

Cunningham, P. and R. Ramlogan (2012), 'Compendium of evidence on the effectiveness of innovation policy intervention: innovation networks', *NESTA and Manchester Institute of Innovation Research*, London and Manchester, available at http://www.innovation-policy.org.uk/share/NESTA_Compendium_Networks_20120516-linked.pdf.

Dopfer, K., J. Foster and J. Potts (2004), 'Micro-meso-macro', *Journal of Evolutionary Economics*, **14**(3), 263–79.

Eickelpasch, A. and M. Fritsch (2005), 'Contests for cooperation: a new approach in German innovation policy', *Research Policy*, **34**(8), 1269–82.

Fritsch, M. and M. Kauffeld-Monz (2010), 'The impact of network structure on knowledge transfer: an application of social network analysis in the context of regional innovation networks', *The Annals of Regional Science*, **44**(1), 21–38.

Gilsing, V.A. (2005), *The Dynamics of Innovation and Interfirm Networks: Exploration, Exploitation and Co-evolution*, Cheltenham, UK and Northampton, MA, USA: Edward Elgar Publishing.

Gilsing, V.A., B. Nooteboom, W. Vanhaverbeke, G.M. Duysters and A. van den Oord (2008), 'Network embeddedness and the exploration of novel technologies: technological distance, betweenness centrality and density', *Research Policy*, **37**(10), 1717–31.

Gould, J. and J. Fernandez (1989), 'Structure of mediation: a formal approach to brokerage in transaction networks', *Sociological Methodology*, **19**, 89–126.

Graf, H. and J.J. Krüger (2011), 'The performance of gatekeepers in innovator networks', *Industry & Innovation*, **18**(1), 69–88.

Hargadon, A. and R. Sutton (1997), 'Technology brokering and innovation in a product development firm', *Administrative Science Quarterly*, **42**(4), 716–49.

Howells, J. (2006), 'Intermediation and the role of intermediaries in innovation', *Research Policy*, **35**(5), 715–28.

Katz, M.L. (1986), 'An analysis of cooperative Research and Development', *RAND Journal of Economics*, **17**(4), 527–43.

Kauffeld-Monz, M. and M. Fritsch (2013), 'Who are the knowledge brokers in regional systems of innovation? A multi-actor network analysis', *Regional Studies*, **47**(5), 669–85.

Lane, D.A. (2009), 'Hierarchy, complexity, society', in D. Lane, D. Pumain, S. van der Leeuw and G. West (eds), *Complexity Perspectives in Innovation and Social Change*, Berlin: Springer, pp. 81–119.

Lane, D.A. and R. Maxfield (1997), 'Foresight complexity and strategy', in W.B. Arthur, S. Durlauf and D.A. Lane (eds), *The Economy as an Evolving Complex System II*, Redwood City, CA: Addison Wesley.

Lynn, L.H., N.M. Reddy and J.D. Aram (1996), 'Linking technology and institutions: the innovation community framework', *Research Policy*, **25**(1), 91–106.

McEvily, B. and A. Zaheer (1999), 'Bridging ties: a source of firm heterogeneity in competitive capabilities', *Strategic Management Journal*, **20**(12), 1133–56.

McPherson, M., L. Smith-Lovin and J.M. Cook (2001), 'Birds of a feather: homophily in social networks', *Annual Review of Sociology*, **27**, 415–44.

Morrison, A. (2008), 'Gatekeepers of knowledge within industrial districts: who they are, how they interact', *Regional Studies*, **42**(July), 817–35.

Nooteboom, B. (1999), *Inter-firm Alliances: Analysis and Design*, London: Routledge.

Nooteboom, B. (2000), 'Learning by interaction: absorptive capacity, cognitive distance and governance', *Journal of Management and Governance*, **4**(1–2), 69–92.

Nooteboom, B., W. Vanhaverbeke, G.M. Duysters, V.A. Gilsing and A. van den Oord (2005), 'Optimal cognitive distance and absorptive capacity', *ECIS working paper* 06-01.

Padgett, J.F. and C.K. Ansell (1993), 'Robust action and the rise of the Medici, 1400–1434', *The American Journal of Sociology*, **98**(6), 1259–319.

Palla, G., I. Derényi, I. Farkas and T. Vicsek (2005), 'Uncovering the overlapping community structure of complex networks in nature and society', *Nature*, **435**(7043), 814–8.

Parolin, L.L. (2010), 'L'innovazione nelle relazioni tra i nodi di un network. Il caso dei fornitori artigiani nell'industria del mobile', *Studi Organizzativi*, 57–76.

Powell, W.W. and S. Grodal (2005), 'Networks of innovators', in J. Fagerberg, D.C. Mowery and R. Nelson (eds), *The Oxford Handbook of Innovation*, Oxford: Oxford University Press, pp. 56–87.

Provan, K.G. and S.E. Human (1999), 'Organizational learning and the role of the network broker in small-firm manufacturing', in A. Grandori (ed.), *Interfirm Networks: Organization and Industrial Competitiveness*, London: Routledge, pp. 185–207.

Rossi, F., A. Caloffi and M. Russo (2013), 'Networked by design: can policy constraints support the development of capabilities for collaborative innovation?', Birkbeck Department of Management, *Working Paper BWPMA* 1305.

Russo, M. (2000), 'Complementary innovations and generative relationships: an ethnographic study', *Economics of Innovation and New Technology*, **9**(6), 517–57.

Russo, M. and F. Rossi (2009), 'Cooperation partnerships and innovation: a complex system perspective to the design, management and evaluation of an EU regional innovation policy programme', *Evaluation*, **15**(1), 75–100.

Russo, M. and J. Whitford (2009), 'Industrial districts in a globalizing world: a model to change, or a model of change?', *WP* 615, Dipartimento di Economia Politica, Università di Modena e Reggio Emilia.

Shohert, S. and M. Prevezer (1996), 'UK biotechnology: institutional linkages, technology transfer and the role of intermediaries', *R&D Management*, **26**(3), 283–98.

Spence, M. (1984), 'Cost reduction, competition and industry performance', *Econometrica*, **52**, 101–21.

Stankiewicz, R. (1995), 'The role of the science and technology infrastructure in the development and diffusion of industrial automation in Sweden', in B. Carlsson (ed.), *Technological Systems and Economic Performance: The Case of Factory Automation*, Dordrecht: Kluwer, pp. 165–210.

Intermediary organizations: a network-based approach to innovation 331

Stark, D. and B. Vedres (2009), 'Opening closure: intercohesion and entrepreneurial dynamics in business groups', *MPIfG Discussion Paper* 09/3.

Tsai, W. (2001), 'Knowledge transfer in intraorganizational networks: effects of network position and absorptive capacity on innovative performance', *Academy of Management Journal*, **44**(5), 996–1004.

Watkins, D. and G. Horley (1986), 'Transferring technology from large to small firms: the role of intermediaries', in T. Webb, T. Quince and D. Watkins (eds), *Small Business Research*, Aldershot: Gower, pp. 215–51.

20. Complexity theory and collaborative crisis governance in Sweden
Daniel Nohrstedt

INTRODUCTION

This chapter considers the relationship between adaptive capacity and the performance of local-level emergency preparedness collaborations in Sweden. Adaptive capacity is a critical resource for communities to prepare for, respond to and recover from various extreme events (Comfort et al., 2010; Norris et al., 2008; Paton and Johnston, 2006). Organizations and networks engaged in crisis planning need to prepare for a range of known and unknown contingencies, which require the development of generic capacities to learn, adjust and change. Emergency preparedness is typically characterized by substantial uncertainty regarding the nature and timing of events, and the viability of established structures and practices for response. In return, managers need to mobilize information and resources from multiple stakeholders in order to increase collective capacities to 'bounce back' and recover from extreme events.

Crisis and emergency management provides a useful case for complexity research. In order to prepare effectively for various contingencies, actors need to cope with at least two sources of complexity related to problems and responses. First, the problems that managers face in this domain cover a range of known and unknown challenges. Most contingencies are routine events but some are complex emergencies, characterized by attributes that are ill-defined and largely unknown (Demchak, 2010; Handmer and Dovers, 2008; Nohrstedt, forthcoming). Second, managers need to continuously monitor organizational issues related to crisis planning, response and recovery. Organizational challenges involve devising structures and practices for coordination, collaboration, communication and leadership. These structures and practices are moving parts that evolve as the result of feedback. Furthermore, crisis management systems are generally embedded in multiple overlapping policy institutions (Lubell, 2013). One of the major challenges of crisis preparedness occurs in the nexus of these two sources of complexity. Ultimately, in order to strengthen capacities to rebound after unanticipated shocks, systems need to be both flexible and adaptive. Achieving this, however, is difficult, not least due to the complexity of governmental systems, which are often resistant to change due to long-term trends that are difficult to resist (Roberts, 2010).

Understanding how systems cope with these sources of complexity is important in order to strengthen capacities to respond to crises. Here loom several research questions that demand attention: how do local-level actors organize across organizational boundaries to collectively prepare for extreme events? What is the ability of these actors to adapt and develop capacities over time? Does adaptation increase the capacity to respond to extreme events? These questions are at the center of this chapter, which focuses on the local level in Sweden.

The study makes two contributions to the literatures on complexity and crisis management. First, the study offers an operationalization of adaptive capacity at the local level. Many studies discuss adaptive capacity at a high level of abstraction and quite frequently as a catch-all explanation for systems' ability to anticipate and respond to extreme events. These studies, however, rarely offer any clear guidance for empirical research (Cutter et al., 2008). Second, the study expands the focus from management processes to outcomes. Although there is considerable interest among policymakers and researchers in the formation and development of inter-organizational collaborations for crisis management, few studies actually document the performance of such systems (Choi and Brower, 2006; McGuire and Silvia, 2010). The study adds to this research by developing an analytical approach to assess outcomes by empirical analysis.

COLLABORATIVE CRISIS GOVERNANCE IN SWEDEN

Collaboration arrangements in themselves are complex systems that consist of interacting autonomous agents that are influenced by local rules, laws and forces (Klijn, 2008). In the Swedish crisis management system, municipalities are responsible for coordinating risk reduction, emergency planning and preparation among authorities and stakeholders at the local level. Since the municipalities have a high degree of autonomy, they have considerable freedom to devise institutional structures and practices for collaboration. As a consequence, these multi-organizational arrangements involve different configurations of actors representing public and private organizations at local, regional and national levels. While some municipalities have adopted arrangements that are consistent with the definition of a 'collaboration' as involving common planning and management by deliberation, other arrangements resemble 'cooperative partnerships' that lack shared decision-making and institutional autonomy (Ansell and Gash, 2008; Gazley, 2010). In the Swedish case as well as in other countries, there are considerable local-level variations in the designs of collaborative arrangements for crisis management. Assessing the effects of these institutional variations is a key research challenge and the topic of this chapter.

In the past decade, several reforms have been made to the Swedish system for crisis management, many of which have aimed at strengthening collaboration across sector and jurisdictional boundaries and levels of authority. In 2006, a new law (SFS, 2006: 544) was enacted that requires local governments to develop crisis planning and to establish crisis committees for coordinating local responses to crisis situations. One guiding idea behind these reforms has been that collaboration is the main instrument to build common capacity for effective action in response to risks and threats (Lindberg and Sundelius, 2013). Meanwhile, so far there has been little discussion about managerial challenges and the potential downsides of increasingly complex patterns of collaboration in this domain. The dominating focus has been to develop institutions that stimulate formation, growth, and development of collaboration. One plausible explanation is that collaboration has been underdeveloped in many municipalities, which has called for continuous policy intervention to foster network growth and expansion. These interventions may have been driven by an overly optimistic perception among policy designers of the benefits of collaborative governance, leaving little room for critical reflection about the

334 *Handbook on complexity and public policy*

challenges encountered at higher levels of networking (cf. Hicklin et al., 2008). Previous research in this domain has identified problems associated with collaboration but these problems have been discussed primarily as barriers to further network integration and collaborative responses to extreme events in specific locations (Palm and Ramsell, 2007; Ödlund, 2010). This chapter takes a broader perspective based on systematic comparisons of adaptive capacity across municipalities.

COMPLEXITY, COLLABORATION AND CRISIS MANAGEMENT CAPACITY

Crises involve remote and unpredictable 'wicked' problems without clearly defined causes, characteristics and solutions, which require managerial responses that cut across different specialized functions, policy sectors, and public and private spheres (Churchman, 1967; Kiefer and Montjoy, 2006). A special type of governance has developed in response to these challenges, characterized by mutual adjustment of organizational policies and procedures towards a common objective and interaction at a low level. The argument holds that wicked problems bring challenges that no agent can handle alone, and therefore which call for multi-organizational collaboration across sectors and levels of government (McGuire, 2006). In fact, disaster research views collaboration as the ideal governance model to strengthen resilience to extreme events but notes that collaboration is difficult in practice. For example, the International Council for Science (2008: 25) notes that 'to facilitate trust and acceptance of decisions reached, governance structures should ideally seek to involve the participation of a wide range of stakeholders. This ideal, however, may often be difficult to achieve in complex environments characterized by inter-group rivalries and with poorly developed institutional frameworks for (e.g. cross-border) negotiation.' To address these difficulties, the Council continues, 'the social sciences will make important contributions to the management perspective and extend into the complexity of the political and social challenges encountered'.

Complexity theory and the literature on collaborative governance offer concepts and assumptions that are a suitable basis to begin unpacking the performance of multi-organizational collaboration in response to wicked problems (Klijn, 2008). These concepts and assumptions enable insight into the implications of complexity for public administration in general and crisis management in particular.

To start with, researchers in complexity research and collaborative studies emphasize the need to shift focus from documenting the formation of collaborations to examining the problem-solving capacity and *outcomes* of multi-governance systems (Duit and Galaz, 2008). That is, even if the complexity of problems and governance systems impose limitations on causal inference, researchers call for more empirical studies to explain performance. According to Shaw (1999: 86, cited in Sanderson, 2000), this involves 'the understanding and explanation of mechanisms operating in a local context, in order that plausible inferences can be drawn regarding other settings, people and interventions that are of interest to policy makers'. Similarly, collaborative governance researchers are increasingly interested in moving from studying the formation and operation of collaboration arrangements to explain outcomes in terms of performance and effectiveness (Kelman et al., 2013).

Complexity theory asserts that changes in social systems occur in a non-linear fashion, which generates variability in system behavior (Duit and Galaz, 2008). In return, system outcomes 'are not precisely predictable or controllable' (Morcöl, 2005: 302). This insight may seem to conflict with the objective to document general explanations of collaborative outcomes; the unpredictability, uncertainty and non-linearity of problems and government systems would simply be inconsistent with an ambition to identify general explanations. Nonetheless, complexity scholars underscore the need for systematic empirical analysis of complex systems to better understand how these are affected by a local context. Complex systems are located in and influenced by local rules, laws and conditions, which in turn mediate the effects of certain events and factors. Understanding the influence of these contextual factors and how they shape outcomes is a legitimate concern for complexity research (Klijn, 2008; Morcöl, 2005). According to Sanderson (2000: 447), understanding the nature and effects of contextual differences calls for 'comparative analyses over time of carefully selected instances of similar policy initiatives implemented in different contextual circumstances'.

The literature offers useful insights to build from in order to study how contextual conditions influence capacities to address wicked problems. One common assumption among complexity theorists, collaborative governance scholars, and adaptive co-management researchers is that problem-solving in the context of complexity is shaped, in parts, by systems' *adaptive capacity*. According to Staber and Sydow (2002: 410), adaptive capacity refers to 'a dynamic process of continuous learning and adjustment that permits ambiguity and complexity'. The assumption – what Duit et al. (2010) coin the 'diversity hypothesis' – holds that the uncertainty of current and future challenges calls for continuous collaboration as a means to mobilize information and knowledge as a basis for action. This insight is supported by crisis management scholars as well. Among others, Comfort (2005) and Wise (2006) suggest that establishing networks of organizations that are committed to continual inquiry, learning and adaptation is a more flexible and robust strategy compared to the standard practice of establishing greater control over risks and threats by administrative structures. Hence, it can be hypothesized that:

H_1: *adaptive capacity has a positive effect on performance*

Hypothesis 1 will be tested using different measures of adaptive capacity and by exploring differences across municipalities. But in addition, the analysis will also test if there is a relationship between the *level of change* in adaptive capacities within units over time and their crisis management performance. Due to the uncertainty associated with crisis events, capacities can be expected to be volatile and susceptible to externally imposed disruption. These characteristics are core attributes of complex adaptive systems. Stated differently, collaborative crisis management can be seen as one example of 'experimenting society' where structures and practices are continuously evaluated and changed in response to feedback (Campbell and Russo, 1999; Oakley, 2000). One can therefore expect that crisis management capacities evolve through trial-and-error; actors are likely to be uncertain about the viability of existing capacities and prone to change given feedback regarding successes and failures (Landau and Stout, 1979; Nohrstedt and Bodin, 2014; van Bueren et al., 2003). Complexity theory assumes that the level of change has implications for the performance of any given system.

336 *Handbook on complexity and public policy*

Adaptation is commonly depicted as a prerequisite for performance in response to complex policy problems. This assumption dominates collaborative governance research, which holds the capacity to adjust to feedback and changing environmental conditions as being a necessary condition for performance (Ostrom, 1990; 2005). While the capacity of networks to innovate and change is generally conceived as an indicator of effectiveness at the network level (Turrini et al., 2010), the effects of networks' adaptive capacity on performance are rarely assessed by empirical analysis. Organizational theory suggests that performance depends on the level of change in any given system over time. Research shows, for example, that organizations that engage in exploitation (refinement of existing structures and practices) are likely to generate predictable returns, while the returns from exploration (experiment with new structures and practices) are much more uncertain (Powell et al., 1996). Meanwhile, March (1991: 71) advocates that 'maintaining an appropriate balance between exploration and exploitation is a primary factor in system survival and prosperity'. This observation is supported by empirical evidence, primarily in financial performance studies (Haveman, 1992; Hoang and Rothaermel, 2010; Lavie and Rosenkopf, 2011; Lichtenhaler, 2009; Macy and Izumi, 1993; Nohria and Gulati, 1996).

These insights correspond to a core assumption within complexity theory, which predicts that the level of 'fitness' of a system (capacity to cope with complex challenges) is related to the level of systemic change (Kauffman, 1995). Going back to March (1991), this implies that effectiveness depends on the ability of systems to strike a balance between minor change (exploration) and major change (exploitation) over time. Systems that only go through minor change are unlikely to improve over time, whereas systems that develop through occasional major leaps and widely different levels of change can perform well at times (reaching so-called 'fitness peaks') but will be highly unstable over time. In return, the optimal level of change is at the intermediate level ('at the edge of chaos') where systems maintain a balance between flexibility and stability (Anderson, 1999; Brown and Eisenhardt, 1998; Weick, 1979). Against this background, this study hypothesizes that, all else being equal:

H_2: *the relationship between changes in adaptive capacity and performance is nonlinear with diminishing returns at higher levels of change*

Hypothesis 2 is tested on the basis of changes in adaptive capacity measures. Thus, while hypothesis 1 focuses on the relationship between adaptive capacity and performance across units, hypothesis 2 adds change in adaptive capacity over time as a plausible explanatory factor. Hypothesis testing is therefore based on difference scores $(Y_t - Y_{t-1})$ as independent variables.

DATA AND MEASUREMENT

Data for this study were collected from a nationwide annual survey administrated by the Swedish Civil Contingencies Agency (MSB) targeting local government public managers in Swedish municipalities. The survey has been conducted by MSB since 2006 as part of a system of mandatory reporting and evaluation established by a 2006 law (SFS, 2006:

Complexity theory and collaborative crisis governance in Sweden 337

544) stipulating the measures that should be taken by municipal and county councils prior to and during extreme events. The survey addressed issues related to civil emergency planning and response management, including organization, planning, risk and vulnerability analysis, collaboration, emergency experience, training and education. The data set for this study includes survey responses from four years (2009–12) with high response rates.[1] Respondents include local-government civil servants with responsibilities for emergency management in Swedish municipalities. The formal positions of the respondents vary across municipalities and include, for example, preparedness coordinators, heads of security and safety, heads of rescue services, and municipality planners. These differences should be kept in mind when interpreting the results of the study.

Survey data has its limitations, however. Self-assessments are generally a source of bias, the most common being the tendency of respondents to consistently overestimate performance (Andrews et al., 2010). Nevertheless, as a source to document outcomes, survey data is a viable alternative to the quantification of emergency management policy (Carreño et al., 2007) or economic indicators of resilience (Rose, 2004), even if the ideal would be to combine multiple indicators (Chen, 2010; McConnell, 2011).

Crisis Management Capacity

Measurement of network outcomes is a widely debated topic in the literature and prior studies offer a range of different empirical approaches and measures. Central issues in this discussion include the level of outcomes (community, network, or organization), and the use of subjective versus objective measures (self-reported capacity vs. real-world impacts). In addition, when dealing with networks in disaster and emergency management, some scholars opt for measures targeting the effectiveness of responses to events as opposed to performance in planning and preparation (McGuire and Silvia, 2010). These different standards make performance measurement a daunting task. In response, some researchers have resorted to measures of self-reported performance, that is, focused on self-perceived capacities (Chen, 2010; Provan and Milward, 2001). In this study, a single measure is used focusing on managers' perceptions of crisis response capacity at the municipality level. Crisis response capacity is included in the survey as one of four predefined performance goals and is coded on a scale from 0 (capacity not achieved) to 2 (capacity fully achieved) with a mean of 1.43.[2]

Adaptive Capacity

To reiterate, adaptive capacity was defined above as 'a dynamic process of continuous learning and adjustment that permits ambiguity and complexity' (Staber and Sydow, 2002: 410). Measurement of adaptive capacity includes multiple empirical indicators of qualities that make any given system more 'error friendly' and conducive to experimentation, improvisation and risk taking (Staber and Sydow, 2002; Yohe and Tol, 2002). According to adaptive governance theory (Innes and Booher, 2010; Keast et al., 2004; Levin, 1998; 2003; Scholz and Stiftel, 2005), three properties are critical in this regard: (1) diversity (inclusion of relevant stakeholders in collaborative arrangements); (2) opportunity for interaction (accessibility of venues for repeated face-to-face dialogue); and (3) methods of selection (capacities and processes to eliminate ineffective

338 *Handbook on complexity and public policy*

strategies and encourage more valued outcomes). These properties provide an operational definition of adaptive capacity for this study.

Diversity is measured by survey responses on the number of reported contacts with other organizations in nine predefined categories (police, county administrative boards, county councils, other municipalities, business, military, religious organizations, voluntary groups, and other organizations). Scores indicate the total number of collaboration partners (range: 0–9) and do not consider organization type or frequency of interaction over time or across units.

Opportunity for interaction is measured using survey responses indicating the number of organizational venues for collaboration exploited by a municipality. Four types of venues are included: local crisis management councils, regional councils, collaboration across municipalities, and other arrangements (open response category). This variable is based on a count on the number of venues used (range 0–4).

Methods of selection are broken down into multiple indicators since effective crisis preparation ideally requires a combination of actions, practices and competencies conducive to learning and adjustment. In addition to diversity and interaction, crisis management research identifies at least five features that are essential to effective crisis planning and preparation: (1) ongoing risk monitoring; (2) continuous education; (3) training through regular exercises; (4) monitoring and adjustment of crisis planning; and (5) learning (Boin and 't Hart, 2010). Risk monitoring is measured by reported annual risk analyses (dichotomy). To measure education, the study includes survey responses about the education of municipality executive boards and staff. Given high correlation (Pearson's $r = .50$) between these variables, a two-item education scale was created (range 0–1). Similarly, training involving different municipality functions (for example political leadership, information officers, crisis management council staff) were combined into a training scale (range 0–1). Reported completion of contingency plans was used as a measure of planning (dichotomy). Finally, learning is measured by difference-scores based on changes in knowledge about risks and vulnerabilities, and range from 0 (no change in knowledge) to 2 (major change).

Controls

As mentioned above, responses to complex policy problems are conditioned by local conditions, which in turn are embedded in regional and national institutions (cf. Lubell, 2013). Prior research indicates that some of these contextual conditions are more important than others in influencing collaborative management and performance. Among these factors are prior hazard experience, population size, fragmentation and political turnover. These factors are included as control variables in this analysis.

First, prior hazard experience can be expected to heighten adaptive capacity in any given municipality over time. This follows from crises as constituting potential triggers for learning and reform (Moynihan, 2008; Birkland, 2006) and the observation that past disaster and emergency experience is positively associated with intergovernmental collaboration (McGuire and Silvia, 2010) as well as with improvements in hazard planning quality (Brody, 2003) and risk reduction policy efforts (Muller and Schulte, 2011). One may thus expect that prior hazard experience influences some of the independent variables (number of partners and venues) included in this study. Prior experience is based on a count of the number of events reported by survey respondents (range 0–11) and includes all types of emergency events. Next, given previous studies suggesting that collaboration is affected by urbanity (Agranoff and McGuire, 2003; McGuire and Silvia, 2010), population size is included as a second control variable.[3] Population size may also be seen as a proxy for administrative capacity, that is the capacity of various organizations to engage in collaboration (Hicklin et al., 2008). While these capacities apply to all organizations (public and private) in any given jurisdiction, the third control variable – fragmentation – is a summary of the number of potential collaboration opportunities with other municipalities and regional government agencies within a county (McGuire and Silvia, 2010). The range for fragmentation (constant within counties) is 1–49. Finally, the analysis controls for political turnover, which refers to changes in the composition of local government. This variable is based on studies linking political turnover to management difficulties in general (McBabe et al., 2008) and in hazards management in particular (Prater and Lindell, 2000). The underlying logic is that a change in government may feed uncertainty among managers, while governmental stability maintains predictability. Turnover is coded as 0 (no change) and 1 (change by election).[4]

DATA ANALYSIS

What are the adaptive capacity factors that influence crisis management capacity and what is the relative importance of each? To test hypothesis 1, which predicts a positive relationship between adaptive capacity and performance, Table 20.1 presents two ordinary least squares (OLS) regression models; one using adaptive capacity variables (Model 1) and one adding the control variables (Model 2). The number of observations drops from 1160 to 822 in both models due to missing data for a total of 338 cases.

The results show that four of the seven coefficients in Model 1 are both statistically significant and correlated with crisis management capacity in the predicted positive direction.[5] The number of collaboration partners, venues, risk monitoring and learning has positive effects on crisis management capacity. The remaining three variables – education, training and contingency planning – did not have statistically significant effects on crisis management capacity. The variance explained by Model 1 is low (8 percent), which is not surprising given that self-perceived crisis management capacities can be attributed to a wide range of factors not accounted for in this Model. Addition of control variables (Model 2) does not change the coefficients or the level of fit. Significant effects from collaboration partners, venues, risk monitoring and learning on crisis management performance seem to hold when controlling for population size, fragmentation, event experience and turnover. These findings give partial support to hypothesis 1, which predicts a positive relationship between adaptive capacity and crisis management capacity.

Turning to hypothesis 2, Table 20.2 presents one linear regression model (Model 1) and two nonlinear regression models (Models 2 and 3) – all using difference-scores as independent variables. Note that these models are tested based on data on the dependent variable in a three-year period (2010–12), which explains why the number of observations drop further in relation to models in Table 20.1.

340 *Handbook on complexity and public policy*

Table 20.1   OLS regression models of adaptive capacity

Independent variables	Model 1		Model 2	
	B	β	B	β
Collaboration partners	.062***	.166	.061***	.164
	(.013)		(.013)	
Collaboration venues	.068***	.128	.062***	.116
	(.018)		(.019)	
Risk monitoring	.221***	.113	.181***	.093
	(.066)		(.066)	
Contingency planning	−.017	−.010	−.032	−.019
	(.057)		(.059)	
Education	−.088	−.054	−.055	−.033
	(.063)		(.064)	
Training	.067	.028	.085	.041
	(.080)		(.080)	
Learning	.191***	.156	.184***	.150
	(.041)		(.041)	
Population size			.000	−.054
			(.000)	
Municipalities per county[a]			.000	−.004
			(.001)	
Event experience			−.010	−.032
			(.012)	
Political regime change (lagged)			.037	.033
			(.038)	
Constant	.712***		.696***	
	(.120)		(.129)	
Adjusted R^2	.08		.09	
Number of municipality years	822		822	

Notes:
OLS regression with unstandardized (B) coefficients and standardized (β) coefficients, and standard errors in parenthesis.
Dummy variables for individual years not reported.
*p < 0.05, ** p < 0.01, ***p < 0.001.
[a] Variable constant within counties.

Only two of the nine coefficients in the linear Model 1 are significant at the .001 level: change in the number of collaboration partners ($p = .002$) and learning ($p = .000$). Both these effects are in the predicted positive direction. The level of fit is low: the combination of independent variables only account for 5 percent in the variance in crisis management capacity, which was expected given that the use of difference scores as independent variables tends to reduce the explained variance (Edwards, 2001). Taken together, these results suggest that the level of change in adaptive capacities over time is a relatively poor predictor of crisis management capacity. At the same time, the results from Model 1 indicate significant and positive effects from change in the number of collaboration partners

Complexity theory and collaborative crisis governance in Sweden 341

Table 20.2 Models of change in adaptive capacity (linear and nonlinear)

Independent variables	Model 1 (linear)		Model 2 (nonlinear – collaboration partners)		Model 3 (nonlinear– learning)	
(Δ2009–12)	B	β	B	β	B	β
Number of collaboration partners	.035*** (.010)	.122	.041*** (.011)	.142	.034*** (.010)	.118
Number of collaboration venues	.013 (.016)	.030	.009 (.016)	.022	.012 (.016)	.028
Risk monitoring	.026 (.051)	.018	.020 (.051)	.014	.022 (.051)	.016
Education	−.085* (.046)	−.070	−.088* (.045)	−.072	−.086* (.046)	−.071
Training	−.054 (.063)	−.033	−.053 (.062)	.032	−.050 (.063)	−.030
Contingency planning	.071 (.047)	.054	.079* (.046)	.061	.072 (.046)	.055
Learning	.188*** (.043)	.156	.344*** (.117)	.285	.349*** (.118)	.289
Collaboration partners squared			−.008*** (.003)	−.103		
Learning squared					−.124 (.084)	−.143
Constant	1.415*** (.026)		1.445*** (.029)		1.409*** (.026)	
Adjusted R²	.05		.06		.05	
F	6.318		6.107		5.806	
Number of municipality years	776		776		776	

Notes:
Linear (Model 1) and nonlinear (Model 2) regression based on difference scores with unstandardized (B) coefficients and standardized (β) coefficients, and standard errors in parenthesis.
Dummy variables for individual years not reported.
*p < 0.05, ** p < 0.01, ***p < 0.001.
Change in Adj R² between Model 1 and Model 2 is statistically significant (p = .006).

and learning on crisis management capacity. Therefore, squared values of collaboration partners (Model 2) and learning (Model 3) are added to test the assumption about diminishing returns (hypothesis 2). A significant positive change in the level of fit between these models and Model 1, and negative significant slopes for the squared terms, would be evidence for nonlinear relationships with diminishing returns (Hicklin et al., 2008).

Results for Model 2 give support for a nonlinear relationship between changes in collaboration partners and crisis management capacity. First, adding the squared term leads to a statistically significant increase (1 percent) in the level of fit between Models 1 and 2. Second, the coefficient for the squared collaboration partner variable is significant

342 *Handbook on complexity and public policy*

and negative, which is evidence for a downward slope. Examination of the marginal effect of changes in the number of collaboration partners on crisis management capacity suggests that the relationship has the shape of an inverted U-curve, with diminishing returns for municipalities with change values above 2. Specifically, a decline in the number of partners is associated with positive effects on crisis management capacity, while an increase in the number of partners above 2 is associated with negative effects. These results lend partial support to hypothesis 2, which predicted a nonlinear relationship between changes in adaptive capacity and performance with diminishing returns at higher levels of change. However, the results for Model 3 do not point in the same direction: the coefficient for the squared learning term is negative but not significant and does not lead to any significant increase in the level of fit compared to Model 1.

In summary, the results reported in Table 20.2 suggest two specifications of the assumption about diminishing returns. First, in this case the nonlinear effect of change in adaptive capacity *holds only for the number of collaboration partners*, while change in the other adaptive capacity variables did not have any statistically significant effects. Hence, changes in the membership structure of collaboration networks appear to be important for performance. Second, the effect of changes in the number of collaboration partners *depends on the scope and direction of change*. Hypothesis 2 predicted diminishing returns at higher levels of change regardless of the direction of change (growth or decline in adaptive capacity). In this regard, the results corroborate the observation by Hicklin et al. (2008) that an increase in networking has positive effects on performance up to a certain point when the effect diminishes or turns negative. What this analysis adds is the insight that changes in adaptive capacity might also be negative – in this case by a decline in the number of partners that managers collaborate with over time. Interestingly, the findings suggest that a decline in the number of partners over time may in fact be good for performance.

CONCLUSION

Complexity theory offers concepts and assumptions regarding the structure, behavior and outcomes of complex social systems. It is argued that complexity theory is a suitable basis to better understand wicked problems such as extreme events (van Bueren, Klijn and Koppenjan, 2003). Complexity theory suggests that effective responses to extreme events depend on resources and capacities that enable systems to adapt, change and learn to cope with uncertainty and surprise (de Bruin et al., 2010). In practice, there is considerable variation between governing systems regarding what solutions they adopt and what measures they take to cope with uncertainty and to increase preparedness. This study takes stock of these variations and investigates how they may affect crisis management capacity.

Assessment of complexity theory as a theoretical basis to better understand – and ultimately cope with – policy problems requires efforts to unpack the notion of complexity and explore its implications by empirical analysis (Cairney, 2012). This study should be seen in that light and poses three empirical questions: how do local-level actors organize across organizational boundaries to collectively prepare for extreme events? What is the ability of these actors to adapt and develop capacities over time? Does adaptation increase the capacity to respond to extreme events?

Regarding the first question, the study confirms that the number of partners and collaboration venues differ between municipalities. These variations were expected, given that the municipalities have considerable freedom to design collaborative arrangements tailored to local conditions. Nevertheless, the evidence opens up new questions regarding the formation and development of interorganizational collaboration at the local level. For example, the study does not provide any answers as to why we see these variations in the first place or why a certain configuration of organizations emerges in any given municipality. One way to better understand network formation would be to systematically examine how managers adapt crisis management capacities to changes in local environments. In response to the second question, the results show that most municipalities adapt over time by changing the set-up of managerial capacities. Most municipalities went through some level of change in collaboration partners, venues and methods of selection. However, these changes were not always in the direction towards more networking; in many cases adaptation was the result of a reduction in the level of collaboration.

To answer the third research question about the relationship between adaptive capacity and crisis management capacity, the study tested two hypotheses derived from complex adaptive systems theory. These suggest that adaptive capacity has a positive effect on performance (hypothesis 1), and that the relationship between changes in adaptive capacity and performance is nonlinear with diminishing returns at higher levels of change (hypothesis 2). Hypothesis testing was conducted using survey data retrieved from the Swedish Civil Contingencies Agency on crisis preparedness work in Swedish municipalities in 2009–12. Both hypotheses are partially supported by the empirical evidence. The study finds a positive relationship between several adaptive capacity variables – including the number of collaboration partners, venues, risk monitoring and learning – and performance in terms of self-reported crisis management capacity. Analysis of the effects of changes in adaptive capacity suggests that changes in the number of collaboration partners (reported collaboration with other organizations) is the only variable associated with crisis management capacity. The results indicate a nonlinear relationship between changes in the number of collaboration partners and crisis management capacity, which lends partial support to hypothesis 2. Meanwhile, this observation only holds for changes in collaboration partners, while changes in the other adaptive capacity factors were not relevant for performance.

While the study confirms that adaptive capacity may contribute to crisis management performance, the findings suggest that some specific capacities are more important than others. These factors include collaboration partners and venues, continuous risk monitoring, and education. However, the analysis does not show why these particular attributes are important in this case or how they influence performance more specifically. One central argument in the literature holds that these attributes provide *resource slack* – reserves of knowledge and pre-established inter-organizational relationships that can be mobilized in response to crisis situations. These resources make social systems more redundant and less vulnerable since information, tasks and relations are distributed across multiple actors (Staber and Sydow, 2002; Comfort et al., 2010).

Future comparative studies should assess in greater detail how different configurations

344 *Handbook on complexity and public policy*

of adaptive systems have responded to extreme events. Such studies would yield important insights regarding the ability to transform preparatory work into more effective response and recovery actions. Focusing on performance in terms of the effectiveness of response and recovery actions would need to go beyond subjective assessment of outcomes (McConnell, 2011; McGuire and Silvia, 2010). A key issue that deserves attention is how investments in adaptive capacity can actually contribute to vigilant coping patterns while avoiding maladaptive actions (Janis and Mann, 1977).

Another issue concerns the extent to which public managers, stakeholders and policy elites can (or even should) actually manipulate adaptive capacity as a basis for improving crisis response capacity. At a glance, it would appear that several adaptive capacity factors highlighted by this study – including collaboration partners and venues, risk monitoring and learning – are relatively easy to manipulate. In this perspective, the results reported here would suggest that performance can be improved by expanding collaborative networks (getting more people involved and finding more venues for interaction), engaging in continuous risk analysis, and by increasing knowledge. However, this view is based on a narrow view of predictability in which uncertainty can be eliminated by planning and control, which stands in stark contrast to the basic premises of complexity theory. Complexity research turns against traditional normative principles about policymaking, which emphasize hierarchical governance, policy planning and mechanical implementation, and proposes an alternative paradigm that sets out from the insight that 'there are no final orders, no happy endings, and no ultimate resting points' (Geyer and Rihani, 2010: 187). Accordingly, policymaking and implementation is characterized by unavoidable uncertainty and unpredictable mistakes, which raise the need for frequent and continuous adaptation at all levels (Rhodes et al., 2011).

The practical advice that follows from this line of reasoning is that policymakers and managers should invest resources to strengthen adaptive capacity but at the same time expand their thinking about constraints imposed by the policy environment (Haynes, 2008; Sanderson, 2000). Concretely, this involves (1) developing a more detailed understanding of the context (actors, institutions, problems) in which any given system is embedded and the history of that context; (2) shifting attention from goals to shared visions; and (3) appointing boundary-spanning individuals that promote more holistic thinking and facilitate interaction among core agents. At the same time, one needs to recognize that political and bureaucratic forces of inertia militate against such changes in mindset (Rhodes et al., 2011). For example, many actors involved in diversified network settings are highly specialized and have little time, capacity and interest to engage in appropriate analysis of the policy setting in which they operate.

NOTES

1. 2009: n = 282 (97 percent), 2010: n = 288 (99 percent), 2011: n = 279 (96 percent), 2012: n = 288 (99 percent).
2. The other goals include planning for risk reduction and emergency response, knowledge about risks and vulnerabilities, and coordination of actors and dissemination of information. The survey operationalizes crisis management capacity based on a number of necessary conditions. To be coded as 'completely fulfilled' (a score of 2), the following four conditions should be met: (1) education and training of local crisis leadership (elected representatives and staff); (2) resources to ensure dissemination to the public and

Complexity theory and collaborative crisis governance in Sweden 345

municipality personnel in times of crisis; (3) a plan for education and training of municipality leaders and administration; (4) technical solutions to ensure power supply, shell protection and information security.
3. Results reported by McGuire (2009) contradict this finding.
4. Changes in local-government coalition by the 2006 elections were coded as 1 for 2009. Similarly, coalition changes by the 2010 election were coded as 1 for 2011.
5. Leaving out non-significant variables from Models 1 and 2 generates little impact on the remaining coefficients or the level of fit. The Durbin–Watson statistic is 1.95 (Model 1) and 1.97 (Model 2), indicating no serial correlation between the residuals. Multicollinearity was checked using variance inflation factors, which are below 1.3 (Model 1) and 2.0 (Model 2), suggesting no or low correlation between independent variables.

REFERENCES

Agranoff, R. and M. McGuire (2003), *Collaborative Public Management: New Strategies for Local Governments*, Washington, DC: Georgetown University Press.

Anderson, P. (1999), 'Complexity Theory and Organization Science', *Organization Science*, **10**(3), 216–32.

Andrews, R., G. Boyne, J. Moon and R. Walker (2010), 'Assessing Organizational Performance: Exploring Differences between External and Internal Measures', *International Public Management Journal*, **13**, 105–29.

Ansell, C. and A. Gash (2008), 'Collaborative Governance in Theory and Practice', *Journal of Public Administration Research and Theory*, **18**, 543–71.

Birkland, T. (2006), *Lessons of Disaster*, Washington, DC: Georgetown University Press.

Boin, A. and P. 't Hart (2010), 'Organising for Effective Emergency Management: Lessons from Research', *Australian Journal of Public Administration*, **69**(4), 357–71.

Brody, S. (2003), 'Are We Learning to Make Better Plans? A Longitudinal Analysis of Plan Quality Associated with Natural Hazards', *Journal of Planning Education and Research*, **23**(2), 191–201.

Brown, S. and K. Eisenhardt (1998), *Competing on the Edge: Strategy as Structured Chaos*, Boston, MA: Harvard Business School.

Cairney, P. (2012), 'Complexity Theory in Political Science and Public Policy', *Political Studies Review*, **10**, 346–58.

Campbell, D. and M. Russo (1999), *Social Experimentation*, Thousand Oaks, CA: Sage Publications.

Carreño, M., O. Cardona and A. Barbat (2007), 'A Disaster Risk Management Performance Index', *Natural Hazards*, **41**, 1–20.

Chen, B. (2010), 'Antecedents or Processes? Determinants of Perceived Effectiveness of Interorganizational Collaborations for Public Service Delivery', *International Public Management Journal*, **13**, 381–407.

Choi, S. and R. Brower (2006), 'When Practice Matters more than Government Plans: A Network Analysis of Local Emergency Management', *Administration and Society*, **37**(6), 651–78.

Churchman, C. (1967), 'Wicked Problems', *Management Science*, **4**, 141–42.

Comfort, L. (2005), 'Risk, Security, and Disaster Management', *Annual Review of Political Science*, **8**, 335–56.

Comfort, L., A. Boin and C. Demchak (2010), *Designing Resilience: Preparing for Extreme Events*, Pittsburgh, PA: University of Pittsburgh Press, pp. 84–105.

Cutter, S., L. Barnes, M. Berry, C. Burton, E. Evans, E. Tate and J. Webb (2008), 'A Place-Based Model for Understanding Community Resilience to Natural Disasters', *Global Environmental Change*, **18**, 598–606.

de Bruin, M., A. Boin and M. van Eeten (2010), 'Resilience: Exploring the Concept and its Meanings', in L. Comfort, A. Boin and C. Demchak (eds), *Designing Resilience: Preparing for Extreme Events*, Pittsburgh, PA: University of Pittsburgh Press, pp. 13–32.

Demchak, C. (2010), 'Lessons from the Military: Surprise, Resilience and the Atrium Model', in L. Comfort, A. Boin and C. Demchak (eds), *Designing Resilience: Preparing for Extreme Events*, Pittsburgh, PA: University of Pittsburgh Press, pp. 62–83.

Duit, A. and V. Galaz (2008), 'Governance and Complexity: Emerging Issues for Governance Theory', *Governance*, **21**(3), 311–35.

Duit, A., V. Galaz, K. Eckerberg and J. Ebbesson (2010), 'Governance, complexity, and resilience', *Global Environmental Change*, **29**(3), 363–8.

Edwards, J. (2001), 'Ten Difference Score Myths', *Organizational Research Methods*, **4**(3), 265–87.

Gazley, B. (2010), 'Linking Collaborative Capacity to Performance Measurement in Government–Nonprofit Partnerships', *Nonprofit and Voluntary Sector Quarterly*, **39**(4), 653–73.

Geyer, R. and S. Rihani (2010), *Complexity and Public Policy: A New Approach to 21st Century Politics, Policy and Society*, Abingdon: Routledge.

Handmer, J. and S. Dovers (2008), *Handbook of Disaster and Emergency Policies and Institutions*, London: Earthscan.

346 *Handbook on complexity and public policy*

Haveman, H. (1992), 'Between a Rock and a Hard Place: Organizational Change and Performance under Conditions of Fundamental Environmental Transformation', *Administrative Science Quarterly*, **37**(1), 48–75.

Haynes, P. (2008), 'Complexity Theory and Evaluation in Public Management', *Public Management Review*, **10**(3), 401–19.

Hicklin, A., L. O'Toole and K. Meier (2008), 'Serpents in the Sand: Managerial Networking and Nonlinear Influences on Organizational Performance', *Journal of Public Administration Research and Theory*, **18**, 253–73.

Hoang, H. and F. Rothaermel (2010), 'Leveraging Internal and External Experience: Exploration, Exploitation and R&D Project Performance', *Strategic Management Journal*, **31**(7), 734–58.

Innes, J. and D. Booher (2010), *Planning with Complexity: An Introduction to Collaborative Rationality for Public Policy*, New York: Routledge.

International Council for Science (2008), '*A Science Plan for Integrated Research on Disaster Risk: Addressing the challenge of natural and human-induced environmental hazards*', available online at http://www.icsu.org/publications/reports-and-reviews/IRDR-science-plan/irdr-science-plan.pdf.

Janis, I. and L. Mann (1977), 'Emergency Decision Making: A Theoretical Analysis of Responses to Disaster Warnings', *Journal of Human Stress*, **3**(2), 35–48.

Kauffman, S. (1995), *At Home in the Universe: The Search for Laws of Self-Organization and Complexity*, New York: Oxford University Press.

Keast, R., M. Mandell, K. Brown and G. Woolcock (2004), 'Network Structures: Working Differently and Changing Expectations', *Public Administration Review*, **64**(3), 363–71.

Kelman, S., S. Hong and I. Herbitt (2013), 'Are There Managerial Practices Associated with the Outcomes of an Interagency Service Delivery Collaboration? Evidence from British Crime and Disorder Reduction Partnerships', *Journal of Public Administration Research and Theory*, **23**(3), 609–30.

Kiefer, J.J. and R.S. Montjoy (2006), 'Incrementalism Before the Storm: Network Performance for the Evacuation of New Orleans', *Public Administration Review*, **66** (special issue), 122–30.

Klijn, E.-H. (2008), 'Complexity Theory and Public Administration: What's New?', *Public Management Review*, **10**(3), 299–317.

Landau, M. and R. Stout (1979), 'To Manage is not to Control: Or the Folly of Type II Errors', *Public Administration Review*, **39**(2), 148–56.

Lavie, D. and L. Rosenkopf (2011), 'Balance Within and Across Domains: The Performance Implications of Exploration and Exploitation in Alliances', *Organization Science*, **22**(6), 1517–38.

Levin, S. (1998), 'Ecosystems and the Biosphere as Complex Adaptive Systems', *Ecosystems*, **1**(5), 431–36.

Levin, S. (2003), 'Complex Adaptive Systems: Exploring the Known, the Unknown and the Unknowable', *Bulletin of the American Mathematical Society*, **40**(1), 3–20.

Lichtenhaler, U. (2009), 'Absorptive Capacity, Environmental Turbulence, and the Complementarity of Organizational Learning Processes', *Academy of Management Journal*, **52**(4), 822–46.

Lindberg, H. and B. Sundelius (2013), 'Whole of Society Disaster Resilience: The Swedish Way', in D. Kamien (ed.), *McGraw-Hill Homeland Security Handbook*, 2nd edn, New York: McGraw-Hill, ch. 53.

Lubell, M. (2013), 'Governing Institutional Complexity: The Ecology of Games Framework', *Policy Studies Journal*, **41**(3), 537–59.

Macy, B.A. and H. Izumi (1993), 'Organizational Change, Design, and Work Innovation: A Meta-analysis of 131 North American Field Studies: 1961–1991', in W. Passmore and R. Woodman (eds), *Research in Organizational Change and Development*, Greenwich, CT: JAI Press, pp. 235–313.

March, J. (1991), 'Exploration and Exploitation in Organizational Learning', *Organization Science*, **2**(1), 71–87.

McBabe, B., R. Feiock, J. Clingermayer and C. Stream (2008), 'Turnover among City Managers: The Role of Political and Economic Change', *Public Administration Review*, **68**(2), 380–86.

McConnell, A. (2011), 'Success? Failure? Something in-between? A Framework for Evaluating Crisis Management', *Policy and Society*, **30**(2), 63–76.

McGuire, M. (2006), 'Collaborative Public Management: Assessing What We Know and How We Know it', *Public Administration Review*, **1**, 33–43.

McGuire, M. (2009), 'The New Professionalism and Collaborative Activity in Local Emergency Management', in R. O'Leary and L. Blomgren Bingham (eds), *The Collaborative Public Manager*, Washington, DC: Georgetown University Press.

McGuire, M. and C. Silvia (2010), 'The Effect of Problem Severity, Managerial and Organizational Capacity, and Agency Structure on Intergovernmental Collaboration: Evidence from Local Emergency Management', *Public Administration Review*, **7**, 279–88.

Morcöl, G. (2005), 'A New Systems Thinking: Implications of the Sciences of Complexity for Public Policy and Administration', *Public Administration Quarterly*, **29**(3/4), 297–320.

Complexity theory and collaborative crisis governance in Sweden 347

Moynihan, D. (2008), 'Learning under Uncertainty: Networks in Crisis Management', *Public Administration Review*, **68**(2), 350–65.

Muller, B. and S. Schulte (2011), 'Governing Wildfire Risks: What Shapes County Hazard Mitigation Programs?', *Journal of Planning Education and Research*, **31**(1), 60–73.

Nohria, N. and R. Gulati (1996), 'Is Slack Good or Bad for Innovation?', *The Academy of Management Journal*, **39**(5), 1245–64.

Nohrstedt, D. (forthcoming), 'Explaining Mobilization and Performance of Collaborations in Routine Emergency Management', *Administration and Society*, early view, 1–29.

Nohrstedt, D. and Ö. Bodin (2014), 'Evolutionary Dynamics of Crisis Preparedness Collaboration: Resources, Turbulence and Network Change in Swedish Municipalities', *Risks, Hazards, and Crisis in Public Policy*, **5**(2), 134–55.

Norris, F., S. Stevens, B. Pfefferbaum, K. Wyche and R. Pfefferbaum (2008), 'Community Resilience as a Metaphor, Theory, Set of Capacities, and Strategy for Disaster Readiness', *American Journal of Community Psychology*, **41**, 127–50.

Oakley, A. (2000), *Experiments in Knowing: Gender and Method in the Social Sciences*, Cambridge: Polity Press.

Ödlund, A. (2010), 'Pulling the Same Way? A Multi-Perspectivist Study of Crisis Cooperation in Government', *Journal of Contingencies and Crisis Management*, **18**, 96–107.

Ostrom, E. (1990), *Governing the Commons: The Evolution of Institutions for Collective Action*, New York: Cambridge University Press.

Ostrom, E. (2005), *Understanding Institutional Diversity*, Princeton, NJ: Princeton University Press.

Palm, J. and E. Ramsell (2007), 'Developing Local Emergency Management by Coordination between Municipalities in Policy Networks: Experiences from Sweden', *Journal of Contingencies and Crisis Management*, **15**, 173–82.

Paton, D. and D. Johnston (2006), *Disaster Resilience: An Integrated Approach*, Springfield, IL, Charles C. Thomas.

Powell, W., K. Koput and L. Smith-Doerr (1996), 'Interorganizational Collaboration and the Locus of Innovation: Networks of Learning in Biotechnology', *Administrative Science Quarterly*, **41**(1), 116–45.

Prater, C. and M. Lindell (2000), 'Politics of Hazard Mitigation', *Natural Hazards Review*, **1**(2), 73–82.

Provan, K. and B. Milward (2001), 'Do Networks Really Work? A Framework for Evaluating Public-Sector Organizational Networks', *Public Administration Review*, **61**, 414–23.

Rhodes, M., J. Murphy, J. Muir and J. Murray (2011), *Public Management and Complexity Theory: Richer Decision-Making in Public Services*, London: Routledge.

Roberts, A. (2010), 'Building Resilience: Macrodynamic Constraints on Governmental Response to Crises', in L. Comfort, A. Boin and C. Demchak (eds), *Designing Resilience: Preparing for Extreme Events*, Pittsburgh, PA: University of Pittsburgh Press, pp. 84–105.

Rose, A. (2004), 'Defining and Measuring Economic Resilience to Disasters', *Disaster Prevention and Management*, **13**, 307–14.

Sanderson, I. (2000), 'Evaluation in Complex Policy Systems', *Evaluation*, **6**(4), 433–54.

Scholz, J. and B. Stiftel (2005), *Adaptive Governance and Water Conflict*, Washington, DC: RFF Press.

SFS (2006), 'Lag om kommuners och landstings åtgärder inför och vid extraordinära händelser i fredstid och höjd beredskap' [Act on municipal and county council measures prior to and during extraordinary events in peacetime and during periods of heightened alert] [In Swedish], 2006:544.

Shaw, I. (1999), *Qualitative Evaluation*, London: Sage.

Staber, U. and B. Sydow (2002), 'Organizational Adaptive Capacity: A Structuration Perspective', *Journal of Management Inquiry*, **11**(4), 408–24.

Turrini, A., D. Christofoli, F. Fosini and G. Nasi (2010), 'Networking Literature about Determinants of Network Effectiveness', *Public Administration*, **88**(2), 528–50.

Van Bueren, E., H.-E. Klijn and J. Koppenjan (2003), 'Dealing with Wicked Problems in Networks: Analyzing an Environmental Debate from a Network Perspective', *Journal of Public Administration Research and Theory*, **13**, 193–212.

Weick, K. (1979), *The Social Psychology of Organizing*, Reading, MA: Addison-Wesley.

Wise, C.R. (2006), 'Organizing for Homeland Security after Katrina: Is Adaptive Management What's Missing?', *Public Administration Review*, **66**(2), 302–18.

Yohe, G. and R. Tol (2002), 'Indicators for Social and Economic Coping Capacity: Moving toward a Working Definition of Adaptive Capacity', *Global Environmental Change*, **12**(1), 25–40.

APPENDIX 1   DESCRIPTIVE STATISTICS

Table A20.1   Descriptive statistics and correlations with crisis management capacity

Variables	r	N	Min.	Max.	Mean	Std Dev.
Crisis management capacity	1.00	1137	0	2	1.43	0.63
Number of collaboration partners	0.19***	1133	0	9	6.64	1.75
Number of collaboration venues	0.15***	1137	0	4	1.62	1.16
Risk monitoring	0.10***	1137	0	1	0.84	0.36
Education	0.05	1137	0	1	0.68	0.40
Training	0.09***	1129	0	1	0.49	0.31
Contingency planning	−0.01	1124	0	1	0.87	0.34
Learning	0.15***	827	0	1	0.32	0.51
Δ Number of collaboration partners	0.13***	833	−8	9	0.16	2.20
Δ Number of collaboration venues	0.07***	834	−4	4	0.29	1.42
Δ Risk monitoring	0.01	869	−1	1	0.08	0.46
Δ Education	−0.08***	813	−1	1	0.03	0.63
Δ Training	−0.03	813	−1	1	0.03	0.57
Δ Contingency planning	0.04	869	−1	1	−0.04	0.49
Δ Learning	0.15***	827	0	1	0.32	0.51
Hazard experience	0.01	1137	0	11	0.88	1.89
Population size	0.02	1160	2421	881235	32581	65380
Fragmentation	0.00	1160	1	49	21.65	14.59
Turnover	0.02	1160	0	1	0.39	0.58

Notes:
*p < 0.05, ** p < 0.01, ***p < 0.001 (two-tailed Pearson's correlation).
Difference scores (Δ) include annual changes in variables ($Y_t - Y_{t-1}$) in 2009–2012.

21. Going for Plan B – conditioning adaptive planning: about urban planning and institutional design in a non-linear, complex world
Gert de Roo

INTRODUCTION: THE URBAN AS A NON-LINEAR PHENOMENON

Cities co-evolve, which means that they do not simply emerge from nowhere or grow linearly. Co-evolution is the non-linear transformation of a city's structures and functions, through which cities change physically, socially and institutionally. Cities are subject to institutional constructs which frame the coherent development of the daily environment, while aiming to create a space for people to live together decently, perhaps even pleasantly. These institutional constructs supposedly mirror a sense of community, either through democratic mechanisms or other mechanisms through which a society's desires can be taken into consideration. Cities emerge over time and new functions, meanings and perspectives consequently co-evolve along with them (Geddes, 1915; Mumford, 1961; Castells and Hall, 1994).

Throughout this evolutionary path the city's institutional reality will repeatedly be redesigned. It is a trajectory of continuously adjusting towards a fit between what is appreciated by communities and the changes that have emerged within the physical environment. This institutional reality represents all the mechanisms which enable a community to express its desires and worries regarding the communal environment, resulting in frameworks of conventions (including those representing conventions of 'good governance').

This includes mechanisms through which the daily environment can be interfered with, influenced by and adjusted to the community's liking (de Roo, 2013). The result is an institutional design which – like any design – is strongly influenced by the societal and scientific perspectives which are being upheld and appreciated at any given moment (Hall, 2002). Nevertheless its design is open to change, and as such, retrospectively reveals evolutionary patterns through time.

While various stages of institutional development can be defined, the awareness of growth was a constant factor for a substantial period (Chapin and Weiss, 1966). The twentieth century in particular showed an exceptional increase in population size (UN, 2013). This resulted in a model of growth of investment in the city's fabric and in its urban qualities. This urban fabric and its qualities were enabled through the expected income generated by housing projects – in which low-cost agricultural land was transformed into expensive housing plots – which were in turn the cause of the transformation of rural into urban land and therefore contributed to the expansion of cities. This city model of growth has had consequences: the city's twentieth century institutional design implicitly embraced a linear perspective on the city's development path. The Western

349

350 *Handbook on complexity and public policy*

world exemplified this linear development route throughout the twentieth century. Since the Industrial Revolution, the population of this part of the world has steadily retreated from the countryside into cities. It is a development which has now affected Asia, though much more quickly and in a shorter period.

This is the traditional context we call Plan A: the assumption of linear growth. This Plan A no longer works well, underpinned as it is by arguments deduced from the 'normal' sciences and their linear understanding of the world we are part of. Functionality, certainty and predictability have proven to be no longer trustworthy, with serious consequences for city planning. It is not easy to venture out beyond current paradigms, as these are entrenched deeply within us, framing our ideas, attitudes, behaviours and actions. Nevertheless, with this contribution we propose a Plan B and what some would call a 'post-normal' science perspective (Funtowicz and Ravetz, 1993), focusing on a non-linear understanding of our world. A non-linear understanding means we consider the world to be undergoing discontinuous change, which is and becomes manifest largely beyond our control. It will have far-reaching consequences for the institutional design used by planners, interacting with and influencing a city's development. Despite this we believe there are strong arguments in support of a Plan B.

THE RISE AND FALL OF FUNCTIONALITY

How does Plan B relate to urban planning? Spatial planners and urban designers have long considered and determined good city forms. In the previous century, 'good' was defined as functional – proclaimed at the fourth CIAM (Congrès Internationaux d'Architecture Moderne) event in 1933, and completed by Le Corbusier's ideas of a minimal and standardized urban design (Mumford, 2000). Functional, minimal and standardized are terms which had social consequences in the sense that these terms advocated an egalitarian society, a society in which the commons are central. Their centrality was not so much meant to enable people to have a say (although democracy functioned – and still does – rather well in quite a number of countries) but to be treated equally. The modernistic era cannot be fully understood if it is considered as merely functional.

The functional has been extremely successful in its own right: if the conditions are 'right' the predictability of a future to come is breathtaking. The way Newtonian science has supported functionality by driving technological innovation forward has been beneficial to every one of us. Nevertheless, we dare to assert that its success has closed our eyes to alternative or additional views. We also believe that the functional is an explicit representation of the idea of the linear, of a world that 'is' and of a 'normal' science using methods based on verification and falsification (Störig, 1950). It is very much object-oriented, clinging to the idea of 'one true' world, a 'real' world perceived from a 'realist' perspective, existing independently from us, and which has – at least to a large extent – a functional sense that is implicitly part of our material world.

The functional can also be regarded as a means of giving effect to the human desire to frame the world we are part of as an ideal within reach. Twentieth-century urban structures were tested against criteria of functionality (McCulloch et al., 1965; Chapin and Weiss, 1966; Shane, 2005; Somer, 2007). Nearness, accessibility and connectivity are such criteria: a favourite location for an industrial site would be near and well-connected to the main road system. A residential neighbourhood needs good connections to a main road and links to a public transport network. It should have several functions to support the community's living environment: a supermarket, a park, a library and so on. A residential neighbourhood is a satellite to a town centre, and a town centre is intended to be a centre of retail trade, entertainment, gastronomy, political debate and so forth, and a stepping stone to the world beyond. With functionality as the paradigm of the spatial planning hierarchy, efficiency and equity became important ideas for the urban.

Subsequently, the process of standardization was embraced with regard to urban growth (Shane, 2005). 'Stamp' planning, through which structures are repeated several times within a space, 'blueprint' planning representing a fixed and controlled design and 'command-and-control' governance were obvious supporters of a linear world view. Various developments can be regarded as confirmations of this linear perspective (Lynch, 1960; Cronon, 1991). Spatial planning was extremely responsive to the strong intellectual belief in a logical-positivist philosophy and the idea that certainty is within reach (Faludi, 1973; Allmendinger, 2009). Obviously, the planner was a part of this process towards utopia, an ideal world within reach, and controlling well defined steps – even processes – towards this ideal was considered the right formula to achieve it (Faludi, 1987).

The rise of closed-systems thinking was very supportive of this idea (McLoughlin, 1969). Closed systems, also known as Class I systems (Kauffman, 1991), framed the 'command-and-control' approach, the obvious route after two world wars: first on account of armies having been so successful at bringing these wars to an end (ignoring the fact that they were responsible for them in the first place, and that half the armies had to lose in ending them), and secondly because planners had to rebuild what these armies had left behind. Planners also had to consider the relatively rapid rise of the urban population as people moved out of the countryside. This growth had to be managed somehow, and functionality was the answer: it resulted in the rise of growth strategies (Chapin and Weiss, 1966), a system which works perfectly so long as the process of urban growth continues. However, growth is no longer self-evident.

QUESTIONING THE LINEAR, THE FUNCTIONAL AND THE ATTITUDE TO CONTROL

The functional paradigm has had an unprecedented impact on both science and global society, including spatial planning and urban design. Spatial planners had never had to reconsider their functional reasoning while the urban continued to expand. Within this linear route of growth the functional paradigm could remain dominant. Here, we challenge this dominance of the functional and the linear regarding the urban, its planning and its governance (Marshall, 2009; Weinstock, 2010).

Furthermore, while the dominance of functional planning is still far from having been stored away in the history books, functionality's position as the dominant criterion for spatial design and urban growth is increasingly coming into question. Questions about the idea of certainty within urban planning had been raised as early as the 1960s and 1970s (Alexander, 1984). There are more drawbacks worth mentioning. The presumed certainties and the belief in their achievability and control fail to deliver increasingly

352   *Handbook on complexity and public policy*

often. The situations being addressed are less universal than planners had thought, and generic methods are not always well equipped for solving ostensibly similar issues.

Perhaps a more fundamental approach is to consider technocratic and functional planning as largely blind to the passage of time, adopting an attitude of trusting linear extrapolations from the here and now and all of its difficulties towards the ideal future (de Roo, 2003). As a result, the blueprint plan has been downsized substantially, no longer valued generically but only particularly with regard to the local land use plan and activities at the construction site. At the strategic level this has largely disappeared, being replaced by progressive planning and feedback planning (Ashby, 1956; Wiener, 1954).

By progressive planning (Forester, 1989) we mean that every so often the strategic proposals of the recent past are critically accessed against societal desires and needs. As a consequence, policy goals are adjusted within the strategic plans. While progressive planning questioned the planning targets, feedback planning is about reconsidering the route towards these targets: is the route still feasible or is an alternative route more desirable? In that respect, the desired targets can also be reconsidered if a once feasible route is no longer manageable (McLoughlin, 1969). Both approaches represent a reasoning that includes the uncertainty of contextual change in the near future, undermining assumed certainties with regard to routes and planning targets. Semi-open systems were introduced within systems theory – also known as Class II systems (Kauffman, 1991) – as a consequence of this lack of certainty, which were embraced not just in planning but in various disciplines supporting alternative views to the world (Jackson, 2003).

From the 1990s onwards, the strategic plan was reconsidered entirely due to a paradigm shift within spatial and urban planning. This change is known as the communicative turn: a turn away from the technical-rational approach and its claim of certainty, towards the communicative approach and its proposal to cope with uncertainty (Forester, 1989). Systems theory understands this step as a move towards open networks. Open networks are systems emphasizing contextual relationships and interactions: the so-called Class III systems (Kauffman, 1991). It became clear that spatial issues were susceptible to public valuation and results were affected by unpredictable dynamics to various degrees. It also means a shift from object-orientation towards intersubjective orientation. The argument was to come to agreements among stakeholders in case of uncertainty. In a sense it is a move away from a factual understanding of reality towards an agreed understanding of reality. This agreement on terms – on the definition of the planning issue, the actions to be taken, the stakeholders participating, the financial commitments, the responsibilities at various stages of the planning process, and so on – frames the planning process. It frames the planning process as facts did before within a functional planning framework, through which – again – certainty has been brought within reach, now by agreement instead of facts.

GOVERNANCE IN A DIFFERENTIATED WORLD

This is fundamental: there is no longer the idea of a single true and undisputed world within which planning issues are all the same and can be dealt with equally. Planning has to consider *a differentiation of planning issues*, resulting in a variety of actions,

Urban planning in a non-linear, complex world 353

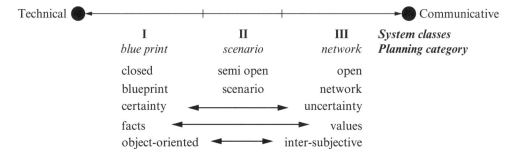

Source: De Roo (2012).

Figure 21.1 A rationality spectrum for spatial planning and its relationship with Class I, II and III systems

approaches and planning behaviour. As was the case with technical-rational approaches, this is also valid for communicative approaches: its success depends on the conditions discussed above framing the best possible fit between situation, issue, action (approach) and consequences. The assumption that planning has developed from one extreme (straightforward and closed situations; issues considered as factual reality; actions based on a technical-rational approach; and the consequences or results being predictable and certain) to another (fuzzy and open situations; issues composed as an agreed reality; actions framed by a communicative approach; and the consequences being unpredictable and uncertain) can be deduced from this reasoning. Between these opposing perspectives on planning we consider a contingency relating the two extremes by assuming a varying degree of certainty (de Roo and Porter, 2006; Zuidema, 2012).

Between the two extremes (see Figure 21.1), a whole range of issues and workable planning options can be positioned, such as progressive planning, scenario planning, area-specific planning, integrated planning and so on (de Roo and Porter, 2006). This differentiation of planning issues comes with consequences: those planning issues that can be dealt with by coordinative government action at a higher level (top-down) are those most frequently encountered, and they can be dealt with as routine. In such situations, planners adopt a technical-rational kind of behaviour, the authorities adopt a 'command-and-control' approach, and a vast majority is probably satisfied with this. At the other extreme, there are planning issues whose essence is not immediately clear due to their fuzzy character, multiple contextual links and the many stakeholders involved, including their varying and opposing interests. These kinds of situations require planners to adopt a communicative approach to the process, with the authorities adopting a facilitating stance within a shared governance environment (bottom-up). There are also an infinite number of planning issues between these two opposing positions. All in all, it means that planning has become 'situational', with an interest in both content and process, with the situation determining which of the two approaches will lead and incorporate the other.

Planning is becoming increasingly situational, while embracing processes of communication, participation, collaboration and interaction (Hillier, 2006). Instead of generic rules and procedures, the uniqueness of a situation is accounted for, considering not only

354 *Handbook on complexity and public policy*

the physical environment but also the people involved and considering them in various ways. Planning is no longer done on behalf of the people, but along with the people, who are now known as stakeholders. These stakeholders no longer depend entirely on political representation, as they increasingly have a direct say in the process, as well as becoming responsible for it.

By and large, the spatial planner functions within governance environments. As such, the interdependency of planners within their governance environment as it undergoes change becomes noteworthy. First, we should consider the shift from government to governance (de Roo, 2003). The top-down model in which a central government takes primary responsibility for policy-making has gradually transformed into governance structures in which authorities participate in facilitating situation-specific initiatives and initiators. As a consequence of this transformation, planners have become less technical experts acting instrumentally by hands-on control, and are increasingly becoming mediators and process facilitators supporting stakeholders (Woltjer, 2000).

THE RISE OF ADAPTIVE PLANNING

At the turn of the millennium the communicative approach within planning saw its heyday. In retrospect it was the beginning of the critical assessment of the conditions under which the communicative approach is an asset to planning. The communicative approach, emphasizing the process of planning, can cause content to be overlooked: if all the parties welcome the outcome of the process, the process is regarded as successful despite the fact that the initial worries which brought everyone together are only partially dealt with, if at all. The communicative approach considers all parties as equal, which is a condition for reaching an effective consensus (Innes, 1995). The parties can all have different, even opposing interests, so long as these interests are equally balanced. Parties with stronger interests, more means or more power than others can easily disrupt the process or create an unbalanced and unsatisfying outcome for some participants. These unsatisfied participants might in turn retreat, and the whole process could easily collapse.

There is a surprising scarcity of research initiatives exploring in theory and in practice into the space between the two ideal approaches to planning, except perhaps for scenario planning. Instead, alternative types of governance are emerging that do not fall between the technical and communicative approaches in the above spectrum. Of particular interest is the growing debate over the issue of adaptive governance. Adaptive governance is a kind of governance which avoids attempting to control our daily environment by purposefully allocating various spatial functions within space, taking the stand that reality will not be created according to a set of planning rules but will evolve more or less autonomously (Hartman et al., 2011). From an adaptive perspective, the role of governance and planning is to guide this evolutionary path, supporting positive and desired consequences while trying to minimize the negative and harmful effects. From this perspective, the spatial planner is neither a technical expert nor a mediator, but can be regarded as a transition manager and 'trend watcher' (de Roo, 2012) and as such is bridging multiple levels of scale. The urban planner guides parties through the processes of change. Moreover, this is quite a reasonable position, given the fact that planners

Urban planning in a non-linear, complex world 355

play a role only when there is change at hand. Adaptive planning is also seen as a logical response to a dynamic, changing and uncontrollable reality.

With the rise of adaptive planning, the essence of planning is brought into question. Why do we plan in the first place? Why does society appreciate spatial planning? Why do communities want to make agreements with each other about their space and place? Why are societies and communities in need of some degree of 'governance'? Why do we have various kinds of 'institutional design'? Simply because we need each other. We want to be able to get somewhere as individuals, while being inextricably connected to one or more communities, and as communities we have to come to agreements. Communal arrangements require governance structures and institutional designs that work, given the cultural, social and physical conditions at a particular time and in a specific space or place. In its essence what we are discussing is not much more than that (Heywood, 2002).

Take, for example, the prevention against floods: every one of us could build their own dyke, but only an overarching strategy to which all agree will result in an effective set of dykes which keep the hinterland dry. Each of us can strive for the best for our own sake, but thanks to Hardin and his famous explanation of the 'tragedy of the commons' (Hardin, 1968) we know only too well that we will all go down the drain together if we do not organize ourselves and restrict ourselves to a set of common rules. If we fail to do this, and rather allow every man simply to toss his own waste over his own fence, for example, to keep his own yard clean and his own living environment pleasant, we will literally perish together in our own filth. In both examples some guiding principles – call them 'institutional design' – are very welcome. Furthermore, there can be no institutional design lacking any form of governance (Parsons, 1995). Governance involves agreeing on a set of common conventions to direct developments (Ostrom, 2005; Geyer and Rihani, 2010). Every kind of governance has its own institutional design represented by a consistent framework of rules, plans, organizations and actors, out of which conditions for interventions emerge.

There are numerous reasons for accepting representative authorities with overarching powers to contribute on our behalf to common or societal interests (Ryan, 2012). However, what would then be socially appreciated, socially acceptable and socially relevant? These questions are no longer easily answered. In fact, the use by authorities of their powers to impose rules which are entirely out of tune with the desires and needs of individuals and communities, even to the extent of constraining developments which would or could be socially desirable, has come under increasing criticism. An example of this is the Dutch authorities, constraining the transition from gas to renewables because gas remains a substantial contributor to the nation's tax revenues. Thus, the attractive notion that the idea of control can be replaced by the idea of reality evolving autonomously and being adaptive to communities' desires arises.

There is yet another notion for planners to consider which has not yet become part of the mainstream debate: the idea of viewing the world around us as an autonomous process of discontinuous change. The very thought of a world that could continue to exist if all planners disappeared is inconceivable to a planner. Of course, perhaps it is fair to consider that a world without an overarching, institutional framework for interference would become less well organized, less functional or even chaotic. However, the implicit belief that the world is the product of planners and the result of purposeful human design is also doubtful. The idea that the world evolves with and without planning, and

356 *Handbook on complexity and public policy*

might be better off if planners did not solely focus on controlling the world as it is, but were instead able to influence the world as it evolves into the future and doing so for the betterment of society, is an interesting one. This idea relates quite well to concepts such as non-linearity, adaptivity, self-organization, emergence and co-evolution (Waldrop, 1992; Allen, 1997; Portugali, 1999; Batty, 2005; de Roo et al., 2012), concepts that we will now try to position within the planning arena.

EMBRACING NON-LINEARITY

How should planning in a non-linear world be considered? Looking into the past, we can see that alternative reasoning, including non-linear ideas, emerged not that long ago. One such example is from the seventeenth century: Ville de Richelieu (Terrien, 2008) in France. Cardinal de Richelieu – known across the world as the evil clergyman who dedicated his life to hunting down the three musketeers – was granted by Louis XIII the right to build a walled city, a park and a château in thanks for his numerous services to the King in his capacity as prime minister. The whole project was structured in accordance with the Fibonacci sequence (Terrien, 2008: 9), a sequence producing self-similar and non-linear structures. The Fibonacci sequence was considered at that time to be an expression of a divine order. Since the ancient past, people have tried to capture a hidden, divine ideal, hoping to be able to address the world using structures, rules or axioms representing a higher 'God-given' order. The whole plan was more or less completed by 1638, being a unique example of urban development with 'planning' rules deduced from Fibonacci's sequence: a divine proportion structuring the world non-linearly.

We are no longer in the seventeenth century. Thinking based on non-linearity will no longer relate planning and decision-making to divine proportions. Nor will non-linearity, planning and 'adaptive' governance emphasize explicit targets (command-and-control governance) or the mediation of processes (shared governance). Instead, efforts are made to consider situational possibilities for change to take place whenever the moment is right; this is known academically as 'possibility space' (Hillier and Abrahams, 2014) and 'windows of opportunities' (Kingdon, 1984). It is not control and reality as it 'is' but change and reality that is 'becoming' that is the common denominator. People are invited to choose their moves and actions in an environment in which conventions do not already dictate the behaviour of those present. This attitude opens up adaptive and self-organizing mechanisms. Adaptive and self-organizing processes have much in common. Adaptivity represents a response to incoming contextual influences triggering an effect of adjusting to a better fit with the environmental context (de Roo and Rauws, 2012). One could say self-organization represents the opposite effect, by which the various parts reorganize in such a way that the whole system becomes better equipped to maintain its identity and to develop within a changing environment (Portugali, 1999). The mechanisms of self-organization are not only internally driven (Maturana, 1980). Depending on the system's context being fairly stable or strongly out of balance, the flow the system is in connects with the inner system's parts (Haken, 1983). The flow the system is in becomes part of or adds to the system's energy or information, which is defined by Prigogine (Prigogine and Stengers, 1984; Prigogine, 1997) as dissipative. In other words, what is regarded as adaptive behaviour seen from a lower level looking

Urban planning in a non-linear, complex world   357

up, can to some extent be seen as a process of self-organization seen from a higher level looking down.

Adaptive and self-organizing processes relate to situations manifest at multiple levels of scale. Such a situation is often perceived explicitly at a particular level (meso), while being highly connected to changing contextual environments, as part of an intensive exchange with developments at higher levels (macro), and which are also 'more than' the sum of highly connected lower-level activities (micro) (Geels, 2005). Such situations exist not as clear and well-defined entities or events but as manifestations of various trajectories at various levels, which link and connect with each other to enable an observer to label or to identify them as a 'situation'. Instead of stable and stand-alone entities or objects, they are emerging properties, also defined as assemblies (assemblages) or arrangements present and manifest at a specific time and place (DeLanda, 2006; Hillier and Van Wezemael, 2012).

These properties, assemblies and arrangements reveal adaptive behaviour and processes of self-organization occurring in situations which are 'out-of-equilibrium' (Prigogine and Nicolis, 1997; Vesterby, 2008). While situations reaching equilibrium are, according to biologists, fairly 'dead' (Lister, 2008), situations which are very much 'alive' are strongly connected to contextual environments which are always in a continuous flow of discontinuous change, like a surfer being carried by waves in the sea, progressing towards an equilibrium but never able to reach it, pushed away again by forces of change, both nearby and far away.

Building on these ideas of adaptivity and self-organization, we are well able to position planning issues within a non-linear and evolutionary perspective: from a problem to be solved definitely (goal driven) via a problem defined by all to be appreciated as a process (consensus driven) to a situation progressing towards a best possible fit with its contextual environment (evolution driven). This evolutionary perspective does come with consequences which fundamentally affect our conceptual framework:

- From being to becoming:
    - From '*in situ*' affairs to 'situations' in a constant state of discontinuous change.
    - From a solid, stand-alone, unchangeable and well defined 'it' to multiple and path-dependent trajectories of becoming.
    - From clearly defined 'splendid isolations' to fuzzy, fluid and vague manifestations of becoming, which suspend between various levels of scale.
- From direct causal relations to a connectedness with varying potentialities depending on time and place.
- From a linear and a-temporal reality to a non-linear reality of change.

If planning is to be considered as non-linear and situational, this means that it should always be regarded as in a state of becoming and in a constant state of discontinuous change. As such, a planning issue is dependent not just on its own configuration but also on its internal and external connections and the potential of these connections to link between the various levels through which the situation becomes manifest. With respect to the planning issue, this requires a perspective which is ultimately different from traditional conceptions. We wish to avoid having to offer a post-modern answer, which would mean having to conclude that this kind of planning issue is unique *per se*

358   *Handbook on complexity and public policy*

and has to be considered as a specific case independent from any preconceived idea of generic label, class or qualities. This would mean that no lessons can be learned, and no generic approaches can be found, which also results in an unworkable imbalance between the generic and the specific actions to be taken by a spatial planner. In that case we would have moved from a predetermined situation to a situation determined solely by its uniqueness.

We feel there has to be an alternative. Our main task therefore is to find commonalities for planning issues which behave non-linearly. Throughout this chapter we have been making links to systems theory. Planning in the early days was driven by a technical-rational approach, which relates rather well to 'Class I systems', also known as closed systems (Kauffman, 1991). The technical-rational approach was succeeded by a kind of scenario planning, which allowed for multiple futures to be considered. Scenario planning relates strongly to 'Class II systems', which are feedback and circular systems. Communicative planning relates strongly to open network systems (Class III systems). It is not that difficult to now expect an increasing interest by planners in 'Class IV systems', which are better known as complex adaptive systems (Kauffman, 1991; Gros, 2008). These complex adaptive systems are central to the complexity sciences. These systems include a conception of time and demonstrate non-linear behaviour (Gleich, 1987; Gros, 2008).

A shift in attention from closed systems to open network systems meant a shift within planning from content to process. However, what would a shift towards complex adaptive systems as the common denominator bring us? The question is how to understand a non-linear evolutionary path for urban space being intrinsically multiple and fuzzy and no longer representing straightforward linear growth. Are we able to identify mechanisms through which the evolution of multiple and fuzzy urban spaces can be influenced? If so, would this give us humans a say in the development of these spaces, without being indoctrinated by the arrogance of considering ourselves the sole creators of space?

We argue that content and process are becoming secondary to *conditions*. As situations undergo processes of change within which structure and function co-evolve, what remain and what continue to give a situation or system its identity over time are the conditions which maintain the situation's or system's balance. A complex adaptive system and the situation it represents are therefore conditional.

A COMPLEXITY UNDERSTANDING OF REALITY

Through complex adaptive systems we can gain a better understanding of evolving spaces. The basis of such systems is internal interactions between dynamics and robustness, as they fluctuate externally between order (uniformity) and chaos (diversity) (Gros, 2008). And it results in quite a different understanding of 'system' from the Class I to III systems, which traditionally consists of nodes and their interactions (Kauffman, 1991). These nodes and interactions are usually represented as dots (the nodes) linked by lines (the interactions).

Here we will focus on the node of the system, elaborating the node conceptually, considering it to be more than 'just' a dot linked to other dots. Here the dot or node is viewed no longer as a black box, but representing a coherent whole of internal and

Urban planning in a non-linear, complex world 359

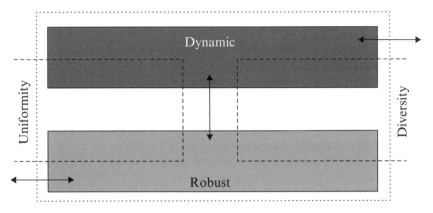

Source: De Roo (2012).

Figure 21.2 The complex adaptive system with its internal and external aspects

external interactions which keep the system balanced and allow it to sustain in processes of change (De Roo, 2012; Neal and Neal, 2013). This coherent whole is responsible for providing sufficient robustness to support dynamic behaviour. Dynamic behaviour without robustness (Cillier, 1998) will lead to nothing but failure.

The dot or node is also open to its contextual environment, through which the system's internal subsystems can be triggered to change. This will probably be a process of self-organization towards a better fit with its environment, to which the system as a whole is also adapting itself. This contextual environment is considered as triggering a complex adaptive system by creating an out-of-equilibrium state. As such, the complex adaptive system floats in its environment, absorbing impulses from outside that affect it internally. These impulses are dissipative in their behavioural nature, adding to or draining the system's energy. These impulses barely affect a system if it is in a more or less stable, uniform and ordered environment, yet affect it severely in a more diverse environment.

This understanding of autonomous, non-linear spatial development (see Figure 21.2) has far-reaching consequences for policy-making, planning and decision-making. Consequently, it also sets conditions for institutional design, through which intentional interference within space and place is initiated. In this chapter we have seen spatial planning going through various phases of thought. It started from a technical (functional) and direct-causal understanding of urban growth, with the planner as the creator of places. During the 1990s, spatial planning found an answer to dealing with fundamental uncertainties by embracing the communicative paradigm. While uncertainty was acknowledged as a fundamental aspect in intervening in urban development, time and non-linearity had not yet been acknowledged as important. Planning continued to have a strong focus on the decision-making moment 't = 0', while the steps that followed after 't = 0' as a consequence of the decision that had been taken were largely ignored.

Adaptive planning is the approach that takes into account the understanding of space and place as an evolutionary process, with non-linear development occurring more or less autonomously. Only recently has this possible route for planning been considered by a small group of theorists (Portugali, 1999; Byrne, 1998; Batty, 2005; de Roo and Rauws,

360 *Handbook on complexity and public policy*

2012; de Roo and Da Silva, 2010). With it, another fundamental aspect has come into the picture: the 'wicked problem'. It is fair to say that it has come *back* into the picture. The concept of the wicked problem was proposed by Horst Rittel and his colleagues, West Churchman and Webber, (Rittel, 1972; Rittel and Webber, 1973) to differentiate problems which can be solved with a definite end ('tamed' problems) from problems which fundamentally cannot be fully understood and which have no definite end. Rittel was a spatial planner, but his suggestions have been ignored for decades by planners. It was within the complexity sciences that proposals were finally made to address wicked problems and the fundamental idea that these cannot be fully understood. Not to solve a problem, but to adapt to the situation, is the response to wicked problems (Rittel, 1972), while adaptivity as an answer to non-linearity is proposed by the complexity sciences. Adaptive planning responds to this proposal and to coping with a world full of discontinuous change and fundamental uncertainty.

Cities are good examples of complex adaptive systems that co-evolve over time in a structural and a functional sense (Batty, 2005; Marshall, 2009; Weinstock, 2010). While co-evolving, the city is undergoing a fundamental transformation in terms of its structure and function (Geels, 2005). This also means a fundamental change of identity. This process of co-evolution and transformation is the result of the system adapting to a new context, through which the system is allowed to achieve a better fit between the system and its environment. During the process of co-evolution, stability decreases while the system's dynamics increase (de Roo, 2012; Hartman et al., 2011). As soon as the system connects effectively with a new contextual environment, stability increases again and the system's identity is likely to have changed radically due to the co-evolution of its structure and function. Cities have gone through various such fundamental changes.

MODELLING NON-LINEAR REALITIES

Here, we are exploring complex adaptive systems as a means of understanding non-linear change. To do so we aim to uncover criteria which characterize a system throughout its life-span, while going through the various transitions and moments of co-evolution. In particular, these criteria could provide complex adaptive systems with a 'sustainable' identity which strongly reflects their specific evolving and self-organizing qualities. If we were able to find such an identity, we could push the idea of complex adaptive systems beyond the metaphor, allowing it to become as 'real' as any other system we use to represent reality. It would give complex adaptive systems a point of reference or markers to adhere to, in order to define them, but also to enable their discussion. Moreover, it would bring a critical assessment of the complex adaptive system within reach.

To arrive at such criteria, these should make reference to a few very fundamental characteristics of the complex adaptive system mentioned earlier in this contribution. The duality of uniformity and diversity is fundamental to representing complex adaptive systems' state in an out-of-equilibrium situation. Uniformity expresses the environmental or contextual 'order' to which a system connects and diversity refers to 'chaos', being the other extreme to which the system contextually relates. The system could progress towards an environment of 'uniformity' in order to find stable ground. However, the

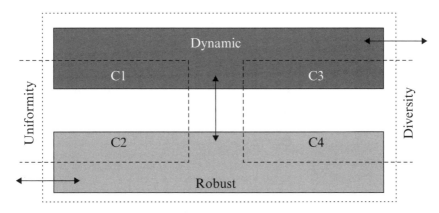

Source: De Roo (2012).

Figure 21.3 The complex adaptive system and its four connectors

system also needs 'diversity', as this opens up possibilities during times of change. Both represent extremes within which the system is able to survive.

Another fundamental characteristic addresses the potential of a complex adaptive system to maintain the structure–function relationship (surviving, not disintegrating) while having the flexibility to adapt (through which structure and function co-evolve). This is due to another and co-existing duality, which is internally oriented. This is the existence of a robust and dynamic relationship. A complex adaptive system has a certain robustness which grounds dynamic behaviour, which could then become a driver for innovation, development and progress, the moment there is a positive fit with its environmental context, as it fluctuates between order and chaos. These dualities represent contrasts that generate what we call 'complexity'.

Without any robustness, a dynamic or flexible attitude to contextual possibilities will have only a small chance for success. System robustness structures the dynamics the system requires to be adaptive to change. The result of this reasoning is complex adaptive systems having *four connectors* (see Figure 21.3; de Roo, 2012): connector 1 (C1) links internal dynamics to contextual uniformity; connector 2 (C2) links internal robustness to contextual uniformity; connector 3 (C3) relates internal dynamics to contextual diversity; and connector 4 (C4) is about internal robustness being linked to external diversity. Obviously, the idea is that a system which is internally and externally fit is a system with all four connectors being responsive to each other in progressing towards a balanced input of impulses and energy (Figure 21.4).

The connectors represent the internal match of various subsystems and their potential to act as a joint force, while oscillating to maintain internal coherence and to respond to external forces, and to adapt outwardly and to self-organize inwardly. Each connector represents a 'gravity' force or 'attractor' (Lorenz, 1996), which balances with the other three connectors, absorbing, oscillating and responding (see Figure 21.4).

How does such a proposal of non-linearity relate to spatial planning? The challenge is to elaborate on this model with proposals that frame a variety of situations as instances of complex adaptive systems within the realm of spatial planning. If the results are

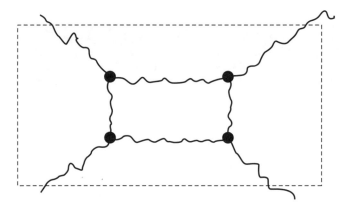

Figure 21.4 The four connectors of a complex adaptive system oscillating as a balanced whole

convincing and fit with an aspect of theoretical discourse, then we can agree on a structural means to identify how the various situations we consider as being planning issues are instances of such systems. This means of identifying systems will no longer focus primarily on content and process, preferring to consider the *conditions* under which the system performs as its prime focus. Moreover, within planning we can now acknowledge the beginning and ending of systems over time, their rise and fall, and the processes of their emergence and co-evolution. Overall, we would have an instrument which gives meaning to complex adaptive behaviour for issues relevant to planning, and we would have a tool from which we could develop a kind of planning with arguments at hand, to intervene adaptively to a world in discontinuous change.

An example to consider is regional development. Regional development relates strongly to spatial change and economic advantages (Atzema et al., 2012). In the context of development and progress, the need to be *competitive* is a given. When relating development to our proposal for a complex adaptive system and its four connectors, it becomes instantly clear that there is more to it than 'just' being competitive. In that respect, competitiveness is one of the four essential conditions of a system to adapt to change. Competitiveness would be the one particular element that specifically relates to internal dynamics and contextual diversity: this would be connector C3 in Figure 21.3. There has to be a second element that relates to internal dynamics. In contrast with 'competitiveness', this element also has to relate to contextual uniformity: this would be connector C1 in Figure 21.3. Instead of 'competitiveness', the idea which could represent internal dynamics and contextual uniformity in spatial-economic development is '*complementarity*' (Hartman et al., 2011; de Roo, 2012). One of the messages we can deduce from this complex adaptive understanding of spatial economic regions is that a competing region will also have to identify its complementary qualities to cooperate with other regions. In other words: it is not just competitiveness that makes a system (a region) adaptive; being complementary as well enables the system (the region) to develop.

The complex adaptive model also informs us when a region is unable to compete and cooperate properly, by showing no robustness in terms of cohesion and compatibility (C2 and C4 in Figure 21.3). Both conditions for regional development – competitiveness

Urban planning in a non-linear, complex world 363

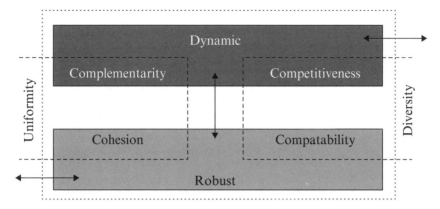

Source: De Roo (2012).

Figure 21.5 Conditions for regional development based on four connectors of complex adaptive systems

and complementarity – cannot be without some kind of cohesion and compatibility. *Cohesion* is understood as a quality of the developing region representing a more or less complete and properly related set of activities which permit the region to function properly. *Compatibility* indicates a situation which is diverse enough to cope with a loss of activities within the region without the region having to collapse, as other activities or functions are able to fill in the gap. We consider competitiveness, complementarity, cohesion and compatibility to be *conditional* to regional development (see Figure 21.5).

This example of regional development is strongly externally oriented. Where a region is said to be ready for change, this means enabling the region to be prepared to relate to macro trends by having a diverse set of potentially strong innovations ready or ongoing. A more internally-oriented example is a neighbourhood which wishes to self-organize to improve by investing in its social capital. Strong social capital is *conditional* for enabling a neighbourhood to maintain or improve its spatial and social quality, to be able to invest and to renovate, and to be an attractive place to live in.

Nienhuis (2014) shows us the importance of the four connectors representing social capital in a neighbourhood in need of internal change (Figure 21.6). This change could be urban or neighbourhood renewal. Social cohesion is one of the conditions (C2) for using social capital successfully. It is the result of the degree to which a neighbourhood community participates in communal life within the neighbourhood. The degree to which people are sympathetic to explicit and implicit rules of behaviour existing within the neighbourhood is of importance to social cohesion. Transcending cultural values in a neighbourhood is another important condition (C1) as it connects the neighbourhood to society and its encompassing socio-cultural attitude, preventing a strong internal orientation which could result in ignorance. Tolerance in a neighbourhood community (condition C4) allows space for dissenters. Where a collectively-felt problem is experienced, it helps to have a community which is not too heavily involved internally, one that is able to maintain a workable balance between upholding a distance and a willingness to support its members if there is a need to do so. Branding (condition C3) refers to the

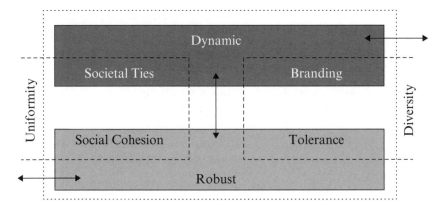

Source: Nienhuis (2014).

Figure 21.6 Conditions relevant to a neighbourhood's social capital during periods of change, represented by four connectors in a complex adaptive system

attractiveness of a neighbourhood within a wider hierarchy of places within the urban area. For example, the position a neighbourhood has in the housing market could be a possible indication. Other 'branding' related criteria could be the perceived quality of the public space available, a neighbourhood's reputation for safety, its historical values and so on. Nienhuis argues that these conditions are important with respect to the resilience of the neighbourhood's community and its willingness to take responsibility for the neighbourhood's qualities. If so, this would make these conditions essential to the neighbourhood's policy and planning. This would not be planning to control or planning to mediate a consensus, but a kind of planning which would facilitate the ambitions that arise from within the community or neighbourhood, which relate to its development within an autonomously transforming world.

These two proposals – regional development and the social capital of a neighbourhood – relate to spatial planning and institutional design. The first example is area-specific and the other relates to a local community. We have shown here that these issues can be considered as complex adaptive systems to be dealt with by an adaptive governance approach. Everywhere in Europe, the issues these proposals relate to would be subject to a political decision-making process. Obviously, it would remain up to politicians and decision-makers whether to follow an adaptive governance approach in these cases of regional development and neighbourhood renewal.

Another possible step is to consider the political process of decision-making itself, and the possibility of framing such a political process within a complex adaptive system's conditions. We expect that socio-political systems also seek balance while undergoing regular processes of sudden change, such as elections. Planning and decision-making have everything to do with politics. To be able to frame planning and decision-making within a socio-political setting would add to the proposals worked out here. This would result in an additional set of conditions to frame the political arena within which the previously presented proposals become subject to decision-making processes. More importantly, it would provide conditions for the fitness of the socio-political system itself.

Urban planning in a non-linear, complex world 365

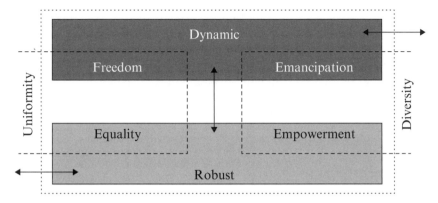

Figure 21.7 Conditions for a political setting based on four connectors of a complex adaptive system

Figure 21.7 attempts to provide an answer. In Figure 21.7 'equality' (C2) is considered crucial as a criterion to be upheld to prevent individuals from dropping out and communities from collapsing. However, if only 'equality' frames policy, planning and decision-making, it will become a burden which constrains all innovation, renewal, creativity, development and progress through its failure to appreciate uniqueness. 'Equality' is therefore balanced by 'empowerment' (C4), expressing the importance of the strength of individuals and the community to support and develop themselves and each other plurally, and to develop themselves differently from others. Equality is supportive to the condition of empowerment, as 'equal opportunity' allows everyone to become aware of their own possibilities, to become self-confident, to stand up and to become politically active.

While social democrats stress the agenda for 'equality', liberals do the same for 'freedom' (C1). Both political attitudes express only a part of a fluid story founded on four (not two) connectors, which seek a common balance as the system responds to change. 'Freedom' stands for the idea that every individual and community should have the right to explore and develop their potentialities. In that respect equality and empowerment are supporting conditions and a robust foundation for making this happen. 'Freedom' represents the individual and the community becoming 'dynamic', with no exceptions and equally for all. If the sociopolitical system only supports 'freedom', imbalance results, as those who are able to progress more quickly and further than others will take the lead and will dominate. Therefore 'emancipation' (C3) is another condition of the socio-political system. Provided that the above is deemed to make sense, equality, empowerment, freedom and emancipation are four conditions which frame the fitness of a political system. As the political system frames spatial planning within a governance environment, equality, empowerment, freedom and emancipations are meta-conditions balancing the process of spatial planning.

This reasoning brings to the fore a non-linear approach to policy and planning (de Roo, 2012). Planning issues relating to contextual change have to be fit to adapt. Adaptive planning would be the route to follow, and *adaptive governance* would represent the attitude of all those involved. This kind of planning is slowly emerging within

366 *Handbook on complexity and public policy*

the European policy arena, due to governments evading their top-down responsibilities. Planning issues which relate above all to the desire to change internally have to be able to allow the processes of *self-organization* to take place. This non-linear approach or argument for choice and decision-making is additional to the traditional approaches to planning: the technical and communicative approaches. While the technical-rational approach is very much *content* related, and the communicative approach underlines the importance of *process*, the non-linear approach stresses the importance of planning being *conditional*.

GOING FOR PLAN B

In this contribution we have observed the main transformations of thought within urban and spatial planning. We started by considering urban and spatial planning as it has evolved after the Second World War: a planning in control, building on the idea of certainty and a 'factual reality', with the planner as the expert. In response to this perspective falling into decline, spatial planning has been enriched since the 1990s by the 'communicative turn'. This 'communicative turn' can be regarded as a response to uncertainties which were hard to ignore within spatial planning processes. The solution to these uncertainties was an 'agreed reality'. Communicative planning and its approaches were in that respect nothing more than another route to certainty.

If the world is considered from a non-linear perspective, this means accepting a world which is not a planner's creation, as it would evolve more or less by itself: a world that is continually evolving in discontinuous ways. This non-linear understanding of reality and the world that surrounds us is increasingly thought of as an interesting idea to explore, not least due to the housing and economic crisis affecting us since 2008. Such a world would need alternative kinds of planning, including those which incorporate alternative understandings of governmental responsibilities. Adaptive planning is the proper response to situations which are highly affected by contextual interference, for example as a consequence of sudden changes or due to local impacts of macro trends. The alternative approaches addressed in this chapter follow a non-linear route of development, which can be explained using a complex adaptive systems model. Such a model positions reality not as an 'it' and an 'it is', but as a highly connected world within which situations become manifest with internal and external links: internally these balance between dynamics and robustness. Externally there is a drive for a system to find the best possible position in its environments or contexts, which can be described as being in flux between uniformity (order) and diversity (chaos), seeking the best fit to develop and to progress.

During such a process of seeking internal and external fitness, the planner's responsibility is as the manager of change. This transition manager tries to enhance the positive effects of non-linear change in support of communities and society as a whole, and to minimize its possible negative effects. For the planner and the discipline of planning to embrace this non-linear perspective would mean a jump from a normal to a post-normal scientific perspective, with planning shifting its attention from content and process to becoming conditional. This would be a major transformation within the discipline of planning: planning going for Plan B.

REFERENCES

Alexander, E.R. (1984), 'After Rationality, What? A Review of Responses to Paradigm Breakdown', *Journal of the American Planning Association*, (1), 62–9.

Allen, P.M. (1997), *Cities and Regions as Self-organizing Systems: Models of Complexity*, Amsterdam: Gordon and Breach Science Publishers.

Allmendinger, P. (2009), *Planning Theory*, Basingstoke: Palgrave.

Ashby, W.R. (1956), *Introduction to Cybernetics*, London: Chapman and Hall.

Atzema, O., T. van Rietbergen, J. Lambooy and S. van Hoof (2012), *Ruimtelijke Economische Dynamiek – Kijk op Bedrijfslocatie en Regionale Ontwikkeling* [Spatial Economic Dynamics – Business Location and Regional Development in Perspective], Bussum: Uitgeverij Coutinho.

Batty, M. (2005), *Cities and Complexity: Understanding Cities with Cellular Automata, Agent-Based Models, and Fractals*, Cambridge, MA: The MIT Press.

Byrne, D. (1998), *Complexity Theory and the Social Sciences: An Introduction*, London: Routledge.

Castells, M. and P. Hall (1994), *Technopoles of the World*, London: Routledge.

Chapin, F.S. and S.F. Weiss (1966), *Urban Growth Dynamics: In a Regional Cluster of Cities*, New York: John Wiley and Sons.

Cillier, P. (1998), *Complexity and Postmodernism – Understanding Complex Systems*, London: Routledge.

Cronon, W. (1991), *Nature's Metropolis: Chicago and the Great West*, New York: W.W. Norton & Company.

DeLanda, M. (2006), *A New Philosophy of Society: Assemblage Theory and Social Complexity*, London: Continuum.

De Roo, G. (2003), *Environmental Planning in the Netherlands: Too Good to be True: From Command-and-control to Shared Governance*, Farnham: Ashgate.

De Roo, G. (2012), 'Spatial Planning, Complexity and a World "Out of Equilibrium": Outline of a Non-linear Approach to Planning', in G. de Roo, J. Hillier and J. Van Wezemael (eds), *Complexity and Planning – Systems, Assemblages and Simulations*, Farnham: Ashgate, pp. 141–76.

De Roo, G. (2013), *Abstracties van Planning* [Abstractions of Planning], Groningen: InPlanning.

De Roo, G. and E.A. Da Silva (eds) (2010), *A Planner's Encounter with Complexity*, Farnham: Ashgate.

De Roo, G. and G. Porter (eds) (2006), *Fuzzy Planning – The Role of Actors in a Fuzzy Governance Environment*, Aldershot: Ashgate.

De Roo, G. and W.S. Rauws (2012), 'Positioning Planning in the World of Order, Chaos and Complexity: On Perspectives, Behaviour and Interventions in a Non-linear Environment', in J. Portugali, H. Meyer, E. Stolk and E. Tan (eds), *Complexity Theories of Cities Have Come of Age: An Overview with Implications to Urban Planning and Design*, Heidelberg and Berlin: Springer-Verlag, pp. 207–20.

De Roo, G., J. Hillier and J. Van Wezemael (eds) (2012), *Complexity and Planning – Systems, Assemblages and Simulations*, Farnham: Ashgate.

Faludi, A. (1973), *Planning Theory, Urban and Regional Planning Series*, Volume 7, Oxford: Pergamon Press.

Faludi, A. (1987), *A Decision-centred View of Environmental Planning: Beyond the Procedural-substantive Controversy*, Oxford: Pergamon Press.

Forester, J. (1989), *Planning in the Face of Power*, Ewing: University of California Press.

Funtowicz, S.O. and J.R. Ravetz (1993), 'Science for the Post-Normal Age', *Futures*, **25**(7), September, 739–55.

Geddes, P. (1915), *Cities in Evolution – An Introduction to the Town Planning Movement and to the Study of Civics*, London: Ernest Benn.

Geels, F. (2005), *Technological Transitions and System Innovations: A Co-evolutionary and Socio-technical Analysis*, Cheltenham, UK and Northampton, MA, USA: Edward Elgar Publishing.

Geyer, R. and S. Rihani (2010), *Complexity and Public Policy: A New Approach to 21st Century Politics, Policy and Society*, London: Routledge.

Gros, C. (2008), *Complex and Adaptive Dynamical Systems: A Primer*, Heidelberg: Springer.

Haken, H. (1983), *Advanced Synergetics*, Heidelberg: Springer.

Hall, P. (2002), *Cities of Tomorrow*, Oxford: Blackwell Publishing.

Hardin, G. (1968), 'The Tragedy of the Commons', *Science*, **162**(3859), 1243–8.

Hartman, S., W. Rauws, M. Beeftink, G. de Roo, D. Zandbelt, E. Frijtes and O. Klijn (2011), *Design and Politics No. 5: Regions in Transition: Design for Adaptivity*, Rotterdam: 010 Publishers.

Heywood, A. (2002), *Politics*, London: Palgrave.

Hillier, J. (2006), *Stretching Beyond the Horizon – A Multiplanar Theory of Spatial Planning and Governance*, Aldershot: Ashgate.

Hillier, J. and G. Abrahams (2014), 'Deleuze and Guattari, Jean Hillier in Conversation with Gareth Abrahams', Exploring Foundations for Planning Theory, *AESOP YA Booklet Series* A-1, Groningen: InPlanning.

Hillier, J. and J. Van Wezemael (2012), 'On the Emergence of Agency in Participatory Strategic Planning',

368   *Handbook on complexity and public policy*

in G. Roo, J. Hillier and J. Van Wezemael (eds), *Complexity and Planning – Systems, Assemblages and Simulations*, Farnham: Ashgate, pp. 311–32.

Innes, J.E. (1995), 'Planning Theory's Emerging Paradigm: Communicative Action and Interactive Practice', *Journal of Planning Education and Research*, **14**(3), 183–9.

Jackson, M.C. (2003), *Systems Thinking, Creative Holism for Managers*, London: John Wiley & Sons.

Kauffman, S.A. (1991), 'The Sciences of Complexity and "Origins of Order"', *SFI WORKING PAPER*: 1991-04-021, Santa Fe: Santa Fe Institute.

Kingdon, J.W. (1984), *Agendas, Alternatives, and Public Policies*, Ann Arbor: University of Michigan.

Lister, N.M.E. (2008), 'Bridging Science and Values: The Challenge of Biodiversity Conservation', in D. Waltner-Toews, J.J. Kay and N.M.E. Lister (eds), *The Ecosystem Approach – Complexity, Uncertainty and Managing for Sustainability*, New York: Columbia University Press, pp. 83–108.

Lorenz, E.N. (1996), *The Essence of Chaos*, Seattle: The University of Washington Press.

Lynch, K. (1960), 'The Pattern of the Metropolis', in L. Rodwin (ed.), *The Future Metropolis*, New York: George Braziller, pp. 103–28.

Marshall, S. (2009), *Cities, Design & Evolution*, London: Routledge.

Maturana, H.R. (1980), *Autopoiesis and Cognition: The Realization of the Living*, Dordrecht: Kluwer.

McCulloch, F.J., P. Brenikov, G.P. Wibberley, C.M. Haar, H.R. Parker, J.P. Reynolds, D.H. Crompton, L. Holford, M. Wright (eds) (1965), *Land Use in an Urban Environment*, Liverpool: Liverpool University Press.

McLoughlin, B. (1969), *Urban and Regional Planning: A Systems Approach*, London: Faber & Faber.

Mumford, E. (2000), *The CIAM Discourse on Urbanism: 1928–1960*, Cambridge: MIT Press.

Mumford, L. (1961), *The City in History*, Harmondsworth: Penguin Books.

Neal, J.W. and Z.P. Neal (2013), 'Nested or Networked? Future Directions for Ecological Systems Theory', *Social Development*, **22**(4), 722–37.

Nienhuis, I. (2014), *Vrijheid, Gelijkwaardigheid & Bevoogding: Over Bewonersparticipatie in het Stimuleren van Sociale Weerbaarheid in (Probleem)Wijken* [*Liberty, Equality & Paternalism About citizens participation in processes stimulating social resilience in (problematic) neighbourhoods*], PhD Thesis, University of Groningen, Groningen.

Ostrom, E. (2005), *Understanding Institutional Diversity*, Princeton, NJ: Princeton University Press,

Parsons, W. (1995), *Public Policy, An Introduction to the Theory and Practice of Policy Analysis*, Cheltenham, UK and Northampton, MA, USA: Edward Elgar Publishing.

Portugali, J. (1999), *Self-organization and the City*, Berlin: Springer-Verlag.

Prigogine, I. (1997), *The End of Certainty*, New York: The Free Press.

Prigogine, I. and G. Nicolis (1997), *Self-organization in Non-Equilibrium Systems*, London: Wiley.

Prigogine, I. and I. Stengers (1984), *Order out of Chaos*, Boulder, CO: New Science Press.

Rittel, H. (1972), 'On the Planning Crisis: Systems Analysis of the "First and Second Generation"', *Bedriftsøkonomen*, (8), 390–96.

Rittel, H. and M. Webber (1973), 'Dilemmas in a General Theory of Planning', *Policy Sciences*, **4**, 155–69.

Ryan, A. (2012), *On Politics – A History of Political Thought from Herodotus to the Present*, London: Allen Lane.

Shane, D.G. (2005), *Recombinant Urbanism – Conceptual Modeling in Architecture, Urban Design and City Theory*, Chichester: Wiley-Academy.

Somer, K. (2007), *The Functional City, CIAM and the Legacy of Van Eesteren*, Rotterdam: NAI-010 Publishers.

Störig, H.J. (1950), *Kleine Weltgeschichte der Philosophie*, Stuttgart: W. Kohlhammer-Verlag.

Terrien, M-P. (2008), *The Ideal City and Château of Richelieu – An Expert Architectural Conception*, Cholet (F): Pays & Terroirs.

UN (United Nations) (2013), *World Population Prospects: The 2012 Revision, Highlights and Advance Tables*, ESA/P/WP.228, New York: Department of Economic and Social Affairs, Population Division.

Vesterby, V. (2008), *Origins of Self-organization, Emergence and Cause, Exploring Complexity: Volume 3*, Goodyear, AZ: ISCE Publishing.

Waldrop, M.M. (1992), *Complexity: The Emerging Science at the Edge of Order and Chaos*, London: Penguin Books.

Weinstock, M. (2010), *The Architecture of Emergence: The Evolution of Form in Nature and Civilisation*, London: John Wiley & Sons.

Wiener, N. (1954), *The Human Use of Human Beings, Cybernetics and Society*, Boston, MA: Houghton Mifflin.

Woltjer, J. (2000), *Consensus in Planning*, Aldershot: Ashgate Publishers.

Zuidema, C. (2012), *Post-Contingency in Planning: Making Sense of Decentralization in Environmental Governance*, PhD Thesis, Groningen: Faculty of Spatial Sciences, University of Groningen.

22. Complexity and health policy
Tim Tenbensel

INTRODUCTION

A much-cited article published in the *British Medical Journal* in 2001 by Paul Plsek and Trisha Greenhalgh has provided a powerful impetus for the application of complexity theory to the domain of health. This introductory article concluded with the following clarion call for complexity thinking:

> This introductory article has acknowledged the complex nature of health care in the 21st century, and emphasised the limitations of reductionist thinking and the 'clockwork universe' metaphor for solving clinical and organisational problems. To cope with escalating complexity in health care we must abandon linear models, accept unpredictability, respect (and utilise) autonomy and creativity, and respond flexibly to emerging patterns and opportunities. (Plsek and Greenhalgh, 2001: 628)

Since the early 2000s, health policy has provided a fertile source of inspiration for those adopting a complexity perspective (Kernick, 2004; Haynes, 2008; Geyer, 2013; Marchal et al., 2013).

In this chapter, I begin by addressing a number of questions that are central to this collection. The first question addressed is whether complexity theory is a coherent theoretical framework or a stock of useful concepts. I then briefly explore whether complexity theory has generated methodological and theoretical advances, and address the question of the nature of advice provided to those involved in policy processes using a complexity perspective.

In the second half of the chapter I develop a critique of one aspect of the way complexity theory is typically applied to health policy, which has particular implications for understandings of policy processes and advice to policymakers. My argument establishes that those who apply complexity concepts to health policy tend to treat the use of hierarchical policy approaches as inherently problematic and antithetical to a complexity-informed understanding of the world. This view, however, fails to recognize that hierarchical policy imperatives are themselves a *product* of complex, dynamic health policy environments. Secondly, I argue that those applying complexity concepts to health have developed an unwarranted habit of assuming that hierarchical policy directives are synonymous with linear, sequential processes of implementation and that more sophisticated attempts to 'command and control' through mechanisms such as health targets often do specifically take into account the importance of local context and contingency. The chapter ends with a call to those who use complexity theory to take hierarchy seriously by engaging with the conceptual repertoires of institutionalism and the broader literature on markets, hierarchies and networks.

369

370 *Handbook on complexity and public policy*

HOW HAS COMPLEXITY THEORY BEEN APPLIED TO HEALTH POLICY?

Complexity theory, as an overarching approach, is remarkably flexible regarding the scales at which it can be applied. In the hands of many scholars, complexity provides a way of understanding macro systemic behaviour. For others, the attraction of complexity theory is its usefulness in unpacking dynamics at a micro, behavioural level. Advocates of this approach note that the complexity lens provides a way of fruitfully linking macro, meso and micro levels of analysis, as core concepts such as feedback loops, strange attractors and emergence not only provide an analytic language for each level, but provide a way of bridging them (Abbott, 2001; Room, 2011).

Health policy is a vast domain and some crude mapping is necessary in order to elucidate the space in which complexity theory can be and/or has already been applied. David Hunter (2003) makes a basic distinction between 'upstream' and 'downstream' health policy. The 'upstream' health policy literature that deploys complexity theory shows a concentration of interest on understanding the complexity of determinants of health, and their effects on health outcomes. In applications of complexity concepts to 'upstream' health policy issues, attention is generally focused on the macro level (Glouberman et al., 2006; Alvaro et al., 2011), though there are examples of analyses at the micro-level (Matheson et al., 2009).

In Hunter's terminology, 'downstream' health policy refers to the broad range of health services that are funded, provided and/or regulated by government. This concerns the organization and delivery of health care services. This has provided even more fertile ground for complexity theory. The health services context, in which clinical autonomy is a crucial contextual feature, is one in which the potential for both inter-organizational coordination and fragmentation abound. At the heart of health services is the role of practitioners, their tacit knowledge and exercise of judgement. In contrast to upstream issues, the majority of applications of complexity theory in studies of downstream health policy tend to focus on 'micro' levels of organization, and broader patterns that emerge from self-organization at the frontline.

Two indicative examples of this emphasis in health services research come from a 2013 special issue of *Social Science & Medicine* on complexity theory and health. In the first of these articles, Lanham and her colleagues investigate two successful instances of 'scale-up and spread' of improvement initiatives – the use of mobile phone messaging to improve adherence to anti-retroviral treatment for HIV in Kenya, and measures to reduce the incidence of MRSA infection in hospitals in the United States (Lanham et al., 2013). Lanham et al. are particularly concerned with the question of how to achieve scale-up and spread across a range of organizations and settings with a wide variety of local contextual features. For these authors, 'understanding self-organization is critical to understanding variation across local contexts', and understanding the role of interdependencies within and between organizations, and the sensemaking (Weick, 1995) of participants are crucial features of what they term 'productive self-organization' (Lanham et al., 2013: 200).

In a second article, Essén and Lindblad provide an account of the 'nationwide spread of and continuous re-invention of an IT-based quality-registry and associated re-invention of rheumatologist practice in Sweden' over a twenty-year period (Essén and Lindblad, 2013).

Complexity and health policy   371

Both articles focus on service-level changes, and may be seen therefore as highly operational in focus. However, these stories delve deeply into the mechanisms by which particular initiatives gained traction, and each of the examples fit squarely into the realm of health policy. In both articles, the key changes over time are how clinicians make sense of what is going on (including how these perceptions and assumptions change), and how actors in organizations (both clinicians and managers) interact within and across service and organizational boundaries. The MRSA example is particularly interesting as it shows how networks within and between hospital clinicians changed – and how 'sense-making activities' at local sites were crucial to interpreting and sharing information. In the Swedish example, patients were also an important source of energy and innovation.

In both these examples, the repertoire of complexity theory concepts helps generate insights that complement policy studies literature that addresses health services and systems. Both stories, essentially, are of bottom-up innovation and diffusion, and both show how small, local changes grow and take off in ways that have significant implications for larger policy developments. With the notable exception of bottom-up policy implementation studies, policy studies have not had the conceptual repertoire to explore these dynamics, given their concentration on identifiable policy actors. The influence and significance of the actions of managers and clinicians only takes on policy significance in retrospect. The focus on 'sensemaking' also fits well with recent developments in health policy implementation literature in which Weick's work has gained prominence (Coleman et al., 2010; Dickinson, 2011).

Another common theme in applications of complexity theory to health is the *failure* and/or unintended consequences of governmental health policy initiatives. Examples from the *Social Science & Medicine* special issue include Xiao et al.'s analysis of pharmaceutical distribution in China (Xiao et al., 2013) and Hannigan's analysis of the introduction of novel mental health service delivery models in Wales (Hannigan, 2013). Other recent examples include an analysis of the failure of performance-based contracting in Uganda (Ssengooba et al., 2012), and policies related to Millennium Development Goals in low- and middle-income countries (Paina and Peters, 2012). These contributions provide some very fine-grained understandings of the dynamics of failures and unintended consequences.

EXPLANATORY THEORY OR REPERTOIRE OF CONCEPTS?

In the opening chapter of this volume, Cairney and Geyer raise important questions about complexity theory's contribution to the study of public policy. Given that health policy has provided fertile ground for the application of complexity concepts, are there particular nuances and development to be found in this domain of policy scholarship?

Research relevant to health policy is frequently highly interdisciplinary, and this can enable considerable cross-fertilization of concepts and metaphors between natural and social sciences. If it is the case that complexity theory has opened up the discussion between these often vastly different research traditions in ways that few conceptual frameworks and paradigms have, then we would expect the arena of health to be at the forefront of this movement. Indeed, this is the case for Plsek and Greenhalgh's *BMJ* piece. Both authors have established careers that span social and natural sciences. Plsek

372   *Handbook on complexity and public policy*

trained as an engineer and is a prominent figure in organizational studies. Greenhalgh trained and practised as a general practitioner before embarking on an academic career that has ranged widely across the social science landscape.

However, this boundary-spanning and cross-fertilization has been the target of critique on the grounds that it is too eclectic. John Paley (2010) took Plsek and Greenhalgh to task on the basis that they misunderstood complexity theory. The heart of his critique is that in the physical sciences complexity theory is applied to *explain* phenomena such as termite mounds using mathematical models.

The weight of complexity scholarship as applied to health policy topics has certainly been towards the 'repertoire of concepts' end of the spectrum, and a cursory analysis of the uses of complexity in health can give an indication of whether there is a solid core to this theoretical enterprise. As an exercise, we take a group of the studies from the last decade that itemize elements of complexity theory that the authors consider to be central to the approach. This is an exercise for indicative purposes, as it is beyond the scope of this chapter to engage in a more systematic approach.

Table 22.1 suggests that no particular concept appears to be core. Some concepts, namely self-organization, emergence, non-linearity, feedback loops, and path dependency, are more prominent than others, but none are common to all. Some complexity concepts are only applied by one set of authors. Other concepts that some proponents of complexity theory would suggest are central – such as strange attractors – do not appear at all.

This need not be a problem. Complexity theory may be most useful to policy studies in providing a stock of metaphors, images and concepts that are useful to think with, rather than providing a coherent theory (Cairney, 2012). As argued above, the studies that use complexity theory to analyse the dynamics of scale-up and spread, or of how localized

Table 22.1   Complexity concepts applied to health policy

Glouberman et al. (2006)	Self-organization; critical nature of local conditions; non-predictability of interventions; dynamic interactions; multiple viewpoints; emergent characteristics
Alvaro et al. (2011)	History (path dependency); feedback loops; critical point; adjacent possibles
Ssengooba et al. (2012)	Non-linearity of implementation
Lanham et al. (2013)	Self-organization; interdependencies; sensemaking
Essén and Lindblad (2013)	Fluctuations; amplifying dynamics; recombination and repurposing; stabilizing dynamics
Paina and Peters (2012)	Path-dependence; feedback; scale-free networks; emergent behaviour; phase transitions
Trenholm and Ferlie (2013)	Self-organization; emergence of novelty; non-linearity; absence of single, formal leader; requisite system variety
O'Sullivan et al. (2013)	Emergence; self-organization; non-linearity; adaptiveness; connectivity

frontline initiatives gain wider traction, make a considerable contribution to policy studies literature. This is because they have reinvigorated and deepened understandings of 'frontline' practice, and widened the focus from resistance to and/or coping with external policy imperatives, to processes which can be seen as actually creating 'public value' (Moore, 1995).

The other issue that arises when any theoretical repertoire is used flexibly is whether the same concepts mean the same things when used by different researchers. Paley, for example, regards the conceptual stretching that has occurred as a result of crossing the science / social science as highly undesirable, preferring a tighter application that is more faithful to the mathematical, explanatory agenda of complexity theory as applied in the physical sciences. The rather pluralistic nature of the adoption of complexity concepts in health policy means that conceptual stretch is highly likely.

Self-organization certainly can mean lots of different sorts of things. Both market and network forms of social coordination could be described as self-organizing, yet they are very different to each other. Emergence is also a potentially wide category, arguably referring to any phenomenon that develops but whose development was not widely anticipated. What is clear is that few, if any, authors make any a priori definition of what these terms do and do not refer to. Indeed, such definitions probably go against the grain of those who are attracted to complexity theory. Once again, this lack of definitional rigour may not matter. After all, any prominent theory of public policy processes (take Sabatier's advocacy coalition framework or Kingdon's multiple streams approach as examples) are also highly prone to conceptual stretch, and arguably more prone the more popular they become. Conceptual development can be thought of as an evolutionary process in which some meanings and applications emerge as particularly useful, while others disappear.

METHODOLOGICAL ADVANCES

Given the above analysis of the use of complexity theory as a reservoir of handy concepts applicable to the field of health policy, any of which can be applied in different ways by different analysts, we would not expect clear methodological developments across the range of applications, even though individual studies may incorporate innovations. Indeed, if complexity theory is primarily a source of conceptual inspiration, then one could argue that there is no particular imperative for a consistent methodological approach.

The most common approach to empirical research across upstream and downstream approaches is the in-depth case study, or small-number comparative case studies. The most common method of analysis is qualitative thematic analysis and/or thick description. In this sense, complexity theory is brought in to provide the conceptual framework to analyse the data, but the same data could just as easily be linked to other systems of interpretation.

There are, however, some applications that draw more on the mathematical modelling aspects of complexity, particularly the analysis of trends in service utilization data developed by Haynes (2008). Mahamoud et al. (2013) use complexity concepts in the service of building simulation models to explore the possible effects of broad policy directions

374 *Handbook on complexity and public policy*

on health outcomes. However, it is not really possible to discern any general methodological trends here.

THEORETICAL CONTRIBUTIONS: IS COMPLEXITY THEORY JUST 'OLD WINE IN NEW BOTTLES'?

In this review of complexity theory as applied to health policy, there is some merit to the argument that complexity approaches are a repackaging of established ideas in social science, and in policy studies in particular.

The concepts of path-dependency, non-linearity and self-organization have been prominent in policy literature for some time now. As noted above, the critique of rationalist, top-down, linear, hierarchical styles of policy implementation, and the need to take notice of local implementation contexts, have been constant themes of policy literature for at least thirty years. Many of the approaches that tell stories of policy failure and/or unintended consequences of implementation processes are part of a lineage in policy studies that can be traced back to Pressman and Wildavsky's analysis of 'how great expectations in Washington are dashed in Oakland' (Pressman and Wildavsky, 1973). If the repackaging of well-established approaches to understanding implementation is all that the complexity lens offers, then perhaps there is no pressing need for yet another approach to explaining policy failure. However, two counter-arguments about the potential novelty of complexity-based approaches can be made.

First, although specific conceptual elements may not be new, it may be the combination of elements that is the most useful contribution of complexity approaches (indeed, it is recombination that is the driver of innovation according to complexity theorists of innovation). To be sure, existing approaches such as Kingdon's multiple streams (1984) and Baumgartner and Jones' punctuated equilibrium framework (1993) also bring together a number of concepts that are akin to complexity concepts. However, I think it really can be argued in some cases that the 'whole' provided by a complexity-inspired analysis of policy process is greater than the sum of the parts. Secondly, complexity concepts have been deployed in combination with elements from other theoretical frameworks in creative and fruitful ways. Lanham et al.'s analysis combines concepts from complexity (self-organization and interdependencies) and Weick's notion of 'sensemaking'. This combination of theoretical concepts enables the authors to explore first how frontline clinicians make sense of new information, and secondly how the diffusion of this sensemaking is pivotal to the spread of new practices.

However, another important consideration is that some of the best characteristics of the old wine may actually be lost in the process of transfer to new bottles. Alvaro et al. (2011) claim that complexity theory tends to be silent about power, and therefore needs to be supplemented by approaches that are able to conceptualize power. If they are right, this is a potential problem with the approach when applied to policy issues (see Graham Room's chapter in this volume). It is also a problem that famously afflicted postwar social theory in which the metaphors and conceptual scaffolding of systems and cybernetics were highly influential (Easton, 1953; Parsons, 1964).

GUIDANCE FOR POLICYMAKERS

Most researchers applying complexity theory see their research as having implications for politicians, public officials and other policy actors, although these are often couched in terms of advice about 'what not to do'.

A key question that arises with the application of complexity theory to health services and policy is that if improvement can come from 'self-organizing, emergent processes', then is it possible that the dynamics of self-organization can be harnessed by policymakers?

One possible response to this question is that harnessing is simply not possible because of the unpredictability of complex systems. In Paley's critique of Plsek and Greenhalgh he also suggests that self-organization is antithetical to intentionality and planning (Paley, 2010). However, few, if any, researchers in health policy subscribe to that particular interpretation, as this would rule out any application of complexity theory to the domain of health policy.

Nevertheless, those who apply complexity theory to policy stories do vary in their judgement of the extent to which complexity-based insights can be intentionally harnessed by health service managers and policy actors. A minimalist position, exemplified by Hannigan, is that key actors in health services appreciate 'the value of carefully considering the possible reverberations of innovation in order that the previously unanticipated becomes expected and planned-for' (2013: 218). Similarly, Xiao et al. highlight 'the importance of performing real-time monitoring and evaluation, with a focus on learning and adjusting policies rather than focusing on simply punishing failure or rewarding success' (2013: 227).

However, some contributors go a great deal further and attribute a more proactive role for health service managers and policymakers in applying complexity concepts. Lanham et al. outline a range of strategies for facilitating sense-making including 'encouraging the inclusion of participants' professional identities in group dynamics, viewing plans as tentative and open to new environmental cues and knowledge updates, encouraging mindful and critical reflection on previous events, and viewing surprises as opportunities to learn' (2013: 200).

Advice to policymakers from those who apply complexity theory often takes the form of a list of precepts. For example, Glouberman et al. (2006: 334–5) provide the following list of recommendations: (1) Gather local information; (2) respect history; (3) consider interaction; (4) promote variation; (5) conduct selection; (6) fine-tune processes; (7) encourage self-organization. Similarly, O'Sullivan et al. (2013: 240–41) advise policymakers to (1) manage complexity as a dynamic context; (2) build situational awareness and connect the dots; (3) dismantle silos and adopt a collaborative lens.

Lists of this nature seem to be popping up frequently wherever academics who apply the complexity lens turn their attention to giving advice; perhaps the best recent example is Room's (2011) list of precepts. This emphasis on contingency, adaptation and flexibility means that advice to policymakers is inevitably vague. Scholars who use complexity theory make little attempt to advise policymakers about what to do. Perhaps, then, the relationship between policy scholar and policy actor is analogous to the relationship between a sports psychologist and an athlete. While the psychologist may stress the importance of 'acting in the moment', for the athlete, what that means

376 *Handbook on complexity and public policy*

at any particular moment is not clear until that moment occurs. Therefore, the best the psychologist can do is to help the athlete reduce the influence of unhelpful extraneous thoughts and habits. So it seems that the role of policy scholars offering advice based on complexity theory is to encourage policymakers to reduce and maybe eliminate unhelpful habits of thought and perception. In the remainder of this chapter, I hone in on one particular piece of advice of this type that is common to most applications of complexity theory to health: the advice that policymakers need to avoid linear thinking and behaviour that results in top-down processes of policy formulation and implementation.

AN IMPLICATION OF COMPLEXITY SCHOLARSHIP: SHOULD POLICYMAKERS ABANDON HIERARCHY?

An emblematic example of this advice can be found in the following quote from David Kernick in his text on complexity and health care organization.

> Policy making is seen not as a set of explicit goals that are engineered from above, but an ongoing maintenance of activities and relationships underpinned by guiding principles. The future state of the NHS is inherently uncertain but emerges from the result of its constituents at a local level. (Kernick, 2004: 101)

In a similar vein, the authors of the Swedish rheumatology study suggest that the moral of the story is that 'policy makers need to get better at monitoring, making sense of and supporting (amplifying) promising practice-driven changes *rather than* initiating and imposing grand changes from above' (Essén and Lindblad, 2013: 211, emphasis added).

Such statements have a long lineage. In public policy theory, there has never been a shortage of theorists and approaches that take aim at reductionism, linearity and 'top-downism'. It is certainly possible to identify explicit proponents of something that resembles such a linear approach (Dror, 1971). The critique of linearity and top-downism in policy studies traces back at least as far as Lindblom (1959), and such arguments appear and reappear in every generation of policy studies, albeit in different conceptual clothing (van Gunsteren, 1976; Fischer, 1990; Parsons, 2004). Overall, the weight of policy studies literature since the 1980s at least, has been highly sceptical of both linear, causal explanations of policy outcomes, and of highly prescriptive, hierarchical processes of implementation. As a central organizing theme, therefore, there is much in common between conventional policy studies approaches to health policy issues, and newer analyses influenced by complexity theory. However, those who have applied complexity theory to health services and policy frequently make an additional move that I believe is problematic in that by doing so, they miss a great deal that may be both interesting and intelligible in terms of complexity theory.

Complexity theory is defined by its proponents as a way of conceptualizing the world (of policy, of health care) that is distinguished from what these proponents regard as an established orthodoxy. This, of course, is what any innovative theoretical approach must do. When applied to health policy (or I suspect any area of policy for that matter), it is a short step to identify aspects of policy that appear to be inspired by 'the old and/ or wrong way of thinking' as inherently problematic. The raw form of this argument in the health services context, is that top-down, hierarchical approaches to policy are based on linear and mechanistic thinking, and as such unnecessarily constrain and simplify the social world such that these policy prescriptions do a great deal of damage and/or are doomed to fail in their own terms. This can be boiled down to a contention that the hierarchical command impulse is the antithesis of the self-organizing. By contrast, self-organization is regarded as natural and therefore virtuous, and hierarchical direction is 'artificial' and therefore to be avoided.

The tendency to distinguish between 'natural' and 'artificial' is a common theme of a large and diverse range of social theory that has influenced the study of public policy. This applies equally to advocates of market mechanisms in policy, many proponents of networks as a superior source of policy innovation, or any of the bottom-up approaches to implementation that emerged in the 1980s. In each of these, the merits of natural, dispersed coordination are juxtaposed against the over-reaching ambitions of planning, control and hierarchy. This stance is problematic for a number of reasons, but a particularly dangerous and self-defeating move within complexity theory. To understand why this is a self-defeating path for advocates of complexity theory to follow, it is necessary to reconnect with some broader themes and traditions in public policy literature.

TREATING HIERARCHICAL IMPERATIVES AS *PRODUCTS* OF POLICY DYNAMICS

First, a complex systems perspective on the dynamics of health policy cannot afford to ignore the broader institutional context of constitutional, electoral democracy. Alongside this, we need to keep in mind the context of health policy in which the system of health services was not built by the state – rather it was the product of a long period of professional consolidation and control of and by medicine (the co-evolution of the medical profession and the hospital). In the UK and its relatively wealthy former colonies such as Canada, Australia and New Zealand, governments entered this space primarily to 'pick up the tab' once the highly inequitable and inefficient consequences of a system built around the needs of the profession became apparent in the first few decades of the twentieth century (Hay, 1989; Tuohy, 1999; Ham, 2000; Klein, 2000). This development occurred when the combination of medical professionalism and market coordination became unsustainable due to the fact that the cost of increasingly specialized, hospital-based medical care escalated to the point that the middle class could no longer afford it. In other European countries, pre-existing institutional mechanisms such as social insurance were re-engineered, such that the state's roles and responsibilities for governing health systems were less to the fore, but nevertheless still important.

The key implication is that in a health system that is required to demonstrate democratic accountability – particularly in systems where health services are funded and provided by government – there are longstanding and legitimate expectations that key policy actors are required to make and implement policy (or at the very least appear that they are doing so). Hierarchical, 'top-down' directives are a product of this dynamic. I'm not arguing that this imperative means that politicians and public officials necessarily know what they are doing, or are skilled in achieving their objectives. The point is that there is a strong imperative for them to follow this script.

This sets up a persistent paradox in the downstream health policy arena dominated

378 *Handbook on complexity and public policy*

historically by provider concerns and interests, and this paradox is particularly strong in countries that have NHS-style systems where the will to control is strong, but the capacity to control is often weak. In this sense, the hyperactivity of successive regimes of ministers, senior public officials and public management fashions, is itself an emergent property of democratic and health service institutions in a tax-funded health system. The hierarchical imperatives that are continually produced can certainly be regarded as self-organizing behaviour, especially when 'making one's mark' is the currency that measures political and managerial success. Similarly, the bureaucratic impulse to audit, monitor and measure can also be regarded as an emergent property of publicly funded services such as health in this context. In this sense, there is nothing artificial about hierarchical imperatives at all; they are a 'natural' consequence of evolution of health systems in specific political institutional contexts.

Secondly, there is a ready stock of conceptual tools that offers a much more nuanced and productive way of understanding hierarchy in a broader context. This stock can be found in policy (and much broader social science) literature on hierarchies, markets and networks/heterarchy (HMN) as distinct ideal types of social coordination (Rhodes, 1997; Tuohy, 1999; Thompson, 2003; Tenbensel, 2005). This is a conceptual framework that shares with complexity theory the virtue of being applicable to macro, meso and micro contexts.

This range of forms is recognizable across a wide variety of social contexts, and they can all be considered 'patterns' of (self-)organizing (Colebatch and Larmour, 1993). When a new group forms for a particular purpose its members may decide that a hierarchical form is the most appropriate for what it wants to do. There may be very good reasons for doing so, including the need to demonstrate accountability to external audiences. Hierarchical features of coordination can and do emerge 'spontaneously' in all sorts of settings, including in health care.

A cornerstone of the HMN literature is that the use of *any* of these forms of coordination can result in dysfunctional consequences, particularly when they fail to take account of the legacies of other modes of coordination (Entwistle et al., 2007). So while 'doses of hierarchy' can have undesirable side-effects, the same can be said for 'doses of markets' and 'doses of heterarchical networks'. In this theoretical framework, there is no reason to single out one type of coordination as being any more or less inherently problematic than any other.

The particular contribution of the HMN literature (and its cousin, cultural theory) is that hierarchies, markets and networks, as ideal types of coordination, only exist in relation and juxtaposition to each other (Rhodes, 1997; Thompson and Ellis, 1997; Jessop, 2000). Manifestations of each mode of coordination constantly rub up against other pre-existing and/or co-evolving coordination styles. In fact, any mode may emerge because of dissatisfaction with the consequences of pre-existing modes. A wonderful metaphor for this, highly consistent with complexity theory, is provided by Christopher Hood when he characterizes virtuosic public management (the deployment of different modes in combination) in terms of a step-dancer's 'ability to shift the balance among a set of ambitious positions no one of which can be sustained for long' (Hood, 2000: 211). This gives us another way of understanding how hierarchical policy imperatives are a *product* of dynamic processes.

There is a third reason why it is a mistake to move hastily from 'linear thinking is

Complexity and health policy   379

problematic' to 'hierarchical governance is problematic'. It is crucial to distinguish between the will to control, and the mechanisms by which governments and governing agencies (and other policy actors) use as part of their attempt to control. There is now a prolific body of literature on metagovernance – the idea that state actors consciously use any or all of the modes (hierarchy, market, network) in their attempts to steer (Jessop, 2003; Sørensen, 2006; Bell and Hindmoor, 2009). Hierarchical policy directives (what governments command) are now typically enacted using a mix of coordination mechanisms that include markets and networks.

Implementation processes are rarely specified in sequential, linear detail, and such a style is highly unlikely in health policy. The linearity associated with hierarchy was simply a background ontological assumption, and the contribution of the early implementation scholars was to make this assumption explicit (Pressman and Wildavsky, 1973; Hogwood and Gunn, 1984). But that simple assumption has been largely replaced over the past 30 years in both policy theory and practice by a precept which has a subtly different character. The hierarchical motif that has emerged over the past 30 years is one of '*we* (the principal) set the *direction, you* (the agent) work out for yourselves *how* to get there'. This certainly requires the use of *authority* (command) but not necessarily *linearity*. One could argue that such a shift in the techniques of hierarchy is completely consistent with the advice of complexity theorists to be sensitive to local context, in that it is perfectly acceptable for different implementing agents to fulfil the same specified policy objective via quite different means.

RE-INCORPORATING HIERARCHY INTO COMPLEXITY THINKING

A good example of this is the use of health targets in New Zealand since the late 2000s. For many commentators on UK health policy, targets are a common *bête noir* (Gubb, 2009; Geyer, 2012). The way in which targets were applied in England in particular provides plenty of grist to the mill of those who see such tough regimes of performance management as crowding out 'softer' forms of information and intrinsic motivations of providers (Radin, 2006), or as stimulating undesirable gaming behaviour (Bevan and Hood, 2006; Radnor, 2008). There is no doubt that targets *can* have such effects. Whether they *do*, however, is very much an empirical question, and not something that can be predetermined simply by the fact that a target is used.

We miss a great deal if the use of targets is *only* understood as a mechanism of top-down control of provider behaviour. One of the New Zealand targets applied to hospital emergency department (ED) waiting times. Here there is a close resemblance to one of the most controversial of the English health targets. In the New Zealand case, the introduction of the ED target was largely attributable to the political pressure from emergency department clinicians who had been trying to get the issue of ED overcrowding on the policy agenda. An incoming government and minister prioritized this issue, and the target of 95 per cent ED admissions seen, treated and/or discharged within six hours was introduced in mid-2009 (Tenbensel, 2009). The story of the target's implementation illustrated many dynamics one could expect when applying a complexity lens. These included resistance to the target from some clinicians, particularly inpatient hospital

380   *Handbook on complexity and public policy*

specialists who resisted encroachments on their clinical autonomy, and some gaming behaviour such as moving patients to acute assessment units in order to stop the target clock (Chalmers, 2014). However, implementation of this target was also characterized by clinician buy-in, particularly where clinicians were able to leverage off the target to secure more resources and staff. Local political alliances emerged between ED clinicians, hospital managers and nurse managers, all of whom showed some commitment to achieving the target. The target served to stimulate a focus on information systems and data capture, and managers and frontline staff built new systems and developed their own service innovations that varied across implementing hospitals (Chalmers, 2014).

Another New Zealand target required local health organizations to ensure that 95 per cent of two-year-old children were fully immunized. This service area was one which had been characterized by a fragmented organizational environment, and a frontline perception that such a target was a worthy aspiration, but impossible to reach (Willing, 2014). This target also stimulated new forms of local networking and cooperation which in turn was partly stimulated by focused accountability on reaching specified targets within each organization, and by an emerging dynamic of 'collegial competition' between general practices, and between primary care organizations. Information sharing between districts also emerged as those who were falling behind sought to learn from those who were closer to achieving the target. As for the ED target, local systems of data capture were tightened, and data was used to inform practice (Willing, 2014). These were all changes that were primarily produced by the external hierarchical stimulus of the target.

The point about these targets is that neither of them was purely hierarchical, and neither involved central government specifying *how* to achieve them in the form of linear prescriptions. The working out of how to achieve them was the product of local processes that drew on practitioner knowledge, experience, micro-political dynamics and networks. In both cases, these processes simultaneously involved hierarchical *and* network coordination. In the case of the immunization target, there was a 'positive feedback loop' in which hierarchical and network dynamics positively reinforced each other.

As such, it is a mistake to fall into the trap of thinking that hierarchical instruments such as health targets are inevitably crude, top-down implements that are incapable of incorporating adaptive feedback and learning at the local, practitioner level. Instead, a more agnostic, open and empirical approach to applying the complexity lens is needed. Aversion to hierarchy is not an inevitable weakness of complexity theory: rather it is something that can be rectified by grafting of complementary theoretical frameworks to the repertoire of complexity theory.

Instead of consigning hierarchical governance to the shadows of complexity theory, it may be more productive to treat hierarchical relations as a particular type of pattern, and hierarchical directives and imperatives as potentially significant 'events' in complex-adaptive systems. In this way they can be analysed as constituent elements of complex systems that interact, recombine and exist in friction with other varieties of coordination (that is, markets, network and community). It may even be fruitful to analyse health policy dynamics in terms of feedback loops that are generated when two or more modes of coordination are present within the same space.

This way of thinking about hierarchy is consistent with the move by Graham Room (2011) to link complexity theory to institutionalist theory. Room canvases both the rational choice institutionalism of Elinor Ostrom and Fritz Scharpf, and the historical institutionalism of Paul Pierson and Kathleen Thelen and argues that these approaches fit well with complexity theory. Using these conceptual frameworks can also help to counteract the tendency in health services applications of complexity theory to focus predominantly on micro, endogenous sources of change rather than macro, exogenous factors.

In health policy scholarship, historical institutionalism has been a very rich source of insight, and some analysts in this tradition also deploy the conceptual repertoire of HMN to explain historical dynamics and/or compare health systems. The best example is Carolyn Hughes Tuohy's *Accidental Logics* (Tuohy, 1999). More recently, analysts of Dutch health policy have used Thelen's concept of 'layering' to describe the dynamic, historical relationships between heterarchical, hierarchical and market mechanisms and incentives (Helderman, 2007; Van De Bovenkamp et al., 2013). These developments are also highly compatible with complexity-inspired approaches.

CONCLUSION

Health policy analysis, situated as it is at the intersection of public policy, management and administration, health services research, epidemiology and medical science may prove to be a very conducive environment for the further development of complexity-inspired concepts and insights. To date, the application of complexity concepts applied to health policy has no clear conceptual core, or clear methodological advances. Rather, complexity theory has provided a relatively novel conceptual repertoire, the contents of which are readily and creatively combined with other theoretical concepts and frameworks.

As complexity analysis matures, we can expect this to change, and a more stable, readily identifiable set of ideas and approaches will probably emerge. However, for complexity theory to stand the test of time in this domain of research, it is crucial that any such theoretical developments continue to be characterized by cross-fertilization with other theoretical traditions such as historical institutionalism at the macro-level and frameworks such as sensemaking at the micro-level. The 'modes of governance' (hierarchies, markets, networks) tradition is another approach that is potentially compatible with complexity theory at any level of analysis. Engagement with theoretical traditions such as these should help to ensure that future growth in the use of complexity theory in health policy scholarship does not come at the expense of analyses of power, or of a more nuanced understanding of the hierarchical imperatives associated with democratic politics and policy implementation.

REFERENCES

Abbott, A. (2001), *Time Matters: On Theory and Method*, Chicago: University of Chicago Press.
Alvaro, C., L.A. Jackson, S. Kirk, T.L. McHugh, J. Hughes, A. Chircop and R.F. Lyons (2011), 'Moving Canadian governmental policies beyond a focus on individual lifestyle: Some insights from complexity and critical theories', *Health Promotion International*, **26**(1), 91–9.
Baumgartner, F.R. and B.D. Jones (1993), *Agendas and Instability in American Politics*, Chicago: University of Chicago Press.

382 *Handbook on complexity and public policy*

Bell, S. and A. Hindmoor (2009), *Rethinking Governance*, Melbourne: Cambridge University Press.

Bevan, G. and C. Hood (2006), 'What's measured is what matters: Targets and gaming in the English public health care system', *Public Administration*, **84**(3), 517–38.

Cairney, P. (2012), 'Complexity theory in political science and public policy', *Political Studies Review*, **10**(3), 346–58.

Chalmers, L. (2014), *Inside the Black Box of Emergency Department Time Target Implementation in New Zealand*, PhD, University of Auckland.

Colebatch, H. and P. Larmour (1993), *Market, Bureaucracy, and Community: A Student's Guide to Organisation*, London: Pluto Press.

Coleman, A., K. Checkland, S. Harrison and U. Hiroah (2010), 'Local histories and local sensemaking: A case of policy implementation in the English National Health Service', *Policy & Politics*, **38**, 289–306.

Dickinson, H. (2011), 'Implementing policy', in J. Glasby (ed.), *Evidence, Policy and Practice: Critical Perspectives in Health and Social Care*, Bristol: Policy Press, pp. 71–84.

Dror, Y. (1971), *Design for Policy Sciences*, New York: American Elsevier.

Easton, D. (1953), *The Political System, an Inquiry into the State of Political Science*, New York: Knopf.

Entwistle, T., G. Bristow, F. Hine, S. Donaldson and S. Martin (2007), 'The dysfunctions of markets, hierarchies and networks in the meta-governance of partnership', *Urban Stud*, **44**(1), 63–79.

Essén, A. and S. Lindblad (2013), 'Innovation as emergence in healthcare: Unpacking change from within', *Social Science & Medicine*, **93**(0), 203–11.

Fischer, F. (1990), *Technocracy and the Politics of Expertise*, Newbury Park, CA: Sage Publications.

Geyer, R. (2012), 'Can complexity move UK policy beyond "evidence-based policy making" and the "audit culture"? Applying a "complexity cascade" to education and health policy', *Political Studies*, **60**(1), 20–43.

Geyer, R. (2013), 'The complexity of GP commissioning: Setting GPs "free to make decisions for their patients" or "the bravest thing" that GPs will ever do', *Clinical Governance*, **18**(1), 49–57.

Glouberman, S., M. Gemar, P. Campsie, G. Miller, J. Armstrong, C. Newman, A. Siotis and P. Groff (2006), 'A framework for improving health in cities: A discussion paper', *Journal of Urban Health*, **83**(2), 325–38.

Gubb, J. (2009), 'Have targets done more harm than good in the English NHS? Yes', *British Medical Journal*, **338**, 130.

Ham, C. (2000), *The Politics of NHS Reform*, London: King's Fund.

Hannigan, B. (2013), 'Connections and consequences in complex systems: Insights from a case study of the emergence and local impact of crisis resolution and home treatment services', *Social Science & Medicine*, **93**(0), 212–19.

Hay, I. (1989), *The Caring Commodity: The Provision of Health Care in New Zealand*, Auckland: Oxford University Press.

Haynes, P. (2008), 'Complexity theory and evaluation in public management', *Public Management Review*, **10**(3), 401–19.

Helderman, J.-K. (2007), *Bringing the Market Back In? Institutional Complementarity and Hierarchy in Dutch Housing and Health Care*, PhD, Erasmus University Rotterdam.

Hogwood, B.W. and L.A. Gunn (1984), *Policy Analysis for the Real World*, Oxford and New York: Oxford University Press.

Hood, C. (2000), *The Art of the State*, Oxford: Oxford University Press.

Hunter, D. (2003), *Public Health Policy*, Oxford: Polity Press.

Jessop, B. (2000), 'The dynamics of partnership and governance failure', in G. Stoker (ed.), *The New Politics of British Local Governance*, Oxford: Oxford University Press.

Jessop, B. (2003), 'Governance and metagovernace: On reflexivity, requisite variety and requisite irony', in H.P. Bang (ed.), *Governance as Social and Political Communication*, Manchester: Manchester University Press, pp. 101–16.

Kernick, D. (2004), *Complexity and Healthcare Organization: A View from the Street*, London: Radcliffe Medical Publishing.

Kingdon, J.W. (1984), *Agendas, Alternatives, and Public Policies*, Boston: Little, Brown.

Klein, R. (2000), *The New Politics of the NHS*, London: Prentice Hall.

Lanham, H.J., L.K. Leykum, B.S. Taylor, C.J. McCannon, C. Lindberg and R.T. Lester (2013), 'How complexity science can inform scale-up and spread in health care: Understanding the role of self-organization in variation across local contexts', *Social Science & Medicine*, **93**(0), 194–202.

Lindblom, C. (1959), 'The science of muddling through', *Public Administration Review*, **19**, 79–88.

Mahamoud, A., B. Roche and J. Homer (2013), 'Modelling the social determinants of health and simulating short-term and long-term intervention impacts for the city of Toronto, Canada', *Social Science & Medicine*, **93**, 247–55.

Marchal, B., S. Van Belle, V. De Brouwere and S. Witter (2013), 'Studying complex interventions: Reflections from the FEMHealth project on evaluating fee exemption policies in West Africa and Morocco', *BMC Health Services Research*, **13**(1).

Matheson, A., K. Dew and J. Cumming (2009), 'Complexity, evaluation and the effectiveness of community-based interventions to reduce health inequalities', *Health Promotion Journal of Australia*, **20**(3), 221–26.

Moore, M.H. (1995), *Creating Public Value*, Cambridge: Harvard University Press.

O'Sullivan, T.L., C.E. Kuziemsky, D. Toal-Sullivan and W. Cornell (2013), 'Unraveling the complexities of disaster management: A framework for critical social infrastructure to promote population health and resilience', *Social Science & Medicine*, **93**(0), 238–46.

Paina, L. and D.H. Peters (2012), 'Understanding pathways for scaling up health services through the lens of complex adaptive systems', *Health Policy and Planning*, **27**(5), 365–73.

Paley, J. (2010), 'The appropriation of complexity theory in health care', *Journal of Health Services Research and Policy*, **15**(1), 59–61.

Parsons, T. (1964), *The Social System*, New York: Free Press.

Parsons, W. (2004), 'Not just steering but weaving: Relevant knowledge and the craft of building policy capacity and coherence', *Australian Journal of Public Administration*, **63**(1), 43–57.

Plsek, P.E. and T. Greenhalgh (2001), 'The challenge of complexity in health care', *British Medical Journal*, **323**(7313), 625–8.

Pressman, J. and A. Wildavsky (1973), *Implementation*, Berkeley: University of California Press.

Radin, B. (2006), *Challenging the Performance Movement: Accountability, Complexity and Democratic Values*, Washington, DC: Georgetown University Press.

Radnor, Z. (2008), 'Muddled, massaging, manœuvring or manipulated?: A typology of organisational gaming', *International Journal of Productivity and Performance Management*, **57**(4), 316–28.

Rhodes, R.A.W. (1997), 'From marketisation to diplomacy: It's the mix that matters', *Australian Journal of Public Administration*, **56**(2), 40–53.

Room, G. (2011), *Complexity, Institutions and Public Policy: Agile Decision-Making in a Turbulent World*, Cheltenham, UK and Northampton, MA, USA: Edward Elgar Publishing.

Sørensen, E. (2006), 'Metagovernance: The changing role of politicians in processes of democratic governance', *American Review of Public Administration*, **36**(1), 98–114.

Ssengooba, F., B. McPake and N. Palmer (2012), 'Why performance-based contracting failed in Uganda: An "open-box" evaluation of a complex health system intervention', *Social Science and Medicine*, **75**(2), 377–83.

Tenbensel, T. (2005), 'Multiple modes of governance', *Public Management Review*, **7**(2), 267–88.

Tenbensel, T. (2009), 'Emergency department waiting time targets', available at http://www.hpm.org/survey/nz/a13/2.

Thompson, G. (2003), *Between Hierarchies and Markets*, Oxford: Oxford University Press.

Thompson, M. and R. Ellis (1997), 'Introduction', in R. Ellis and M. Thompson (eds), *Culture Matters*, Boulder: Westview Press.

Trenholm, S. and E. Ferlie (2013), 'Using complexity theory to analyse the organisational response to resurgent tuberculosis across London', *Social Science & Medicine*, **93**, 229–37.

Tuohy, C.H. (1999), *Accidental Logics: The Dynamics of Change in the Health Care Arena in the United States, Britain and Canada*, New York: Oxford University Press.

Van De Bovenkamp, H.M., M. De Mul, J.G.U. Quartz, A.M.J.W.M. Weggelaar-Jansen and R. Bal (2013), 'Institutional layering in governing healthcare quality', *Public Administration*, **92**(1), 208–23.

Van Gunsteren, H. (1976), *The Quest for Control: A Critique of Rational-Central Rule Approach in Public Affairs*, London: John Wiley and Sons.

Weick, K. (1995), *Sensemaking in Organizations*, Thousand Oaks: Sage.

Willing, E. (2014), *Understanding the Implementation of New Zealand's Immunisation Health Target for Two-year-olds*, PhD, University of Auckland.

Xiao, Y., K. Zhao, D.M. Bishal and D.H. Peters (2013), 'Essential drugs policy in three rural counties in China: What does a complexity lens add?', *Social Science & Medicine*, **93**(0), 220–28.

23. A case study of complexity and health policy: planning for a pandemic
Ben Gray

INTRODUCTION

The health sector has achieved many things by applying a reductionist approach to understanding health problems. People who can see following cataract surgery, who have been cured of TB with anti-tuberculous drugs or saved from smallpox because it was eradicated by immunization, have a lot to be grateful for. As a result of these successes the provision of healthcare in most Western countries is dominated by a reductionist approach of relying on 'specialization' to provide services, valuing detailed knowledge about a narrow area of human function over an understanding of the whole. There is increasing emphasis on 'Evidence Based Medicine' of which the gold standard evidence is the meta-analysis of multiple randomized controlled trials (Sackett et al., 1996). This is a method that relies on trying to eliminate all variables except one to study the outcome. There is also a focus on 'universal' ethical principles for doctors to follow, at least in the majority of cases. Medical practitioners are encouraged to adopt the same solution to individual problems based on the evidence and universal ethical principles.

Yet the results of these trials and principles do not provide an adequate set of rules for practitioners to follow. They may be useful to address simple problems in which there is a universal solution, but not complicated or complex problems which defy simple solutions. In this chapter, I present this argument from the perspective of a busy general practitioner in healthcare, faced with limited information, complexity, and the need to negotiate medical decisions with a wide range of colleagues and patients. First, I expand on the difference between simple, complicated and complex problems in healthcare. Second, I present a critique of the influence of pharmaceutical companies and other factors, which produce a focus on simple problems, treating individuals with medicine rather than considering the wider context of their health. Third, I identify the limitations to a reduction of ethics into four key principles. Fourth, I provide a detailed case study of my involvement in a response to a pandemic, to identify the large number of (often unpredictable) factors that have to be taken into account when planning for and/or responding to a healthcare event. I argue that, if healthcare systems are complex systems, it is impossible to apply a reductionist logic in which there is 'one best way' or a clear solution to problems. One alternative is to seek partial solutions which generate a degree of consensus within a population. Healthcare is about social relations and community building as much as health science.

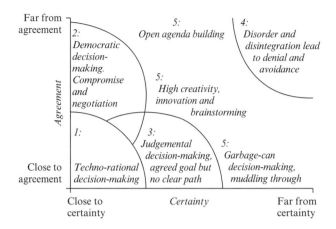

Figure 23.1   Stacey diagram

HOW DOES HEALTH POLICY FIT INTO COMPLEXITY THEORY?

The Stacey diagram (Figure 23.1) (Geyer and Rihani, 2010) allows us to see that the health policy successes of reductionist thought are all found in Zone 1 in the region where 'problem and response are clear and actors agree on how the problem should be solved'.

This is described by Glouberman and Zimmerman (2002) (Table 23.1) in slightly different terms by separating problems into 'simple' 'complicated' and 'complex'. Simple problems are reliably solved with the use of a recipe/protocol; complex problems benefit from expertise but each situation is unique so protocols have limited application.

Table 23.1   Simple, complicated and complex problems

Following a recipe (simple)	Sending a rocket to the moon (complicated)	Raising a child (complex)
The recipe is essential	Formulae are critical and necessary	Formulae have a limited application
Recipes are tested to assure easy replication	Sending one rocket increases assurance that the next will be OK	Raising one child provides experience but no assurance of success with the next
No particular expertise is required But cooking expertise increases success rate	High levels of expertise in a variety of fields are necessary for success	Expertise can contribute but is neither necessary nor sufficient to assure success
Recipes produce standardized products The best recipes give good results every time	Rockets are similar in critical ways There is a high degree of certainty of outcome	Every child is unique and must be understood as an individual Uncertainty of outcome remains
Optimistic approach to problem possible	Optimistic approach to problem possible	Optimistic approach to problem possible

386 *Handbook on complexity and public policy*

In developing health policy it is vitally important to determine whether the problem being addressed is simple, complicated or complex. If it is simple then a protocol (recipe) can be drawn up with a high chance of success, but if it is complex then such a protocol would probably fail. For example trying to stop a measles epidemic from expanding within a community that believes in the efficacy of immunization is at worst a complicated problem: there is close to certainty about the intervention (immunization for measles works) and agreement in the community means that implementation will be able to proceed (within Zone 1 of the Stacey diagram). If vaccine efficacy was low and there was substantial opposition within the community then the problem would become a complex one (Zone 5 in the Stacey diagram).

One of the difficulties in developing health policy is that there are factors that assert that health problems are more simple than they really are.

Bioethics/Values

Bioethics is a field that is well explained by looking through a complexity lens. Ethical problems are by definition ones that do not have an agreed answer (or we would not be discussing what the ethical answer should be). There is usually incomplete information: how effective might this treatment be; what the prognosis of the condition is; what other people affected by the decision think. There are often conflicting values, for example between the relative importance of increased longevity versus quality of life, and the costs of treatment often need to be considered. However, arguably the most dominant text in bioethics (Khushf, 2004), Beauchamp and Childress' *Principles of Biomedical Ethics* (2009), asserts that for most ethical problems there is a correct solution. They declare that there is a 'Common Morality' (p. 2) which is a 'set of norms shared by all persons committed to morality' (p. 3). They then declare four principles as being 'general guidelines for the specification of more general rules' (p. 12): autonomy, justice, non-maleficence and beneficence. From the general norms, 'specifications' can be made to further describe the norm. They acknowledge that there is sometimes a moral dilemma. Their position is that there is nearly always a single right way to do things and that this is discovered by rational thought applying the principles. They believe that these principles transcend culture and should apply to all people.

There is no acknowledgement that their account has inevitably been developed from within a particular culture, based on the values of that culture. Their approach is akin to that discussed by Taylor (2003) in relation to the medical profession which she described as the 'Culture of no Culture': doctors are right and everyone else has a 'cultural' opinion. Beauchamp and Childress place great emphasis on individual rights and state: 'It is extremely difficult for morally compelling social objectives to outweigh basic rights' (Beauchamp and Childress, 2013: 16).

Hofstede et al. (2010) have pointed out that the USA is the most individualistic country and that the majority of the world's population live in countries with a much more collective orientation.

Collective societies, such as Maori in New Zealand, would want to add 'the well-being of the collective'; environmentalists would add 'the well-being of the planet', but, despite much criticism, the seventh edition of the book published in 2013 (Beauchamp and Childress, 2013) retains just the four original principles. This approach to bioethics

Complexity and health policy: planning for a pandemic 387

directs the practitioner along a path of considering the problem and rationally deciding on *the* (reductionist) answer, by contrast with a public health ethics (systems complex) approach described below that puts much greater emphasis on dialogue with stakeholders to develop an answer that is acceptable to those involved.

Best Practice Guidelines

Practice guidelines are an essential tool with which to synthesize the available evidence and provide guidance to coalface clinicians who cannot possibly keep up to date with the literature. For simple problems such guidelines are invaluable. The World Health Organisation Surgical Checklist (Gawande, 2010) is a good example. The checklist was developed in response to the high levels of morbidity and mortality from surgery. Whilst surgery is a complex activity that cannot be done according to a protocol and requires experience to do well, there are elements of the task that are simple. These are elements that are in Zone 1 of the Stacey diagram, where there is close to certainty that the intervention is successful and close to agreement on the action involved. For example, one element of the checklist is to ensure that no surgical instruments are left behind in the abdominal cavity after an operation. However, because medicine does not have a wide appreciation of complexity science, such guidelines are also applied to complex problems. There are best practice guidelines on breast cancer screening (Willett, 2010), stroke (Donnellan et al., 2013) and a major source of advice to New Zealand GPs is the 'Best Practice Advisory Centre' (BPAC.org.nz). Without a clear understanding of the difference between a protocol for a simple problem (close to certain evidence with agreement from all parties) and a guideline for a complex problem (uncertain evidence without agreement from all parties), there is a risk of presuming that complex problems are simpler than they really are.

Commercial Pressures

Best practice is determined from the evidence. The literature on Evidence Based Medicine provides a hierarchy of research methods according to the likelihood of reaching truthful conclusions (Guyatt and Rennie, 2002). At the top of the hierarchy are randomized controlled trials (and meta-analysis of these). These aim to simplify the problem by controlling all variables except one. As a methodology, it is best suited to the investigation of pharmaceuticals, and much more difficult to apply to other interventions. As a result, high quality evidence is dominated by treatments amenable to randomized controlled trials, particularly pharmaceuticals. Doing medical trials is very expensive, so only organizations that stand to make money out of the proposed treatment are likely to invest in such trials. There is some evidence, for example, that exercise is an effective treatment for depression (Rimer et al., 2012) but Rimer et al.'s review notes that more research is needed. Such research is only likely to be done by publicly funded bodies that have a much smaller promotional budget than drug companies to promulgate the results. Designing a study on the efficacy of exercise is more complicated than designing a study on drug treatment. It is hard to do placebo exercise; ensuring compliance with the exercise regime is more complicated than compliance with a pill, and it is very likely that within a study population there will be variation in the amount of exercise already being done, leading

388 *Handbook on complexity and public policy*

to confounding of the results. As a result, doctors are much more likely to prescribe an anti-depressant than to encourage exercise. The pharmaceutical industry exerts significant financial pressure and is very successful. In 2002 the combined profits for the ten drug companies in the Fortune 500 in the United States (US$35.9 billion) exceeded the profits of the other 490 businesses put together (US$33.7 billion) (Washington, 2011). Some of these pharmaceuticals are of great value for conditions which are simple, for example vitamin D for rickets. However, many treat conditions that are better described as complex problems: psychiatric illness, cardiovascular disease, effects of diabetes, and there is a big risk that funding on simple solutions from the evidence of randomized controlled trials interferes with the search for more complex ways forward. For example Reeve et al. (2013) argue that it is likely that Expert General Practice is a better approach than increased pharmaceutical use to managing patients with multiple co-morbidities, but that establishing the evidence to support this assertion is difficult.

Individual Patient Care versus a Public Health Approach

As a rule, commerce argues strongly for the individual patient model of healthcare and patient choice. The prototypical example of this is the tobacco industry which, for decades, promoted the simple analysis of the problem: that it is people who choose to smoke. It has been clearly shown (World Health Organization, 1998) that merely relying on persuading people to stop smoking is an ineffective strategy for controlling this epidemic, and that a complex approach is required that involves individual patient approaches, as well as controls on marketing, taxes on the product, and public health campaigns to change social acceptability. A similar scenario is playing out with the obesity epidemic, where the food industry that sells unhealthy foods focuses on patient choice as the problem, and obesity as a problem of the individual being unable to control themselves. There is no evidence that this approach will make any difference to the epidemic, and an approach similar to that adopted for tobacco will need to be developed (Nestle, 2000).

The combination of the idea of a 'cure' for everything, seeing all problems as residing in the individual, and the embedded view of medical ethics that there is a right solution to every problem, which underpins the proliferation of Best Practice Guidelines, encourages people to think that the problem they are addressing is within Zone 1 of the Stacey diagram and is a simple problem with a single solution.

Long-term Conditions are Complex Problems

In addition to the bias within the sector to seeing problems as being in Zone 1 of the Stacey diagram, a majority of the work of the health sector is addressing problems that are undoubtedly complex and fall within Zone 5. Around one in four people live with a long-term condition; the majority with more than one (Barnett, 2012; van Weel and Schellevis, 2006); 70 per cent of health service spending in the UK, for example, is on dealing with long term conditions (Kings Fund, 2012; UK Department of Health, 2012). These are problems for which there is not a solution.

Public Health Practice and Complexity

Public health is the medical discipline most attuned to the complexity paradigm: it inevitably focuses on systems. Tobacco control is an excellent example of an issue requiring this focus. Thomas and Gostin's (2013) recent review of the endgame begins by saying 'There are complex legal and ethical trade-offs involved in using intensified regulation to bring smoking prevalence to near-zero levels' (p. i55), and follows with an analysis of the conflicting results of varying policies and the difficulty of knowing which is the right thing to do. Nonetheless this does not stop people from trying to assert that these are simple problems amenable to solution by the use of protocols. Dunne et al. (2013) describe an audit of cancer control coalitions in the USA to see whether they complied with Best Practice for tobacco control strategies.

An instructive area of public health policy to examine to see how complexity as a paradigm is understood, is the management of pandemics, particularly the Severe Acute Respiratory Syndrome (SARS) epidemic in 2002–03 (Wang and Chang, 2004) and the H1N1 influenza pandemic of 2009 (Moreno et al., 2009). Both of these outbreaks involved novel organisms. At the outset of the epidemic there was significant uncertainty around all the important parameters: degree of infectivity, likely mortality rate, most vulnerable populations and mode of transmission. After the SARS epidemic many jurisdictions developed or updated pandemic plans (Wilson and Baker, 2009), which were then utilized when the H1N1 pandemic arrived.

H1N1 INFLUENZA CASE SCENARIO

The primary care clinic I work in had the first case of H1N1 flu in our city, Wellington, New Zealand. We knew that the epidemic had arrived in Auckland and particularly that Melbourne (Australia) was affected. One of our patients came to the clinic with flu-like symptoms and a history of having just arrived from visiting relatives in Melbourne. The patient was isolated and swabbed and when the swab came back positive the public health authorities were contacted.

Planning for pandemics can be divided into five stages (Immunisation Advisory Centre New Zealand, 2013): planning for it; border management; cluster control; pandemic management; and recovery. New Zealand at the time were controlling the border to try to keep H1N1 out and trying to control clusters of outbreak by quarantining, so the public health workers came and instructed the staff who had had closest contact with the index patient to go home and avoid going out as much as possible. All the patients who were in the waiting room when the patient was there were contacted and told to stay at home, and the index patient was sent home to be cared for by her family. The plan incurred significant difficulties for us. We did not have locum receptionists available so that half of our receptionists were off work despite being well. The index family spoke limited English and rather than stay at home, family members went to the Emergency Department, worried about flu, making the risk of spread higher than if they had not been contacted. One of the patients in the waiting room (Ms B) had a history of significant personality disorder and on being told that she had been exposed to H1N1 and

390 *Handbook on complexity and public policy*

asked to stay at home, promptly went to the Emergency Department and announced to all present that she had swine flu.

The factors described in the first section affected the response in this case.

Shared Values

A presumption behind pandemic planning (congruent with conventional bioethics) is that everyone would want to act vigorously to contain and control a pandemic and thus put effort into developing a pandemic plan. This may not be true on the following levels:

1. Individual level: Ms B had little ability to focus on anyone but herself.
2. A local level: it might be that poor communities would be more focused on problems of good housing, food and access to urgent routine medical care. The focus of their values is more on immediate problems than future problems. The migrant family with the index case wanted medical attention for themselves and their family and were not concerned about the risk of spreading the infection to others.
3. The international level: The World Health Organization (WHO) has put much effort into pandemic planning and most people from developed countries would think that this is a high priority, as pandemics affect the whole world. However, not all countries see this the same way. Sedyaningsih et al. (2008) document the reasons why Indonesia ceased to provide samples of H5N1 influenza virus to the WHO despite understanding that this would impair WHO ability to respond to a possible outbreak from this organism. They made this decision because of two episodes. The first was that a scientific paper was written about a virus they supplied that did not include their scientists as authors, contrary to the existing agreements on virus sharing. The second was their virus was passed on to a vaccine manufacturing company without their consent and again contrary to the regulations. Their view was that this highlighted global inequities in that the vaccine would be manufactured using the virus that they provided at no cost.

> Disease affected countries, which are usually developing countries, provide information and share biological specimens/virus with the WHO system; then pharmaceutical industries of developed countries obtain free access to this information and specimens, produce and patent the products (diagnostics, vaccines, therapeutics or other technologies), and sell them back to the developing countries at unaffordable prices. (Sedyaningsih et al., 2008: 486)

Given that many developing countries have no capacity to respond to a pandemic, and that they already have high levels of morbidity and mortality to other infectious diseases, it is not surprising that they do not share the same concern about a pandemic as developed countries.

Following the Pandemic Plan/Guideline

The public health authorities were treating this as a Zone 1 problem with high agreement and high certainty. The plan they were following had been developed following the SARS epidemic, an organism that had high case fatality. At the time of the first case it was unknown how infectious or fatal H1N1 would be, so they acted on the

Complexity and health policy: planning for a pandemic 391

basis of a worst case scenario, whilst acknowledging the uncertainty. We did not agree with their response. If we had been consulted rather than directed we would have advised of the potential problems with Ms B, and that attempting to control her behaviour was likely to be counterproductive, and that it would have taken much more effort and significant use of interpreters to prevent the index family from going to the hospital. By following the plan to the letter rather than consulting, the risk of spread was increased.

Individual Patient Care versus a Public Health Approach

Pharmaceutical treatment of infectious diseases highlights the individual approach as compared to approaches such as vaccination, quarantining of infectious people, eliminating vectors, and environmental change (draining swamps to lessen malaria risk, improving housing or drainage for respiratory or gastrointestinal infections). If there had been no effective pharmaceutical to treat H1N1 then the only approaches available would have been public health approaches. Unlike previous influenza pandemics for the H1N1 pandemic, Oseltamivir (Tamiflu®) was available. On the presumption that it worked, those individuals who did not want to participate in public health measures could buy themselves some tablets (if they could afford them) and feel some security in continuing to behave as they wished. Unfortunately, as discussed below, Oseltamivir has not been proven to be effective, so its widespread use probably undermined public health efforts by instilling some complacency that for example quarantining was not as important because people had taken the medication.

Commercial Pressures

The story of Oseltamivir (Tamiflu®) during the H1N1 pandemic highlights the theoretical problem of the bias introduced by powerful commercial enterprises. The H1N1 pandemic led to huge profits for Roche, the manufacturer of Oseltamivir (Tamiflu®). 'Prior to the global outbreak of H1N1 influenza in 2009, the United States alone had stockpiled nearly US$1.5 billion dollars worth of the antiviral' (Doshi et al., 2012). It was understood that Oseltamivir was effective at limiting the severity of H1N1 infection and if given to contacts of infected people it could limit the spread of the virus. Doshi et al. (2012) have reviewed the evidence on which these claims are based and found that the United States Federal Drug Administration (FDA) did not accept company evidence that the drug limited complications from infection, nor did it prevent further transmission. The FDA sent Roche a warning letter (FDA, 2000) asking Roche to desist from stating that Tamiflu has been shown to reduce serious complications of influenza, as Roche had not presented evidence to substantiate this claim. Unlike the wider public, the FDA had access to the full clinical reports. Despite considerable pressure and assurances that they would release information, at the time of Doshi et al.'s paper Roche had failed to release the full clinical study reports publicly. This information was released shortly after publication (PM Live, 2012). Those reports that they have released cast doubt on the validity of the claims made for Oseltamivir prior to the H1N1 pandemic.

392 *Handbook on complexity and public policy*

PLANNING FOR A PANDEMIC: A COMPLEXITY APPROACH

Planning for a future pandemic is hard for a number of reasons:

1. Knowledge about the infectious organism will be uncertain and changing; how infectious? How fatal? H1N1 had a low case fatality of 0.3 per cent (Tuite, 2010) whereas SARS had a much higher case fatality rate of 21 per cent (Karlberg, 2004).
2. Knowledge is not shared by all in the community.
3. Decisions are inevitably based on particular values that may not be shared by all.
4. Whilst there is legal power to enforce decisions, in practice, compliance with decisions is voluntary.
5. The risks and costs of responding to a pandemic need to be added to the risks and costs of business as usual.

Prior to the SARS epidemic there was little in the literature addressing the values (ethics) surrounding decisions made during a pandemic (Thompson et al., 2006) but the SARS epidemic highlighted the difficulty of responding to such an episode without an ethical framework to work from. They summarized their findings:

- Good pandemic planning requires reflection on values because scientific information alone cannot drive decision-making.
- The development of an ethical framework for hospital pandemic planning calls for expertise in clinical, organisational and public health ethics.
- Stakeholder engagement is essential for the ethical framework to be relevant and legitimate.
- The ethical framework contains procedural and substantive ethical values to guide decision-making.
- Three key elements of integration of ethics in to pandemic planning are (1) sponsorship from senior hospital administration; (2) vetting by stakeholders and; (3) decision review processes.
- An ethical framework is robust to the extent that pandemic influenza planning decisions are seen to be ethically legitimate by those affected by them.
- In order to increase the robustness of pandemic planning in general, timely public debate about the ethical issues is essential. (Thompson et al., 2006: 11)

An important feature of this summary is that it is predicated on one of the tools that Geyer and Rihani (2010) describe as an essential tool for working with complex problems: 'real stakeholder engagement'. 'The traditional steep hierarchy that separates stakeholders from decision-makers by layers of bureaucrats and 'experts' is obviously not fit for complex situations' (p. 69). It is hard to envisage what this might mean in practice but Torda (2006) gives some good examples of values that need some agreement in planning for a pandemic that highlights how difficult this process might be.

Reciprocity

> Society should recognise the burden on health workers and others involved in protecting the public during a pandemic, and support them appropriately. (Torda, 2006: s74)

The burden of a pandemic falls unevenly. With SARS there was a high case fatality rate of 21 per cent (Karlberg, 2004) so clinical staff risked their lives going to work. Why

Complexity and health policy: planning for a pandemic 393

should they do that and what supports would make them more likely to go to work? This can be addressed by appealing to professionalism and the duty of care ... but we also need cleaners to go to work. Within the Wellington Hospital most clinical staff effectively have unlimited sick leave. Thus anyone needing to be off work would get paid. By contrast, the cleaning staff are all contracted with limited sick leave and some workers may have had no sick leave left. Particularly with this infection, where the need for people to stay home was often to prevent spread in the institution (rather than because of being too sick to work), the burden is paid by the worker with the benefit going to the rest of the community. If prior to the pandemic industrial relations with the cleaners had been poor (as stakeholders they were not listened to and had little power) then the likelihood of the cleaners happily carrying the extra burden is less.

During the peaks in Wellington, general practice clinics, the emergency department and public health authorities worked particularly hard. People asked to stay at home to prevent spread lost income and students had their assessments disrupted. A few people became seriously ill. For most in the community things continued as normal. Those of us working hard because of flu could have our commitment enhanced or deflated. For us our commitment was enhanced when the Health Board offered to recompense the clinic for extra costs incurred because of the pandemic (staff off with flu, quarantined because of contact, and extra staff needed for high patient loads). Our clinic was significantly affected as the epidemic was more prevalent in poorer communities. This was a small amount of money but the gesture was very important. By contrast, the University told students that if they had the slightest respiratory symptoms then they should not sit their exams. However, all students who did not sit the exam were required to present a medical certificate to apply for an *aegrotat* pass. This led to an extra bulge of patients needing to be seen when we were busy seeing sick people. This deflated us. Pandemics are not all or nothing. The decision as to when a pandemic is bad enough, for example, to waive the need for students who miss their exams to get a sick note is easy if there is widespread disruption, but if it is not that bad then a good decision will only be made if there is the ability of 'stakeholders' (like our clinic) to be able to have input to such decisions. The difficulty of achieving this is compounded by our inability to do anything except seeing the deluge of patients presenting, there is little time to attend meetings. Real stakeholder involvement is not easy to achieve.

Proportionality

> Measures to protect the public should not overly restrict the liberty of individuals and these measures should be proportional to the threat. (Torda, 2006: s74)

If this virus had been universally fatal and very infectious then it would have been reasonable to call the police to ensure that Ms B stayed at home. If this had been done in this case there would have been public outrage. When is it bad enough to justify cancelling an important sporting event or a much anticipated school trip or cancelling staff holidays? When is the harm caused by cancelling elective services to cover for influenza greater than the harm caused by the influenza? There is no easy way for these decisions to be made, but if decision makers get significantly out of step with their community then trust is lost. If trust in authority is undermined it then becomes very hard to do anything.

Stewardship, Trust and Solidarity

> Stewardship, trust and solidarity are closely interwoven. Stewardship refers to the notion that decision makers need to act ethically and in the interests of the individual and the population. This requires good, informed decision making and high levels of trust from both those involved in providing care and the public. (Torda, 2006: s74)

As a GP I don't have time to read all about H1N1. I need to trust my public health colleagues to make good decisions. My trust was waning because I wasn't sure that they were taking into account all the consequences of their decisions; they were not adequately linked in to all the stakeholders. Trust cannot suddenly be manufactured when a pandemic appears. If the bureaucracy does not trust me to do the right thing they require auditing and reporting which takes time without any benefit to the patient. Unnecessary bureaucracy undermines trust. When I was finally allowed to prescribe funded Oseltamivir (Tamiflu®) for affected patients I had to fill out a form for every patient and fax it to the public health authorities because:

> The Ministry of Health requires clear documentation of the person's illness and that you confirm they meet the eligibility criteria. The form sent to RPH meets this requirement. Please note that pharmacies will be sending a copy of the prescription to Regional Public Health and that DHBs or the Ministry of Health may audit this process. (New Zealand Ministry of Health, 2009)

When the waiting room was full of patients, spending time doing this task was not welcome. This highlights that 'real stakeholder involvement' is a prerequisite of effective management, not a desirable add on.

At the end of the first week after our patient had been diagnosed I was stressed. We were bearing the burden (insufficient reception cover, greatly increased numbers of patients) and getting little support and I felt no one wanted to hear of our troubles: I was losing trust in the authorities. The thought went through my mind that if another patient turned up then maybe I would not report them. This is what solidarity is about. If I had acted in that way (and others had done the same) then all ability to control the pandemic would have been lost.

WHAT USE ARE CHECKLISTS/PROTOCOLS?

With the benefit of hindsight several of the public health decisions made in response to this pandemic resulted in our task of responding being more difficult. The plans developed following SARS, which had a much higher case fatality rate than H1N1, coloured the public health response to being more interventionist than in retrospect was necessary. A better understanding of complexity could have led to developing a protocol or checklist that covered the simple elements of action required and a process and list of stakeholders to address the complex elements. This distinction is clearly described by Gawande (2010) in his case study of the use of checklists in building large buildings. He described the complicated schedule that was used to coordinate the work of all the trades. This was developed in advance and was effectively a complicated recipe. However, they had a different checklist for addressing the unexpected. In his example of a skyscraper building, the floors on the upper storeys of the partially constructed building were not flat and it was not certain why or how to fix it. This was a complex problem that was unique to this particular building and had not been anticipated in advance. This second checklist was called a 'submittal schedule'. Unlike the schedule that described all the *construction* tasks that needed to be done and in what order, this specified *communication* tasks to ensure that all trades who needed to be consulted about a particular unanticipated problem were consulted. For a pandemic, the protocol (recipe) might, for example, include the logistics of transporting extra masks and gloves and how to declare 'code red' in hospitals to progressively cancel non-essential services. The equivalent of his 'submittal schedule' would be a list of stakeholders who needed to be consulted about particular decisions. A plan that does not distinguish between these two sorts of checklists will inevitably lead to either under-reaction or over-reaction, depending on whether the new organism turns out to be less or more virulent than the one on which the plan was based. This is usually not clear at the time decisions are being made. Many of the decision points hinged on values that were not shared. If the health authorities had proposed action that was too far from the accepted values of those affected (for example calling the police out to force Ms B to stay at home), there could have been widespread non-adherence and loss of trust in the authorities, with a possible catastrophic outcome from an uncontrolled epidemic.

A plan for managing a pandemic using a complexity perspective would contain the following elements:

1. Determine what elements of the plan involve Stacey diagram Zone 1 elements: simple elements for which there is widespread agreement about the need for the action and close to certainty on the benefit of the action. Develop a plan to cover these elements.
2. Develop a list of values that are pertinent to this problem and this community (National Ethics Advisory Committee, 2007; Torda, 2006; Thompson et al., 2006), and try to develop some consensus in the community as to what they think about each value.
3. Develop a list of stakeholders and a detailed process of how communication with them might happen. Prior to the pandemic much could be resolved.

Much work can be done in advance of the pandemic. Many stakeholders may be happy to delegate their input (for example to representatives from professional colleges, or particular community groups) so this can be sorted out in advance. Relationships can be built, preferably using synergies with other planning. For example there would be a lot of overlap with other disaster planning, such as earthquakes. Many of the decisions will be particular to the next pandemic so a process of ensuring stakeholder input in real time is essential.

POLITICS

This process will help with a number of elements of planning but an outstanding problem is what is to be done if agreement on values cannot be reached between the stakeholders. If the values required are not shared prior to the crisis then there is a high chance that

396 *Handbook on complexity and public policy*

they will not suddenly become shared values during a crisis. It is facile to note that none of the lists of values cited above include the making of a profit out of a pandemic. Clearly Roche did make a significant profit from the pandemic and at the time of the pandemic did not release the information on which their application for FDA approval (which was not granted) was based (Jefferson and Doshi, 2014). Their approach did not accord with the value of transparency, as they did not provide all relevant information prior to the pandemic to allow planners to make good judgements regarding where to put resources.

As Geyer notes when discussing soft systems methodology (Geyer and Rihani, 2010) 'problems and solutions have to be revisited over and over again as time goes on' (p. 69). If we really listened to stakeholders we might put the 'problem' of pandemic planning on hold and widen our focus to look at international equity, as suggested by the Indonesians. The problems that get chosen for attention are those that matter most to those with more power (nationally or internationally). It is hardly surprising that Indonesia and other developing countries are not as worried about pandemic flu as much as Western countries, given that large numbers of their populations die from other infectious diseases all the time. If we do adopt a complexity approach to addressing problems, it is inevitable that an element of the process will have to move out of the realm of implementing public policy and into the realm of political decision making, which is the process we use to decide what to do when there are conflicting values.

CONCLUSION: COMPLEXITY AND HEALTH POLICY

There are important barriers to the application of the complexity paradigm to health policy. There are vested interests who would like us to believe there are simple solutions, particularly the pharmaceutical industry. There are also others who profit from the status quo: bariatric surgery will never be an effective solution for the obesity epidemic but there are many surgeons who have a booming business providing this operation, diverting attention from more effective strategies. These commercial interests tend to characterize problems as being within the individual (and thus amenable to treatment with pills or surgery) rather than acknowledging the importance of public health and society-wide approaches. They may actively lobby against a public health approach (the tobacco industry) to maintain their commercial power.

An ethical approach that believes that there is such a thing as the right way to do everything conflicts directly with a complexity paradigm and contributes to proliferation of best practice guidelines that attempt to turn complex problems into complicated problems that can be addressed by the use of protocols.

Public health as a discipline has a good understanding of the complexity paradigm and this is well illustrated by the approach needed to address a pandemic. The case study illustrates that an important area where the level of complexity is irreducible is in the values that sit behind decisions relating to a pandemic. A diversity of values within the community is an element of complexity that will always be present, perhaps more so in health than other areas of public policy. This highlights that the public policy is interwoven with the political process, which is how we make decisions on the values of our community. Any policy development that is not well linked in to some form of political process is unlikely to reach a successful conclusion.

Complexity and health policy: planning for a pandemic 397

Finally it needs to be said that by discussing 'health policy' I am guilty of approaching human well-being from a reductionist viewpoint. It is well known that many of the determinants of health lie outside the purview of Ministries of Health. Housing, social welfare, education, utilities infrastructure, penal policy, environmental policy and many others can have an equal, if not larger, impact on the health of the population. It could be that the best way to improve the health of the population would be to significantly cut expenditure within the health sector and instead put the money into a different area of expenditure. Grasping the whole is always difficult and in addressing complex problems one effective strategy is to consider the problem in manageable parts. The trick is to maintain a sense of the whole.

REFERENCES

Barnett, K. (2012). Epidemiology of multimorbidity and implications for health care, research, and medical education: a cross-sectional study. *The Lancet* (UK edn), 380, 37–43.

Beauchamp, T.L. and J.F. Childress (2009). *Principles of biomedical ethics*, New York: Oxford University Press.

Donnellan, C., S. Sweetman and E. Shelley (2013). Health professionals' adherence to stroke clinical guidelines: a review of the literature. *Health Policy*, 111(3), 245–63.

Doshi, P., T. Jefferson and C. Del Mar (2012). The imperative to share clinical study reports: recommendations from the Tamiflu experience. *PLoS Medicine*, 9, e1001201.

Dunne, K., S. Henderson, S.L. Stewart, A. Moore, N.S. Hayes, J. Jordan and J.M. Underwood (2013). An update on tobacco control initiatives in comprehensive cancer control plans. *Preventing Chronic Disease*, 10.

Food and Drug Administration (FDA) (2000). Warning letter to Roche [online]. Accessed 27 August 2013 at http://www.fda.gov/downloads/Drugs/GuidanceComplianceRegulatoryInformation/Enforcement ActivitiesbyFDA/WarningLettersandNoticeofViolationLetterstoPharmaceuticalCompanies/UCM166329. pdf.

Gawande, A. (2010). *The checklist manifesto: how to get things right*, London: Profile Books.

Geyer, R.R. and S. Rihani (2010). *Complexity and public policy: a new approach to twenty-first century politics, policy and society*, Abingdon: Routledge.

Glouberman, S. and B. Zimmerman (2002). Complicated and complex systems: what would successful reform of Medicare look like? *Changing Health Care in Canada: The Romanow Papers*, 2, 21–53.

Guyatt, G. and D. Rennie (2002). *Users' guides to the medical literature: a manual for evidence-based clinical practice*, Chicago: AMA Press.

Hofstede, G.H., G.J. Hofstede and M. Minkov (2010). Cultures and organizations: software of the mind: intercultural cooperation and its importance for survival. 3rd edn, New York: McGraw-Hill.

Immunisation Advisory Centre New Zealand (2013). New Zealand Pandemic Planning [online]. Accessed 30 August 2013 at http://www.guidetools.com/IMAC_influenza/pandemic_planning_NZ_pandemic_plan. html.

Jefferson, T. and P. Doshi (2014). Multisystem failure: the story of anti-influenza drugs. *BMJ: British Medical Journal*, 348.

Karlberg, J. (2004). Do men have a higher case fatality rate of Severe Acute Respiratory Syndrome than women do? *American Journal of Epidemiology*, 159, 229–31.

Khushf, G. (2004). *Handbook of bioethics*, Dordrecht: Kluwer Academic Publishers.

King's Fund (2012). *Long term conditions and ultimorbidity* [online]. Accessed 11 August 2013 at http:// www.kingsfund.org.uk/time-to-think-differently/trends/disease-and-disability/longterm-conditions-multi-morbidity#morbidity.

Moreno, R.P., A. Rhodes and J.-D. Chiche (2009). The ongoing H1N1 flu pandemic and the intensive care community: challenges, opportunities, and the duties of scientific societies and intensivists. *Intensive Care Medicine*, 35, 2005–2008.

National Ethics Advisory Committee (2007). *Getting through together: ethical values for a pandemic* [online]. Accessed 11 August 2013 at http://neac.health.govt.nz/system/files/documents/publications/getting-through-together-jul07.pdf.

Nestle, M. (2000). Halting the obesity epidemic: a public health policy approach. *Public Health Reports (1974)*, 115, 12.

398 *Handbook on complexity and public policy*

New Zealand Ministry of Health (2009). Instructions to GPs on prescribing Oseltamivir (personal communication).

PM Live (2012). Roche offers compromise in Tamiflu debate 2012 [20/2/15]. Available at http://www.pmlive.com/pharma_news/roche_offers_compromise_in_tamiflu_data_debate_451786.

Reeve, J., T. Blakeman, G.K. Freeman, L.A. Green, P.A. James, P. Lucassen, C.M. Martin, J.P. Sturmberg and C. Van Weel (2013). Generalist solutions to complex problems: generating practice-based evidence: the example of managing multi-morbidity. *BMC Family Practice*, 14, 112.

Rimer, J., K. Dwan, D.A. Lawlor, C.A. Greig, M. McMurdo, W. Morley and G.E. Mead (2012). Exercise for depression. *Cochrane Database of Systematic Reviews*, 7.

Sackett, D.L., W.M. Rosenberg, J. Gray, R.B. Haynes and W.S. Richardson (1996). Evidence based medicine: what it is and what it isn't. *BMJ: British Medical Journal*, 312, 71.

Sedyaningsih, E.R., S. Isfandari, T. Soendoro and S.F. Supari (2008). Towards mutual trust, transparency and equity in virus sharing mechanism: the avian influenza case of Indonesia. *Annals Academy of Medicine Singapore*, 37, 482.

Taylor, J.S. (2003). Confronting "culture" in medicine's "culture of no culture". *Journal of the American Association of Medical Colleges*, 78(6), 555–9.

Thomas, B.P. and L.O. Gostin (2013). Tobacco endgame strategies: challenges in ethics and law. *Tobacco Control*, 22, i55–i57.

Thompson, A., K. Faith, J. Gibson and R. Upshur (2006). Pandemic influenza preparedness: an ethical framework to guide decision-making. *BMC Medical Ethics*, 7(1), E12.

Torda, A. (2006). Preparing for an influenza pandemic. *Medical Journal of Australia*, 185, S73–6.

Tuite, A.R. (2010). Estimated epidemiologic parameters and morbidity associated with pandemic H1N1 influenza. *Canadian Medical Association Journal*, 182, 131–6.

United Kingdom Department of Health (2012). *Long Term Conditions* [online]. Available at https://www.gov.uk/government/publications/long-term-conditions-compendium-ofinformation-third-edition.

Van Weel, C. and F.G. Schellevis (2006). Comorbidity and guidelines: conflicting interests. *The Lancet*, 367, 550–51.

Wang, J.-T. and S.-C. Chang (2004). Severe acute respiratory syndrome. *Current Opinion in Infectious Diseases*, 17, 143–8.

Washington, H.A. (2011). *Deadly monopolies: the shocking corporate takeover of life itself, and the consequences for your health and our medical future*, New York: Doubleday.

Willett, A.M. (2010). *Best practice diagnostic guidelines for patients presenting with breast symptoms*. London: UK Department of Health.

Wilson, N. and M.G. Baker (2009). Comparison of the content of the New Zealand influenza pandemic plan with European pandemic plans. *Journal of the New Zealand Medical Association*, vol. 122.

World Health Organization (1998). *Guidelines for controlling and monitoring the tobacco epidemic*. Geneva: WHO.

24. How useful is complexity theory to policy studies? Lessons from the climate change adaptation literature
Adam Wellstead, Michael Howlett and Jeremy Rayner

INTRODUCTION: WHAT CAN COMPLEXITY THEORY LEARN FROM THE CLIMATE CHANGE ADAPTATION LITERATURE?

The use of metaphors is widespread in policy studies (Morgan, 1980; Dowding, 1995). These root metaphors provide a central theme to a policy framework and allow analysts a starting point in advancing their understanding of policy phenomena (Mio, 1997). But not all metaphors are as useful as others in informing research, knowledge and action. As Zashin and Chapman (1974) pointed out, a long-standing problem in political studies, for example, is the constant issue whereby much relevant experience and accumulated knowledge of political processes and phenomena is 'excluded from the mainstream of the discipline by its commitment to the use of a vocabulary modeled on that of the natural sciences'.

This is true of complexity theory, viewed as the application of a metaphor from system thinking applied to the study of public policy. When metaphors such as complexity are used in social science research, the 'empirical referents, more explicitly their connections with the experience of real people, seems even more tenuous than those of the traditional theoretical concepts' such as arguments, interests and positions (Zashin and Chapman, 1974: 292. In place of these older concepts – and traditional political theory constructs such as rights, power, authority or legitimacy – the use of cybernetic metaphors such as equilibrium, feedback, input, transactions, games, and the structural-functional models they often entail, have limitations when it comes to analysing policy-related activity and behaviour.

This chapter critically assesses the use of complexity metaphors in the policy literature in this light and recommends future research directions. In doing so we highlight significant problems recently uncovered in the climate change adaptation scholarship which serves as the example, *par excellence*, of the application of complexity maxims and concepts to policy-making. We argue that much of this adaptation theory's failure to have much impact upon policy-making (Wellstead et al., 2013a and 2013b; Howlett, 2014) can be attributed to its overlooking the lessons of conventional policy literature which emphasizes meso- and micro-level policy and governance relationships, relying too heavily on macro-level eco-system-based concepts and constructs. We note a similar issue with complexity theory in general and suggest it requires complementing by existing insights from the policy sciences into the actual processes of policy-making in order to better explain and inform policy studies.

400 *Handbook on complexity and public policy*

THE COMPLEXITY APPROACH TO PUBLIC POLICY

What is the complexity approach to policy study? Geyer's research on issues affecting European integration in the early 2000s argued, 'the traditional linear view of science has been giving way to a growing non-linear or complexity framework and since the 1980s has been spilling over into the social sciences' (Geyer, 2003: 16). Such a framework, based on complex interactions between policy-making elements, non-linearity and emergent properties, he argued, 'does not eliminate or solve the problem of complexity. However, it provides a new and intriguing ontological and epistemological foundation for addressing the problem of complexity' (Geyer, 2003). Similarly, Sanderson (2009) found that complexity theory provided insights into a policy environment that is dynamic and non-linear, and an unstable system where disequilibrium is the norm.

Complexity theory has its own vocabulary and metaphors, however, which bring potential new insights and directions to policy research, but which can also mislead (Lissack, 1999). First, complex systems are characterized as those containing a large number of elements with non-linear interactions. These interactions usually having a fairly short range. The environment in which they occur is open, and the whole of the system is said to be greater than its parts (Geyer, 2003; Cairney, 2012a). Complex systems are also argued to be often in a state of disequilibrium and to have a history of interactions and states which matter in understanding how the system and its parts (co) evolve. Each element in the system is unaware of the behaviour of the system as a whole (Geyer, 2003) and 'self-organization' is a central complexity principle whereby agents change their actions endogenously, without external pressures to do so (Teisman and Klijn, 2008). Teisman and Klijn (2008) argue that such 'actions systems can consist of one single actor, but can also be a group, an organization or a set of organizations. In fact only single actors can act but they do so as members of a larger system, which also affects this system' (p. 344). Cairney (2012a) also highlights the role of emergence and 'strange attractors' as complex systems shift between states of activity. These complexity theory metaphors provide policy research with a distinct vocabulary and grammar and a number of helpful avenues for inquiry. There have also been a number of applications of complexity theory in the public policy field. They include Geyer and Rihani's (2010) discussion of health, international relations, development, and terrorism issues using tools of complexity such as 'cascades of complexity', 'complexity mapping', 'fitness landscapes', and the use of Stacey diagrams. Geyer (2012) also applied the notion of such a 'complexity cascade' to an evaluation of UK education and health policy.

Cairney (2012a; 2012b) has noted how the concepts and ideas behind complexity theory resonate with some of those already developed within the policy sciences, such as punctuated equilibrium, historical institutionalism, implementation concerns, and approaches to policy focusing on instruments and tools. However, complexity theory is currently pitched at a very high macro and abstract level which make it somewhat problematic when applied to a down-to-earth subject of day-to-day political activity such as policy-making. Missing from most contemporary applications of complexity theory to policy-making are the key political variables – 'interests', conflicts, bargaining, trade-offs, deal-making and others – which animate more traditional policy studies and it is argued below that scholars who use the complexity metaphor should not abandon these traditional policy frameworks in developing and applying complexity concepts

Complexity theory and policy studies: lessons from climate change 401

to policy-making. Rather they should give them new life within complexity theory as 'they continue to play an important role but within the wider context of complexity' (Wellstead, 2007: 147). This means that more sophisticated accounts of policy than those currently used by complexity theorists are required; these must include factors such as 'the impact of policy makers' ideologies, about the nature of decision making, [and] upon the conduct and outcomes of the various stages of the policy process' which shape policy outcomes (Smith and May, 1980: 156). How this can be done is illustrated below using the example of the use of complexity constructs in policy analysis of climate change. This is probably the single area in which complexity constructs have been most used, and the strengths and weaknesses of these applications are revealing of those of complexity theory as a whole when applied to policy studies.

ADAPTATION TO CLIMATE CHANGE: TAKING THE ECOSYSTEM METAPHOR TOO FAR

Adaptation to climate change has become a particularly pressing issue for all levels of government. This is true in both developed and developing countries faced with an onslaught of high-profile climate-related impacts and disasters. In response, governments and NGOs have invested considerable levels of resources developing voluminous vulnerability assessments and ambitious adaptation frameworks. A flourishing adaptation to the climate change research industry has also emerged, led by a new generation of climate change-oriented social scientists resulting in an impressive output of articles diagnosing policy problems and recommending management solutions found in multidisciplinary journals such as *Climate Change, Ecology and Society, Global Environmental Change* and *Mitigation and Adaptation Strategies for Global Change*. For both on-the-ground efforts and scholarly endeavours, climate change impacts and their associated vulnerabilities in ecosystems and socio-economic systems have been well chronicled and attention has turned to determining the contours, and recommending the content, of climate change adaptation policy.

Many of these studies have consciously or unconsciously applied a complexity theory framework to their analyses. Emison (2008), for example, examines changes to the US Clean Air Act within the context of complexity theory, linking this to adaptive systems theory, and many other studies have similarly been heavily influenced by complexity concepts and precepts. A growing number of national and sub-national exercises in particular draw heavily upon the Intergovernmental Panel on Climate Change (IPCC) in its Third and Fourth Assessment reports (Preston et al., 2010). The Third Assessment report, in turn, draws upon Smit et al.'s (1999) climate change assessment framework, which was heavily influenced by complexity theory concepts and arguments. As a result, this general approach has been emulated at a number of sectoral and governmental levels. This trend continues in the Fifth Assessment, with complexity highlighted as a major theme (IPCC, 2014).

Figure 24.1 from Füssel and Klein's (2006) well-known and oft-cited assessment of climate change adaptation policy processes typifies the main components of these approaches to adaptation studies (see Figure 24.1). The authors draw their analysis from earlier contributions, namely Smit and Wandel (2006), as well as resilience studies such

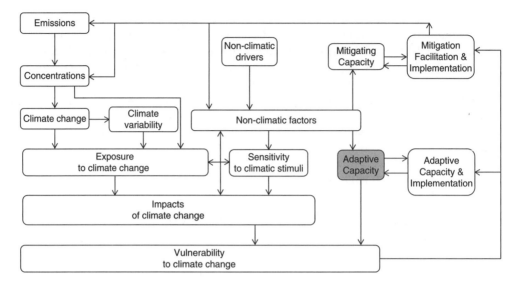

Source: Füssel and Klein (2006).

Figure 24.1 Füssel and Klein's model of adaptation policy processes

as Folke (2006), Gallopin (2006), Adger (2006) and Nelson et al. (2007). As Figure 24.1 shows, this framework is pitched at a very high level of generality and attempts to model policy relationships based on the metaphor of ecosystem dynamics. Complexity concepts such as exposure, sensitivity, impact, adaptive capacity, vulnerability and adaptation are all featured in this work. The systems-level thinking behind such analyses is clear even in the definition of 'adaptation' used in these models, referring to 'the adjustment in natural or human systems in response to actual or expected climatic stimuli or their effects, which moderates harm or exploits beneficial opportunities'. Similarly adaptive capacity is concerned with 'the ability of a system to adjust to climate change (including climate variability and extremes) to moderate potential damages, to take advantage of opportunities, or to cope with the consequences' (Füssel and Klein, 2006: 18).

Misplaced Functionalism in Complexity-inspired Policy Analysis

In their work, Füssel and Klein (2006) identify two important adaptation-related functions that they argue governments usually perform: facilitation and implementation. Facilitation refers to activities that enhance adaptive capacity, such as scientific research data collection, awareness raising, capacity building, and the establishment of institutions, information networks and legal frameworks for action. Implementation refers to activities that actually avoid adverse climate impacts on a system by reducing exposure or sensitivity to climatic hazards, or by moderating relevant non-climatic factors.

Other leading scholars in the climate change adaptation field echo this approach to explaining government actions but only where government appears as a 'dependent variable' influenced by larger system-level concerns. For example Nilsson et al. (2011) note that the role of institutions and governance processes 'needs to be considered' along

Complexity theory and policy studies: lessons from climate change 403

with the physical production and social variables that compose the system in applying this model to policy-making and outcomes. Brooks et al. (2005) similarly develop a suite of governance proxies for national-level vulnerability to climate change (for example, political stability and rule of law) but the specification of exactly how these variables affect policy dynamics 'on-the-ground' is missing. Plummer and Armitage (2007) also identify capacity and capacity building, institutions, social capital and networks, learning, and vulnerability and livelihoods as critical in their assessment framework, and argue that these influence environmental governance, but without any details on how such processes actually work in either theory or practice. Climate change scholarship postulating governments as independent variables are exceedingly rare. Adger et al. (2007), for example, note that adaptive capacity is influenced by 'the nature of governance structures' while Smit and Wandel (2006) state only that 'improvements in institutions' may lead to increased adaptive capacity. Neither clarifies the conditions under which this is likely to occur.

A good example of the limitations these models have for understanding actual policy-making processes on the ground can be found in the work of Engle and Lemos (2010). They note the importance of 'governance and institutional mechanisms' as determinants in characterizing adaptive capacity and rank a suite of governance and institutional indicators they think are important in affecting policy-making. The key political and governance considerations that would explain, for example, how the indicators will be used to coordinate activities towards adaptation goals in the absence of political commitment from government and how that commitment emerges, are missing.

In such cases, the analysis jumps quickly and uneasily between high-level abstraction and micro-level policy recommendations, skipping over the 'missing middle' of the meso-level governance variables that are critical to joining the macro and micro levels together in practice (Voss and Borneman, 2011; Nilsson et al., 2011). Activities like public policy-making, law-making and legislative and administrative behaviour require causal and intentional modes of scientific explanation which take seriously the activities of policy-makers and the political and social forces which drive them.

This literature thus typically notes the importance of political institutions in addressing adaptation and adaptive capacity, but models these variables only in very general terms, with a lack of specifics with regard to the precise mechanisms and relationships involved in deriving policy recommendations and instrument choices (Daedlow et al., 2011). In these complexity theory-inspired studies 'politics', 'governance' and 'policy-making' are understood in a 'functional' way: as a kind of input variable promoting 'necessary' adaptation 'functions' in response to system-level changes and needs (Holling, 1973; 2001; Folke et al., 2002). The logic is one which simply assumes that governance activities will be performed in specific ways due to system-level prerequisites (Cummins, 1975), but ignoring the policy process itself and the possible non-performance of 'mission critical' tasks (Howlett et al., 2009a; Wu et al., 2010; Weible et al., 2012).[1]

The 'Black-box' Problem

Even when some socio-political variables are incorporated into a climate change vulnerability assessment framework, a second problem arises in these studies due to the lack of specificity about the mechanisms and internal workings of institutional and other components of political systems and policy sub-systems. This is the so-called 'black-box' problem of unspecified process variables and mechanisms, which also plagued early work in the policy and political sciences that attempted to describe and model decision-making and other policy processes. Concerns with the limitations of high-level systems-theoretic models when applied to policy-making surfaced more than forty years ago when these models first emerged in the social sciences, and these same concerns are features of today's climate change assessment frameworks and other complexity-theory-inspired studies (Black, 1961; Gregor, 1968; Landau, 1968; Stephens, 1969).

Like their contemporary climate change counterparts, 1960s-era political scientists such as Gabriel Almond (1965) and David Easton (1965) and many others suggested that high-level systems-based metaphors could describe much political behaviour and help explain outcomes. Following general systems theory scholars such as von Bertalanffy (1969), they argued that the political system exists in an environment that inputs resources and demands into that system which then produces outputs (decisions and supports) operating with feedback loops back to the environment and into the system as new inputs. As Figure 24.2 shows, this model described government or a political system as a simple feedback system in which a 'black box' (government) converted inputs into outputs which, in turn, fed back into the environment to generate

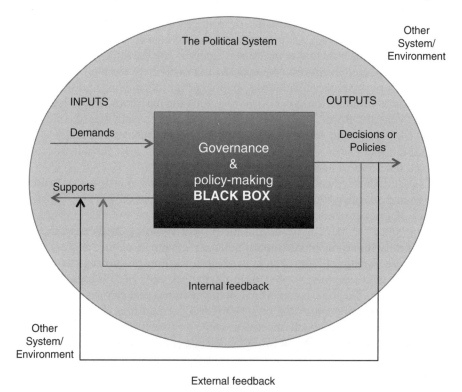

Source: Adapted from Easton (1965).

Figure 24.2 The 'black box' model of a political system following Easton (1965)

Complexity theory and policy studies: lessons from climate change 405

new inputs and so on. This logic is identical to that contained in Füssell and Klein's framework (Figure 24.1).

Like earlier generations of social-cybernetic or system-theories, much systems-inspired modelling in the climate change adaptation area continues to rely too heavily upon a flawed natural sciences metaphor. As early as the 1970s this overly abstract and general conception of a political system as a resource conversion mechanism had already been largely discredited in political science and policy studies as providing an insufficient and misleading view of government and its activities. For example, Lilienfeld (1978) labelled systems theory an 'ideological movement' because of its tendency to assume that systems maintain themselves in a state of equilibrium and concluded that it contained little relevance to the real world where actors actively sought and produced change, and even less practical application. Similarly, Chilcote (1994) found 'black-box' systems-level frameworks did little to explain political or policy change, yielded few testable hypotheses, and presented a strong ideological underpinning that sought to downplay political conflict and promote a technocratic understanding and approach to political life. Thorson (1970) lamented the whole enterprise was futile so long as the 'black-box' of real political processes remained unopened and unexamined. In general, Groth (1970) found that 'structural-functionalism black-box models have run aground trying to specify its model of the social system untangled by monumental ambiguities and values in the guise of survival considerations'. These models also failed convincingly to establish at least some underlying social and political relationships as 'behavioural universals for these allegedly goals of survival and adaptation' (Groth, 1970: 499). As Elster argued, opening up the black box and showing the cogs and wheels of the internal machinery was necessary to reveal a continuous and contiguous chain of causal or intentional links between the explanans and the explanandum (in Hedström and Ylikoski, 2010).

The ultimate purpose of most climate change assessments/frameworks is to accurately inform policy-makers of the feasible directions and procedures through which climate change adaptation can be accomplished. It is here in the provision of practical advice to policy-makers that the failure to attend to governance arrangements and institutions within complexity theory is most detrimental (Koliba et al., 2010). The meso dimensions of policy and governance missing from complexity theory are the most critical in affecting on-the-ground policy change and implementation. Ascher (2001), for example, noted that implementation of a number of resource management doctrines informed by complexity-inspired adaptation studies would lead, at best, to a range of poor results due to unintended consequences, the promotion of perverse incentives and other kinds of policy failures linked to the adoption of infeasible policy alternatives (Marsh and McConnell, 2010; McConnell, 2010).

While complexity theory-inspired studies such as the kinds of climate change vulnerability assessment frameworks cited above can provide a useful heuristic for understanding system-level impacts on policy-making and policy outputs, the assumptions inherent in these approaches leave much to be desired in terms of understanding or accurately characterizing policy processes, and are of little use to actual policy-makers. Further development of key concepts, such as governance and policy, is required in order for complexity-inspired analyses to accurately model meso-level political variables and escape this misplaced structural-functionalism. Complexity theory-influenced climate

406 *Handbook on complexity and public policy*

change adaptation frameworks await incorporation of the results of other studies which specifically focus on the meso-level of analysis (Peters, 1999; Hall and Taylor, 1996).

THE NEED TO INCORPORATE POLICY AND GOVERNANCE VARIABLES INTO ADAPTATION STUDIES AND COMPLEXITY THEORY

The neglect of meso-level variables in complexity-inspired climate change studies is not unique. Nilsson et al. (2011) found similar shortcoming in high-level energy systems studies, which share many similar characteristics based on complexity theory precepts. As they put it:

> The hitherto superficial treatment of institutions and politics in energy future studies is somewhat surprising. Many literatures concerned with systems-technical change recognize the importance of institutions in shaping (and interacting with) technological systems. These insights have emerged not only in economic history, sociology and political science, but also prominently in innovation systems studies, evolutionary and institutional economics, socio-technical systems, and even more recently in transitions management and elsewhere. These perspectives share several insights about institutions, what they are, why they are relatively stable and how they change. (pp. 1117–18)

In order to enter the realm of 'feasibility' and achieve practical relevance, adding meso-level variables to climate change adaptation studies, and more generally to complexity theory itself, is a prerequisite. This involves moving beyond the abstractions inherent in contemporary system-level complexity theory and instead seeking to incorporate knowledge of formulation and implementation into complexity-inspired models of political and policy interaction (Treib et al., 2007). This involves the need for a better dialogue between complexity theory and governance studies.[2]

As Frechette and Lewis (2011) have argued:

> in order to develop a comprehensive understanding of the dynamics of change, analysts require a meta-theoretical approach that not only provides complementary insights into how rules change over time, but also pushes the boundaries of conventional analysis to consider the constitutional arrangements that structure collective action and the subsequent performance of forest governance structures. (p. 582)

More sophisticated analyses are required, for example, that capture governance-related impacts of even such basic policy-making structures as federalism or the territorial division of powers between governments, the so-called basic multi-level or 'polycentric' nature of contemporary governance (Ostrom, 2008; 2009; Enderlein et al., 2011; Aligica and Tarko, 2011; McGinnis, 1999). Governance studies have shown, for example, that governments, lacking the knowledge or the mandate to govern alone, have increasingly chosen to try to construct policy consensus through more engaged and interactive forms of policy-making and to allow non-state actors to implement those policies within a broad framework of incentives, benchmarking and private governance (Sprinz and Vaahtoranta, 1994; Zito, 2007), a phenomenon which originates at the meso-level and affects micro- and macro-level behaviour and systemic conditions.

Complexity theory and policy studies: lessons from climate change 407

Three specific aspects of policy-making need to be explicitly modelled in new complexity frameworks. The first is to examine the structure and pervasiveness of policy networks (Howlett, 2002; 2011). In this network dimension, the number and diversity of actors (state and non-state) that exert some degree of power or influence over the outputs of the governance arrangements is a key facet of policy-making (Knoke and Kuklinski, 1982; Knoke, 1987). The concern of the analyst is to identify where political power lies in relation to society and the state (Lukes, 1974; Lindblom, 1977; Katzenstein, 1978; Offe, 1984) and the analytical challenge is to determine whether, and to what extent, in specific sectors and issue areas, the state or its agents are directly dictating the outcomes that emerge from the governance arrangement, more loosely 'steering' the arrangement, or alternatively whether ultimate power to determine outcomes rests with non-state actors (for example corporations, unions, environmental civil society organizations and so forth) (McCool, 1998).

This echoes findings in other sectors which have also pointed out the advantages to practitioners of incorporating governance into macro-level systems thinking. For health care, for example, Gómez (2011) has argued that:

> the practitioner community stands to gain from applying these theoretical approaches to their analysis of the institutional aspects of health governance and health system governance. Instead of merely measuring the presence of elite stewardship, strategic vision, responsiveness, and the like, this alternative approach suggests that practitioners begin their analysis by specifying the following issues: political and bureaucratic elite beliefs, interests, and the supportive coalitions that motivate elites to become stewards, visionaries, and to pursue institutional change. In contrast to the existing literature, this approach therefore sees elite interests and coalitions as key independent variables while the aforementioned health governance and health system governance indictors are treated as outcomes to be explained. (p. 210)

The second dimension has to do with modelling the rigidity of institutional policy-related arrangements – namely their formal or informal nature. The institutional aspects of governance arrangement can be assessed in terms of factors such as precision (how closely government constrains private action); obligation (the 'bindingness' of government commands); and delegation (the extent to which the power to adjudicate and enforce these obligations is retained by a regulator or delegated to an independent third party) (Tollefson et al., 2012). This also involves taking seriously the complex multi-level or 'polycentric' nature of contemporary policy-making and governance. Recently, Doelle et al. (2012) usefully explored these dimensions in a study of climate change-based forest governance arrangements in Canada, New Zealand and the US.

Finally, incorporating the third dimension, the policy process, is needed to understand the dynamic features of governance arrangements. The policy cycle (Hill, 2007; Howlett et al., 2009b; Pal, 2010; Wu et al., 2010) and policy change frameworks developed in the policy sciences, such as the advocacy coalition framework, institutional rational choice and structural choice (Moe, 1984; Ostrom, 1991; Schlager and Bloomquist, 1996; Sabatier and Jenkins-Smith, 1999), are all well-developed approaches that draw upon the network and institutional dimensions of policy-making to provide a fine-grained, more empirical lens on understanding the complexity and challenges of governance.

Complexity theorizing which incorporates these elements is hard to find but such an analysis would help to overcome the unrealistic functionalism of much applied complexity theory as it stands, which in practice often assumes that governance will

408 *Handbook on complexity and public policy*

simply 'get done' as a kind of system maintenance activity. Recent advances in the natural resource governance literature, in particular, have sought to capture these governance dynamics and their effects on policy-making (Howlett et al., 2009b, Tollefson et al., 2008).[3] Incorporating such logics and findings would help move complexity approaches forward. As Voss concluded in his study of adaptive management which takes meso-level variables seriously (Voss, 2011):

> Politics cannot be escaped or bypassed, nor eliminated or completely controlled by governance designs, but they can be analyzed and reflected on in order to devise more robust design strategies for new reflexive forms of governance. This is what we hope to encourage and support with the provision of this framework and sketching of avenues for further research.

CONCLUSION: COMPLEXITY THEORY AND THE POLICY PROCESS

Informed by complexity theory, existing climate change adaptation frameworks and vulnerability assessments suffer from conceptual weaknesses which limit their accuracy and policy relevance. But policy scientists who choose to apply complexity theory using only its systems-based focus on feedbacks, attractors and emergence risk the same problems found in the climate change adaptation field. These studies follow an implicit 'structural-functionalist' logic, which treats governments as a 'black-box' and policy-making as an undifferentiated and unproblematic output of system-level dynamics and requisites.

The absence of considerations of meso-level governance or societal steering activities and capacities in the framework literature partly explains the lack of impact on the ground that existing climate change adaptation studies and strategies, for example, have had among and upon policy-makers. The complexity models on which they draw were developed for other reasons, such as ecosystem impact modelling and studies of community resilience (Walker and Cooper, 2011), and they are not suited to policy analysis without significant modification. Although currently in vogue in many geography and natural resource management programmes, such frameworks are not well suited to the development of *feasible* policy prescriptions since they ignore or downplay the actual practices of policy-making, where the issues of political power, unequal resource distribution and institutional legacies noted in many case studies are very central concerns (Skodvin, 2010; D'Alessandro et al., 2010).

The arguments presented here are intended to further the efforts to improve complexity theory in its application to policy-making by highlighting the need to adequately model and account for governance arrangements and policy-making processes rather than relying upon outmoded and inaccurate models redolent of political and sociological theory of the 1950s and 1960s. A focus on 'macro' ecological and social systems-level variables has ignored or minimized the key role played in public policy decision-making by 'meso' or middle range variables such as constitutional structures, electoral and administrative considerations. More accurate modelling of micro-level variables related to the nature of public policy decision-making processes in democratic states is required for complexity theory to move into the mainstream of policy analysis (Voss and Bornemann, 2011; Nilsson et al., 2011).

Complexity theory and policy studies: lessons from climate change 409

NOTES

1. Jon Elster (1986) has noted this kind of functionalism, in the social sciences, is a 'puzzling and controversial' mode of explanation in general because, unlike other scientific modes such as causal or intentional explanations (where the intended consequences occur earlier in time), early events are explained by another event later in time (p. 31). Thus, in a functional explanation, 'we cite the actual consequences of the phenomenon in order to account for it' (p. 31). Feedbacks loops are the essential mechanism in functional reasoning because they provide 'a causal connection from the consequences of one event of the kind we are trying to explain to another, later event of the same kind' (p. 32). However, in social and political situations, as Elster further argued, such explanations are 'only applicable when a pattern of behaviour maintains itself through the consequences that benefit some group, which may or may not be the same group of people displaying the behaviour' (p. 32). That is, an institution or a behavioural pattern X is explained by its function Y for group Z if and only if: (1) Y is an effect of X; (2) Y is beneficial for Z; (3) Y is unintended by the actors producing X; (4) Y (or at least the causal relationship between X and Y) is unrecognized by the actors in Z; (5) Y maintains X by a causal feedback loop passing through Z (p. 28). Most attempts to use functionalism in social and political explanations fail because they are missing one or more of these five features (Elster, 1985). And, as Elster further noted, in political life there are many examples of singular, non-recurring events that produce unintended policy consequences (such as wars, riots and rebellions), while feedback loops are often postulated or tacitly assumed when they do not in fact exist (Elster, 1986). Hence, explanatory theories must move beyond simple functional modes of identifying and linking variables together to predict or model outputs and their impacts and effects. Elster (1985) argues instead for causal or intentional forms of explanation in the social sciences because functionalism is only applicable in biology and ecosystems.
2. 'Governance' is a term used to describe the different possible *modes* of government coordination of non-governmental actors (Rosenau, 1992; Rhodes, 1996; de Bruijn and ten Heuvelhof, 1995; Kooiman, 1993; 2000; Klijn and Koppenjan, 2000). That is, governments control the allocation of resources between social actors, providing a set of rules and operating a set of institutions setting out 'who gets what, where, when, and how' in society and managing the symbolic resources of state legitimacy which are crucial for the attainment of any policy goal, including but not limited to climate change adaptation.
3. This is true, for example, in many areas where efforts have been made to develop 'integrated strategies' such as forestry and coastal marine eco-system management (Howlett and Rayner, 2006a; 2006b), and similar efforts are typical in both climate change mitigation and adaptation efforts (Voss et al., 2006). In these new governance modes, the lines between public and private have become blurred (Gatto, 2006): from a mode of coordination based on hierarchical top-down, command and control by government actors or their agents, governments have increasingly experimented with new modes of governance that rely on the incentives provided by markets and by the sharing of information in governance networks.

REFERENCES

Adger, W.N. (2006), 'Vulnerability', *Global Environmental Change*, **16**(3), 268–81.
Adger, W.N., S. Agrawala, M.M.Q. Mirza, C. Conde, K. O'Brien, J. Pulhin, R. Pulwarty, B. Smit and K. Takahashi (2007), 'Assessment of Adaptation Practices, Options, Constraints and Capacity', in M.L. Parry et al. (eds), *Climate Change 2007: Impacts, Adaptation and Vulnerability. Contribution of Working Group II to the Fourth Assessment Report of the Intergovernmental Panel on Climate Change*, Cambridge: Cambridge University Press, pp. 717–43.
Aligica, P. and V. Tarko (2011), 'Polycentricity: From Polanyi to Ostrom, and Beyond', *Governance*, **25**(2), 237–62.
Almond, G. (1965), 'A Developmental Approach to Political Systems', *World Politics*, **17**(2), 183–214.
Ascher, W. (2001), 'Coping with Complexity and Organizational Interests in Natural Resource Management', *Ecosystems*, **4**(8), 742–57.
Black, M. (1961), 'Some Questions about Parsons' Theories', in M. Black (ed.), *Social Theories of Talcott Parson*, Englewood, NJ: Prentice Hall.
Brooks, N., W.N. Adger and P. Mick Kelly (2005), 'The Determinants of Vulnerability and Adaptive Capacity at the National Level and the Implications for Adaptation', *Global Environmental Change*, **15**(2), 151–63.
Cairney, P. (2012a), 'Complexity Theory in Political Science and Public Policy', *Political Studies Review*, **10**(3), 346–58.

410 *Handbook on complexity and public policy*

Cairney, P. (2012b), *Understanding Public Policy: Theories and Issues*, Basingstoke: Palgrave Macmillan.

Chilcote, R.H. (1994), *Theories of Comparative Politics: The Search for a Paradigm Reconsidered*, 2nd edn, Boulder: Westview Press.

Cummins, R. (1975), 'Functional Analysis', *The Journal of Philosophy*, **72**(20), 741–65.

Daedlow, K., V. Beckmann and R. Arlinghaus (2011), 'Assessing an Adaptive Cycle in a Social System under External Pressure to Change: The Importance of Intergroup Relations in Recreational Fisheries Governance', *Ecology and Society*, **16**(2), 3.

D'Alessandro, S., T. Luzzati and M. Morroni (2010), 'Energy Transition Towards Economic and Environmental Sustainability: Feasible Paths and Policy Implications', *Journal of Cleaner Production*, **18**(6), 532–9.

de Bruijn, J.A. and E.F. ten Heuvelhof (1995), 'Policy Networks and Governance', in D.L. Weimer (ed.), *Institutional Design*, Boston: Kluwer Academic Publishers, pp. 161–79.

Doelle, M., C. Henschel, J. Smith, C. Tollefson and A. Wellstead (2012), 'New Governance Arrangements at the Intersection of Climate Change and Forest Policy: Institutional, Political and Regulatory Dimensions', *Public Administration*, **90**(1), 37–55.

Dowding, K. (1995), 'Model or Metaphor? A Critical Review of the Policy Network Approach', *Political Studies*, **43**, 136–58.

Easton, D. (1965), *A Systems Analysis of Political Life*, New York: Wiley.

Elster, J. (1985), *Ulysses and the Sirens*, Cambridge: Cambridge University Press.

Elster, J. (1986), *An Introduction to Karl Marx*, New York: Cambridge University Press.

Emison, M. (2008), 'The Potential for Unconventional Progress: Complex Adaptive Systems and Environmental Quality Policy', *Duke Environmental Law and Policy Forum*, **7**(167), 167–92.

Enderlein, H., S. Wälti and M. Zürn (2011), *Handbook on Multi-Level Governance*, Cheltenham, UK and Northampton, MA, USA: Edward Elgar Publishing.

Engle, N. and M. Lemos (2010), 'Unpacking Governance: Building Adaptive Capacity to Climate Change of River Basins in Brazil', *Global Environmental Change*, **20**(1), 4–13.

Folke, C. (2006), 'Resilience: The Emergence of a Perspective for Social–Ecological Systems Analysis', *Global Environmental Change*, **16**(3), 253–67.

Folke, C., S. Carpenter, T. Elmqvist, L. Gunderson, C.S. Holling and B. Walker (2002), 'Resilience and Sustainable Development: Building Adaptive Capacity in a World of Transformations', *AMBIO: A Journal of the Human Environment*, **31**(5), 437–40.

Frechette, A. and N. Lewis (2011), 'Pushing the Boundaries of Conventional Forest Policy Research: Analyzing Institutional Change at Multiple Levels', *Forest Policy and Economics*, **13**(7), 582–9.

Füssel, H.-M. and R.J.T. Klein (2006), 'Climate Change Vulnerability Assessments: An Evolution in Conceptual Thinking', *Climate Change*, **75**(3), 301–29.

Gallopin, G. (2006), 'Linkages between Vulnerability, Resilience, and Adaptive Capacity', *Global Environmental Change*, **16**, 293–303.

Gatto, A. (2006), 'The Law and Governance Debate in the European Union', Discussion Paper, Geneva: International Institute for Labour Studies.

Geyer, R. (2012), 'Can Complexity Move UK Policy Beyond "Evidence-Based Policy Making" and the "Audit Culture"? Applying a "Complexity Cascade" to Education and Health Policy', *Political Studies*, **60**(1), 20–43.

Geyer, R. (2003), 'Beyond the Third Way: The Science of Complexity and the Politics of Choice', *The British Journal of Politics & International Relations*, **5**(2), 237–57.

Geyer, R. and S. Rihani (2010), *Complexity and Public Policy: A New Approach to Twenty-first Century Politics, Policy and Society*, London: Routledge.

Gómez, E.J. (2011), 'An Alternative Approach to Evaluating, Measuring, and Comparing Domestic and International Health Institutions: Insights from Social Science Theories', *Health Policy*, **101**(3), 209–19.

Gregor, A.J. (1968), 'Political Science and the Uses of Functional Analysis', *American Political Science Review*, **62**, 425–39.

Groth, A. (1970), 'Structural Functionalism and Political Development: Three Problems', *Western Political Science Quarterly*, **23**, 485–99.

Hall, P.A. and R.C.R. Taylor (1996), 'Political Science and the Three New Institutionalisms', *Political Studies*, **44**(5), 936–57.

Hedström, P. and P. Ylikoski (2010), 'Causal Mechanisms in the Social Sciences', *Annual Review of Sociology*, **36**, 49–67.

Hill, M. (2007), *The Policy Process*, 4th edn, Harlow: Pearson.

Holling, C.S. (1973), 'Resilience and Stability of Ecological Systems', *Annual Review of Ecology and Systematics*, **4**(1), 1–23.

Holling, C.S. (2001), 'Understanding the Complexity of Economic, Ecological, and Social Systems', *Ecosystems*, **4**(5), 390–405.

Complexity theory and policy studies: lessons from climate change 411

Howlett, M. (2002), 'Do Networks Matter? Linking Policy Network Structure to Policy Outcomes: Evidence from Four Canadian Policy Sectors 1990–2000', *Canadian Journal of Political Science*, **35**(2), 235–68.

Howlett, M. (2011), *Designing Public Policies: Principles and Instruments*, New York: Routledge.

Howlett, M. (2014), 'Why are Policy Innovations Rare and so Often Negative? Blame Avoidance and Problem Denial in Climate Change Policy-Making', *Global Environmental Change*, available at doi:10.1016/j.gloenvcha.2013.12.009.

Howlett, M. and J. Rayner (2006a), 'Globalization and Governance Capacity: Explaining Divergence in National Forest Programmes as Instances of "Next-Generation" Regulation in Canada and Europe', *Governance*, **19**(2), 251–75.

Howlett, M. and J. Rayner (2006b), 'Convergence and Divergence in "New Governance" Arrangements: Evidence from European Integrated Natural Resource Strategies', *Journal of Public Policy*, **26**(2), 167–89.

Howlett, M., M. Ramesh and A. Perl (2009a), *Studying Public Policy: Policy Cycles and Policy Subsystems*, 3rd edn, Ontario: Oxford University Press.

Howlett, M., J. Rayner and C. Tollefson (2009b), 'From Government to Governance in Forest Planning? Lesson from the Case of the British Columbia Great Bear Rainforest Initiative', *Forest Policy and Economics*, **11**, 383–91.

IPCC (2014), 'Summary for Policymakers', in C.B. Field et al. (eds), *Climate Change 2014: Impacts, Adaptation, and Vulnerability. Part A: Global and Sectoral Aspects. Contribution of Working Group II to the Fifth Assessment Report of the Intergovernmental Panel on Climate Change*, Cambridge, UK and New York, USA: Cambridge University Press, pp. 1–32.

Katzenstein, P. (1978), *Between Power and Plenty: Foreign Economic Policies of Advanced Industrial States*, Madison: UWP.

Klijn, E.H. and J.F.M. Koppenjan (2000), 'Public Management and Policy Networks: Foundations of a Network Approach to Governance', *Public Management*, **2**(2), 135–58.

Knoke, D. (1987), *Political Networks: The Structural Perspective*, Cambridge: Cambridge University Press.

Knoke, D. and J.H. Kuklinski (1982), *Network Analysis*, Beverly Hills: Sage.

Koliba, C., J. Meek and A. Zia (2010), *Governance Networks in Public Administration and Public Policy*, New York: CRC Press.

Kooiman, J. (1993), 'Governance and Governability: Using Complexity, Dynamics and Diversity', in J. Kooiman (ed.), *Modern Governance*, London: Sage, pp. 35–50.

Kooiman, J. (2000), 'Societal Governance: Levels, Models, and Orders of Social–Political Interaction', in J. Pierre (ed.), *Debating Governance*, Oxford: Oxford University Press, pp. 138–66.

Landau, M. (1968), 'On the Use of Functional Analysis in American Political Science', *Social Research*, **35**(1), 48–73.

Lilienfeld, R. (1978), *The Rise of Systems Theory*, New York: John Wiley & Sons.

Lindblom, C.E. (1977), *The Policy-Making Process*, New Haven: Yale University Press.

Lissack, M.R. (1999), 'Complexity: The Science, its Vocabulary, and its Relation to Organizations', *Emergence*, **1**, 110–26.

Lukes, S. (1974), *Power: A Radical View*, London: Macmillan.

Marsh, D. and A. McConnell (2010), 'Towards a Framework for Establishing Policy Success', *Public Administration*, **88**(2), 564–83.

McConnell, A. (2010), 'Policy Success, Policy Failure and Grey Areas In-Between', *Journal of Public Policy*, **30**(3), 345–62.

McCool, D. (1998), 'The Subsystem Family of Concepts: A Critique and a Proposal', *Political Research Quarterly*, **51**(2), 551–70.

McGinnis, M. (ed.) (1999), *Polycentricity and Local Public Economies: Readings from the Workshop in Political Theory and Policy Analysis*, Ann Arbor: University of Michigan Press.

Mio, J.S. (1997), 'Metaphor and Politics', *Metaphor and Symbol*, **12**(2), 113–33.

Moe, T. (1984), 'The New Economics of Organization', *Journal of American Political Science*, **28**(4), 739–77.

Morgan, G. (1980), 'Paradigms, Metaphors, and Puzzle Solving in Organization Theory', *Administrative Science Quarterly*, **25**(4), 605–22.

Nelson, D.R., W.N. Adger and K. Brown (2007), 'Adaptation to Environmental Change: Contributions of a Resilience Framework', *Annual Review Environmental Resources*, **32**, 395–419.

Nilsson, M., L. Nilsson, R. Hildingsson, J. Stripple and I. Eikeland (2011), 'The Missing Link: Bringing Institutions and Politics into Energy Future Studies', *Futures*, **43**(10), 1117–28.

Offe, C. (1984), *Contradictions of the Welfare State*, Boston: MIT Press.

Ostrom, E. (1991), 'Rational Choice and Institutional Analysis: Towards a Complementarity', *American Political Science Review*, **85**(1), 237–43.

Ostrom, E. (2008), *Polycentric Systems as One Approach for Solving Collective-Action Problems*, SSRN eLibrary, available at http://papers.ssrn.com/sol3/papers.cfm?abstract_id=1304697.

412   *Handbook on complexity and public policy*

Ostrom, E. (2009), *A Polycentric Approach for Coping with Climate Change*, SSRN eLibrary, available at http://papers.ssrn.com/sol3/papers.cfm?abstract_id=1494833.

Pal, L. (2010), *Beyond Policy Analysis: Public Issue Management in Turbulent Times*, 4th edn, Toronto: Nelson Education.

Peters, B.G. (1999), *Institutional Theory in Political Science: The 'New Institutionalism'*, London: Pinter.

Plummer, R. and D.R. Armitage (2007), 'Charting the New Territory of Adaptive Co-management: A Delphi Study', *Ecology and Society*, **12**(2), 10.

Preston, B., R. Westaway and E. Yuen (2010), 'Climate Adaptation Planning in Practice: An Evaluation of Adaptation Plans from Three Developed Nations', *Mitigation and Adaptation Strategies for Global Change*, 1–32.

Rhodes, R.A.W. (1996), 'The New Governance: Governing Without Government', *Political Studies*, **44**(4), 652–67.

Rosenau, J. (1992), 'Governance, Order, and Change in World Politics', in J.N. Rosenau and E.-O. Czempiel (eds), *Governance without Government: Order and Change in World Politics*, Cambridge: Cambridge University Press, pp. 1–29.

Sabatier, P. and H. Jenkins-Smith (1999), 'The Advocacy Coalition Framework: An Assessment', in P. Sabatier (ed.), *Theories of the Policy Process*, Boulder: Westview.

Sanderson, I. (2009), 'Intelligent Policy Making for a Complex World: Pragmatism, Evidence and Learning', *Political Studies*, **57**, 699–719.

Schlager, E. and W. Bloomquist (1996), 'Three Emerging Theories of the Policy Process', *Public Research Quarterly*, **49**(3), 651–72.

Skodvin, T., A. Gullberg and S. Aakre (2010), 'Target-group Influence and Political Feasibility: The Case of Climate Policy Design in Europe', *Journal of European Public Policy*, **17**(6), 854.

Smit, B. and J. Wandel (2006), 'Adaptation, Adaptive Capacity and Vulnerability', *Global Environmental Change*, **16**(3), 282–92.

Smit, B., I. Burton and J.T.K. Richard (1999), 'The Science of Adaptation: A Framework for Assessment', *Mitigation and Adaptation Strategies for Global Change*, **4**(3–4), 199–213.

Smith, G. and D. May (1980), 'The Artificial Debate between Rationalist and Incrementalist Models of Decision Making', *Policy & Politics*, **8**(2), 147–61.

Sprinz, D. and T. Vaahtoranta (1994), 'The Interest-Based Explanation of International Environmental Policy', *International Organization*, **48**(1), 77–105.

Stephens, J. (1969), 'The Logic of Functional Systems Analysis in Political Science', *Midwest Journal of Political Science*, **13**(1), 367–94.

Teisman, G.R. and E.-H. Klijn (2008), 'Complexity Theory and Public Management: An Introduction', *Public Management Review*, **10**, 287–97.

Thorson, T. (1970), *Biopolitics*, New York: Holt, Rinehart and Winston.

Tollefson, C., F. Gale and D. Haley (2008), *Setting the Standard: Certification, Governance and the Forest Stewardship Council*, Vancouver, BC: University of British Columbia Press.

Tollefson, C., A.R. Zito and F. Gale (2012), 'Symposium Overview: Conceptualizing New Governance Arrangements', *Public Administration*, **90**, 3–18.

Treib, O., H. Bähr and G. Falkner (2007), 'Modes of Governance: Towards a Conceptual Clarification', *Journal of European Public Policy*, **14**(1), 1–20.

von Bertalanffy, L. (1969), *General System Theory: Foundations, Development, Applications*, revised edn, New York: George Braziller.

Voss, J. and B. Bornemann (2011), 'The Politics of Reflexive Governance: Challenges for Designing Adaptive Management and Transition Management', *Ecology and Society*, **16**(2), Art. 9.

Voss, J., D. Bauknecht and R. Kemp (eds) (2006), *Reflexive Governance for Sustainable Development*, Cheltenham, UK and Northampton, MA, USA: Edward Elgar Publishing.

Walker, J. and M. Cooper (2011), 'Genealogies of Resilience: From Systems Ecology to the Political Economy of Crisis Adaptation', *Security Dialogue*, **42**(2), 143–60.

Weible, C., T. Heikkila, P. deLeon and P. Sabatier (2012), 'Understanding and Influencing the Policy Process', *Policy Sciences*, **45**(1), 1–21.

Wellstead, A. (2007), 'Filling in the Gaps: Complexity and the Policy Process', in R. Geyer and J. Bogg (eds), *Complexity, Science and Society*, London: Radcliffe Press, pp. 141–7.

Wellstead, A., M. Howlett and J. Rayner (2013), 'The Neglect of Governance in Forest Sector Vulnerability Assessments: Structural-Functionalism and "Black Box" Problems in Climate Change Adaptation Planning', *Ecology and Society*, **18**(3).

Wellstead, A., J. Rayner and M. Howlett (2013a), 'Beyond the Black Box: Forest Sector Vulnerability Assessments and Adaptation to Climate Change in North America', *Environmental Science & Policy*, **35**, 109–16.

Wu, X., M. Ramesh, M. Howlett and S. Fritzen (2010), *The Public Policy Primer: Managing Public Policy*, London: Routledge.

Zashin, E. and P.C. Chapman (1974), 'The Uses of Metaphor and Analogy: Toward a Renewal of Political Language', *The Journal of Politics*, **36**, 290–326.
Zito, A. (2007), 'European Union: Shifting Environmental Governance to the Supranational Level', in A. Breton, G. Brosio, S. Dalmazzone and G. Garrone (eds), *Environmental Governance and Decentralisation*, Cheltenham, UK and Northampton, MA, USA: Edward Elgar Publishing, pp. 140–72.

25. Agent Based Modelling and the global trade network
Ugur Bilge

INTRODUCTION

In the aftermath of globalization, the global economic system and global trade have become ever more important. This was highlighted further after the recent economic crisis which spread amongst nations like a viral epidemic. The global economy presents one of the most complex systems we know, and the global trade network is a complex system for several reasons. First, the global economy is a network of autonomous agents with multiple levels of properties, characteristics or dimensions. Some of these dimensions are geographic, social, political and cultural. As a general rule, it might be natural to think that geographic neighbours are also strong economic partners with tightly coupled economies, but there are many exceptions to this rule for historical reasons. Global economy has a long history and is path-dependent like many complex physical systems that depend on initial conditions that are difficult to measure accurately. The global economy is an adaptive system, as it has many agents such as sovereign nations and multinational companies that are able to observe their state and adapt accordingly. It can be defined as self-organized, as the global economy has no central controller that can turn levers and change the course of the global economic system; even though there are several global institutions, they have not been able to prevent crises emerging frequently in the recent century. Global trade links between nations provide positive as well as negative feedback, creating a far from stable network. The global economy is by nature a non-equilibrium system that suffers from collapses, and extreme events such as economic depressions and crises are emergent properties of this complex system. They can be triggered by weather or climate events as well as social political reasons to the point that it is often difficult to establish what causes which. For these reasons the global economy can best be described as a complex adaptive social system consisting of a network of interacting autonomous and interdependent agents.

As always, crises and extreme events can prompt innovative research. Here a research project is described that looked into the global trade network as an investigation of potential game changers and extreme events for the next twenty years. Our focus was on the global economy and in particular Gross Domestic Product (GDP) growth and the global trade network, formed of import and export links amongst nations. One of the problems in exploring complex systems is that there are no generic tools in this domain, and therefore it is hard to model the global economy, and capture its extreme behaviour using linear tools. The Global Trade Network Simulator (GTNS) described in this chapter attempts to model one aspect of the global economy by focusing on the global trade network. The Agent Based Modelling (ABM) approach was used in developing GTNS, which is probably the most appropriate technique for exploring complex networks of agents such as trading nations. Agent Based Modelling is a computer simulation technique involving autonomous agents each equipped with their own data, connections and rules. ABM philosophy can be summarized as 'the whole is greater than the sum of its parts', and the modelling effort is worthwhile as it tries to explore unexpected and extreme behaviour (see the chapters by Astill and Cairney, and Morçöl, for a broader discussion of ABM and complexity). GTNS was commissioned as a subtask in the Game Changers (GC) project. It was built with ABM to model the global trade network and trade-related GDP growth and run expert-driven or machine-generated what-if scenarios for investigating extreme events in the next decades.

THE GAME CHANGERS PROJECT AND EXTREME EVENTS

The Game Changers project was based at the International Institute for Applied Systems Analysis (IIASA) in Laxenburg, Austria, as part of the X-events in Human Society initiative. The project took place between 2009 and 2011 and the main partners of the project were the Finnish and Scottish governments, supported by researchers and consultants mainly from Finland, Austria and the UK. The goal of the Game Changers project was to identify drivers that could have a major impact on the global economy and detect potential shocks, and measure their impact on the specific economies in question. These drivers or events, also called X-events (or extreme events), are by definition rare events with a large impact on the global economic system.

> An extreme event is unlikely but potentially significant. This refers not only to the immediate trigger event but also to the following sequence of linked events. The event is conditional on the surrounding context. Events typically have varying unfolding and impact times, and outcome is determined depending on the mood and expectation of the system. (Casti, 2012)

The Game Changers project examined the global economic system including social and political dimensions and asked the question: what are the main drivers which might cause a major transformation in the global economic system of 2030? The project used qualitative and quantitative research methods. As the project was mainly funded and supported by Finnish and Scottish institutions, it had a strong emphasis on the analysis and impact of game changers on these economies, with a particular focus on forest industry, communications technologies, food and drink, life science and digitalization. Several brainstorming sessions narrowed down the game changers into natural events, human events and economic events. Extreme natural events were classified as climate change and extreme weather, a potential ice age, volcanoes, plant diseases and bio-system collapses. Social and human system game changers were listed as loss of faith in the political system, problems with workings of the patent system, trust in technology, integration of commercial and political power, changing roles of private and public sectors and the emergence of global order. A number of scenarios were discussed regarding the new world order and changing bloc structure in the world geopolitical scene as well as the impact of global organizations such as the United Nations. Economic game changers included the future of globalization and centralization, credit crunch, development of virtual money and exchange, zero cost energy and supercomputing and singularity. These game changers were analysed in a systematic and qualitative fashion.

416 *Handbook on complexity and public policy*

The qualitative analyses of the Game Changers project and the brainstorming sessions showed that most shocks could be linked with GDP, and so impacts and shocks would be measured as changes in GDP growth. GDP growth is probably the most overrated measure of a nation's performance in the global arena. It is a single measure but is widely used for comparing the relative performance and wealth of nations. GDP growth rate (percentage change over previous period, usually year or quarter) is analogous to the existence or non-existence of economic activity, and is used for comparison with other nations as well as previous periods. Why some nations are rich yet others remain poor is an interesting question (Kay, 2003), and answers may lie in a number of seemingly unrelated domains such as geography, natural resources, education, skills, social infrastructure, technology and innovation. Several GDP models have previously been built and long-term forecasts have been made using these systems. One of the forecasts of the global GDP made by the Club of Rome (COR) back in the 1960s was remarkably accurate. In 1972 the COR published a report called 'The limits to growth' and argued for the dangers of unchecked economic and population growth. The report used a computer simulation called World3, which modelled global food and industrial systems, population, non-renewable sources and pollution.

Another computer simulation which is called International Futures (IFs) (Hughes, 1999) also modelled global demographic, economic, agriculture and food, energy, environment, technological and socio-political change. The system can run user-defined scenarios, generate forecasts with provided new assumptions and compare results with baseline assumptions. The IFs simulator incorporates simple growth models such as population dynamics, basic negative and positive feedback loops and equilibrium dynamics in economy (Price, Production, Demand, Inventories and Capacity Utilization). But it is not a network-based tool and does not model network dynamics and complex what-if scenarios involving network interactions.

After examining these models a decision was made not to model an economy or GDP, but instead the GTNS was proposed to model global trade and its impact on GDP which most contributors believed was important in the post-globalization era. The building of the GTNS fitted nicely into the objectives of the Game Changers project. As GDP growth and the global trade network pose a complex problem, the Agent Based Modelling technique was used in constructing the GTNS. In the first version which is presented here, a coarse-grained approach was adopted, and only the most significant connections for Finland were modelled as the agents of the simulation. The simulator provided visualization of global trade, and facilitated what-if scenario analysis between 1990 and 2030. Users could provide their text-based scenarios using GTNS syntax, including GDP and trade shocks for one or more countries, force countries to change their trading partners, play out currency wars and observe contagion. Results are then compared with baseline scenarios where agents of the simulation follow their trend line for growth and trade. The GTNS does not calculate GDP but simulates changes on GDP as a result of changes in trade patterns. Later the GTNS calculates the impact of changes in global economy and trade on specific business sectors and industries in the focus country (in this case Finland).

AGENT BASED MODELLING

As opposed to classic computer simulations, the ABM approach is a 'bottom-up' modelling technique, sometimes resulting in unexpected, so-called 'emergent' behaviour. ABM is suitable for real world problems where there are a number of autonomous, heterogeneous and interacting agents. Often the development of an Agent Based Simulation (ABS) is in itself useful and educational where the end product does not necessarily predict the future, but is a virtual laboratory to experiment with ideas, and to test possible real-life scenarios in a safe environment.

Unfortunately, there are no off-the-shelf ABM packages available which fit every problem. Early ABS environments built in the USA, such as Swarm, are platforms for looking into the theoretical aspects of ABS. Later work on social systems, trading and the spread of culture is described in SugarScape (Epstein and Axtell, 1996). Most simulation environments such as Swarm and NetLogo are highly abstract, as they model agents as nodes on graphical grids. The idea of using computer simulations and ABM as laboratories for many real-life problems has been gaining momentum (Casti, 1997). TRANSIMS, the first of its kind, involved a fine grain simulation (at the level of each car and household) of Albuquerque's traffic (Casti, 1997) and was a dedicated system with the purpose of simulating and researching air pollution levels for the city of Albuquerque in New Mexico, USA. In the UK, SimStore supermarket simulator was developed by SimWorld Ltd as a business tool where supermarket customers are modelled as autonomous agents fulfilling their shopping lists by moving in a virtual store (Casti, 1999). ABM can be used for social, organizational and business applications (Bonabeau, 2002; Banks, 2002). ABM could also be used as a tool for decision support in organizational policy-making where standard tools fail (Lempert, 2002). Today ABM in the social context is often used together with network analysis techniques for visualizing and simulating social agents in organizations (Barabási, 2002; Buchanan, 2002).

ABM has several advantages; it promotes decentralized thinking and it can run what-if scenarios and evaluate outcomes from multiple perspectives. It encourages users to look for emergent and unexpected properties of system behaviour even when agents of the system behave mechanically, obeying simple rules. It also allows the possibility of agent adaptation and rule change, and can give the user new insights, if not an outright prediction. ABM can work in the past, which can be used for validation, and in the future for testing scenarios. Statistical analyses of results could give a level of confidence in predictions.

The Global Trade Network promised to be a productive domain to apply Agent Based Modelling and ideas from Network Theory. With ABM, we would not need to know the overall equations defining the global economy, but would focus on trade between nation states and model from the bottom-up. Instead of trying to forecast the global economy in twenty or thirty years' time from now, we could focus on simulating, observing and running what-if scenarios with an open mind in order to learn from the simulations, generate insights and understand the mechanisms of global trade connections. We could use recent ideas from network analysis techniques to identify cliques, most utilized nodes and links.

A standard modelling practice was not adopted because it would not have provided us with anything useful, as our main objective was to understand economies during crises and under extreme conditions. As Cecchetti (2006) wrote: 'Standard modelling strategies provide virtually no information about the behaviour of the economy when it is under

418 *Handbook on complexity and public policy*

stress.' For the global trade network, ABM philosophy was adopted from the beginning, and expert as well as machine-generated scenarios could be used, and their outcomes would be discussed for further investigation.

BUILDING THE GTNS FOR FINLAND

In the Game Changers project we were primarily interested in extreme events that might impact on the global economy and Finland in particular. We established early on that whatever events happen would somehow show up in the GDP, and that trade was a significant part of GDP. Global trade, contagion of economic crises (as well as prosperity), changing trade links and global economic conjuncture were design considerations. One of the first priorities for the simulation was to model Finland's economy in a global trade context, so as to be able to run what-if scenarios into the next twenty years with a graphics environment for generating insights to stimulate discussions. So a user-friendly graphics user interface and self-explanatory visualization became one of the early requirements for the simulation tool. The programming environment was another consideration, and the Java programming language was chosen in the development of GTNS as it provides a strong graphical user interface and platform-independent execution on Java-enabled Internet browsers.

One of the most important considerations in model building is what to model. After an early analysis, a total of 22 countries were chosen as the agents of simulation. In the selection two considerations were taken into account. These were to be a sufficiently large economy and to have strong trade links with Finland. The selected agents of the simulation are: Finland, Sweden, Denmark, Belgium, Holland, Germany, France, Spain, Italy, USA, Mexico, Canada, Brazil, China, India, Japan, Indonesia, UK, Norway, Russia, Turkey and South Africa. A trade-off has been made in the selection of the subset of countries: a larger simulation would have presented cumbersome data modelling and a more crowded network map with little benefit to projections of Finland's GDP, as the relatively small nations would have little impact on results in the twenty years' simulation horizon. Still, the software has been tailored to be able to include an arbitrary number of countries in the simulation should we need to run it for a different country in the future.

Each agent of the simulation has a GDP, GDP growth rate, Imports, Exports and Currencies. Data between 1990 and 2009 were collected for display purposes and the simulation scope was set between 2010 and 2030. A mixture of open access web data sources were used, such as the CIA World Factbook archive, World Bank (WB), and International Monetary Fund World Economic Outlook (WEO). The major economies' GDP is given in local currency in constant prices from IMF/WB. All growth rates data were taken from IMF/WB (in constant prices in local currency) and the CIA World Factbook data is used for displaying major trading partners for each country between 1990 and 2008. For 2009 Finland's export breakdown for each country is taken from Finland's official dataset for sector exports. For some countries we used 2009 currency conversion rates as the basis for 2009–30 but they can be modified by user-defined currency events. The Global Trade Network Simulator displays GDP both in local currency and USD, and maintains currency rates throughout simulation.

Agent Based Modelling and global trade 419

Trade-driven GDP Growth

The basic formula for Gross Domestic Product is:

$$GDP = Consumption + Investment + Government\ Spending + \textbf{Exports} - \textbf{Imports}$$

If we focus on the changes in trade patterns we would be able to calculate the changes in GDP in terms of changes in exports and imports. So we would need to calculate the year-on-year difference in exports and imports, and calculate the difference in GDP growth resulting from these two variables. Then we use this difference in growth rate as a percentage to compare it with the baseline growth rate for projecting into the future.

In summary, the simulator does not attempt to calculate GDP; it has a baseline GDP growth rate, and assumes it will continue unless trade links change. The GTNS calculates the change in the growth rate as a result of changes in exports and imports. So increased exports have a positive impact on the growth rate and increased imports have a negative impact. The GDP formula above also gives us insights about countries' dependency on exports; for example, exports in the USA have a much lower impact on GDP than exports in Germany or China, because these countries generate huge income from exports, while in the USA consumption is a major part of its GDP.

An economic shock such as a country's GDP going down for various reasons spreads on the network through the export links towards this country. To give an example: if the USA GDP goes down by 10 per cent, countries that are exporting to the USA will be affected by a drop in their GDP (depending on their exports to the USA and how much their GDP depends on exports). The shock ultimately affects all countries on the network. We assume that any effect works in a symmetrical fashion, and the opposite of a drop also affects countries, this time positively spreading through the network.

Baseline Projections

Baseline projections simply take the average growth rates for a specified number of years (N) and project into the future using the same rate, GDP growth rate, export growth rate and import growth rate. By default we use N = 8 years. We can then calculate and visualize GDP for all countries in the network. Growth rates can be changed by user-defined what-if scenarios and by ABM rules changing trade connectivity. Taking N as 8 years we find the following values for Finland in 2009: GDP Growth Rate is 1.69 per cent, Export Growth Rate is 6.53 per cent, and Import Growth Rate is 10.06 per cent.

Export Links Matrix

Links of the global trade network are based on exports to other countries, for example in 2009 Finland's export breakdown was: Germany (13.1 per cent); Sweden (9.3 per cent); Russia (7.4 per cent); UK (7 per cent); USA (6.7 per cent); China (4 per cent); Italy (3.6 per cent); France (3.5 per cent); Netherlands (3 per cent); Spain (2.6 per cent); Belgium (2.6 per cent); Norway (2.5 per cent).

The list above consists of the export destinations that form more than 2 per cent of

420 *Handbook on complexity and public policy*

Finland's exports. There are 13 export partners for Finland, while Mexico and Canada heavily rely on exports only to the USA.

- Mexican export breakdown: USA (81 per cent), Canada (3.4 per cent).
- Canada export breakdown: USA (79 per cent), UK (2.8 per cent), Japan (2.3 per cent).

As a result Canada and Mexico show a strong dependency on the USA in terms of their exports (81 per cent and 79 per cent respectively) and are therefore more vulnerable to shocks from their large neighbour.

The simulator is provided with the 22 countries' exports distribution in an export links matrix. Once GTNS reads the Export Links Matrix, it is able to display the links in a graph (Figure 25.1).

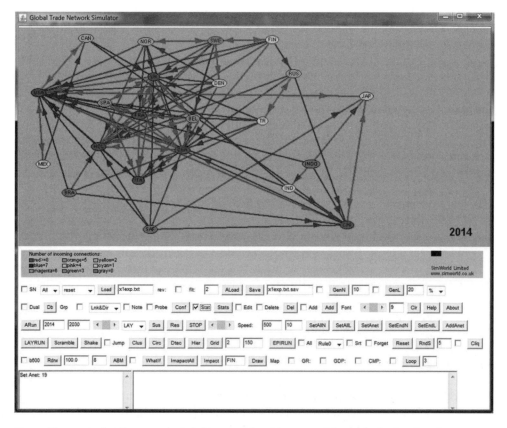

Note: Please note that figures are included in greyscale, without use of the original colour key, for illustrative purposes in order to give the reader a feel for the simulator's interface.

Figure 25.1 Global trade network simulator

Network Analysis: Identifying Cliques

In network analysis, cliques are two or more nodes with mutually directed connections. In the trade network, cliques are defined as groups of trade partners with more than 2 per cent of exports. Using clique-searching algorithms the GTNS finds and visualizes all cliques in the network for a given trade network. The emergence of new cliques can indicate new trade blocs and disappearing old cliques can flag up the breakdown of old alliances.

In the Finland model there were 15 cliques altogether in 2009, but most of these were obvious groups, as countries in these cliques are members of the European Union such as Germany, France, Netherlands, Belgium and Italy; others were known big traders such as USA, Germany and China, as well as USA, China and Japan, and there was an interesting group consisting of Russia, Italy and Turkey. These cliques were displayed on the network view as shown in Figure 25.2.

In addition to the network view, we can use the Map View (Figure 25.3) to probe into the individual country's data. Map View also provides a score table for the GDP of the countries. We can run what-if and ABM scenarios and see outcomes in the score table. Future projections are compared with current assumptions that use the mean values of the last N years' values as the basis for the future GDP, import and export growth rates. As the system is driven by variable percentage growth rates, it displays exponential GDP growth in the period between 2010 and 2030. Growth rates can be changed by user-defined what-if scenarios and by ABM rules changing trade connectivity. Map View also displays the GDP ranking, Growth Rate ranking, or the difference from baseline projection for each year between 1990 and 2030.

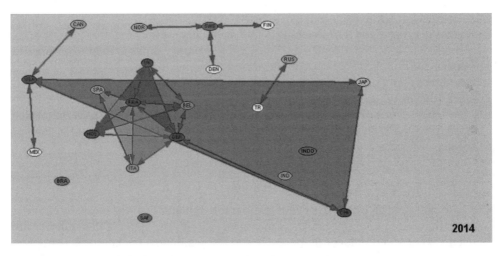

Note: Please note that figures are included in greyscale, without use of the original colour key, for illustrative purposes in order to give the reader a feel for the simulator's interface.

Figure 25.2   Trade cliques

422 *Handbook on complexity and public policy*

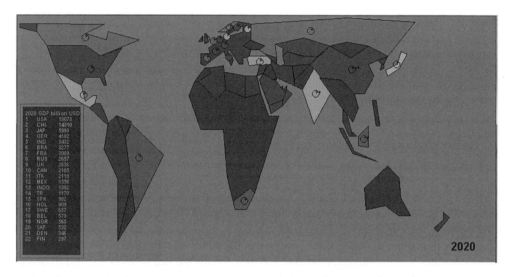

Note: Please note that figures are included in greyscale, without use of the original colour key, for illustrative purposes in order to give the reader a feel for the simulator's interface.

Figure 25.3   Map View

Impact of GDP Shocks

As mentioned above, using a breadth-first search technique we can track export links and establish the impact of a GDP shock on each country of shocks by all others in the network, carried by the exports links matrix. Here we use the breadth-first search algorithm to calculate the impact on the network.

Finland's exposure to GDP shocks in 2009 was the highest to Germany (2.34 per cent), followed by USA (1.45 per cent) and Canada 0.3 per cent), and the total exposure was 12.88 per cent.

This shows that given the above network structure, Finland's GDP can go down by 2.34 per cent (from its baseline) if the German GDP goes down by 10 per cent from its baseline levels. This dependency can change only by reducing exports to Germany and increasing exports to another country such as Japan or Mexico. Also other trade changes on the network influence these impact numbers.

Currencies

Currency rate changes also trigger changes in trade. In fact this is the basis for currency wars. Nations often apply a number of measures to lower the value of their currency to stimulate exports and dampen imports. The GTNS simulator includes this feature by allowing the currency of a country to go up or down a certain percentage, causing an increase/decrease in exports and imports. As all import and export figures are provided in US dollars, USD is chosen as the benchmark currency, so all devaluations and revaluations are carried out as percentage change against US dollars.

Agent Based Modelling and global trade 423

Table 25.1   Sector breakdown for exports

	Sector 1	Sector 2	Sector 3	Sector . . .
Germany	10	20	30	5
USA	40	20	10	10
UK	10	20	40	10

Countries can belong to blocs or with no blocs. Currently there is only one bloc that has a common currency (euro) which can value or devalue currencies of a group of countries against the USD. Countries can also leave the euro and devalue currency. We assume that the currency rates stay the same as in their 2009 values unless changed by what-if scenario events such as devaluations.

Exports and Sectors

A good insight about what to model came from Finnish colleagues, as they suggested including Finland's exports at sector level, and modelling the impact of change in exports on these sectors. This made a lot of sense, and zooming into sectors provided information about how exports varied in their constitution depending on the target country. So, a drop or increase in exports to certain countries would have an impact on sectors and their employment levels, which was also important for other game changers such as the social mood in Finland.

We used data for 42 different sectors (or industries); see Table 25.1. Exports to Germany consist of 10 per cent in sector 1, 20 per cent in sector 2 and so on all the way to sector 42. We assume the sector export breakdown stays the same unless changed by what-if scenarios or ABS runs.

What-if Scenario Rule Syntax

GTNS enables users to run their own scenarios using a number of keywords. These are GDP, EXPORT, IMPORT, LINK, CHANGE and CURR rules.

GDP Rule

This rule creates GDP growth events, originating from a country or group of countries in a given year, with number of years of duration and amplitude in terms of GDP growth rate. There are two ways of providing the GDP growth rate, and these are in absolute terms (GDPD) or relative to the baseline growth rate (GDP). For example: in 2014 Russia's growth will be -5 per cent for one year, and the World economy will shrink by -1 per cent for 3 years, while eurozone GDP will shrink by -2 per cent (see Table 25.2).

424   *Handbook on complexity and public policy*

Table 25.2   GDP rule syntax

Year	Duration	Source	GDP	Per cent
2014	1	RUS	GDP	−5
2014	3	World	GDP	−1
2014	2	Euro	GDP	−2

Table 25.3   EXPORT and IMPORT rule syntax

Year	Duration	Source	Exp/Imp	Per cent	Destination
2015	5	FIN	EXPORT	10	RUS
2015	5	FIN	IMPORT	−5	RUS
2016	3	USA	EXPORT	−10	World
2020	3	CHI	EXPORT	5	Euro

EXPORT and IMPORT Rules

This rule specifies year to year change in exports from one country to another. For example, between 2015 and 2020 Finland's exports to Russia will go up by 10 per cent, but imports from Russia will go down by −5 per cent. In 2016 USA exports to the world will go down by −10 per cent, and in 2020 for 3 years China exports to the eurozone will go up by 5 per cent (see Table 25.3).

LINK and CHANGE Rules

This rule assigns user-defined export connections to a country. In 2022 Finland's export links will be 30 per cent to USA, 30 per cent to UK, and 10 per cent to Russia. Once connections change they will stay changed to the end of simulation. Change rule simply reassigns a relative increase or decrease in export links. Once connections change they will stay changed to the end of simulation.

CURR Rule

Currency rule enables users to play out currency war scenarios. The simulator interprets currency devaluations by increasing exports to the rest of the world, and reducing imports to that country with a given ratio. This ratio is a coefficient which takes into account the export dependency of that country. For example in 2015 Finland could leave the eurozone, and devalue its currency by −10 per cent. Assuming this is achievable, the simulator calculates exports increasing by 2 per cent, and imports going down by 2 per cent (see Table 25.4).

Agent Based Modelling and global trade 425

Table 25.4   CURR rule syntax

Year	Duration	Country	CURR	Per cent
2015	1	FIN	CURR	−10

Table 25.5   A test what-if scenario

Year	Duration	Country	Source	Event per cent	Destination
2014	1	RUS	GDP	−5	
2016	1	USA	IMPORT	−10	World
2020		USA	EXPORT	−10	UK, GER
2022	1	FIN	LINK	100	USA,30,UK,30,RUS,10
2025	3	World	GDP	−20	
2029	1	FIN	CHANGE	100	USA,−1,UK,1,FRA,1

A Test Scenario

Here is a user-defined sample what-if scenario for testing purposes:

In 2014 Russia's GDP will go down by 5 per cent.
In 2016 USA imports from the rest of the world will go down by 10 per cent.
In 2020 USA exports to UK and Germany will go down by 10 per cent.
In 2022 Finland's export links are USA 30 per cent, UK 30 per cent and Russia 10 per cent.
In 2025 a worldwide economic crisis causes 20 per cent drop in GDP in 3 consecutive years.
In 2029 Finland's exports to USA go down by 1 per cent, to UK go up by 1 per cent, and to France they go up by 1 per cent.

This translates easily to the what-if scenario code required by the simulator shown in Table 25.5.

The simulator is able to run this what-if scenario and show outcomes in all countries in terms of GDP, growth rate, imports and exports for all years until 2030. Now using the syntax described above users can try their own scenarios.

GTNS WHAT-IF SCENARIO PROCESS

Use of scenarios for planning in geopolitical settings is common. Multinational companies and governments use this methodology where experts from different backgrounds and disciplines gather to generate future scenarios, and sometimes play out against each other in the form of war games (Rickards, 2012). But these exercises often fail miserably when predicting extreme events. Most expert-generated scenarios are constrained by the

426   *Handbook on complexity and public policy*

past experience, personal biases of experts and available data. Sometimes these scenarios are combined (Bremner and Keat, 2009) to eliminate these shortcomings but still future extreme events remain unpredictable by human experts.

One of the problems with the GTNS was that what-if scenarios in the real world are expressed in narrative form whereas the simulator requires numbers and instructions presented in a very strict syntax to 'understand' and process these scenarios and present numerical and graphical outputs. These outputs in turn need to be processed by economists and interpreted back into narrative form. So later in the project we adopted a three-stage procedure to process scenarios.

The first step involved starting with a question which has relevance in terms of policy, strategy or the game changer. In the second step this question is worded as a qualitative narrative. In the third step, the narrative scenario is converted into GDP and trade terms and written in the syntax that the simulator can process. Then the simulator is run several times with the scenario and neighbouring scenarios to assess sensitivity of the numbers provided to the simulator. Results are assessed graphically and numerically, and agent performances are ranked and colour coded for easy interpretation. Once there is confidence in the numerical, graphical outputs, an interpretation of the results is written in a qualitative, narrative form. Finally these results are reported back to the problem owners for further interpretation and assessment of the relevance for the game changer in question.

Game Changer Scenarios for the Next Two Decades

In the Game Changers project when future extreme events were discussed, the main focus was always on USA–China dynamics and its potential impact on other countries, particularly on Finland. The areas of contention and cooperation between USA and China were outlined in a seven-page document; they are summarized here in a number of subheadings:

- Global Economy
  - trade and exchange rates
  - holding of USA Treasuries
  - demand for energy
- Geopolitical Issues and Global Security
  - narrowing competitive power gap
  - Taiwan
  - future of North Korea
  - competition for key resource region
- Ideological and Philosophical Issues
  - democracy
  - human rights
  - religious freedom

The USA–China scenarios are loosely based on Chimerica, a term coined by Niall Ferguson to describe the dynamics between USA and China, as a marriage in heaven, 'East Chimericans did the saving and West Chimericans did the spending' (Ferguson,

Agent Based Modelling and global trade 427

2009). How this marriage would end was a matter for much speculation in the Game Changers project. The project prepared a number of narrative Chimerica scenarios (Casti and Ilmola 2012). Here is an example:

> The US is still the world's sole superpower, while China is rapidly emerging as a new one and is the only country that can realistically challenge the US for global dominance. Meanwhile, their economic ties are now so important that China and the US have since 2007 been described as 'Chimerica' – two sides of a single economy that comprises a third of global GDP. Therefore, any deterioration in this relationship would be detrimental for global political, economic, and financial stability.

Flavours of China Scenarios

This exercise reveals a number of possibilities for USA versus China. If we focus on China challenging USA dominance, this challenge could come in three different flavours:

- Scenario 1: China is increasingly assertive, belligerent even, in all areas of conflict and contention (the 'hard' path). In this scenario the wildest card is actual military confrontation between China and the US.
- Scenario 2: China is still assertive but exercises its assertion in a more subtle way, often through diplomatic channels, denial of resource exports and the like (the 'soft' path).
- Scenario 3: China weakens due to internal stresses of an economic, political and social nature, while the USA somewhat miraculously re-emerges as the dominant global power. This scenario involves progressively slower Chinese growth, while the USA regains confidence and influence.

In order to convert the foregoing verbal scenarios into a language the simulator understands, we must engage in a kind of 'simspeak', in which a somewhat more detailed story is told in numbers rather than words.

Translation of Scenarios

Scenario 1: Between 2010 and 2015, an aggressive China increases trade in Asia. The Asian partners then turn their back on the Chinese in 2015, and return to the USA as the main trading partner until 2030. (This is now modified as: Between 2012 and 2017 China is aggressive, later for 10 years China Exports to Japan, India and Indonesia shrink, and Chinese GDP growth is 5 per cent, USA Exports to Japan, India and Indonesia increase.) (See Table 25.6.)

When we run the above scenario on GTNS, in 2030 China is by far the worst off country, as it achieves 40 per cent less than its baseline. Finland also suffers, being off the baseline by 3.7 per cent. All Finland sectors do badly: the worst-performing sectors are Paper Products and Publishing exports (down by USD 0.2 bn), Machinery and Equipment exports (down by USD 0.18 bn) and Electronic Equipment exports (down by USD 0.16 bn).

Scenario 2: This so-called 'soft path' scenario suggests that between 2012 and

428   *Handbook on complexity and public policy*

Table 25.6   Scenario 1

Year	Duration	Source	Event	Per cent	Destination
2012	5	CHI	EXPORT	10	IND
2012	5	CHI	EXPORT	10	INDO
2012	5	CHI	EXPORT	10	JAP
2012	5	CHI	GDPD	12	
2017	10	CHI	EXPORT	−30	IND
2017	10	CHI	EXPORT	−30	INDO
2017	10	CHI	EXPORT	−30	JAP
2017	10	USA	EXPORT	30	JAP
2017	10	USA	EXPORT	30	IND
2017	10	USA	EXPORT	30	INDO
2017	10	CHI	GDPD	5	

2022 China slowly increases exports to Asia (5 per cent per year to Japan, India and Indonesia) and achieves 11 per cent GDP growth every year, following a careful, yet determined strategy to push the USA out of the Asian import market.

This scenario run on GTNS gives the following results: by 2030 all economies do well, particularly China achieving 12 per cent higher GDP than in the baseline scenario. Finland also does well, showing an increase of 0.06 per cent. As for sectors: Paper products and Publishing exports are up by USD 0.038 bn; Machinery and Equipment exports are up by USD 0.035 bn; and Electronic Equipment exports are up by USD 0.031 bn.

Scenario 3: GTNS results show by 2020 China underperforms 3 per cent below its baseline while the USA grows 1 per cent above its baseline, then after 2020 China underperforms by 5 per cent for the next 8 years. Mexico, Canada and USA all do well. Finland does 3.27 per cent worse than the baseline. China was normally expected to overtake USA by 2021 (baseline forecast) but this scenario delays this event by 5 years to 2026.

Similar China scenarios have been discussed elsewhere (Rickards, 2012); as the Chinese GDP composition is the mirror image of the USA, should China decide to increase consumption (at the expense of increasing imports from the USA), both USA and China should be able to increase their GDP growth. Maybe optimum time spans or paths for such scenarios can also be investigated by the use of simulations.

Machine-generated ABM Scenarios

In addition to the user-defined scenarios, the GTNS is able to generate and test machine-generated scenarios using a number of simple rules. These rules are mainly designed to enable each agent (country) to modify its trade links. There are five types of agent rules and a further three meta rules (acting on agent rules). In the real world, agent rules can correlate to policy changes by governments; the GTNS can systematically or randomly activate these rules to generate thousands of scenarios, and results of these scenario runs can be evaluated by experts. Since the project was completed in 2012, the ability to change trade connections was observed by several countries, and currency wars are a hot topic in the global competition for increasing exports and reducing imports.

Agent Based Modelling and global trade 429

- Rule 1. Increase trade with nearest N neighbours, and decrease trade with furthest N neighbours by X per cent.
- Rule 2. Increase trade with richest N agents, and decrease trade with poorest N agents by X per cent.
- Rule 3. Increase trade with poorest N agents, and decrease trade with richest N agents by X per cent.
- Rule 4. Increase trade with fastest growing N agents, and decrease trade with slowest growing N agents by X per cent.
- Rule 5. Increase trade with the same bloc N agents based on nearest neighbours, and decrease trade with other blocs N by X per cent.

Complex adaptive behaviour of agents can best be described by agent-level rule changes. Both in the real world and in the simulation, agents monitor their past performance and change rules following a number of higher level rules called meta rules. Below are three meta rules that are used to guide policy change by adapting a new agent rule.

- Meta Rule 1: Measure agent's GDP growth performance for last year, and compare it with the rest of the agents. If the performance is good then stick to the same rule as last year. If the performance is not good then flip a coin, and depending on the outcome, either choose a new rule randomly, which must be different from last year's rule, or adopt the best performing agent's rule.
- Meta Rule 2: Same as Meta Rule 1, but assess performance for the past 1, 2 or X years.
- Meta Rule 3: If the agent's performance is good, continue the same as last year, otherwise invert the current rule, that is if the agent is using the 'trade with richest N agents' rule, then this time use the 'trade with the poorest N agents' rule.

The Global Trade Network Simulator eliminates human bias by machine-generation of scenarios, and by facilitating the assessment of scenarios from a multi-agent perspective and the global economy as a whole, in numerical and graphical detail.

ABM functionality enables us to test agent-level strategies such as Finland (or any country or all countries) increasing trade with its nearest neighbours, the richest, the poorest, the fastest growing or same bloc nations. Agents can also be initialized with randomly assigned rules, and we can check their performance every year; if they are not doing well, then they can automatically switch to another rule. ABM also can be used in conjunction with given user-defined what-if scenarios. This joint set-up provides insights into the complexity of the global economy, and the difficulties in making correct decisions in a complex, unpredictable world.

The simulator can also loop by randomly selecting a subset of N countries creating GDP shocks by all going down by 10 per cent in the same year. When randomly selecting N countries every year between 2010 and 2030, every run produces different results and a further analysis enables us to search for extreme or unexpected outcomes. We later highlight and display the difference from the baseline scenario and display GDP, growth rates and sector breakdown for exports. We then need to analyse results to find what combinations of events can impact Finland's economy and its sectors.

430 *Handbook on complexity and public policy*

DISCUSSION

Complexity is all around us, and it is very easy to produce complex behaviour using simple rules. Probably the simplest way of generating chaotic complex behaviour is using a positive feedback loop. Networks with positive and negative feedback pathways can display complex behaviour. Social networks of agents have an added level of complexity as they can adapt and change their behaviour. In that sense the global economy and global trade network is not only complex, but a complex adaptive system that displays adaptive behaviour. ABM is a good modelling technique for complex and complex adaptive systems, as it imitates real world agent behaviour in computer memory. The ABM approach presented here provided a suitable modelling technique for global economy, from a non-predictive perspective, mainly to explore, understand and generate insights about future shocks and extreme events for the global economy.

In our project we faced several challenges. Probably the most important of these was in combining qualitative, narrative scenarios with a quantitative computer simulation approach. We believe they need to be used together to achieve the best results. In social, geo-political and socio-economic domains, narrative-based scenario analysis is often the standard course of action in the search for potential futures. Complementing this process with computer modelling and adding numbers into it provides more credibility and a broader perspective. The ABM approach used in GTNS adds network dimension to what-if scenarios, and enables researchers to look into a problem from a multi-agent perspective and the global system as a whole. In our opinion, ABM also strengthens the analysis process by randomly generating unthinkable scenarios for exploring 'known unknowns'. This could be the major advantage compared to classical scenario-based workshops where participant bias due to similar backgrounds, or confirmation bias where experts like to agree, often limits thinking outside the box.

An interesting post analysis showed that machine-generated scenarios are not that unrealistic. In the real world too, countries can and do change their trading partners in a matter of years, by increasing trade with their neighbours or targeting fastest-growing economies. A recent example is Turkey, which established stronger trade links with Iraq, Iran and UAE as a result of the recent crisis in Europe. According to CIA Factbook 2012, Iraq has now become the second major market to Turkey, with 7.1 per cent of exports going to this country. Figures indicate that in a matter of three to four years, a country can change its trade connectivity to a great degree to adapt to changes in the global economy and react to economic shocks. Turkey seems to have found new export partners in the Middle East to replace the declining demand in Europe as a result of the belt-tightening measures after the 2008 crisis.

Since 2009 some countries are making a big effort to devalue their currencies or at least prevent them from appreciating, so in a way currency wars are here for more exports and fewer imports. The UK and USA were initially successful in depreciating, and there has been pressure on China to appreciate the Yuan (Roubini and Mimh, 2011). But there are always risks that currency wars can turn into trade wars, which could have a bigger negative consequence. According to OECD the collapse in global trade seen in 2009 was a result of the credit crunch that occurred in 2007 and 2008, 'as the recession intensified global exports also fell in large numbers, and the credit crunch encouraged trade wars in

Agent Based Modelling and global trade   431

a tit-for-tat fashion. This continued and intensified in the aftermath of the 2009 collapse, and did not help the recovery efforts' (Roubini and Mimh, 2011).

Changes in the global trade may not be an early indicator of shocks in the global economy. It is possible that the global debt links are a more accurate indicator in the detection and spread of shocks in the global economy. As the recent crisis showed, the complex web of cross-holdings of European sovereign debt (Roubini and Mimh, 2011) helped shocks spread quickly over the debt network, and triggered a collapse in trade. So, a global debt network simulator of 'who owes who' connections could also be worth investigating in a similar way to the trade network simulation.

One of the issues with the Finland GTNS model is that it would not be able to capture the so-called butterfly effect – that is how a small country in Latin America can cause a global crisis – as it only models the large and seemingly relevant countries for our investigation. But this is only a data issue and in fact all countries can be added to a global trade network simulation for searching the potential butterfly effect. GTNS is by no means complete and it can be improved further and extended to a larger set of countries, as well as running with smaller time steps, such as quarterly, or trade and GDP figures for detecting and reacting to new shocks in the coming decades.

REFERENCES

Banks, S.C. (2002), 'Agent Based Modelling – A Revolution?', *Proceedings of the National Academy of Sciences of the United States of America*, **99**(3), 7199–200.

Barabási, A.L. (2002), *Linked: The New Science of Networks*, Cambridge, MA: Perseus Publishing.

Bonabeau, E. (2002), 'Agent-based Modeling: Methods and Techniques for Simulating Human Systems', *Proceedings of the National Academy of Sciences of the United States of America*, **99**(3), 7280–87.

Bremner, I. and P. Keat (2009), *The Fat Tail: The Power of Political Knowledge for Strategic Investing*, Oxford: Oxford University Press.

Buchanan, M. (2002), *Small World: Uncovering Nature's Hidden Networks*, London: Weidenfeld & Nicolson.

Casti, J.L. (1997), *Would-be Worlds: How Simulation is Changing the Frontiers of Science*, New York: John Wiley & Sons.

Casti, J.L. (1999), 'Firm Forecast', *New Scientist*, **2183**, 24 April, 42–6.

Casti, J. (2012), *X-Events: Complexity Overload and the Collapse of Everything*, New York: HarperCollins.

Casti, J. and L. Ilmola (2012), *7 Shocks and Finland*, IIASA Project Report, available at http://www.aka.fi/Tiedostot/Tapahtumat/IIASA2012/IIASA_2012_report.pdf.

Cecchetti, S.G. (2006), 'Measuring the macroeconomic risks posed by asset price booms', NBER Working Paper Series, No. 12542, National Bureau of Economic Research.

Epstein, J.M. and R.L. Axtell (1996), *Growing Artificial Societies: Social Science from the Bottom Up*, Washington, DC: Brookings Institution Press.

Ferguson, N. (2009), *The Ascent of Money: A Financial History of the World*, New York: Penguin Books.

Hughes, Barry B. (1999), *International Futures: Choices in the Face of Uncertainty*, 3rd edn, Boulder, CO: Westview Press.

Kay, J. (2003), *The Truth about Markets: Why Some Nations are Rich but Most Remain Poor*, London: Penguin Books.

Lempert, R. (2002), 'A New Decision Sciences for Complex Systems', *Proceedings of the National Academy of Sciences of the United States of America*, **99**(3), 7309–13.

Rickards, J. (2012), *Currency Wars: The Making of the Next Global Crisis*, New York: Portfolio/Penguin.

Roubini, N. and S. Mihm (2011), *Crisis Economics: A Crash Course in the Future of Finance*, New York: Penguin Press.

26. The international financial crisis: the failure of a complex system
Philip Haynes

INTRODUCTION

This chapter will evaluate the ongoing global international financial crisis since 2007 using a complex systems perspective. The global financial system has the attributes of a complex social system outlined in the introduction to this book. As a result, it requires a systemic application of policy interventions. The chapter explores the relevance of complex systems methodology and concepts to theorizing about the international political economy. The international and national economies are argued to be systems that display the behaviour previously identified by scientists and social scientists as symptomatic of a complex structure. At the core of this are dynamic and unpredictable interactions and feedback. A number of key concepts used previously by social scientists who apply complexity theory, and by the editors of this volume, are discussed with reference to their possible explanation of economic behaviour. In particular, this chapter applies the seminal work of the environmentalist and systems theorist, the late Donella Meadows, and seeks to develop further her conceptual articulation of how policy makers can intervene in economic systems and the current crises. Key concepts include reinforcing feedback, balancing feedback, and self-organization. These concepts are operationalized with specific reference to the economic policies applied during the financial crisis by major international governments. The policies analysed include: austerity, fiscal expansion, financial repression, and quantitative easing (QE). Meadows' (2009) argument for systems change posited that a change in the value paradigm is the most powerful method for achieving macro change. This is relevant to the application of policy change post the financial crisis, as otherwise continuation of previous policies with their preference for the continuing privatization of money and markets will lead to further manifestations of instability, inequality and eventual environmental chaos. The exploration of how to develop policy in such a dynamic system uses Meadows' (2009) seminal approach to complex systems interventions. The chapter builds on previous research and evidence about the values and beliefs that led to the financial crisis and how governments initially responded (Pettifor, 2006; Haynes, 2012).

The emerging debate about Full Reserve Banking (Benes and Kumhof, 2012) as a new financial system 'rule book' is one of a number of contemporary policy interventions evaluated. The role of public values as a method to create stability in any complex system transformation is argued to be a key element of any future policy success. Based on the evidence of the complex systems analysis, the chapter makes a normative argument for a future model based on a greater scale of intervention and global cooperation.

The international financial crisis: failure of a complex system   433

THE COMPLEX SYSTEMS APPROACH: A NEW METHOD FOR THEORIZING POLITICAL ECONOMY

This chapter synthesizes the global financial system as a complex system using the lens of complex systems theory as promoted in the introduction to this volume by Geyer and Cairney.

The total financial system is greater than the sum of its parts. This is illustrated by the impact of globalization on nation states, with transnational corporations and individuals increasingly able to move money across the world following financial deregulation (Dicken, 2011). The enhanced ability of the global financial market to create new credits in the electronic banking system and to move and invest these credits has been a defining feature of twenty-first-century macroeconomics. Bankers and policy makers cannot predict the future outcomes of such dynamic complexity using complicated computer-based probability models. Intervening in such a system is a qualitative decision judgement by policy makers about systemic patterns and their evolution (Haldane, 2012).

Many actions in the global economy are by their nature self-reinforcing and potentially exponential in their growth. This market reinforcement, if not checked by balancing policy actions, creates instability (Mintsky, 2008). Some actions in complex systems are dampened (balancing feedback) while others are amplified (reinforcing feedback). Conventional monetary policy – like the setting and adjustment of central bank interest rates – is a balancing feedback. The setting of interest rates is one method that central banks have used to stabilize the instability of the wider economy. If prices are rising too fast due to reinforcing demand and lack of supply, central bank interest rates are one policy lever to increase the cost of credit and thereby decrease demand. Similarly, unconventional monetary policy became a new form of balancing feedback after the global financial crisis, whereby some central banks purchased financial assets to increase credit liquidity and inflation, and thereby prevent a rapid reinforcement of falling prices and deflation. Monetary policy is a method for balancing market feedback to attempt to induce stability in the economy.

The tension between instability and stability is also inherent in the value contradictions at the heart of the political economy of the financial system. Bankers and speculators profit from the instability and growth of money through buying and selling monetary-based assets. This happens despite the fact that sovereign currency is owned by nation states and is a public good. The ability of banks to create credit without close regulation by a central bank leads to inflated salaries and bonuses as these private institutions are able to retain a proportion of the money produced. Profits are made on interest rates that are considerably higher than central bank interest rates.

Complex systems evolve between stability and instability. They demonstrate periods of relative stability. For example, in the period from 2003 to 2007 the majority of OECD countries experienced GDP growth, increases in the proportion of the population in paid employment and had low consumer price inflation (CPI) (Haynes, 2012). Nevertheless, after 2007 these patterns become unstable. It has been argued that it was the collapse of house prices in the US and the related collapse of the secondary mortgage investment market that tipped the global economy to 'the edge of chaos', and from stability into instability (FCIC, 2011). The selling of mortgage credit to American households had found its way into large institutional investments in financial products related to the

434 *Handbook on complexity and public policy*

repackaging of mortgage products. Another example of longer-term instability is the increasingly creative use of electronic money after the financial and capital deregulation in the US, UK and other countries in the 1980s (Krugman, 2008; Stiglitz, 2010). This was followed by market reinforced credit growth from financial institutions and the ability to inflate asset markets (Ryan-Collins et al., 2012).

Complex systems are particularly sensitive to initial conditions that later produce a long-term momentum or 'path dependence'. An example of this feature in global finance is the tendency for investment and innovation to lock in at an early historical point in the evolution of the market (Arthur, 1989). Once the baseline product or technology is accepted by a critical mass of customers, it is locked into services and production that feed the reinforcing demand. Examples are fixed term mortgage investments, credit insurances and Microsoft software. Geyer and Cairney remind us that the existence of path dependency does not mean that complex systems are structurally determined. Paradoxically 'complex systems exhibit "emergence"'. This behaviour results from interaction between elements at a local level rather than via central direction. An example of this in the global financial system would be the growing proportion of women entering education and employment. Consequently, they earn financial wages independent of any male partner and become a consumer and active agent in the market place. This change was not driven by government policy (although governments have responded to the phenomenon). The change is driven by changing social attitudes often related to the secularization of society and the political activity of women. An increasing female labour force has injected a high value supply of additional labour into the world economy (that is competitively priced when compared to male labour, due to structural inequalities). The evidence that this behaviour has evolved differently, and at different speeds in different countries, illustrates the fact this feature has 'emerged', rather than being hierarchical and structurally driven from the top down. Instead, it has emerged from the interaction of local elements like neighbourhood changes in family and community life, where values associated with these forms of existence have changed.

In addition to emergence at the local level, complex systems have part of their structural complexity defined by issues of level and scale. Scalability refers to the fact that patterns of structure and behaviour can be observed in similar forms at both the macro and micro level. For example, as stated above, the Financial Crisis Inquiry Commission (FCIC, 2011) showed the roots of the financial crisis were in the increase of sub-prime mortgage lending to low-income families in the US. The scalability of this irresponsible lending was that large international financial institutions developed combined packages of mortgage debt and resold them to institutional financial investors across the globe. At the core of this scalability were values and beliefs about profit making and a belief that the housing market and house prices would continue to rise. Taken together, these issues of emergence, levels and scales help us to understand the interactive nature of complex systems and that they are dynamic rather than static. This questions the economic concept of equilibrium: that financial markets will return to a natural balance of supply and demand. Complex interactions make this rather impossible, but a feature of these dynamic systems is that they still possess some elements of stability. The global financial economy, despite its instability, does exhibit stable and predictable times and places.

As Geyer and Cairney stated in the introduction, complex systems may 'demonstrate extended regularities of behaviour with short bursts of change'. Many aspects of the global financial system are remarkable for their continuity and stability. Examples are the discovery and use of energy forms, like oil with its key derivatives such as petrol. Cheap oil was a major element in the continuation and expansion of the twentieth-century industrialization and technological revolution. However, at certain periods of history, the supply of oil became confined to geographical regions where production was at its cheapest and supply in abundance. The law of limited natural resources dictates that such periods will not last beyond a few decades. In the 1970s, the world economy suffered a major crisis and significant period of change due to a short war in the Middle East that disrupted supply and exponentially increased oil prices. One of the results was that exploration started to occur in difficult environments such as the cold-water oceans. A rise in price meant that exploration in such areas was economically viable. In time, this allowed the market to stabilize again, as there was more diverse oil supply and investments in other energy forms. There has been a similar mix of periods of stability and instability in the world gas market in the last forty years. So one manifestation of the mix of stability and instability is periods of regularity, although then interrupted by bursts of change.

Attractors

If complex systems are not dominated by instability, and these are occasional features, what is it that creates stability? It is attractors that provide order in complexity and help bring back stability at times of instability (Byrne and Callaghan, 2013: 26–30; Haynes, 2012: 10–11; Geyer and Rihani, 2010: 38–9). The policy of setting central bank interest rates during a period of conventional monetary policy provides an attraction to order, as interest rates restrain price growth, but it is recognized that a time lag will exist before a rise in interest rates has an effect on prices. Setting interest rates does not have a stand-ard, uniform policy outcome, but will likely create a pattern of price changes. This is an example of an attractor.

No single attractor can ever provide complete and uniform order and stability. Attractors create patterns of similarity, but not identical patterns and linear predic-tion. Values and beliefs are also key attractor points in social systems. Value systems create boundaries around expected behaviour patterns in given social groups and social contexts, but not in uniform and highly predictable ways. Shared public values bring stability.

APPLYING MEADOWS' FRAMEWORK FOR INTERVENING IN COMPLEX SYSTEMS

Meadows (2009) proposed a number of key elements in a complex system that can be used to both diagnose the cause of problems and to design policy solutions. For the purposes of this chapter, these are summarized here as eight elements: resources (such as stocks and flows); information; rules; reinforcing feedback; balancing feedback; self-organization; goals; and paradigm shift. Each element of intervention is considered in turn with regard to intervening in the global financial system.

436 *Handbook on complexity and public policy*

Resource Management: Stocks and Flows

Prior to the 2007–08 crisis there were considerable imbalances in the monetary and capital stocks and flows in the global financial system. International data (Furceri et al., 2011) showed that by the start of the crisis in 2007, Foreign Direct Investments (FDI) had grown from less than 1 trillion US dollars in 2003 to a peak of 2.5 trillion in 2007. Foreign currency reserve stocks held by governments doubled between 2003 and 2007 from less than half a trillion US dollars to in access of 1 trillion. International capital flows in the form of debt portfolios peaked in 2006 at over 2 trillion dollars, but then started to decline with the crisis in the US and European housing and mortgage security markets. The unprecedented build-up of foreign reserves and increased flows of international investments created much instability in the global financial system, as these movements and stocks were subject to little regulation (Kay and Vickers, 1988) and used by companies, or nations, to protect their own interests. Both Krugman (2008) and Stiglitz (2010) have written of the disruptive effects in recent history on newly developing economies like Mexico, Argentina, Indonesia and Thailand, caused by the instability of global flows.

While newly developing nations' governments built foreign reserve stocks of US dollar-based investments as a defence against global market turbulence, in contrast many banks in the developed world had depleted stocks of money capital. They developed a financial model that involved borrowing on the secondary money markets to minimize their need for capital. They had decreased their capital to liquidity ratios. For example, this is what caused the collapse of the Northern Rock Bank in the UK, in 2007, so that it required nationalization by the UK Government in 2008 (Darling, 2011).

Meadows (2009) argued that policy makers use stocks and flows to intervene in complex systems. Her approach recognizes the periodic inconsistencies of complex systems and their ebb and flow in periods of relative stability and instability, with intervention often needed to prevent the further reinforcement of feedback that creates instability. Stocks of resources such as credit provide a cushion that can be utilized in times of instability. Witness the lack of capital financial stocks in the banking sector in the great financial crisis of 2007–08 and all the effort expended since trying to improve the stocks of capital in ratio to lending (European Banking Authority, 2013). The long time taken to build capital stocks in banks has, however, decreased their ability to lend credit into the economy in the short term and has been a key factor in delaying economic recovery. Rather differently, China has used its formidable US dollar reserves to prevent its currency from appreciating too rapidly when it believed this might create a decline in its export-based economy. This in turn has probably slowed the recovery and competitiveness of manufacturing production in other countries (Wolf, 2010: Rajan, 2010). One policy intervention available to governments and central banks is their use of financial stocks and reserves and the rate at which these flow into the financial system.

Information

The provision of information and its presentation are a powerful form of policy intervention. Governments seek to change corporate, household and state administrative behaviour by the presentation and use of information. The developed economies also

The international financial crisis: failure of a complex system 437

seek to share information definitions and analysis through their collaboration in international organizations like the World Bank, International Monetary Fund (IMF) and Organisation for Economic Cooperation and Development (OECD). The presentation of statistics about growth, employment and economic confidence feeds back into the system and affects the way the system responds. Developed governments seek to instil confidence in the use of such information by encouraging independent audit by international bodies. These policy methods deliver confidence and bring stability. The lack of confidence in Greece's use of official statistics after it became a member of the European Monetary Union was cited as a major reason for its rapid spiral of financial problems in 2010 and subsequent decline (Lynn, 2011).

Rules

Policy rules can operate as attractors in complex systems and are designed to create patterns of order in social and economic behaviour. All governments use rules as forms of policy interventions. These rules include policy statutes and legislation, or less formal mechanisms such as requirements placed on government departments and officials. Rules can limit instability and provide stability. In a normative sense, they may provide security and protection of the vulnerable from the extremes of market behaviour. The minimum wage used in many developed countries is a policy rule that seeks to ensure the poor have the ability to survive in the labour market. Competition rules and regulations try to prevent markets evolving towards monopolies that give a few suppliers control that disadvantages the public good. One feature of market deregulation after 1980 was that regulatory bodies, set up to restrict uncompetitive commercial mergers and acquisitions, had fewer formal powers to intervene in market processes. As a result the financial sector saw a number of large mergers and acquisitions before the crisis, and governments felt compelled to intervene to save these institutions because they were 'too big to fail'. Despite their damaging behaviour, the consequence of letting them go bankrupt was too catastrophic.

Governments can use capital controls and exchange controls as policy rules to try to limit the impact of global financial instability on their populations. In 2013, Cyprus implemented capital controls to try to stabilize its banking crisis and to prevent too much euro-denominated capital leaving the country. The liberalization of the global economy since the 1980s saw a reduction in financial rules and regulations and a corresponding increase in the growth and success of transnational corporations to move financial resource around the globe to increase their profitability. A declining use of such policy rules has resulted in an increased instability in global finance and a rise in global inequalities of wealth and income (Krugman, 2008; Stiglitz, 2010).

Reinforcing Feedback

The most obvious example of a reinforcing feedback is a financial bubble where prices of a particular asset, service, product or commodity rise exponentially. Such behaviour is often the source of problematic instability for social systems, as witnessed in the 2007–08 Financial Crisis. At that time house prices rose rapidly in the US and some European countries and then collapsed, having risen much more rapidly than average earnings in

438 *Handbook on complexity and public policy*

the preceding years. By using policy mechanisms that seek to reinforce behaviours that are believed to be socially and economically beneficial, governments may contribute to price bubbles. Home ownership has previously been seen as socially optimal, and for this reason, policies that encourage home ownership, such as deregulation of planning permissions to encourage developers to borrow and build, can further exacerbate a property price bubble.

Another major reinforcing bubble before the Financial Crisis was the liberalization of credit supply created by the deregulation of banking systems, so that across the world, credit loans grew rapidly from the early 1970s and at an exponential rate into the 'noughties' (Ryan-Collins et al., 2012). This growth in credit fed the debt crisis of 2007–08 with household debts at record levels in major economies like the US and UK. Unregulated credit expansion led to artificially high commodity and share equity prices as money was cheaply and readily available to put into such investments.

Balancing Feedback

Government policy has to consider interventions that will balance feedbacks and as Meadows (2009) observed, the key judgement is how to achieve an appropriate check against the strength of feedback so that an intervention is relative to the impacts the government wants to correct. While the government may want to dampen selling and asset price growth, they rarely want to close a market and stop trading. In addition, such market exuberance is linked to 'feel-good factors' and used by politicians to win electoral support. The problem is that policy analysts may see exuberance as irrational when taking a longer-term view. As a result, a policy judgement about how much of a policy lever to pull and check reinforcement, and how much influence to exert, becomes challenging for those in government and central banks. Often governments are dependent on periodic intervention events like annual reviews of tax rates. Interventions that are adjustable and can evolve rapidly might be preferred, but are more difficult to design and implement. Examples might be the monthly setting of central bank rates. This allows the policy maker to adjust intervention in relation to dynamic feedback such as falling or rising prices. The most recent example of an attempt at balancing feedback used in response to the crisis was QE, as a form of unconventional monetary policy. This has been used where interest rate setting no longer influences the demand for credit because economic confidence is so low and the stock of credit depleted. QE tries to balance this fall in demand by making liquid money available. The central bank purchases financial assets, thus placing new money into the bank accounts of those who sell the assets. This introduces velocity into the financial system and some of the resulting financial activity will stimulate demand. A very different argument about balancing feedback has come from the political lobby that supports so-called 'austerity' policies. The proponents of this policy are acting to stop the reinforcing behaviour of governments borrowing to finance public expenditure that limits unemployment and relative poverty. The supporters of austerity policies see the long-term consequences of this reinforcement as high government debt that will drive up taxes and reduce economic growth. These examples show how policy interpretations of reinforcement, its causes and checks, is in part dependent on normative political value judgements and interpretations, and not a scientific judgement based on economic and statistical analysis alone.

Self-organization

The ability for local and micro self-organization to modify, disrupt or improve the operation of a system is a key interaction in complex systems (Ricaurte-Quijano, 2013). Meadows (2009) observed that government policy could seek to understand this self-organization and in some circumstances seek to encourage and facilitate it, where self-organization was judged to have an optimal social and public benefit. One example is the ability of small Non-Government Organizations (NGOs) to innovate in public policy and to generate collaborative behaviours based on non-economic values of religion, humanity and neighbourliness. The growth of small-scale community-based credit agencies, such as Credit Unions and Cooperatives, is an example of a self-organizing response to the financial crisis. Credit Unions offer lower interest rates to the poor and vulnerable than private profit making lenders. Similarly, small-scale community energy groups can share their investments to generate micro renewable energy while sharing the benefits. Governments can support such self-organization by setting a favourable regulatory environment and/or giving tax incentives. Self-organization can also be manifest in socially problematic forms and is not always a universal public good for policy makers. For example, inquiries into the causes of the Financial Crisis (FCIC, 2011) concluded that some bankers and investors had self-organized to deliver mortgage-backed securities and investments that offered low quality, high risk, and fraudulent products to the wider market place.

Changing Goals

Meadows (2009) argued that all social and economic organizations benefit from shared goals and that the articulation of these goals through democracy and leadership is an important aspect of intervention. The articulation of goals seeks to coordinate behaviours and priorities, thereby changing the dynamic of the social and economic system over time. Goals are linked to an articulation of values, and a prioritization about what values are believed to be of importance in a particular historical period. They are a potential source of stability. Policy is disrupted by unclear or contradictory goals and this is symptomatic of a dissonance in values and beliefs. The financial crisis of 2007–08 represented a key event in the period of economic history that Kaletsky (2010) has described as 'Capitalism 3.0', an ideological period after the late 1970s that has been characterized by excessive market liberalism pervading all aspects of social and public values. The increasing dominance of market values over public life has failed to deliver the social outputs that many hoped were possible through technological progress. Income and wealth inequality increased and many countries failed to make a substantial impact on collective aspirations like sustainability and market security for essential items like food, housing and energy. Goals bring politics into the macro-political economy and link economic behaviour to social values. The articulation of such values is the attraction to order that unstable systems need if they are to achieve periods of stability. Consideration of banking reform has included much-needed discussion about public values and ethics.

440 *Handbook on complexity and public policy*

Paradigm Change

Similar to goal setting and its ability to redefine social and economic values and result-ant behaviour, Meadows (2009) observes that the ultimate in goal setting is paradigm change. The priority for Meadows et al. (1972) was the 'Limits to growth thesis' with its central belief that a continued global model of growth based on materialism and con-sumption would create major destructive instabilities, not least pollution and scarcity of resources in the eco-system. For Meadows, the paradigm needed to change urgently to reflect the external reality, but she was aware that such large-scale international change could only be achieved if the beliefs and values of the most powerful actors, and wider public, changed. The paradigm shift she required was that environmental protection and sustainability would become the driving ethos of the global world, above ideas of individualism, material consumption and profit maximization. Meadows argued that paradigm change was much more likely to have a large-scale impact on a complex social system than smaller-scale interventions like changing stocks and flows of money.

DEFINING CURRENT MACROECONOMIC POLICY IN A COMPLEX SYSTEMS ENVIRONMENT

This section of the chapter scopes the current macro policy environment and the analyti-cal approach offered by a complex systems evaluation.

Global Political Economy as a Complex System

The definition of globalization used in this chapter is that it is characterized by a domi-nance and evangelism of Western market liberalism and associated values. This has resulted in a convergence of economic ideas, but political conflict and resistance remain. There has been a growing power of transnational corporations as economic entities and the tendency of national governments to fear the power of these corporations and to feel subjected to their needs and requirements over those of their civil and public interests. The globalization of finance of the last 30 years was linked to the privatization of credit growth in international banks (Jackson and Dyson, 2012). These institutions increasingly had an international focus with only light touch regulation from national governments. They therefore had extraordinary financial power to move resources and investments globally, as exchange and capital controls were relaxed.

The global financial system is argued in this chapter to have many of the attributes of a complex system where the precise future of the system is largely unknowable due to many interactions and feedbacks between its dynamic components (Haldane, 2012). At best, analysts can diagnose repetitive patterns in certain places in the system that might give policy makers some ideas about how to forecast future behaviour and therefore how to manage the system, but precise prediction and long-term forecasting are extremely challenging. It is not surprising therefore, that during certain periods economists have become focused erroneously on certain components in the system, believing that some variable relationships can be used to manage much of the system. While these scientific theories are rooted in some empirical evidence and have some predictive and managerial

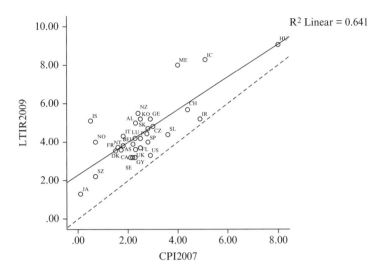

Source: www.oecd.org 2013.

Figure 26.1 The relationship between annual percentage consumer price inflation (2007) and future long-term interest rates (2009), 31 OECD countries

value, they fail over the longer term, sometimes with devastating consequences, because they become part of the system's dysfunctional feedback and amplify certain types of feedback at key moments in time while ignoring other interactions.

An example of this was 'conventional monetary policy': the method for controlling market supply and demand through the setting of central bank interest rates (NEF, 2013). Many economists and central bankers like Mark Carney (2013) began to argue that the management of consumer price inflation had largely been solved by the raising of central bank interest rates to reduce price inflation. Thus, Figure 26.1 shows a strong association amongst OECD countries between a nation's consumer price inflation and long-term interest rates.

Due to financial liberalization in many developed countries, post 1980, and the privatization of credit supply, central bank rates had a reducing impact on credit supply and credit control (Ryan-Collins et al., 2012; Pettifor, 2006). Asset inflation (that is, house and equity prices) was all too easily ignored while central banks focused on CPI. In Europe, there was the additional challenge of the merging of numerous international currencies that increased the power of the European Central Bank (ECB) and resulted in a lack of a national flexibility regarding interest rate setting (Krugman, 2013). Here the strength of the German economy meant that when the euro currency was launched, euro-based credit was too readily available in countries like Spain, Portugal, Italy and Ireland (Lynn, 2011).

Behind the orthodoxy of central bank interest rates setting, as an intervention for inflation and credit control, the expanding and merging private banking sector generated billions of dollars of new electronic credit, much of it converted to property debts, but also lent against equities and commodities (Ryan-Collins et al., 2012). Such speculation

442   *Handbook on complexity and public policy*

created asset 'bubbles' (asset inflation). For example, house prices rose exponentially in the period from 2003–07 in most developed countries, as did energy commodity prices. In the short term, consumers and households in the developed world were provided with a limited degree of protection because manufacturing exports from the newly emerging economies in Asia continued to fall in price and therefore neutralized price inflation. In sum, this imbalance in the global market system between bank credit allocation, rising asset prices and competitive supply side prices for consumption resulted in a very large credit bubble within many developed countries.

POLICY INTERVENTIONS

The Financial crisis of 2007–08 has become a longer and more sustained period of economic stagnancy. For Western countries, in particular in Europe, it appears to have been the longest period of economic contraction, as measured by lost output and reduction in GDP, since the Great Depression of the early 1930s. In this next section, the chapter focuses on the predominant national policies (Table 26.1) that have been used, giving an assessment of their likely systemic effect and wider international influence on the global system.

Austerity

Austerity policies were adopted in a number of countries, especially in Europe and the eurozone area. Governments pursuing austerity problematize the crisis as being more about government debt than private corporate and household debt. They seek to rebalance the feedback of increasing government debt. They acknowledge that additional government expenditure and investment has been necessary to save financial institutions and prevent a breakdown in the market economy, but believe the imperative for economic recovery is for it to come from the private sector and for public sector intervention to be limited to crisis intervention and emergency support only. Austerity governments seek urgent reductions in the annual government deficit (the percentage of GDP borrowed each year by the government to balance public expenditure) and seek reductions in total government debt as a percentage of GDP. They therefore reject Keynesian arguments for generating recovery via government investment into large infrastructure projects and believe that high levels of public debt limit growth (Reinhart and Rogoff, 2010). Infrastructure projects should come from private sector investment. Austerity governments seek to reduce welfare 'dependency' and to incentivize work, including subsistence level, part-time and casual work. This they see this as generating a dynamic and flexible labour market that will stimulate more rapid economic recovery.

When considering a systems analysis of austerity policy it is noticeable that it is values driven (with regard to goals and paradigm) with a strong belief in individual and market values, with the competitive market seen as providing the optimal moral basis for system stability. The growth of the state and public sector is viewed as providing negative values such as dependence on others and inefficiency in organization.

The extent to which austerity policies are successful is contested. Economic growth has been very sluggish in the countries pursuing austerity. Private sector job creation has

Table 26.1 Policy interventions: the financial crisis

Policy	Problem definition	Govt policy intervention	Current situation	Current outcome	International interaction	Complex system feature
Austerity (UK, much of EU)	Public debt more problematic than private debt High taxation and govt borrowing will crowd out private investment	Reduce public borrowing and expenditure Reduce employment in public sector and privatize services	Increased private household debt Increased relative poverty Low growth	Private job creation is part-time and low paid Austerity reduces demand	Combined austerity effect in Europe amplifies recession and unemployment	Balancing feedback is to prevent increased govt borrowing Attraction to market values
Fiscal expansion (US, Japan)	Lack of demand and investment, contributing low growth, employment and consumption	Government borrows at low rate in recession to invest in infrastructure projects for employment and growth	Public investment that has long-term benefits (i.e. roads, rail) This stimulates private activity	Limited success in US and Japan Unclear how to fine-tune with monetary policy	Possible migration of labour to expansion areas	Balancing feedback is increase in govt spending to prevent deflation Intervention with stocks and flows
Unconventional monetary policy 1 Monetary QE (US, UK, Japan)	Low interest rates are not promoting demand Encourage spending by making liquid money available	Central bank purchases assets to increase liquidity IR kept low by this intervention	Inflation risks Inefficient distribution of liquidity so limited growth	Low IR Mildly inflationary Decline in value of savings	Capital outflows to developing economies Fear of capital flights, if QE is 'unwound'	Emergence of new credit and liquidity Intervention focus is on system stocks and flows, new flow emerges

Table 26.1   (continued)

Policy	Problem definition	Govt policy intervention	Current situation	Current outcome	International interaction	Complex system feature
Unconventional monetary policy 2 Strategic QE (China, Taiwan, South Korea)	Target productive investment to stimulate long-term growth and social stability Imbalances in financial activity need addressing	Central bank targets credit at productive areas of economy	Increased employment Growth based on job growth Productive investment	Has tended to work in newly emerging countries focused on investment in exporting industries	Labour migration	Intervention by credit rules Focus is on system stocks and flows, with central bank directing flows
Financial repression UK	Need to reduce all forms of debt Inflationary policy needed so prices rises are greater than IR	Negative real interest rates	Total value of debts is reduced Investment favours inflation proof assets, i.e. property, commodities	Risks from credit expansion and inflation	Capital flight – outward investment	Balancing feedback to reduce debt Inflate flow of money to reduce stock of debt
Competitive devaluation China	Growth needs to be driven by exports so maintain relative low value of currency	Central bank purchases foreign reserves, maintains low interest rates Banking policy favours credit to export orientated businesses	Export-led growth	Internal industrial and labour market imbalances Global trade imbalances	Instability in exchange rates Reduction in capital flows and FDI Depression if too many countries pursue	Balancing feedback stops the market inflating domestic currency value

Fractional reserve banking	Fractional reserves promote credit expansion Financial crisis was caused, in part, by over relaxation of capital to lending ratios	Improve banks' capital stocks so they can cope with crisis and rebalance	Banks' lending into economy is reduced while they build up stocks of monetary reserves	Govts support bank lending in the short term, to keep economy active	Bank recapitalization in many countries reduces credit supply on a global scale	Balancing feedback is capital ratios (as break on credit growth) Paradigm sees market values as best way to expand and allocate credit
Full reserve banking	Fundamental economic problems are about credit supply, i.e. excessive growth in supply of credit and resulting instabilities	Implement full reserve banking so that credit growth and inflation are closely controlled by central bank	Concerns about increased power of central bank and how to make central bank accountable	Not implemented, but fears that innovative credit supply will cease	Reduction in FDI Scale effect of more stability in international finance	Balancing feedback is full reserve requirement Paradigm believes money is public good, not market commodity A rules approach to stock and flow

446   *Handbook on complexity and public policy*

had some success, but is not providing increases in household income. Many new jobs appear to be low paid, on casual contracts and part-time, thus failing to lead to growth because wage levels are too low to stimulate economic activity, and those in these jobs remain dependent on some government support, such as benefits to help pay market rents. There are also concerns that austerity policies are adding to the private household debts of the poor (because of wage rises being below inflation) and wealth and income inequalities are continuing to rise as a result. In terms of the system scale effect, at the international level, austerity policies in the eurozone combined to amplify a lack of demand, prolonging recession and rising unemployment.

Fiscal Expansion

The idea of fiscal expansionism comes from Keynesian economics and argues much the opposite of austerity. Government expenditure on investment, achieved by public borrowing, is used in a recession to increase demand and get an economy growing. The best-known proponent at the present time is Paul Krugman (2013). In terms of a complex system analysis, the policy is less driven by a fundamental belief in the collective benefit of public activity, but more in a theory of 'stocks and flows'. In a recession, stocks of money (liquidity) must come from the state and central bank, and investment must flow via the government who can target productive projects that will create employment and growth and provide long-term benefits to the economy. Obvious examples of such projects would be social housing, renewable energy production, roads, rail and airports, as they provide benefits to society over many future years.

Fiscal expansion has not been popular in Europe, largely because of money stocks and flows overseen by the transnational ECB in partnership with the IMF and this has seriously restricted the ability of national governments to use monetary expansion to generate fiscal investment. In historical terms, this is a significant scaling up of system rules. It is argued that there need to be more strategic investment decisions and more fiscal monies flowing from the rich areas of Europe to the poor areas (Krugman, 2013).

Japan has also attempted fiscal expansion in its financial crisis and recessions. As a result, public net debt has risen to over 200 per cent of GDP, the highest in the developed world. Fiscal investment has not succeeded in kick-starting the economy, but neither has high public debt bankrupted the country, and many of the host population still invest in government bond issues. It is argued that Japan's prolonged crisis, which predates the world financial crisis, illustrates the need for coordination of political, monetary and fiscal interventions in a complex economy, and fiscal expansion alone is insufficient (Turner, 2008).

The US has attempted fiscal expansion as one element of its policy repertoire since the crisis of 2007–08. President Obama announced hundreds of billions of dollars of investment soon after his election, but Krugman (2013) argued it was an insufficient percentage of total GDP to have much influence on economic growth and employment growth, especially as the Government was also implementing some contradictory austerity on local based state services. An important feature of US policy has also been monetary QE. There is a policy argument that if countries fiscally expand at the same time as easing monetary policy, this can have a multiplier effect on the global economy.

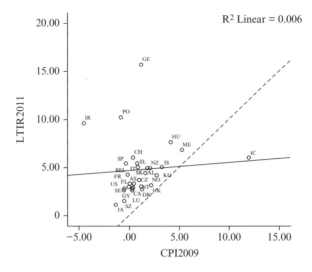

Source: www.oecd.org 2013.

Figure 26.2 The challenge to conventional monetary policy: a weaker relationship between annual percentage consumer price inflation (2009) and future long-term interest rates (2011), 31 OECD countries

Unconventional Monetary Policy 1: Quantitative Easing

Some of the world's biggest economies, the US, Japan and UK, have moved into a period in recent years of unconventional monetary policy referred to as QE. This is the antithesis of the conventional monetary approach where the setting of central bank interest rates is an effective monetary policy for controlling price inflation and dealing with cycles of supply and demand. One of the features of national and international banking crises is that central bank interest rates hit theoretical zero and fail to stimulate demand or relate to consumer price trends. At this point, the 'conventional' idea about monetary system stocks and flows needing to respond to consumer prices has stop working (see Figure 26.2). This situation was first documented in Japan before the 2007–08 international crisis. The central bank sought to ease the supply of liquid money by purchasing assets and thereby creating money by indirect rather than via direct creation. The core balancing aim of QE policy is to check the system feedback of price deflation (Joyce et al., 2011; 2012).

Since the crisis, US and UK central banks have invested hundreds of billions of dollars/pounds in QE. It is argued (NEF, 2013; Haynes, 2012; Joyce et al., 2011; 2012) that while QE prevented deflation and rising insolvencies and unemployment, by keeping interest rates low, it also raised inflation when wage levels were flat (creating poverty) and contributed to bubbles in asset prices, including via outward international flows. It is argued to be unfair on savers and pensioners.

448   *Handbook on complexity and public policy*

Unconventional Monetary Policy 2: Strategic Quantitative Easing

Given the partial failures of QE, with it creating asset inflation and increases in unstable global credit flows, other unconventional approaches to liquidity and credit creation have been proposed that would target credit growth more at productive investment, and the kind of investment that would stimulate employment and stable economic growth. Perhaps the best recent example of such a policy framework was the document, *Strategic Quantitative Easing: stimulating investment to rebalance the economy* published by the UK New Economics Foundation (NEF). Strategic QE is a government-directed monetary stimulus, probably via the central bank, where credit investment is targeted at lending to strategic productive areas in the national economy. Examples of similar policies to date would be the Brazilian development bank and Germany's system of regional investment banks (see NEF, 2013: 2). The strategic targeting of credit can also be linked historically with the rapidly developing economies of Asia in the last 40 years. Some likely challenges of strategic QE for countries using it are that it will reinforce inward skilled migration rather than increase local employment. If geared towards exporting, it is dependent on global market trends that are unpredictable, if there are no global policies on capital flows and exchange rate stability. Strategic investment that focuses on domestic markets is liable to cause difficult planning debates, that short term politicians prefer to avoid, especially in highly densely populated countries. This is evidenced by political debates in the UK about the possible development of transport and energy infrastructure.

Financial Repression

Financial Repression is an economic policy to reduce debt whereby a government and central bank maintain interest rates below inflation (so real interest rates are negative) in order to reduce the total level of debt. Politicians rarely advocate financial repression explicitly, as it could be seen as a position of national weakness that undermines a currency. At worst, this could create a vicious cycle of feedback and panic of currency selling on the international money markets. Nevertheless, after World War Two, it is argued that countries with high national debts, like the UK, effectively dealt with their high debt-to-GDP ratios by repression (Reinhart and Sbrancia, 2011). If too many countries try to repress their currency values and reduce debts at the same time, the global repercussions of repression are potentially damaging. For national governments, repression also runs the risk of capital outflows and an inability to encourage inward investment, and savings for pensions. Repression is likely to encourage asset-based investment and inflation, such as property or gold, as a hedge against inflation. Given that repression is a monetary-based intervention that seeks to realign monetary and debt stocks and flows in the system, it is prone to the unintended consequences of what other nations and transnational corporations do in response.

Competitive Devaluation

The deliberate devaluation of a currency has some similarities to financial repression, but also some important differences. While repression has the consequence of devaluing

The international financial crisis: failure of a complex system   449

and debasing a national currency to reduce debts and enable growth, devaluation is almost exclusively directed at maintaining or expanding an export economy. This is a balancing feedback intervention, designed to prevent the market reinforcement of an increase in currency value. The best known recent example is China, with its purchase of US dollar reserves and attempt to keep its export markets secure after the global crisis of 2007–08 (Rajan, 2010). As a result, China faced political pressure from the US to let its currency appreciate upwards, so that manufacturing growth could occur in the US. China's approach involves purchasing foreign currency to keep its own currency weak. These foreign reserves give the country market flexibility at times of global instability.

Competitive devaluations were a global feature of the 1930s Depression and an indication of a lack of global coordination from the nations involved. In sum, they reduced growth and expansion. Competitive devaluation, while attractive to individual nations, can therefore lead to reductions in investment and consumption and price deflation. As a national economic policy, the irony is that the outcome depends on what other nations do. Such policies illustrate the dynamic and strongly interactive element of complex systems.

Fractional Reserve Banking

There has been much international discussion since the financial crisis of 2007–08 about how the current dominant world system of banking, fractional reserve banking (reserves are a fraction of credit), can be made more stable. The focus for reform has been on capital-to-liquidity ratios and how to increase capital stocks. It is acknowledged that capital stocks fell too low in the 'noughties' as credit expanded and this was a major cause of the crisis and the insolvency of the banks (Kay, 2010). The rebuilding of capital stocks has provided a short-term problem in that improving capital ratios has reduced the amount of credit in the financial system and this limits economic activity. This current banking policy seeks to build system stocks and reserves as a future buffer against instability.

Previous rapid credit growth and reinforcement supported risky and unethical lending and so banking reform also focused on the ethics and rewards of banking. Rebuilding ethics and values as an attraction to new order could be a paradigm shift in terms of Meadows' account. Fractional reserve banking would then be underpinned by better values and would be more closely accountable and regulated.

There remain concerns that the system of fractional reserve banking is an erroneous method of managing stocks and flows because it places too much emphasis on private credit creation away from the central bank (with limited public values) and that it has failed as a paradigm. At worst, such critics say fractional reserve banking is a dysfunctional form of adaptive self-organization and monetary emergence from private organizations.

Full Reserve Banking

The theoretical alternative to fractional reserve banking is full reserve banking (reserves equal credits). This alternative is a radically different paradigm for the monetary system. It proposes different core values and ideas about the stocks and flows of money. Full

450   *Handbook on complexity and public policy*

reserve banking, as a paradigm, would give more regulatory and strategic control to the central bank, as all credit creation would be linked to central bank deposits or stable and predictable loans. The Chicago economist, Irving Fisher, proposed this paradigm after the financial crash of 1929. He advocated full reserve banking and believed it would allow much more stability of credit supply in the financial system, greatly reduce the possibility of bankruptcy by financial institutions, and reduce total net debts, including public debt. In 2012, economists (Benes and Kumhof, 2012) at the IMF revisited Fisher's plan and concluded it would have substantial positive effect on the US economy if implemented. Similarly, in the period after the financial crisis of 2007–08, some academics and campaign groups have argued for full reserve banking as part of the post-crisis banking reforms. For example, Positive Money in the UK have published a detailed plan of how full reserve banking could be implemented to reduce debts, arguing that the growth of credit and resulting record stocks of debt are a continuing and fundamental problem that the financial system faces until the banking paradigm is changed (Jackson and Dyson, 2012). The main criticism levelled against full reserve banking is that it will restrict credit supply due to punitive credit controls and thereby will reduce economic activity. It is clear that the detail of how full reserve banking is implemented is vital in this respect, with regard to the independence and accountability of the central bank and their regulatory control of the private banking sector. Nevertheless, given the scale of the financial crisis, the persistent instability of money credit creation and flows since global deregulation, full reserve banking is an interesting new paradigm to reduce debt and bring financial stability to the markets. By offering the potential for monetary stability and the better management of credit, it also offers the chance for governments to build public values in the market system by preferring the types of credit expansion that are in the public interest. Such a monetary paradigm also offers the indirect chance to deal with the crisis of rising wealth and income inequalities and need for investment in sustainable energy production and use.

CONCLUSION

The complex systems analysis of national policy interventions and their likely outcome in a global financial system illustrates the interconnectivity within the global system. Given the extent of globalization, national policy outcome has become increasingly dependent on what happens in other countries. Transnational corporations also illustrate this as major players in national and international economies, and they have increasingly shifted their focus from the national to international. National governments have been cautious to limit this, fearing a detrimental effect for their citizens rather than a positive one. While the financial crisis has enabled governments to rethink their ability to take some control over their domestic public and banking policies, the dominant international logic has not shifted its paradigm.

Table 26.2 shows two international policy paradigms. Although both are active and followed to a limited extent, financial liberalization remains the dominant paradigm. This is illustrated in the central belief that open markets are the most likely avenues for delivering growth and associated benefits and it is better for private banks to innovate with credit than public institutions. There is an emphasis on the further removal of trade

The international financial crisis: failure of a complex system 451

Table 26.2   Globalization policy: post the financial crisis

Policy	Assumption	Intervention	Likely outcome
Financial liberalization	Growth in open and free global market will bring rapid growth and benefits	Removal of trade barriers such as import duties and capital controls Paradigm values driven by markets	Instability in global economy, increasing regularity of boom, bust and crisis Growth of manufacturing in Africa and South America and high tech and services in Asia Further industrialization and urban poverty Growing inequality of wealth and income Increase in environmental damage
Global governance and cooperation	Countries must negotiate common good and agree priority values and actions	Global security by consensus Carbon and pollution controls Paradigm values driven by human rights and their relationship to welfare, i.e. health, education Sanctions against offender countries	Developing global forms of cooperation and global institutions Impossibility of achieving a form of democracy on a global scale, so continental and regional collaborations persist

barriers, such as import duties and capital and exchange controls. Given the outcomes of the last 25 years, it is not difficult to forecast the likely continuing overall pattern of such a paradigm: growth in manufacturing in Africa and South America, with the continuing and rapid urbanization of those continents. There would be some growth of high tech and services in Asia, and an up-scaling for innovation and services in the developed West. This would be accompanied by growing inequality of wealth and income, both on a global scale and national level, as some are much more able to adjust and adapt to an increasingly competitive world than others. With a dominant focus on growth that gives largely material and/or monetary rewards, we must expect more environmental damages. As Meadows herself predicted, such chaotic complex system growth would surely implode at some point as basic resources can no longer sustain the system. Fundamental collapse, whether in 50, 100 years or 150 years, becomes inevitable.

An alternative paradigm that stabilized the system might still be achievable. Many of the core ideas exist already, but they do not dominate and restrain behaviour enough. An alternative system paradigm is one based on more stable banking, and global governance and cooperation. The financial market becomes subservient to global security by consensus. Carbon and pollution controls are established and enforced, and basic human rights

452 *Handbook on complexity and public policy*

underpin all activities and sanctions against offender countries. All this depends on a strengthening of global forms of governance facilitated by international institutions. The nature of complex systems and the emergence of order dictates that countries will find it easier to first cooperate on a continental scale (evidence already suggests this), and that an appropriate lower national/regional level of democracy exists alongside higher global decision-making bodies. Such levels are an important aspect of a stable complex system.

Better global governance requires a global monetary system of some description. The ending of the Gold Standard has been characterized by an open system where the only stabilizing feature is the strength and belief in the value of the US dollar. The evidence for this was that the dollar remained popular after the 2007–08 crash. While this was something of a paradox when much of the crisis had originated in the US, if the dollar had failed, the crisis would most likely have been worse. The dollar will not remain the global currency of default forever. It would be better that an alternative policy for an international monetary reserve is planned rather than having to be implemented in response to a major economic crisis. Future stability depends on more monetary stability between the world's largest currencies like the dollar, euro, yuan and yen. Some international method of a collaborative and shared reserve 'stock' type protection is one such requirement (Wolf, 2010).

In conclusion, national financial policies can still have significant and important contributions to make for their welfare of their peoples, but any improvement in the preciseness and predictability of the application of types of public and economic policy would appear to depend on effective international cooperation. At the core of this is credit creation, monetary stock reserves and the international flows of credits and monies. At the heart of any agreement needs to be a values-based paradigm. Although complex systems approaches in the social sciences have been characterized by their attempt to use a scientific approach that borrows from physical systems like the weather and biological process that behave in indeterminate and highly interactive and dynamic ways, it is vital that social scientists do not ignore the normative imperative in social science. The use of complexity theory in political economy shows that values and norms can be included as part of the system dynamic, and borrowing from Meadows' historical work, we can see that ensuring political debate is active in financial policy, can be at the core of a complexity theory-based understanding.

REFERENCES

Arthur, B. (1989), 'Competing Technologies, Increasing Returns, and Lock-in by Small Historical Events', *Economic Journal*, **99**, 116–31.
Benes, J. and M. Kumhof (2012), *The Chicago Plan Revisited*, IMF Working Paper, New York: IMF, available at http://www.imf.org/external/pubs/ft/wp/2012/wp12202.pdf.
Byrne, D. and G. Callaghan (2013), *Complexity Theory and the Social Sciences: The State of the Art*, London: Routledge.
Carney, M. (2013), 'Dr Mark Carney: Response to the Treasury Committee's questionnaire', House of Commons Treasury Select Committee, 18 April 2013, London: UK Houses of Parliament, accessed 6 August, 2013 at http://www.publications.parliament.uk/pa/cm201213/cmselect/cmtreasy/944/944we03.htm.
Darling, A. (2011), *Back from the Brink: 1,000 days at Number 11*, London: Atlantic Books.
Dicken, P. (2011), *Global Shift: Mapping the Changing Contours of the World Economy*, 6th edn, London: Guilford Publications.

The international financial crisis: failure of a complex system   453

European Banking Authority (2013), *Risk Assessment of the European Banking System*, Luxembourg: European Banking Authority, accessed 21 August 2013 at http://www.eba.europa.eu/-/the-eba-risk-assessment-report-july-2013.

Financial Crisis Inquiry Commission (FCIC) (2011), *The Financial Crisis Inquiry Report. Final Report of the National Commission on the Causes of the Financial Crisis and Economic Crisis in the United States*, New York: Public Affairs.

Furceri, D., S. Guichard and E. Rusticelli (2011), *Medium-Term Determinants of International Investment Positions: the Role of Structural Policies, Working Paper No. 863*, Paris: OECD, accessed 20 July 2012 at http://www.oecd-ilibrary.org/economics/medium-term-determinants-of-international-investment-positions_5kgc9kzsm19x-en.

Geyer, R. and S. Rihani (2010), *Complexity and Public Policy: A New Approach to 21st Century Politics, Policy and Society*, London: Routledge.

Haldane, A.G. (2012), 'The Dog and the Frisbee', speech to the Federal Reserve Bank of Kansas City's 366th economic policy symposium, 'The changing policy landscape', Jackson Hole, Wyoming, 31 August, accessed 7 March 2013 at http://www.bis.org/review/r120905a.pdf.

Haynes, P. (2012), *Public Policy Beyond the Financial Crisis: An International Comparative Study*, London: Routledge.

Jackson, A. and B. Dyson (2012), *Modernising Money*, London: Positive Money.

Joyce, M., M. Tong and R. Woods (2011), 'The United Kingdom's Quantitative Easing Policy: Design, Operation and Impact', *The Bank of England Quarterly Bulletin*, Q3, **51**(3), 200–12.

Joyce, M., D. Miles, A. Scott and D. Vayanos (2012), 'Quantitative Easing and Unconventional Monetary Policy – An Introduction', *The Economic Journal*, **122**(564), 271–88.

Kaletsky, A. (2010), *Capitalism 4.0*, London: Bloomsbury.

Kay, J. (2010), 'Should we have "Narrow Banking"?', in A. Turner et al. (eds), *The Future of Finance and the Theory that Underpins it*, London: London School of Economics, accessed 11 March 2014 at http://www.futureoffinance.org.uk/.

Kay, J. and J. Vickers (1988), 'Regulatory Reform in Britain', *Economic Policy*, **3**(7), 286–351.

Krugman, P. (2008), *The Return of Depression Economics and the Crisis of 2008*, London: Penguin.

Krugman, P. (2013), *End this Depression Now*, New York: W.W. Norton and Co.

Lynn, M. (2011), *Bust: Greece, the Euro, and the Sovereign Debt Crisis*, Hoboken, NJ: Bloomberg Press.

Meadows, D. (2009), *Thinking in Systems: A Primer*, London: Earthscan.

Meadows, D., G. Meadows, J. Randers and W.W. Behrens III (1972), *The Limits to Growth*, New York: Universe Books.

Mintsky, H.P. (2008), *Stabilising and Unstable Economy*, Maidenhead: McGraw-Hill.

New Economics Foundation (2013), *Strategic Quantitative Easing: Stimulating Investment to Rebalance the Economy*, London: NEF, available at http://www.neweconomics.org/publications/entry/strategic-quantitative-easing.

Pettifor, A. (2006), *The Coming First World Debt Crisis*, London: Palgrave Macmillan.

Rajan, R. (2010), *Fault Lines: How Hidden Fractures Still Threaten the World Economy*, Princetown, NJ: Princetown Press.

Reinhart, M. and K. Rogoff (2010), 'Growth in a Time of Debt', *American Economic Review*, **100**(2), 573–8.

Reinhart, M. and M.B. Sbrancia (2011), 'The Liquidation of Government Debt', *NBER Working Paper* No. 16893.

Ricaurte-Quijano, C. (2013), *Self-organisation in Tourism Planning: Complex Dynamics of Planning, Policy-making, and Tourism Governance in Santa Elena, Ecuador*, PhD thesis, Brighton: University of Brighton, available at http://ethos.bl.uk.

Ryan-Collins, J., T. Greenham, T.R. Werner and A. Jackson (2012), *Where Does Money Come From? A Guide to the UK Monetary and Banking System*, London: New Economics Foundation.

Stiglitz, J. (2010), *Freefall: Free Markets and the Sinking of the Global Economy*, London: Penguin.

Turner, G. (2008), *The Credit Crunch*, London: Pluto Press.

Wolf, M. (2010), *Fixing Global Finance*, London: Yale University Press.

CONCLUSION

27. Conclusion: where does complexity and policy go from here?

Paul Cairney and Robert Geyer

After reviewing all of the chapters, trying to put them in a coherent order and then getting them edited and organized (with the help of our excellent Research Assistant, Nicola Mathie), what have we learned about the 'state of complexity and policy thinking', the nature of the 'complexity dilemma' that confronts academics and policy actors working in this area and some potential future directions?

THE STATE OF COMPLEXITY AND POLICY THINKING

As is well known, complexity-inspired approaches and interpretations in social science have been growing since the early 1990s. More than twenty years later, applying complexity to policymaking is now well established. Complexity thinking has been used in a number of governmental research and policy applications in countries such as the UK, Australia and USA and by international organizations like the OECD and EU. There are a number of academic departments, research institutes, academic journals, organizations and funding bodies that are working on complexity and policy. Typing 'complexity and policy' into an Amazon or Google search box generates a bewildering array of scholarly publications and areas to explore. The Edward Elgar book series on complexity (of which this book is a part) is just another general indication of how 'mainstream' complexity has become. Clearly, our edited book is just a small part of this larger wave.

Nevertheless, did we notice anything in particular about the contributors, chapters and book that may give us some pointers about the current state of complexity and policy? Looking down the list of contributors, three points immediately come to mind: they are predominately academic, UK/US/English language oriented, and male. The first observation is not surprising. When we sent out our announcements looking for contributors we sent them to academic and policymaker networks. Nevertheless, it is obviously very difficult for busy policy actors to take the time to write an unpaid chapter contribution. A key question for future efforts for complexity and policy thinkers is: can we find a format for encouraging policymakers to contribute to, and take part in, the development of complexity thinking, and make it worth their time? If we really want to engage with policymakers, and believe that complexity improves policymaking, do we need some form of more concise interactive formats to broaden the reach/appeal of complexity? Can academic books be interwoven with blog posts, websites, social media and other communication tools? This is clearly something that we need to think about and act upon in the future if we want to take complexity further out of the realm of academia.

Second, most of the chapters, with some key exceptions, were heavily oriented to the English-speaking world. This is an obvious bias that comes from who we are

457

458   *Handbook on complexity and public policy*

(English-speaking academics working in the UK), the language that we work in, and our key reference points. Despite our desire to make the book as international as possible we were very limited in what we could achieve. Nevertheless, the few cases we were able to include, such as policing in Brazil, emergency preparedness in Sweden, firm networking in Italy, and migration policy in China, point to the possibility of further comparative and international areas that could be explored. Larger projects that are able to bring together and assess the development of complexity thinking and activities in the non-English-speaking world would be particularly fascinating and something we hope to see in the future. Moreover, despite some recent growth in complexity-oriented international academic organizations, work on complexity and development is still rare. How complexity relates to the developing world is a huge area of interest and potential. As Graham Room explored in his chapter on power, complexity does have a great deal to say about the global inequalities and the power imbalances that maintain them. Exploring this issue would clearly make complexity more relevant to academics and policy actors in those parts of the world.

Third, why is the book so male dominated? Again, we admit our obvious biases: we are both male, and the networks which we contacted were predominately male. In general, and as in many areas of social science such as political economy, complexity seems currently to be a largely male dominated academic area. We feel that this is partially a reflection of its links to the natural sciences, where men tend to be numerically dominant, and to the general dominance of men in the upper levels of academia. Nevertheless, it is an obvious weakness in a book applying complexity to policy because so many of the actors in policy sectors are women. Finding a strategy and language to reach out to these policy actors, to 'polish the gem of complexity' as Catherine Hobbs argues in her chapter on local policy, is a major challenge for complexity thinkers and researchers.

One small part of the solution may lie in our attempts to identify links to the current literature. We have identified the ties between the complexity theory literature, as applied to policy, and well-established literature in the social sciences. By identifying this broader knowledge, we can reflect on the extent to which we routinely draw on, and cite, the work of women. For example, complexity has potentially strong ties to new institutionalism, and rational choice variants such as the Institutional Analysis and Development framework pioneered by Elinor Ostrom. We may also reflect on the links between complexity theory, institutionalism and feminist research, since they have a common focus on the routine rules and biases that reflect and reinforce inequalities. In the face of a complex world and a limited ability to understand it, people often behave in quite predictable ways and produce regular patterns of behaviour when they interact. They produce simple rules to deal with complexity. They deal with an almost infinite amount of information, and ways in which to understand it, by relying on cognitive short cuts, to decide what information to process and how. These short cuts can include individual habits and social norms, many of which are difficult to change when they become established. People develop beliefs and groups of people often develop shared ways of thinking, which can involve the power to establish dominant ways of thinking and acting, to benefit some and often marginalize others. Governments operate in a comparable way, breaking their policy responsibilities into a series of departments and units, each of which has some potential to develop its own 'standard operating procedures', to decide where to seek information and how to process it to make decisions. In that context, the study of policymaking is about identifying those rules, how stable they are, and how they might change. If we can draw further on feminist and related research, our studies may focus increasingly on how institutions and patterns of behaviour should be challenged.

A PRAGMATIC APPROACH TO THE COMPLEXITY DILEMMA

As academics who have been working with complexity and policy for over a decade, we are often struck by what can be called the 'complexity dilemma' or 'complexity tension'. Fundamentally, this means that we live in societies that are infused with the traditional rationalist and orderly 'scientific' framework. In this tradition, one is continually tempted to try and break a complex world into a simple model, and use a small set of simple rules, based on linear cause and effect, to help control the policymaking system and solve policy problems. From this perspective, with enough human effort, knowledge or evidence, the problem can be understood, responsibilities allocated to key actors, policy outputs applied and outcomes measured. The obvious difficulty for those working in the field of complexity is, as Derrida is claimed to have said, 'if things were that simple, word would have gotten round'!

Instead, we have to be pragmatic when faced with perhaps insurmountable problems regarding our knowledge and control of the world. Pragmatism involves a recognition and acceptance of the limits of our knowledge and understanding, and ability to gather evidence, developing models when we know that they only tell us part of the story, and adapting to policymaking situations that are often beyond our control. This does not mean that there is no such thing as progress. Human knowledge is progressing all of the time and at an incredible rate. The problem is that human interaction and social development, as well as its interaction with the natural world, continues to change. Hence, there is always more to learn. Complexity never dies unless the system becomes a closed linear one and we would certainly not want to live in such a system.

The difficulty is that such an uncertain and 'wishy-washy' approach is difficult to defend in the context of a policy process that demands clear evidence and solutions from academics and holds elected officials and policymakers to account for outcomes over which they have limited control. Again, this does not mean that complexity thinking implies that there should be no policy accountability or responsibility or that all policy audits/targets are a burden. What it demands is a balanced and pragmatic approach to the strengths and weaknesses of all types of policy. Again, this is difficult to 'sell' to pressured policy actors working in an open political and media context that demands policy action X should lead to policy outcome Y. Moreover, although this dilemma can be found in almost all areas covered by the book, the meaning and nature of the pragmatic response to this dilemma varies markedly by topic and according to the audience with which we engage. So, in the conclusion, we use these themes of complexity dilemmas and pragmatism to sum up a range of debates on complexity research, as they apply to theory and practice.

460 *Handbook on complexity and public policy*

THE COMPLEXITY DILEMMA AND SCHOLARSHIP: THE IMPORTANCE OF CHALLENGING 'REDUCTIONISM'

It is in this context of dilemma that we should understand the often bold challenge to 'reductionism' in science. The pursuit of complexity science has parallels in the historical pursuit of science and the modern scientific method. The latter's merits were stated so boldly in the past because they represented a rejection of paradigmatic religious belief as the basis for understanding the world. Instead, 'positivist' science was based on gathering evidence through induction, with your own eyes, instead of relying on faith. Further, since the social stakes were so high, it was difficult for scientists to express uncertainty about their methods and results. Now, in a period of scientific ascendancy, the challenges of uncertainty and complexity can be articulated and addressed in a more open way.

This complexity challenge is often made in a polemical way, by setting up complexity science as a fundamental rejection of the assumptions and methods of the past, and with reference to a stylized form of naïve 'positivism'. This tradition is continued in the chapter by Givel, who provides a critique of current policy science, particularly as it is practised in the US. Yet our broader aim is to reintroduce a sense of uncertainty about what science can achieve and how certain we should be, by rejecting the idea that we can understand complex systems by breaking them down into their component parts and determining the relationships between them. Rather, complex systems are greater than the sum of their parts, and elements interact in ways that are difficult to measure.

When applied to policymaking, in theory and practice, this broad complexity debate has two distinct elements. First, the key tenets of complexity may not represent such a radical rejection of social or political science because, while some 'positivist' methods and approaches may be more prominent in academic journals, there is not a sense of 'one best way' to do research. Ever since the rise and success (particularly during the industrial revolution) of linear rational science based on reductionism, causality, predictability and determinism, a whole range of scientific, philosophical and policy-related actors have been challenging its dominance. So, the issues and concerns we raise often chime with earlier debates – from the concerns of 'instrumentalist' policy thinkers in the 1950s and 1960s, the philosophical Pragmatists of the early twentieth century, and all the way back to Kant who argued against a 'mechanical' vision of nature and humanity. Of course, each policy area tends to have its own unique historical developments and language – see, for example, the chapter on the post-WWII development of complexity and planning by de Roo. Nevertheless, the general trends are remarkably similar.

Complexity theorists can therefore find elements of their arguments already rehearsed in debates on ontology and epistemology that have always been a feature of social science. For example, Morçöl draws on discussions of positivism and critical realism, while Little draws on discussions of reality, while both examine debates (which both relate to Bent Flyvbjerg) on the relative merits of in-depth case study versus broader quantitative analysis. Further, as we discuss in the introduction, we can trace many complexity themes to well-established studies of punctuated equilibrium, path dependence, governance and implementation studies. Complexity theory may represent a new package of ideas, described in a new way, with far greater parallels in the natural sciences than most theories, but many elements are already well understood in policy studies.

Second, however, as scholars we operate in a system that often seems to favour

Conclusion 461

particular academic activities. We are increasingly asked to reinforce a simple understanding of the world by demonstrating the 'impact' of our work on it, in a linear manner which suggests very simple mechanisms of cause (we provide new knowledge) and effect (it has an important impact on the way that other people think and act). Without this clear 'evidence base', and the identification of linear causal relationships, impact is difficult to demonstrate, and scholarship may be deemed invalid and become rejected or go unfunded. To compete in this context, complexity thinkers may be tempted to make large claims for their work. The difficulty is that complexity thinking is fundamentally about uncertainty and emergent processes, and complexity theorists face a choice of betraying their fundamental positions or continually losing out in the battle for influence. Though it sounds trite, complexity does force one to recognize that scholarship is both a political and academic enterprise.

How might we respond to such a dilemma? As a whole, the chapters in this book provide a mix of suggestions, from engaging in scholarly debates, providing the tools to have a direct impact on policymakers and organizations, and providing new analyses of case studies to better understand the real world. For example, the theoretical chapters provide various ways in which to respond to a sense of novelty and continuity, with some authors challenging (Givel) and others reinforcing the value of existing theories in their disciplines (Wellstead et al., Tenbensel), and some challenging their disciplines to combine established and complexity ideas (Webb, Little, Morçöl, Room). Others (particularly Mitleton-Kelly and Price and Haynes) argue that complexity does provide an overarching framework and practical tools that can make a clear impact on policy. They demonstrate this through the use of their policy toolkits and the way in which they have applied them to various policy situations. Similarly, the various chapters on modelling (in particular: Johnson, Edmonds and Gershenson, and Hadzikadic, Whitmeyer and Carmichael) explored the potential of complexity-inspired modelling tools to add to our understanding of particular policy areas and situations.

A collection of chapters demonstrate the descriptive and prescriptive aspects of complexity thinking, showing how policymakers and organizations deal with complexity and how they can reasonably be held to account. For example, Little argues that people and organizations deal with complexity by developing distinctive, simple understandings of the world, producing their own sets of rules about how to process information and act. These understandings and rules are based on a snapshot of time, and quickly become dated, but they also endure in common knowledge and institutions which are difficult to challenge and change. Policymaking therefore involves the interaction between a wide range of people, institutions, and new events and knowledge, in often unpredictable ways, partly because people respond to information using dated rules. Little's contribution connects well with Haynes' study of the economic crisis, which largely represents the outcome of multiple institutions furthering a very partial and often dated understanding of the global economic system, and struggling to adapt to key changes. It also informs Gray's critique of healthcare responses to pandemics. If we combine their insights, we can begin to understand why people and organizations fail to adapt quickly to quickly-changing circumstances: they process information through well-established lenses, which often distorts their understanding of rapid and unpredictable change. While these insights perhaps entail a sense of hindsight bias – we can only really understand these events after they have occurred and, by then, the world has

462   *Handbook on complexity and public policy*

moved on – they also prompt us to recognize the important limitations of established institutions and ways of thinking.

THE COMPLEXITY DILEMMA IN THE REAL WORLD: WHAT IS A PRAGMATIC PRACTITIONER?

If such ways of thinking are limited, and potentially damaging, what might we replace them with? How should people think and act pragmatically when they engage in complex policymaking systems? In the area of political practice, 'pragmatic' can mean very different things for each audience. For example, Wellstead et al.'s critique of complexity theorists in the natural sciences is that they identify complexity in the natural world but not the political process, which is often treated as a 'black box' in which to inject knowledge and expect a direct and proportionate response. In that sense, pragmatism refers to the need for scientists to understand the limited impact that they may make simply by raising problems and expecting knowledge to have a direct impact on policymaking.

Rather, they should adapt to the policymaking world in which they seek influence, by seeking to identify where the 'action' is in a political system, engage with the right people and organizations, form coalitions with like-minded actors, find the right time to identify problems, find the right way to 'frame' problems for powerful audiences – and, perhaps most importantly, recognize that, in democratic political systems, many other people have an as-legitimate role as providers of knowledge and opinions. In other words, they could learn a lot from interest groups which often maintain multi-level lobbying strategies, either directly or as part of networks.

In that context, many of our modelling chapters come into play, as a collection of strategies that can be used to engage with policymakers. Most notably, many chapters identify the role of agent based modelling, which can generate scenarios and outcomes that can be used to inform policymaking. The models do not tell us what will happen, or what we should do. Rather, they help provide some clarity by identifying the outcomes of a large number of interactions between people following simple rules, and encouraging policymakers to think about those outcomes and set priorities.

The dilemma for elected politicians and civil servants is more complicated, since there is often a large gap between how they can act and how they must account for their actions. So, on the one hand, complexity thinking can be used to reject the idea that power is concentrated in the hands of a small number of people in central government, or that a government can control policy outcomes. Instead, they can adapt to their policy environments and seek to influence various parts of it, while accepting that the interaction between large numbers of people and institutions makes control of a whole policymaking system impossible.

On the other hand, elected policymakers have to justify their activities with regard to well-established accountability mechanisms – such as in the UK where government ministers are accountable to the public via ministers and Parliament. The media and public expect ministers to deliver on their promises, and few ministers are brave enough to admit their limitations. Civil servants may also simultaneously receive policymaking training which encourages them to think about complexity and the limits to their influence, and management training to encourage them to use simple rules and techniques to

Conclusion 463

exert control over their policymaking tasks – again, because their knowledge of what is possible rubs up against what is expected of them.

Being pragmatic in this context is not easy. It is a topic which can be informed by complexity thinking, but only if we move some distance from our original polemical stance. In other words, it may be sensible to produce a range of measures based on a more realistic policymaking philosophy, and potential strategies including: relying less on centrally driven targets, and punitive performance management, in favour of giving local bodies more freedom to adapt to their environment; trial-and-error projects, that can provide lessons and be adopted or rejected quickly; and, to teach policymakers about complexity so that they are less surprised when things go wrong. Yet, as Tenbensel makes clear, these strategies should not be selected simply because we reject a caricature of top-down policymaking. Rather, we should consider how complexity thinking can be compared to what actually happens in government, which forms relationships with organizations using a mix of government, market and network solutions. We should also reflect on the limited role of outcomes-based measures of policy success, since they involve outcomes that span decades and are therefore difficult to reconcile with elections that appear every four to five years.

Practitioners at the 'street' level do not operate in the same kinds of conditions. Indeed, Michael Lipsky's classic study of street-level bureaucrats already highlights a degree of pragmatism in some professions, which recognize their inability to fulfil all government objectives and, instead, fulfil an adequate number while maintaining a degree of professional morale (see our introductory chapter). To do so, they draw on a simple set of rules generated through professional practice and experience. In this case, complexity thinking has relatively little to offer. In contrast, it may be profoundly useful to examine cases in which things go profoundly wrong, when their simple rules contribute to catastrophic events. An excellent policy example of what a complexity oriented approach can do is found in the 2011 *UK Munro Review of Child Protection* (https://www.gov.uk/government/publications/munro-review-of-child-protection-final-report-a-child-centred-system). Following a string of high-profile child abuse cases, Professor Eileen Munro was asked to carry out a wide-ranging and in-depth review of UK child protection policy. Inspired by systems and complexity thinking, Munro produced an impressive document that highlighted the failings of the former well-intentioned but misguided approach that resulted in a tick-box culture and a loss of focus on the needs of the child. These weaknesses were further amplified by a media and public culture which demanded that 'lessons must be learned' and some individual or process must take responsibility/blame.

The core problem, which the Review made clear, was that in highly complex situations there are no simple solutions, lines of responsibility or easy targets to blame. What made this situation even worse was a knee-jerk governmental response that demanded ever-growing targeting and audit regimes to show that 'lessons' had really been learned. The difficulty, as the Munro Review aptly demonstrated, was that this did little for the actual protection of children, while greatly complicating the policy process of child protection. Hence, one of the key conclusions of her report was that there needed to be a radical reduction in central prescription in order to help social workers move from a compliance to a learning culture, and that we had to recognize that the larger societal pressures to find 'someone to blame' (amplified by the mass media) misshaped the policy response to child protection.

464 *Handbook on complexity and public policy*

Finally, the existence of complexity presents a dilemma for the public: it should accept that it cannot simply blame a small number of elected ministers for the ills of government. On the other hand, it should not absolve government entirely; complexity should not be an excuse used by policymakers to take no blame for their actions. In this case, Room's chapter develops insights regarding our ability to hold groups of people to account for the ways in which they act, and interact with each other, to produce outcomes that disadvantage other groups. He explores the argument that 'emergence' does not necessarily represent an unpredictable outcome that no one can control or be held to account for. Rather, people benefit from the outcomes that emerge from the simple rules maintained by some powerful groups. Accountability is about, for example, focusing on the ability of policymakers to challenge the simple rules that benefit some at the expense of others.

We have more work to do to produce clear and practical advice to policymakers that they can use and defend. Getting people to agree that policymaking systems are complex is easy. Working with them to produce pragmatic strategies, to adapt to complexity, is hard. Getting them to prioritize these strategies, in the face of media, public and parliamentary pressures to hold them to account for their decisions, may often seem impossible. Yet, in several chapters, we can see that these conversations are taking place. Price et al. use complexity thinking to engage with a range of local policymakers to further the spread of ideas. Mitleton-Kelly goes one step further, using complexity thinking to study and help reorganize policymaking organizations. Hobbs makes a clear argument for the role of complexity in local government. Meanwhile, Tenbensel and Gray identify applications to health, while de Roo makes similar arguments in the field of planning.

FUTURE DIRECTIONS, CURRENT CHALLENGES

In many ways, the challenges faced by complexity scholars now are similar to challenges faced in the past. First, it is difficult to get a sense of the 'state of the art' in complexity theory, to establish what we know and still need to know. We can, to some extent, address this question in edited volumes, but this Handbook highlights the diversity of understandings and approaches. It does not represent a 'systematic review', which seeks to identify a comprehensive list of sources and code the literature's main themes and conclusions – and it remains to be seen just how possible it is for a review to produce a consistent set of ideas. It also only scratches the surface (in an admittedly biased way!) of political practices across the globe, and we may need to go beyond the academic book, to invite a large number of people to tell their 'complexity stories', in theory and practice, before we know just how far the concept reaches, and how useful it is, as a source of a common language to describe policymaking.

Second, it is not clear just how far we have come in generating a language of complexity that everyone understands and shares. It is clear, even from the chapters of this book alone, that different people understand complexity, and seek to apply its insights, in very different ways. This can be a useful outcome, when the language of complexity is used widely, to challenge existing approaches in a range of disciplines, prompting scholars to modify their arguments to engage with the language of their audience. Terminological spread can be a good thing, if good ideas spread to many audiences, prompting new ideas to develop. However, as Edgar Morin noted in his 2005 essay on 'restricted and general

Conclusion   465

complexity', this outcome can also reflect a tendency to treat complexity primarily as a metaphor (or shorthand for complicated), which undermines our ability to tell if these arguments can be pulled together, or if we can accumulate insights in a meaningful way. This problem is magnified when we seek to combine insights from the natural and social sciences: we use the same language of complexity and emergence, but to refer to very different processes. As Morçöl describes, the social sciences seek to understand the operation of complex individuals in complex worlds, adding an extra layer to explanation. In this respect, this book represents a microcosm of that problem: each author engages, to a greater or larger extent, with key terms, focusing perhaps most on the idea of 'emergence', but not presenting detailed or common arguments about what it means and what its implications are.

Putting these ideas together often seems like an impossible task. Take, for example, a case study of mental health and policy. The instant problem is that we can identify *multiple* complex systems. When we apply these insights to individuals, we find that the brain is a complex system, in which thoughts, feelings and actions result from the interplay between nodes and neurons. That person may have 'complex needs', which refers to a wide variety of social and institutional responses to their demands on social and public services. Public organizations and institutions may operate within a complex policymaking system, in which mental health only appears very infrequently on the high-level political agenda, and in which policy is often made locally in the relative absence of central direction. The continually emerging and evolving cultural position of mental illness within a society adds another layer of complexity. Policymakers also operate within complex international systems, in which governments and organizations form networks, share information, and coordinate action. Even in this short example, we have identified multiple interrelated complex systems and types of complexity. Finding a language in which to study these issues as a whole, and a method that transcends each element of study, may seem like an insurmountable task, prompting us, again, to be pragmatic about the extent to which we can understand and influence our object of study.

In the end, we do not pretend to have the answers to these problems. Rather, our hope is that by continually adding 'bits and pieces' to the academic debates, working with policymakers on particular problems and teaching our students about complexity, small changes will continue to build and basic attitudes and beliefs will shift in a complexity direction. Whether you are a complexity 'convert', 'dabbler' or just curious about it, we hope that you have enjoyed this book, that it will spark your interest and help to add to the debate and larger process.

Index

Achorn, E. 160
Actionable Capability for Social and
   Economics Systems (ACSES) 229–39
  clustering 233–6
  Cobb–Douglas production function and
    implementation 231–3
  coercion theory 232–3
  expected outcomes (questions answered,
    policies evaluated) 230–31
  followership theories 232
  initial agent configuration for all simulation
    runs 234
  legitimacy theory 232–3
  motivation weights instantiating different
    social theories 233
  number of pro-government citizens as
    function of surge level 238
  project description 229–30
  representative theory 232–3, 236
  repression theory 232–3
  results after 50 time steps 234
  results after 200 time steps 235
  results from Run 2 235
  results from Run 3 236
  social influence theories 232, 236
  surges 236–8
  surprise 233–8
  threshold response 236–8
adaptation 3, 39–40, 224, 228, 375
  see also adaptive action; adaptive capacity;
    adaptive systems
adaptive action and violent crime measures in
  Brazil 11–12, 284–97
  Brazilian Institute of Social Research (IBPS)
    293
  changing patterns of adaptive action 289–93
  Civil Police 289–90
  confrontation 286–7, 288, 294
  containment 286, 288
  eroding patterns of progress: recent
    problems 293–5
  illegal drugs market and drugs gangs 286–7,
    288, 291–2, 293
  image of the police 292–3
  institutional problems 285
  linearity 285–7
  Military Police 285, 287, 289–90, 292–3
  organized crime 286, 294

  police death squads 294
  Primeiro Comando da Capital (PCC) 286–7,
    288
  structural problems 285
  Unidades Policiais Pacificadoras (UPPs –
    Pacifying Police Units) 284, 291–5
adaptive capacity and collaborative crisis
  governance in Sweden 12, 332–45, 348
  adaptive capacity 337–8
  changes in adaptive capacity and nonlinear
    performance (hypothesis 2) 336, 339,
    343
  collaboration partners 339, 340–41, 342,
    343–4, 348
  collaboration venues 339, 340–41, 343–4, 348
  contingency planning 338, 339, 340–41, 348
  controls 338–9
  cooperative partnerships 333–4
  crisis management capacity 334–6, 337, 348
  data analysis 339–42
  data and measurement 336–9
  descriptive statistics 348
  diversity 338
  education 338, 339, 340–41, 343, 348
  event experience 340
  fragmentation 338–9, 348
  interaction opportunity 338
  learning 338, 339, 340–41, 343–4, 348
  models of change (linear and nonlinear) 339,
    341
  multi-organizational collaboration 334
  municipalities per county 340
  OLS regression models 339, 340
  political turnover 338–40, 348
  population size 338–9, 340, 348
  positive effect of adaptive capacity on
    performance (hypothesis 1) 335–6, 339,
    343
  prior hazard experience 338–9, 348
  resource slack 343
  risk monitoring 338, 339, 340–41, 343–4, 348
  selection methods 338
  self-perceived crisis management capacities
    337, 339, 343
  Swedish Civil Contingencies Agency (MSB)
    336, 343
  training 338, 339, 340–41, 348
adaptive governance 354, 356

Index 467

adaptive planning 354–7, 359–60
adaptive systems 33, 175–6, 414, 430
see also complex adaptive systems
Adger, W.N. 402, 403
advocacy coalition framework theory 69, 373
agent-based modelling (ABM) 8–9, 13–14,
19–22, 165–7, 197, 200, 214–17, 462
conceptualization 165
design and development 165–7
effective model-building considerations 167
individual level 167
modelling complexity 154, 161, 207
policy outputs and outcomes in United
States 65
political science 144, 147
power, complexity and policy 27
public administration policy 83, 85–6, 87, 88
system-wide collective level 167
verification and validation 166
see also agent-based modelling (ABM) and
complex adaptive systems (CAS)
agent-based modelling (ABM) and complex
adaptive systems (CAS) 9, 221–39
adaptation 224, 228
Afghanistan 229
agents 223, 227
attribute-based rules of behaviour 223–4,
227
Defence Advanced Research Project Agency
(DARPA) 229, 2388
design of agent-based modelling (ABM)
223–6
emergence 221, 223, 226, 229, 239
environment (open or closed) 224, 228
fitness function 228
IF Conditions – THEN Actions 224
locality principles 225, 228–9
Markov models 230
procedural rules of behaviour 224, 227–8
utility function model 230
see also Actionable Capability for Social and
Economics Systems (ACSES)
agent-based modelling (ABM) and the global
trade network 13–14, 414–31
agent rules 428–9
baseline projections 419
Chimerica 426–7
China scenarios 427
coarse-grained approach 416
crisis and extreme events 414
CURR rule 424–5
currencies 422–3
EXPORT and IMPORT rules 424
export links matrix 419–20
exports and sectors 423

fine-grained approach 417
Game Changers (GC) project and extreme
events 415–16, 418, 426–7
GDP rule 423–4
Geopolitical Issues and Global Security
426
Global Economy 426
Global Trade Network Simulator (GTNS)
414–15, 416, 418–29, 430–31
Gross Domestic Product (GDP) growth 414,
416, 418, 419, 421, 422
'hard' path scenarios 427, 428
International Institute for Applied Systems
Analysis (HASA) (Austria) 415
LINK and CHANGE rules 424
machine-generated scenarios 428–9, 430
Map View 421, 422
meta rules (acting on agent rules) 428–9
narrative-based scenario analysis 430
network analysis: trade cliques identification
421–2
'soft' path scenarios 427, 428
translation of scenarios 427–8
user-defined scenarios 428
what-if scenarios 416, 417, 418, 421, 423,
425–9, 430
X-events in Human Society initiative 415
Alam, S.J. 217
Albert, R. 85
Almond, G. 404
Alvaro, C. 372, 374
American Legal Realism 53–5
Ansell, C.K. 328
Armitage, D.R. 403
Arthur, W.B. 21, 25–6
Ascher, W. 405
Ashby, W.R. 254–5
Astill, S. 8
attractors 14, 107, 408, 435
socially complex phenomena 174, 181, 186
see also strange attractors
Axelrod, R. 82, 86, 104, 106
Axtell, R. 86, 88

backcasting 178, 182
Baderin, A. 33
Bajardi, P. 217
Bar-Yam, Y. 153
Barabási, A.-L. 85
Bateson, G. 251
Batty, M. 275
Baumgartner, F.R. 4, 374
Beauchamp, T.L. 386
Benington, J. 247
Berger, P.L. 21, 30

468 *Handbook on complexity and public policy*

Berman, M. 216
best fit relationship 133
best practice 274, 387, 388, 396
Bhaskar, R. 69, 105
Bichard, Lord 246
Bilge, U. 13–14
'black swans' 275
'black-box' problem 403–6, 408, 462
Boone, M. 275
Boons, F. 98
bottom-up policy 13, 22, 95, 98
    agent-based modelling (ABM) 165, 417
    health policy 371, 377
    power, complexity and policy 25, 27, 28
    realism 43, 45
    urban planning and non-linearity 353
bounded rationality 280
Bovaird, T. 11, 246
Brazil *see* adaptive action and violent crime
    measures in Brazil
Brighton Complex Systems Toolkit 7, 92, 95,
    96, 101, 107
Brizola, L. 286
Brooks, N. 403
Buchanan, M. 221
Burt, R.S. 30, 318, 319
butterfly effect 14, 135, 431

Cabral, S. 291
Cairney, P. 8, 14, 300, 371, 400, 433–4
Caloffi, A. 12
Cameron, K. 175
Cano, I. 293
Cardia, N. 285
Carmichael, T. 9
Carneiro, L.P. 285–6, 294
case studies 42, 85, 136–9, 373, 460
    public administration policy 83, 85
Casti, J.L. 85, 151–2, 415
causality, transformative 176–7
cause-and-effect chains/mapping 261, 264, 265,
    266–7, 270, 276, 277, 279
Cecchetti, S.G. 417–18
cellular automata (CA) 24, 213–14
CFinder software 322
chaos/edge of chaos 5, 100–103, 107, 140–41,
    151, 275
    adaptive capacity and collaborative crisis
        governance 336
    agent-based modelling (ABM) and global
        trade network 430
    Critical Legal Studies movement (CLS) 49
    information theory 213
    international financial crisis 433, 451
    local government design skills 250

policy outputs and outcomes in United
    States 74
political science 135, 140–41
socially complex phenomena 172, 174, 175
urban planning and non-linearity 358,
    360–61, 366
Chapman, J. 249, 255, 288
Chapman, P.C. 399
Chilcote, R.H. 405
Childress, J.F. 386
China *see* education inequality in China
Church, M. 176
Churchman, C.W. 254
Cilliers, P. 38–40, 44–5, 50
climate change adaptation 13, 399–409
    adaptive capacity 402–3
    adaptive management 408
    adaptive systems theory 401
    'black-box' problem 403–6, 408
    command-and-control 409
    ecosystem dynamics metaphor 401–6
    facilitation 402
    functionalism 402–3, 409
    governance 403, 409
    implementation 402
    institutional mechanisms 403
    Intergovernmental Panel on Climate Change
        (IPCC) 401
    policy and governance variables 406–8
    policy processes 401–2
    political system 404–5
    socio-political variables 403
    structural-functionalism 405, 408
    systems-level approach 402–5, 406, 408
closed-systems thinking (Class I systems) 351,
    353, 358
Club of Rome 'Limits to Growth' model
    215–16, 416
co-evolution 9, 184–5, 201
    health policy 377
    power, complexity and policy 28
    problem space 111–12, 114–15, 118, 119–20,
        122–6
    public administration policy 83
    socially complex phenomena 186
    urban planning and non-linearity 360
Cobb–Douglas production function 231–3
cognitive mediation roles of models 192–5,
    199–201
cognitive models 192–201
Cohen, D. 104, 106
Cohen, J. 173
collaborative crisis governance *see* adaptive
    capacity and collaborative crisis
    governance in Sweden

Collier, J. 176
Comfort, L. 335
command-and-control approach 351, 353, 356
communicative approach 352–3, 354, 358, 359, 366
complex adaptive systems (CAS) 104, 151, 358, 359, 361–2, 365
see also agent-based modelling (ABM) and complex adaptive systems (CAS)
complex adaptive systems (CAS) and public outcomes improvement by Birmingham City Council 11, 26–82
'Be Healthy' 270
'blue skies' innovation 276
Business Development and Innovation 276
cause-and-effect chains 261, 264, 265, 279
cause-and-effect mapping 261, 266–7, 270, 276, 277
'Cleaner, Greener, Safer City' 270
conceptual design of model (phase 1) 266–8
dependencies 266
disruptive innovation 280
dynamic modelling 280
innovation 276–8
input–output–outcome relationships 264, 270, 273
intervention holders 268–9, 270
knowledge domains 272–5
KPMG 265
Marketing and Attracting Investment programme 269, 270, 271, 274, 281
Marketing Birmingham 266
meta-planning approach 281–2
Modelling Birmingham (first phase) 265, 276, 278, 279, 280, 281
modelling process 368–71
outcomes-based approach 261, 267, 270, 271
partial quantification over a period of time 276
pathways to outcomes approach 261, 279–80
quantification of links in strategy maps (phase 3) 270, 271
rolling out and expanding modelling approach (phase 4) 270
strategy maps development (phase 2) 265, 266, 268–70, 274
strengths and limitations 272
'Succeed Economically' outcome area 261–3, 270
Total Quality Management (TQM) 278
trade-offs 270
University of Birmingham 265
University of Warwick 265
Value for Money models 272

'what works' centres 265
worklessness programme 280
complex ethics 177–8, 182–3, 185–6
complexity modelling 8, 150–68
agent-based models (ABMs) 165–7
benefits 158–60
challenges and limitations 160–61
complex adaptive system (CAS) 151
complex system 150–51
complexity science 151
complicated system 150
context and asking the right questions 157–8
definitions 150–53
methods and tools 161–4
network analysis 161–4
rationale for necessity for new research tools 153–6
research modelling perspectives 157
simple system 150
traditional research versus complexity research 155–6
complexity toolkit 7, 92–107
Brighton Complex Systems Toolkit 7, 92, 95, 96, 101, 107
Community Research and Evaluation Gateway 94, 95
Complex Practice Toolkit 92, 94, 107
empiricism, data and evaluation 105–6
facilitation, distribution, direction and rules 96–7
hard (restricted) complexity 92, 93
heuristics and rules of thumb 99–100
information scanning and resource management 101–4
intervening in the system 95–6
policy experiments portfolio 104–5
policy system definition 95–6
policymaker leadership 97
self-organization encouragement 98–9
soft (metaphorical or general) complexity 92, 93
complicated system 150
computer programming languages 161, 166, 418
Conn, E. 95, 105–6
connectivity 118, 123, 126
Connolly, W. 34
contingency planning 6, 375
adaptive capacity and collaborative crisis governance in Sweden 338, 339, 340–41, 348
Critical Legal Studies movement (CLS) 52–4, 58–60
convergent rules 181–2, 184, 186

470 *Handbook on complexity and public policy*

coordination 290, 378–80
creative pattern-based and puzzle-solving
thinking 139–41
crises or catastrophes 74–5
Critical Legal Studies movement (CLS) 6–7,
48–62
American Legal Realism 53–5
anti-stability argument 54
causal intuition 52
compositional fallacy 51
conceptual fallacy 57
contingency 52–4, 58–60
'the Crits' 48, 55–6, 58, 60
democracy 51
destabilization 55–8, 61
distortion and delusion 58
divisional fallacy 51
emergence 50–52, 55–8, 61
equilibrium 55–6
Formalism 53, 54, 56
indeterminacy 54, 57, 58–60, 62
law 49–50
Law and Economics movement 55, 56
Law, Science and Policy school 53
Law and Society Association 53
Legal Realism 49
neo-formalism 59
objectivity 54
perpetual critique 54
realism 56, 59
reductionism 50–52, 56–7, 60–62
rule-making 50–51
self-criticality 52
self-reflexivity 59, 60–61
stability/instability 55–6
stagnation 56, 57–8
suppression 57–8
critical naturalism 69–70
critical realism 7, 66, 67, 69–70, 105, 460
Cui Yongyuan 312
Cunningham, D.J. 200
Cunningham, P. 314
CYNEFIN framework 11, 273, 279

Dahl, R.A. 24, 30
Darking, M. 7
Darwin, C. 30
data mining 65, 161
Dawe, A. 20–21
de Medici, C. 328
de Roo, G. 12–13, 460, 464
de Souza, L.A.F. 287
decoding process 194, 201–2
deduction/deductive inference 79–80, 85–9
defining policy system 95–6

Deleuze, G. 49
Denyer-Willis, G. 285, 287, 288, 295
dependant variables 132–5, 156, 402
dependence *see* inter-dependence; path
dependence
Derrida, J. 459
Descartes, R. 89
design component of model 195, 200–201
destabilization 55–8
Dilthey, W. 89
diversity hypothesis 335
Doelle, M. 407
Dopfer, K. 21–2, 27
Doshi, P. 391
double hermeneutic 84
Dreyfus paradox 84
Duarte, T. 293
Duit, A. 335
Dunne, K. 389
Dupuy, J.-P. 193
Dworkin, R. 54
dynamical system modelling 161, 163

Easton, D 404
Edmonds, B. 9, 218
education inequality in China 12, 299–312
Area C achievement and struggles 306–9
Area C background 306
centralized political system 311
compulsory, high-school and higher
education 302–3
Convention on the Rights of the Child 312
'creative complexity' and educational policy
309, 312
Education Committee 308
education system 302
financial issues of migrant children's
education 302, 303, 308, 311
GAOKAO examination process 299, 302,
304, 308–9, 310, 311
government's dilemma 308
higher-level education entrance examination
310, 311
illegal schools (Tongxin School) 307–8, 312
main responsibilities 310–11
mass media coverage 307
migrant children 309–10, 311
Ministry of Education 307, 312
National Law on Education (1986) 302
policy in practice and challenges faced in
Area C 307–9
President's (Ju Jintao) Remark 2012 301, 306
problems encountered 309–10
responses from migrant families 308
social and cultural factors 311

Two Main Responsibilities Policy ('Decision on Basic Education Reform and Development' 2001) 300–301, 303, 304, 305, 306–9, 311
updated version (2006) 301
Ehrlich, P. 82
Eichelberger, C.N. 224
Einstein, A. 72, 82
Elder-Vass, D. 22–3, 25, 26, 27, 29
elementary statistics 132–9
Elias, N. 97, 252
Elster, J. 405, 409
emergence 2–3, 6, 8, 96, 153, 155, 464–5
  agent-based modelling (ABM) and complex adaptive systems (CAS) 221, 223, 226, 229, 239
  agent-based modelling (ABM) and global trade network 414, 417
  climate change adaptation 400, 408
  complex adaptive systems (CAS) and public outcomes improvement 274
  Critical Legal Studies movement (CLS) 50–52, 55–8, 61
  health policy 370, 372–3, 375
  information theory 213
  international financial crisis 434
  local government design skills 250
  policy outputs and outcomes in United States 65, 72–3, 75
  political science 136
  power, complexity and policy 21–3, 27
  problem space 114–15, 120, 124, 125, 126
  public administration policy 82, 83, 86, 89
  realism 33
  regular (simplexity) 173, 185
  socially complex phenomena 173, 176
  super (complicity) 8, 173, 185
  urban planning and non-linearity 357
emergency preparedness *see* adaptive capacity and collaborative crisis governance in Sweden
Emison, M. 401
empiricism 105–6
employment in an Arctic Community (agent-based modelling (ABM)) 216–17
encoding process 194, 201–2
Engle, N. 403
entropy 70–71, 75
Eoyang, G. 284, 289
epistemological determinism 40
epistemology 34, 35–41, 45–6, 80–81, 88
Epstein, J.M. 86, 88, 157, 218
equilibrium 55–6
  see also punctuated equilibrium framework
escape dynamics 145

Essén, A. 370, 372
ethics 177–8, 182–3, 185–6, 384, 386–7, 396
EUROMOD tax and benefit micro-simulation model 178
explanatory variables 132, 136
exploitation 104, 336
exploration 104, 336
external modelling loops 194–5, 200–202

Faber, E.M.H. 201
facilitation, distribution, direction and rules 96–7
falsifiable hypotheses 68
Farnsworth, R. 221
feedback/feedback loops 152, 161
  adaptive capacity and collaborative crisis governance 335–6
  balancing 432–3
  climate change adaptation 404, 408, 409
  health policy 370, 372
  'Limits to Growth' (Club of Rome) 215
  political science 143
  reinforcing 432–3
  system dynamics 212
  see also negative feedback; positive feedback
feminist and related research 458–9
Ferguson, N. 426
Ferlie, E. 372
Fernandez, J. 318
Fernandez, S. 251
Fisher, I. 450
Flyvbjerg, B. 43, 44, 81, 84, 85, 87, 89, 460
Folke, C. 402
followership theories 232
formality of models 205–7, 210–11
Foucault, M. 30, 41, 44
Frechette, A. 406
Fuchs, C. 176
Fuerth, L.S. 201
functionality 192–5, 350–51
Füssel, H.-M. 401–2, 405
future directions and future challenges 464–5
futures studies 178
futures thinking 248

Galston, W. 33, 35
Game Changers (GC) project and extreme events 415–16, 418, 426–7
Game of Life 213–14
Gawande, A. 394
general complexity 40
generalized capacity 20, 28, 29, 30
generative mechanisms 22, 25, 26
Gershenson, C. 9
Geuss, R. 33, 35, 37–8

472 *Handbook on complexity and public policy*

Geyer, R. 14, 249–50, 253, 300, 309, 312, 371, 392, 396, 400, 433–4
Giddens, A. 21, 30, 84
Gilsing, V.A. 319, 328
Gioja, L. 221
Givel, M. 7, 460
global trade network *see* agent-based modelling (ABM) and the global trade network
Glouberman, S. 372, 375, 385
Gómez, E.J. 407
Gostin, L.O. 389
Gould, J. 318
Gray, B. 13, 461, 464
Greenhalgh, T. 369, 371–2, 375
Greiling, D. 252
Griffin, D. 97
Groth, A. 405
Guattari, F. 49
Guzzini, S. 32

H1N1 *see* health policy for pandemic planning
Hadzikadic, M. 9, 224
Hallsworth, M. 250
Hannigan, B. 371, 375
hard (restricted) complexity 40–41, 92, 93
Hardin, G. 355
Harré, R. 22–3
Harrison, N.E. 221
Hart, H.L.A. 53–4
Hatch, M.J. 253
Haynes, P. 7, 14, 373, 461
Hayward, C.R. 23–4, 25, 27, 28
health policy 369–81
    downstream approaches 370, 373, 377–8
    guidance for policymakers 13, 375–6
    hierarchical approaches 376–7
    hierarchical imperatives as products of policy dynamics 377–9
    hierarchies, markets and networks/heterarchy (HMN) 378–9, 381
    hierarchy re-incorporation into complexity thinking 379–81
    'layering' 381
    methodological advances 373–4
    theoretical contributions 374
    upstream approaches 370, 373
    see also health policy for pandemic planning
health policy for pandemic planning 13, 384–97
    autonomy, justice, non-maleficence and beneficence 386
    best practice guidelines 387, 388, 396
    bioethics/values 384, 386–7, 396
    collective societies 386–7

commercial pressures 387–8, 391, 396
common morality 386
complexity approach 392–4
Evidence Based Medicine 384, 387
following pandemic plan/guideline 390–91
H1N1 influenza case scenario (New Zealand) 217, 389–91
individual patient care versus public health approach 388, 391
individualistic societies 386
long-term conditions as complex problems 388
Oseltamivir (Tamiflu®) 391, 394
politics 395–6
proportionality 393
protocol (recipe) 386, 395, 396
public health practice and complexity 389
randomized controlled trials (and meta-analysis) 384, 387–8
reciprocity 392–3
SARS 390, 392, 394
shared values 390
simple, complicated and complex problems 385
Stacey diagram 385–6, 387, 388, 395
stakeholder involvement 392, 394, 395
stewardship, trust and solidarity 394
universal ethical principles 384
values (ethics) 392, 396
World Health Organisation 390
World Health Organisation Surgical Checklist 387
Heckman two-stage probit (probit model with sample selection) 323–6
Heclo, H. 3
Hegel, G.W.F. 176–7
heterogeneity and difference distinction 39–40
heuristics and rules of thumb 99–100
Hicklin, A. 342
hierarchical approach 374, 376–81
hierarchical control 409
hierarchical level 159
Higgs boson: 'near discovery' and its relevancy 79–80
Hirschman, A.O. 25–6, 27, 28, 29, 30
HIV spread and social structure model (South Africa) 217
Hjern, B. 4–5
Hobbs, C. 11, 458, 464
Hofstede, G.H. 386
Holladay, R.J. 284, 289
Hood, C. 378
Houchin, K. 100
Howells, J. 316, 328
Howlett, M. 13

Hu Jintao 301
Hudson, R. 103
Human, O. 38–40, 44–5
Human, S.E. 316–17
Hunter, D. 370

independent variables 133–5, 156
indeterminacy 54, 57, 58–60, 62
individual-based modelling 214–15
Indonesia *see* problem space: case study in a government agency in Indonesia
information scanning 101–4
initial conditions 71, 72–3
Innes, J.E. 221
innovation 105
  as a complex adaptive system (CAS) 276–8
  disruptive 279–80
  policies *see* network-based approach to innovation policies in intermediary organizations
instability *see* stability/instability
institutionalism 4, 69, 380–81
inter-connectivity 113, 120, 123
inter-dependence 4, 113, 118, 120, 123, 125, 126, 370, 374
intermediary organizations *see* network-based approach to innovation policies in intermediary organizations
internal modelling loops 194–5, 200–202
international financial crisis 26, 432–52
  asset bubbles (asset inflation) 442, 447
  attractors 435
  austerity policies 438, 442, 443, 446
  capital-to-liquidity ratios 449
  central bank interest rates setting 441–2
  competitive devaluation 444, 448–9
  consumer price inflation and interest rates 441, 447
  conventional monetary policy 441
  credit agencies 439
  current macroeconomic policy 440–42
  electric money 434
  feedback balancing 438
  feedback reinforcement 437–8
  Financial Crisis Inquiry Commission 434
  financial liberalization 450–51
  financial repression 444, 448
  fiscal expansion 443, 446
  fractional reserve banking 445, 449
  Full Reserve Banking 432, 445, 449–50
  global governance and cooperation 451–2
  global political economy 440–42
  goals, changing 439
  house prices collapse in United States 433–4
  information 436–7

'Limits to growth' thesis 440
  market reinforcement 433
  Meadows framework 435–40
  non-governmental organizations (NGOs) 439
  paradigm change 440
  policy interventions 442–50
  political economy theorizing 433–5
  Positive Money 450
  quantitative easing (QE) 438, 443, 444, 446–7, 448
  regularities of behaviour 434–5
  resource management: stocks and flows 436
  rules 437
  scalability 434
  self-organization 432, 439
  stability/instability 433–5
  stocks and flows 446, 449
  unconventional monetary policy 443, 444, 447, 448
  values and beliefs 435, 439, 449, 452
International Futures (IFs) computer simulation 416
Italy *see* models in policy-making

Jackson, P.M. 252
Jenks, C. 87
Johnson, L. 8
Jones, B. 4, 374
Josilyn, C. 193

Kaletsky, A. 439
Kant, I. 69, 460
Kauffman, S. 82, 85, 176
Kenny, R. 11
Kernick, D. 376
Keynes, J.M. 30
King, G. 81, 82, 84, 89
Kingdon, J.W. 373, 374
Kirsten, A. 286
Klein, R.J.T. 401–2, 405
Klijn, E.-H. 3, 400
knowledge representation, roles of 193
Krugman, P. 436, 446
Kuhn, T. 81, 190, 256

Laming, H. 100
Landman, T. 42
Landuyt, N. 253
Lanham, H.J. 370, 372, 374, 375
Lasswell, H. 68
law and policy *see* Critical Legal Studies movement (CLS)
Le Corbusier 350

474 *Handbook on complexity and public policy*

leadership 96–7, 103, 250–51
  distributed 96–7, 120–21, 123–4
Lederman, L. 78
Lehmann, K. 11–12, 288
Lemos, M. 403
Levin, S.A. 221–2
Lewis, N. 406
Lewis, O. 131
Lichtenstein, B.B. 95, 96
Lilienfeld, R. 405
Lindblad, S. 370, 372
Lindblom, C. 5, 376
linearity
  adaptive action and violent crime measures
    in Brazil 285–7
  and cause and effect 7, 461
  climate change adaptation 400
  health policy 374, 376–7, 379
  policy outputs and outcomes in United
    States 65–74
  political science 133–6
  public administration policy 78
Lipsky, M. 4, 463
Little, A. 6, 460, 461
local, national and international policy 9–14
locality principle 225, 228–9
lock-in 20, 28
Luckmann, T. 21, 30
Lukes, S.M. 23–4, 27, 28, 29, 30

Macal, C.M. 160
MacLean, D. 100
macro-level behaviour 20, 24–5, 28, 157, 160
  climate change adaptation 399, 400, 403,
    406–7, 408
  health policy 370, 378, 381
  international financial crisis 434
  urban planning and non-linearity 357,
    366
Mahamoud, A. 373
Mandelbrot, B. 103
Mandelbrot set 141, 143
March, J. 336
Maroulis, S. 222
Marris, P. 30
Marten, G.G. 221
mathematical equation-based models 154
Mathie, N. 457
McEvily, B. 318
Mead, G.H. 97
Meadows, D.H. 95–6, 101–2, 107, 432, 435–40,
    449, 451–2
mental models 196
Menzies Lyth, I. 100
meso-level behaviour 157, 315

climate change adaptation 399, 403, 405–6,
    408
health policy 370, 378
urban planning and non-linearity 357
meta-analysis 384, 387–8
micro-level behaviour 24–5, 124, 157, 160,
    314–15
  climate change adaptation 399, 403, 406
  health policy 370, 378, 381
  international financial crisis 434
  urban planning and non-linearity 357
Millemann, M. 221
Miller, J.H. 152, 222
Mitleton-Kelly, E. 7–8, 464
mixed methods approach 8, 42, 44, 137–9,
    279–80
modelling complexity 9, 137, 143–5, 205–19
  abstraction of a phenomenon 205–6
  adaptive model 208–9
  cellular automata (CA) 213–14
  Club of Rome 'Limits to Growth' model
    215–16
  employment in an Arctic Community
    (agent-based modelling (ABM)) 216–17
  explanation 207
  exploration 207
  formality of models 205–7, 210–11
  Game of Life 213–14
  generality 210–11
  goals 211
  H1N1 influenza: pandemic preventive
    measures evaluation model 217
  HIV spread and social structure model
    (South Africa) 217
  individual- and agent-based modelling
    214–15
  information theory 213
  limitations 217–18
  network theory 212–13
  predictive models 207–8
  prospects 217–18
  racial segregation model (Schelling) 216
  relevant interactions 209
  representational model 208–9
  simplicity 210–11
  specificity of manipulation 206
  specificity of reference 206
  system dynamics 212
  tools and approaches 212–15
  use of models 207–9
  validity 210–11
models in policy-making 9, 190–202
  abstractions (theoretical bases) 196
  action domain 196
  barriers 198–9

CityDev model 197
cognitive models 192–5, 196, 197, 198, 199–201
communication function 200
contextual issues 199
cultural inertia 198–9
decision-making 191–2, 196–7
design component of model 195, 200–201
functional roles 192–5
goals 199
ICT-human mediation 190–91, 195, 198, 201
intelligent packages 195–6
Italian case study (IRES and Florence University) 195–7, 199
knowledge levels 196
knowledge representation roles 193
language gap 199
mental models 196
model of the observable 196
new requirements 198–201
opportunities 198–9
representational (semantic) component of model 195, 198, 200–201
social issues 196
stylized model approaches 196
syntactic component of model 195, 198, 200
users' expectations and experience 200
see also agent-based modelling; complexity modelling
Monjardet, D. 287
Moore, M.H. 247
Morçöl, G. 7, 460, 465
Morin, E. 32, 93, 464–5
Moynihan, D.P. 253
Mulgan, G. 249
multiple streams approach 69, 373, 374
Munro, E. 463

nature of policymaking 70–72
negative feedback 2, 4, 5, 126
    agent-based modelling (ABM) and global trade network 414, 430
    policy outputs and outcomes in United States 65, 71–2
    political science 132
    problem space 115, 123
Nelson, D.R. 402
nested policy systems 71, 72–3
NetLogo software 158, 166, 417
network analysis 30, 65, 161–4, 212–13
    average degree 162
    betweenness 162
    closeness centrality 162
    clustering coefficients of connectivity 162
    connectivity measures 162

degree 162
degree centrality 162
degree distribution 162
nodes and links 162
size 162
see also agent-based modelling (ABM) and the global trade network; network-based approach; networks; social network analysis
network-based approach to innovation policies in intermediary organizations 12, 314–29
    bridging positions 318
    brokerage index 315, 322, 329
    brokerage positions 319–21, 326
    brokers 318–20, 322, 323, 328, 329
    cognitive distance 319
    data and methodology 320–26
    descriptive characteristics of agents' characteristics 324–5
    empirical results 326–7
    empirical strategy 321–6
    exploitation processes 319, 328
    exploration processes 319, 328
    features of intermediaries 316–18
    innovation centres 320
    intercohesion measurement 315
    intercohesive agents 318–21, 322, 326, 328, 329
    intercohesive nodes 318, 322, 323
    intermediary positions and innovative contexts 319–20
    k-cliques and *k* nodes 322, 329
    knowledge and competencies 328
    knowledge and information dissemination 319
    knowledge spillovers 314–15
    knowledge-intensive business service providers (KIBS) 320–21, 328
    learning and innovation processes 319
    local associations 326, 328
    local government 326, 328
    many-to-one relationships 317
    nodes 318, 322
    one-to-one relationships 317
    overlapping cohesive subgroups 319
    participating organizations by type 321
    policies 314–15
    process innovations 321
    Regional Programmes of Innovative Actions (RPIA-ITT) 320, 329
    regression results 327
    Single Programming Document (SPD) 320
    small and medium-sized enterprises (SMEs) 320

476 *Handbook on complexity and public policy*

Social Network Analysis (SNA) 315, 317, 322, 329
'space' services 328
structural holes 318, 319
'switch' services 328
technological environments 321, 323
Tuscany Region (Italy) 320–27
Virtual Enterprises (RPIA-VINCI) 320, 329
networks 212–13
 adaptive 212
 adaptive capacity and collaborative crisis governance 342
 climate change adaptation 407
 conceptualization 163–4
 function of systems 212
 health policy 377, 380
 informal 117
 local government design skills 253, 254
 natural 212
 preferential attachment 162
 problem space 122
 relational 176
 small world 146–7, 163
 structure of systems 212
 types 162–3
new methods toolbox 139–45
new order 113, 124–5
New Zealand *see* health policy for pandemic planning
Newton, I. 66, 72, 89, 173
Nienhuis, I. 363–4
Nilsson, M. 402–3, 406
Nohrstedt, D. 12
non-linearity 151, 159
 agent-based modelling (ABM) and complex adaptive systems (CAS) 223
 climate change adaptation 400
 health policy 372, 374
 policy outputs and outcomes in United States 67
 political science 132, 135
 public administration policy 83
novelty 104, 105, 155, 275, 461
nowcasting 178, 183

Obama, B. 446
observations 133–4, 138
Occelli, S. 9
ontology 35–8, 80–81, 88
open networks (class III systems) 352, 353, 358
opening up spaces of the possible 173–4, 181–2, 184, 185
order (uniformity) 21, 213, 358, 360–61, 366
Ormerod, P. 105

Ornstein, R. 82
Ostrom, E. 380, 458
O'Sullivan, T.L. 372, 375
out-of-equilibrium state 357, 359–60

Padgett, J.F. 328
Page, S.E. 152
Pagels, H.R. 168
Paina, L. 372
Paley, J. 372, 373, 375
Palla, G. 322
pandemic planning *see* health policy for pandemic planning
Parson, T. 20, 21, 23, 25, 30
Pascale, R.T. 221
path dependence 2, 4, 5, 20, 460
 agent-based modelling (ABM) and global trade network 414
 health policy 372, 374
 international financial crisis 434
 power, complexity and policy 25, 28
 realism 33, 38
Peake, S. 222
Peat, F.D. 159
Peirce, J. 288
Pender, H. 160
period doubling 140
Peters, D.H. 372
Philp, M. 32, 33, 35
phronesis 41–6
Pierson, P. 4, 381
plausibility test 279
Plsek, P.E. 369, 371–2, 375
Plummer, R. 403
pluralism 67, 68–9
polarization 23–5
policy diffusion theories 69
policy experiments portfolio 104–5
policy outputs and outcomes in United States 7, 65–75
 complexity model of public policy 75
 context of policy process 71, 72, 73
 critical realism 67
 early public policy theory 67–8
 initial conditions 71, 72–3
 linear models 65–74
 modern policy theories 68–70
 nesting 71, 72–3
 policy system change over time 71, 74
 policymaking, nature of 70–72
 positivism 66–9
 post-positivism 66, 67, 69
 space-time 71, 72
 strange attractors 71, 73–4
political 'gaming' 279–80

Index 477

political science 8, 131–48
  case study analysis 136–7
  case study process 138–9
  chaos 135, 140–41
  comparison of methods 147–8
  creative pattern-based and puzzle-solving thinking 139–41
  descriptive statistics 132
  elementary statistics 132–9
  escape dynamics 145
  linearity 133–6
  Mandelbrot set 143
  modelling and deduction 137
  modelling and simulation 143–5
  multiple methods 137–9
  network (social network) analysis and relations 145–7
  new methods toolbox 139–45
  non-identical coupled logistic maps 142
  qualitative methods 135, 136–7, 139, 145, 147
  quantitative methods 132–9, 145
  small world networks 146–7
Porter, D. 4–5
positive feedback 2, 4, 5, 20, 126
  agent-based modelling (ABM) and global trade network 414, 430
  health policy 380
  policy outputs and outcomes in United States 65, 71–2
  political science 132
  power, complexity and policy 28
  problem space 115, 123
positivism 7, 32, 44
  naïve 460
  policy outputs and outcomes in United States 66–9
  political science 131
  public administration policy 80
post-positivism 7, 66, 67, 69
Potts, J. 21–2, 27
power, complexity and policy 6, 19–30, 253, 374
  economic development 25–7
  emergence 21–3, 27
  non-zero-sum power 20, 25, 28
  segregation and polarization 23–5
  self-organization 27–8
  social systems 19–20
  zero-sum concept 20, 23, 26, 28, 29, 30
pragmatic approach 459, 462–4
precepts list 375
predictive models 207–8
Pressman, J. 374
Price, J. 7

Prigogine, I. 82, 85, 86–7, 356
problem space: case study in a government agency in Indonesia 7–8, 111–27
  addressing a complex problem 119–20
  authoritarian mindset 118
  common themes 116–17
  communication, increasing capacity of 122
  corruption and nepotism 116, 119, 125, 126
  cultural barriers to communications 117
  EMK methodology 7–8, 111, 116
  Enabling Environment preparation 120–23
  ethnic and religious groupings or 'gangs' 111, 119, 124–5, 126
  induction and mentoring 122–3, 125
  insights gained 124–5
  interview findings 116–18
  Jakarta Workshop (2013) 118–19
  leadership 120–21
  learning environment development to facilitate co-evolution 122–3
  local culture 119
  multi-disciplinary pilot activities that bridge different divisions 121–2
  opportunities and challenges faced 125
  ownership and responsibility, building a sense of 121–2
  personal integrity development as basis for organizational integrity 121
  planning process should make appropriate use of staff 123
  principles of complexity, application of 123–4
  respect and status, loss of 117
  roles, allocation of and responsibilities 116–17
  roles and functions of government agency 121
  'single fighters', prevalence of 117
  strong leader, crucial importance of 116
  'survival' culture leading to individualistic tendencies rather than organizational resilience 117
  technical and management skills 123
  ten principles of complexity theory 112–15
  tensions/dilemmas 117
  transparency 121
  underlying assumptions 117–18
procedural rules of behaviour 227
projective approach 191
Provan, K.G. 316–17
public administration policy 7, 78–89
  context 84–5, 87
  deduction/deductive inference 79–80, 85–9
  epistemological framework 80–81, 88

478 *Handbook on complexity and public policy*

European Organization for Nuclear
     Research (CERN) 78, 79, 86, 88
Higgs boson 78–89
ideal theory 83–4
induction 80
methodological issues 80–81, 83
New Public Administration movement 81
Occam's razor 81–2
ontological framework 80–81, 88
qualitative methods 81, 86–9
quantitative methods 80–81, 86–9
simplicity versus complexity 81–3
standard model of particle physics 79, 86, 89
theory of everything 83, 89
universal generalizations 83–5, 86, 88
public outcomes improvement *see* complex
     adaptive systems (CAS) and public
     outcomes improvement by Birmingham
     City Council
public service (local government) design skills
     11, 245–56
  Accenture 248
  Beacon Scheme 253
  challenging assumptions 251–3
  Chartered Institute of Public Finance and
     Accountability 248
  civic enterprise and enterprising councils
     246
  collaboration 251
  Commission on the Future of Local
     Government (Leeds City Council) 246
  Communities of Practice networks 251
  Complexity and Policy Analysis
     international seminar 249
  creative councils initiative 248
  Delphi survey 247
  fundamental reform and public value 245–7
  future insights 253–6
  futures thinking 248
  integrative leadership 251
  International Research Society for Public
     Management 249
  local vision initiative 247
  Munro Review of Child Protection 250
  Nesta: Public Services Lab development
     with Local Government Association
     248
  Political and Constitutional Reform
     Committee 247–8
  Principles for 2020 Public Services (Royal
     Society of Arts) 247
  Public Management and Policy Association
     248
  strategic systems leadership 250
  System Stewardship 250

systems approaches 248
punctuated equilibrium framework 2, 4, 69,
     374, 460

Qian Liu 12
qualitative methods 65, 86–9, 101–3, 155,
     157–8
  agent-based modelling (ABM) and global
     trade network 415–16, 430
  complex adaptive systems (CAS) and public
     outcomes improvement 274
  health policy 373
  political science 135, 136–7, 139, 145, 147
  problem space 111–12
quantitative methods 86–9, 101–3, 136, 155,
     156, 157–8, 460
  agent-based modelling (ABM) and global
     trade network 415, 430
  complex adaptive systems (CAS) and public
     outcomes improvement 265, 270, 274,
     279
  political science 132–9, 145
  problem space 111
Quinn, R. 175

Raadschelders, J.C. 87
racial segregation model (Schelling) 216
Radin, B. 80
Ramlogan, R 314
random agent networks (RANs) 212
randomized controlled trials 384, 387–8
randomness *see* chaos
ratchet effects 20, 28
rational choice models 69, 87
Rayner, J. 13
realism/realist approach 6, 22, 30, 32–46, 191,
     460
  American Legal Realism 53–5
  constitutive challenge 38–41
  Critical Legal Studies (CLS) 56, 59
  detachment realists 33
  epistemology 38–41, 80–81
  mono-realism 35
  new debate 34–5
  ontology 35–8, 80–81
  phronesis as new form of real political
     analysis 41–3
  real, towards complex economy of 43–5
  transcendental 69–70
  urban planning 358–60
  see also critical realism
reductionism/reductionist approach 2, 40,
     151–2, 209, 460–62
  Critical Legal Studies movement (CLS)
     50–52, 56–7, 60–62

health policy 376, 384, 387
political science 131, 138
Reeve, J. 388
reflexive complex policymaking 175–6
regulated autonomy 176, 182–3
relational model of complexity 38, 40
representational model 208–9
representational (semantic) component of
    model 195, 198, 200–201
Resnick, M. 222
resource management 101–4
restricted complexity 40–41, 92, 93
Ricaurte, C. 7
Riccucci, N.M. 81
Richardson, L. 286
Rihani, S. 249–50, 253, 309, 312, 392, 400
Rimer, J. 387
risk management and planning 102
risk monitoring 338, 339, 340–41, 343–4,
    348
Rittel, H. 360
Robb, J. 222
Room, G. 4, 6, 375, 380–81, 458, 464
Rossi, F. 12
Rössler, O.E. 85
Ruhl, J.B. 49, 50–52
Russo, M. 12, 317

Sabatier, P. 68, 106, 373
Sageman, M. 222
Sanderson, J. 106, 252, 300, 334–5, 400
Sawyer, R.K. 221
scenario modelling techniques 65, 161, 358
    see also what-if scenarios
Scharpf, F. 380
Schelling, T.C. 21, 24, 25, 27, 28, 82, 207, 208,
    216
Schön, D.A. 175
Schram, S. 41–2
second order complexity 253
Seddon, J. 103, 252
Sedyaningsih, E.R. 390
segregation 23–5
self-organization 7, 20, 97–100, 159, 163
    adaptive 449
    adaptive action and violent crime measures
        in Brazil 284, 288–9
    agent-based modelling (ABM) and complex
        adaptive systems (CAS) 221, 223,
        225–6, 229, 239
    agent-based modelling (ABM) and global
        trade network 414
    climate change adaptation 400
    Critical Legal Studies movement (CLS) 50,
        54

health policy 372–4, 375, 377, 378
information theory 213
international financial crisis 432,
    439
policy outputs and outcomes in United
    States 70–71, 75
power, complexity and policy 27–8
problem space 113–15, 117, 120, 123, 124,
    125, 126
productive 370
public administration policy 83
socially complex phenomena 176, 180
urban planning and non-linearity 356–7,
    359, 366
self-reflexivity 59, 60–61
semantic component of model 195, 198,
    200–201
Semboloni, F. 9
semi-open systems (Class II systems) 343, 352,
    358
Senge, P. 96
sense-making 279, 370–71, 374–5, 381
sensitivity analysis 161
Shank, G. 200
Shannon, C.E. 213
Shapiro, I. 42
Sharkansky, I. 82
Shaw, I. 334
signs, general theory of 200
Simon, H.A. 80
simplicity versus complexity 81–3
simplification 40
SimStore 417
simulation 143–5, 154–61, 166, 195–6, 215
small world networks 146–7, 163
Smircich, L. 252
Smit, B. 401, 403
Smith, J. 87
Snow, C.P. 43
Snowden, D.J. 275
social constructionism 21, 69–70
social investment approach 183–6
social naturalism 105
social network analysis (SNA) 12, 145–7, 164,
    315, 317, 322, 329
    agent-based modelling (ABM) and global
        trade network 430
    ego-centered 164
    health policy for pandemic preventive
        measures 217
    local government design skills 252
    political science 145–7
    public administration policy 83, 85, 87,
        88
social systems 19–20

480 *Handbook on complexity and public policy*

socially complex phenomena 8–9, 171–86
  Active Inclusion 179
  adaptive and reflexive complex policymaking 175–6
  child poverty 180
  Children and Young People's Policy Framework 179
  complex ethics 177–8, 182–3, 185–6
  complicit and valued rules for educational advantage 181–2
  Department of Children and Youth Affairs (DCYA) 179
  driver of social change, complexity as 172–4
  educational disadvantage and educational attainment 180–83
  Europe 2020 strategy 186
  Investing in Children: Breaking the Cycle of Disadvantage 184
  National Action Plan for Social Inclusion 2007–2016 (NAPinclusion) 178–9, 186
  National Social Target for Poverty Reduction 179
  OECD Report: *Doing Better for Children* 179
  opening up spaces of the possible 173–4, 181–2, 184, 185
  social complexity as theoretical complexity 174
  social inclusion/exclusion 180, 183–5
  social investment approach 183–5, 186
  Social Investment Package 179, 184
  transformative causality 176–7
  'work in progress' 178–80
soft (metaphorical or general) complexity 92, 93
soft modelling 274
Sornette, D. 222
space of possibilities, exploration of 95, 97, 105–6, 113–15, 120, 124–6
space–time 71, 72
Squazzoni, F. 22, 27
Ssengooba, F. 372
Staber, U. 335
stability/instability 14, 103
  Critical Legal Studies (CLS) 55–6
  international financial crisis 433–5
  policy outputs and outcomes in United States 74–5
  political science 141
  urban planning and non-linearity 360
Stacey diagram 385–6, 387, 388, 395
Stacey, R. 97, 175, 176
stages heuristic theory 68, 70
Stark, D. 318

Steinmo, S. 131
Stenger, V. 79, 88–9
Stengers, I. 85, 86–7
Stewart, I. 173
Stiglitz, J. 436
Stokey, E. 82
strange attractors 2, 71, 73–4, 100
  climate change adaptation 400
  complex adaptive systems (CAS) and public outcomes improvement 275
  health policy 370
  socially complex phenomena 174
Strogatz, S.H. 146
Stroud, J. 7
structuration 21
stylized model approaches 196
SugarScape 417
Sweden *see* adaptive capacity and collaborative crisis governance in Sweden
Swift, A. 32, 34–5
Sydow, B. 335
syntactic component of model 195, 198, 200
systems-level focus 402–5, 406, 408
systems-theory approach 59, 159, 352

Taleb, N.N. 103, 155–6
Taylor, J.S. 386
Teisman, G.R. 400
temporal networks 212
Tenbensel, T. 13, 463, 464
Teresi, D. 78
Tesfatsion, L. 221
Thelen, K. 381
Thomas, B.P. 389
Thorson, T. 405
Tian, Jianrong 312
Tierney, J. 285, 287, 288, 295
tipping points and thresholds 159, 182–3
Titmuss, R.M. 29
top-down approach 13, 22, 95–6
  agent-based modelling (ABM) 166
  climate change adaptation 409
  health policy 374, 376–7
  power, complexity and policy 25
  urban planning and non-linearity 353–4
Torda, A. 392–4
trade networks *see* agent-based modelling (ABM) and the global trade network
tragedy of the commons 355
TRANSIMS 417
transparency 119, 125
Treadwell Shine, K. 8–9
Trenholm, S. 372
triangulation approach 155

Tsoukas, H. 253
Tuohy, C.H. 381
Turchin, V. 193
Tushnet, M.V. 55

Ucinet software 322
United States *see* policy outputs and outcomes
    in United States
universal generalizations 83–5
urban planning and non-linearity 12–13,
    349–66
  adaptive governance 354, 356
  adaptive planning 354–7, 359–60
  arrangements 357
  assemblies (assemblages) 357
  branding 363–4
  chaos (diversity) 358, 360, 361, 366
  closed-systems thinking (Class I systems)
      351, 353, 358
  cohesion 363
  command-and-control approach 351, 353,
      356
  communicative approach 352–3, 354, 358,
      359, 366
  compatibility 363
  competitiveness 362–3
  complementarity 362–3
  complex adaptive systems (Class IV systems)
      358, 359, 361–2, 365
  consensus driven 357
  control 351–2
  emancipation 365
  emerging properties 357
  empowerment 365
  equality 365
  evolution driven 357
  feedback planning 352
  Fibonacci sequence 356
  freedom 365
  functionality 350–52
  goal driven 357
  governance and differentiation of planning
      issues 352–4
  gravity force or attractor 361
  interactions 358
  interdependency of planners 354
  linearity 351–2
  nodes 358–9
  non-linearity 356–8, 359, 360–66
  open networks (class III systems) 352, 353,
      358
  order (uniformity) 358, 360–61, 366
  out-of-equilibrium state 357, 359–60
  Plan A 350
  possibility space 356

progressive planning 352
rationality spectrum for spatial planning
    353
reality 358–60
regional development 362–4
robust and dynamic relationship
    361
scenario planning 358
self-organization 356–7, 359, 366
semi-open systems (Class II systems) 343,
    352, 358
shared governance 356
situational planning 353–4
social capital in a neighbourhood
    363–4
social cohesion 363–4
societal ties 363–4
socio-political systems 364–5
strategic plan 352
structure–function relationship
    361
technical-rational approach 353, 358,
    366
tolerance 363–4
Ville de Richelieu 356
windows of opportunities 356

Vedres, B. 318
violent crime measures in Brazil *see* adaptive
    action and violent crime measures in
    Brazil
von Bertalanffy, L. 404
Voss, J. 408
vulnerability assessment framework 403,
    405

Waldo, D. 80, 81
Waltz, K. 32
Wandel, J. 401, 403
Wangel, J. 178
Warren-Adamson, C. 7
Watts, D.J. 146
Webb, T. 6–7
Webber, M. 360
Weick, K. 371, 374
Wellstead, A. 7, 13, 462
Westminster model 10
what-if scenarios 215, 416, 417, 418, 421, 423,
    425–9, 430
White, S. 32, 34–5
Whitford, J. 317
Whitmeyer, J. 9
'wicked' policy problems 28–9, 104–5, 334,
    342, 360
Wildavsky, A. 374

Williams, B. 32, 33, 35
Wise, C.R. 335
Wolf-Branigin, M. 151, 154
World3 computer simulation 416

Xiao, Y. 371, 375

Yang, K. 80

Zaheer, A. 318
Zashin, E. 399
Zeckhauser, R 82
Zimmerman, B. 385